Advanced Robots: Design, Control and Application—2nd Edition

Advanced Robots: Design, Control and Application—2nd Edition

Guest Editor

Ioan Doroftei

Basel • Beijing • Wuhan • Barcelona • Belgrade • Novi Sad • Cluj • Manchester

Guest Editor
Ioan Doroftei
Mechanical Engineering,
Mechatronics and
Robotics Department
"Gheorghe Asachi" Technical
University of Iasi
Iasi
Romania

Editorial Office
MDPI AG
Grosspeteranlage 5
4052 Basel, Switzerland

This is a reprint of the Special Issue, published open access by the journal *Actuators* (ISSN 2076-0825), freely accessible at: www.mdpi.com/journal/actuators/special_issues/4L234RU4NV.

For citation purposes, cite each article independently as indicated on the article page online and using the guide below:

Lastname, A.A.; Lastname, B.B. Article Title. *Journal Name* **Year**, *Volume Number*, Page Range.

ISBN 978-3-7258-2756-5 (Hbk)
ISBN 978-3-7258-2755-8 (PDF)
https://doi.org/10.3390/books978-3-7258-2755-8

Contents

About the Editor

Ioan Doroftei

Prof. Ioan Doroftei is a graduate of the Faculty of Mechanical Engineering at the "Gheorghe Asachi" Technical University of Iaşi, 1986. He obtained the Ph.D. title in 1998. Since 1990, he has been with the Department of Mechanical Engineering, Mechatronics, and Robotics at the "Gheorghe Asachi" Technical University of Iaşi, serving as the head of this department since 2016. He has been a Ph.D. supervisor in the field of mechanical engineering since 2008. Prof. Doroftei has specialized in robotics at the Université Libre de Bruxelles, Belgium. He has also been a visiting professor at Université Blaise Pascal, Clermont-Ferrand, France. Prof. Doroftei has been president of the Romanian Association for the Science of Mechanisms and Machines (ARoTMM), affiliated with IFToMM, and the chair of Romania Member Organization in IFToMM since 2013 till the end of 2022. He is a member of the Technical Committee for Linkages and Mechanical Controls since 2019 and a member of the Technical Committee for Robotics and Mechatronics since 2019 in the same federation. He is the first vice president of the Romanian Robotics Society (SRR), affiliated with the International Federation of Robotics (IFR) since 2021, and the president of the Iasi Branch of SRR since 1996. Prof. Ioan DOROFTEI is a member of the International CBRNE Institute, Belgium, Explosives Knowledge Centre (EKC) since 2018, and the International Measurement Confederation (IMEKO, TC17)—Measurement in Robotics since 2019. He has published, as the sole author or co-author, 14 books and manuals and over 260 articles in national and international journals and conference proceedings. He is the editor of 13 volumes (principal editor for 11 of them), including 3 published by Springer and 6 by other international publishers. He is the author or co-author of 13 patents. He has contributed to solving over 24 national and international scientific research grants. He is a reviewer for over 30 international journals.

Preface

Research into the design, control, and application of advanced robots has increased over the past few decades, with many different and interesting projects being developed. Advanced robots have many promising applications in various areas of modern society. These robots could yield significant positive impacts on society, but they also carry the potential to cause negative impacts. Therefore, these impacts should be considered and discussed from the perspectives of not only technical solutions but also relevant social issues that concern safety, law, ethics, psychology, and philosophy.

This Special Issue, entitled "Advanced Robots: Design, Control, and Application—2nd Edition", showcases the most recent contributions in the theoretical and experimental research from fields related to the design, control, and application of advanced robots (particularly human–robot interactions (HRI) and social robotics); healthcare and medical applications; service and assistance; soft robotics; autonomous robots; etc.

We would like to express gratitude to all the authors who contributed to this Special Issue and to the staff at MDPI for their excellent technical and editorial support.

Ioan Doroftei
Guest Editor

 *actuators*

Article

"You Scare Me": The Effects of Humanoid Robot Appearance, Emotion, and Interaction Skills on Uncanny Valley Phenomenon

Karsten Berns *[ID] and Ashita Ashok [ID]

Department of Computer Science, University of Kaiserlautern-Landau, 67663 Kaiserslautern, Germany; ashita.ashok@cs.rptu.de
* Correspondence: karsten.berns@cs.rptu.de

Abstract: This study investigates the effects of humanoid robot appearance, emotional expression, and interaction skills on the uncanny valley phenomenon among university students using the social humanoid robot (SHR) Ameca. Two fundamental studies were conducted within a university setting: Study 1 assessed student expectations of SHRs in a hallway environment, emphasizing the need for robots to integrate seamlessly and engage effectively in social interactions; Study 2 compared the humanlikeness of three humanoid robots, ROMAN, ROBIN, and EMAH (employing the EMAH robotic system implemented on Ameca). The initial findings from corridor interactions highlighted a diverse range of human responses, from engagement and curiosity to indifference and unease. Additionally, the online survey revealed significant insights into expected non-verbal communication skills, continuous learning, and comfort levels during hallway conversations with robots. Notably, certain humanoid robots evoked stronger emotional reactions, hinting at varying degrees of humanlikeness and the influence of interaction quality. The EMAH system was frequently ranked as most humanlike before the study, while post-study perceptions indicated a shift, with EMAH and ROMAN showing significant changes in perceived humanlikeness, suggesting a re-evaluation by participants influenced by their interactive experiences. This research advances our understanding of the uncanny valley phenomenon and the role of humanoid design in enhancing human–robot interaction, marking the first direct comparison between the most advanced, humanlike research robots.

Keywords: uncanny valley; humanoid robots; emotional responses; human–robot interaction; social robots

Citation: Berns, K.; Ashok, A. "You Scare Me": The Effects of Humanoid Robot Appearance, Emotion, and Interaction Skills on Uncanny Valley Phenomenon. *Actuators* **2024**, *13*, 419. https://doi.org/10.3390/act13100419

Academic Editor: Alessio Merola

Received: 15 August 2024
Revised: 13 September 2024
Accepted: 23 September 2024
Published: 16 October 2024

1. Introduction

Within the premise of humanoid robots in human–robot interaction (HRI), one finds the pervasive phenomenon of uncanny valley. The uncanny valley hypothesis, proposed by Masahiro Mori, states that the humanlike perception of social humanoid robots (SHRs) enhances their likeability and familiarity [1]. However, this hypothesis was prefaced with a warning that excessive humanlikeness triggers a shift of the affine perception from positive feelings to a sense of eeriness [2]. This shift in perception is referred to as the uncanny valley effect (UVE) which stems from either aberrant behaviour or lack of realism, posing a significant challenge in HRI [3].

Recent advancements in SHRs have emphasized humanlikeness in both design and behavioural interaction. From a hardware perspective, robotic agents include a pair of eyes, a mouth to communicate, a body, two hands, legs (mostly immobile), etc. [4]. Meanwhile, from a behavioural perspective, SHRs have been personalized with unique names, backstories, voices, gaze algorithms, turn-taking, clothing, hair, etc. [5–8]. In social robotics, robot personality has been used to either establish a connection with human interlocutors or to make similar-looking agents differentiable [9,10]. Studies have focused on robot personality-induced uncanniness and have implied that robots that depict emotional stability tend to reduce feelings of eeriness [11,12].

To achieve humanlike communication is the primary goal of HRI [13]. The acceptability of robots as social agents underpins a display of rich anthropomorphic behaviour that is fluid, mutually regulated, affective, intuitive, and natural [14]. The impact of a robot's appearance and behaviour on the human disposition is deemed significant, particularly in tasks requiring social interaction [15]. This interaction dynamic was explored in a groundbreaking piece of research by Herbert Clark and Kerstin Fischer, who examined the intriguing question of why humans perceive social robots as sentient beings while, at other times, they are perceived as mere mechanical constructs or machinery [16].

Significant research has been conducted with the appearance and movement of robots as variables to assess how these factors influence the UVE. Previous research has primarily focused on how different levels of humanlikeness affect user perception by using robot images as stimuli [17]. A study with a similar focus on university students' perception of a tutor robot's appearance found the preference to be a mix of both mechanical coupled with a humanlike look and few facial features [18]. Analysing robot stereotypes on the dimensions of communion and agency, a study found humanoid and android robots (rated high on agency and communion) to evoke positive emotions such as pride and admiration, typically not triggered by discriminatory tendencies such as passive or active harm, which exists in the case of low-agency communion robots [19]. Additional findings reported that humanoid robots were assumed to be a part of the "in-group" based on the Stereotype Content Model and the Behaviours from Intergroup Affect and Stereotypes Map, aligning with societal robotics. At the same time, a discrepancy arises due to said anthropomorphism, prompting human normative expectations that do not translate to robots. Extending this narrative, a study exploring the influences of robot decisions found that more humanlike robots were favoured less and attributed as being deficient, highlighting a moral uncanny valley effect [15]. Although most studies have found that humans show aversion to highly humanlike robots within a few seconds of viewing photo stimuli, indicating a robot's perceived humanness [2,20,21], one study established a preference for realistic humanlike visualization of humanoid home assistants [22]. Studies have also highlighted the importance of robot appearance coupled with response in relation to their contribution to SHR's perceived social presence regarding natural HRI [23]. The above-mentioned studies carried out robot evaluations through photos, videos, or live interaction stimuli, but one prior research suggests that movement (shown in GIF or video) is crucial to reducing the eeriness, but this creates privacy concerns due to robot eye movements [24,25]. Additionally, studies have highlighted that the emotional expression of humanlike faces inhibits the UVE in HRI but this, of course, has understandable limitations due to the texture of robot skin used [26].

The aspect of human likeness has been considered in different research activities within our research group robotics research lab (RRLab) (https://rrlab.cs.rptu.de/ (accessed on 15 July 2024)) since 2006, in which we developed the first humanlike head able to express Ekman's six basic emotions. The results indicated that the missing degree of freedom (DOF) in the robot face (e.g., eye and eyelid movement) leads to inadequate recognition compared to human facial expressions. In addition, it was found that the speed of the facial expressions, which is essential for the quality of the recognition process, was also unsatisfactory. In Hirth 2012, we defined an optimal speed function, which could be parameterized for the different facial expressions [27]. With the development of the humanoid upper-body system ROMAN, the extension of the body motion and hand and arm gestures only partially lead to a pronounced humanlike robot appearance [28]. The synchronization and the adaptation of various robot body motions according to the interaction situation made the holistic implementation solution very difficult. The use of our behaviour-based control architecture iB2C [29,30] enables an overlap of different body motions by using adequate fusion mechanisms [31]. Not only the movement of the system but also the social behaviour of the robot influences humanlikeness. With the robot's automatic change in emotional states, we developed a humanlike adaptation of the robot's behaviours in typical interaction scenarios. With an "emotion-based architecture", an appraisal system adjusts the social reaction based on the satisfaction level of motives and the observed situation around the humanoid robot [27]. For humanlike robot behaviour, it is essential that the robot perceives the intention of the interaction partner. We additionally developed a detailed communication model, in which descriptors for verbal and non-verbal communication are separated [28]. Besides recognizing mimics, hand and body gestures, and feedback cues, we determine the personality of the interaction partner [32,33]. As a result, a better understanding of the interaction partner has been reached, which allows a better adaptation of the social behaviour of the robot system. In addition, personalised verbal communication also plays a vital role in attributing humanlikeness [34].

The current study aims to survey the perception of SHR Ameca among university students as student companions. A study with a similar focus on the perception of tutor robot's appearance by university students found the preference to be a mix of both mechanical coupled with a humanlike look and few facial features [18]. For this purpose, two studies with Ameca in a university setting were devised: (a) Study 1, assessing the expectations from SHRs in hallway/corridor; (b) Study 2, a follow-up study comparing three humanoid robots on their humanlikeness. The technical system, the EMAH robotic system, is developed for and implemented on Ameca (hereon referred to as EMAH), an upper-torso enabled humanoid robot as shown in Figure 1.

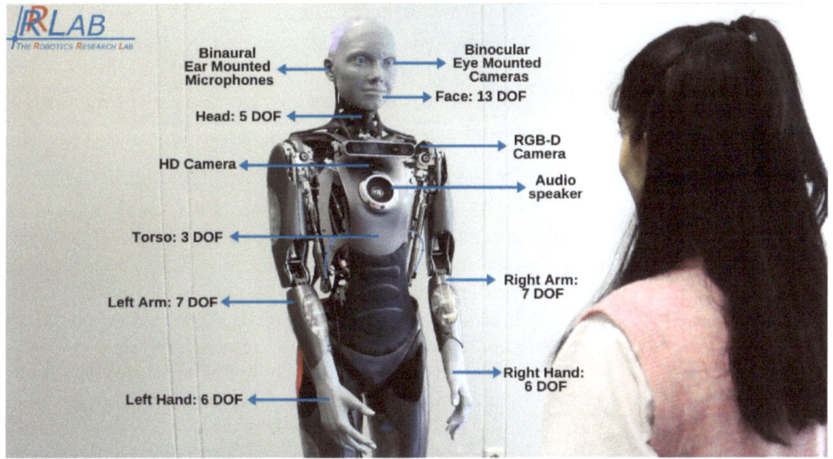

Figure 1. EMAH (Empathic Mechanized Anthropomorphic Humanoid) robotic system implemented on Gen 1 Ameca robot from Engineered Arts.

2. Study 1: Robot in University Hallway

The growing demand for advanced HRI in public settings, where social robots can provide knowledge and companionship, highlights the relevance of this study. The core objective is to assess how SHRs can seamlessly integrate into a university corridor environment and effectively initiate and sustain meaningful interactions. To develop and navigate SHRs in companion contexts, one must comprehend human–robot collaboration and establish meaningful social interactions with human interlocutors [35]. A fundamental property of social robots is their capacity to interact with humans by following social cues and regulations [36]. Additionally, social robots' bidirectional communication and emotional human–robot interaction are crucial for their efficiency in social settings [37]. The "open social environment" in HRI refers to a setting where robots interact with humans and other agents while being aware of human interactional structures like turn-taking (venues such as shopping malls, museums, cultural sites, wide corridors, etc.) [38].

2.1. Research Question

R1: What robot features are important for hallway HRI?

2.2. Experimental Design

The study aimed to identify characteristics EMAH should offer for hallway interaction and acquire essential insights in a university setting. This study used an in-person field study and an online questionnaire to obtain student feedback about SHRs. Participants were notified of anonymizing their data and provided their informed consent. The presented user study was approved by the Ethics Committee of the Department of Social Sciences, University of Kaiserslautern (RPTU Kaiserslautern).

2.2.1. Field Study

The Wizard of Oz technique [39] was utilized where the investigation included a researcher controlling EMAH situated at the university corridor of RPTU Kaiserslautern. The language of interaction was English and German. The SHR was positioned at the university corridor, replicating real-life interactions. The main focus of the study was to examine how people react to the presence of the humanoid robot and to explore three specific research inquiries. This experimental design enabled the investigation of human perceptions and responses to a robot in a realistic university corridor setting. EMAH initiated interaction by saying "Hello, my name is EMAH. Do you have time to answer three short questions?" and if the participant agreed, it proceeded to ask the following questions:

- What are your feelings towards me as a robot?
- What tasks or information would you find useful from me?
- Would you interact with me if you saw me again?

All responses were noted down using real-time Google speech-to-text API in English and German.

2.2.2. Online Study

A within-subjects empirical study was carried out in English. The online form creator involve.me (https://www.involve.me/ (accessed on 20 February 2024)) was employed. A six-part mixed methods questionnaire was surveyed, including demographic factors (age, gender, residence history, language, and field of study). The second part explained EMAH and 'Social Robots'. Third, a brief video of EMAH at the RPTU Kaiserslautern university hallway along with four perspectives was shown with the instruction "You are going to watch a short video where EMAH is depicted in the university hallway of RPTU Kaiserslautern. Also, shown below, there are four perspectives of EMAH in the hallway. In the next three pages, we will ask you questions regarding this robot scenario. Please watch the video, view the images, and focus on the robot and your feelings toward the robot.". A 5-point Likert scale [40,41] was used to respond to 11 statements with 1 = Not at all, 2 = Slightly, 3 = Moderately, 4 = Very, 5 = Extremely. Participants were asked to express their ideas and sentiments about EMAH honestly. Participants were asked 14 yes–no questions about EMAH's physical appearance and functionality. Finally, participants were asked to respond to further open-ended questions, including elaborating and suggestive answers about their particular needs and expectations.

2.3. Participants

2.3.1. Field Study

The study involved 24 participants with different characteristics. The age range of the participants spanned from 18 to 55 years, encompassing a wide variety of age groups. The study included male and female participants. In addition, the participants had diverse language backgrounds, encompassing German, French, and English.

2.3.2. Online Study

A diverse and representative sample of 25 people (14 female, 10 male, and 1 other) from numerous nationalities volunteered for the study. The participants spoke German, French, English, etc. The average age of participants was 26.08 years (range 20–32 years, SD = 3.21). Participants were recruited in several ways, including online questionnaire distribution via department-based instant messaging, social media, and email invitations. This study targets overseas students at our university having access to a computer/laptop, smartphone, headphones, and internet connection. The average completion time of the online questionnaire was 16.58 min. Before participating, all subjects gave written consent, assuring informed contribution to the study.

2.4. Stimuli

2.4.1. Field Study

As shown in Figure 1, the main focus of our research is the Ameca robot in HRI, with a stature of 187 cm and a weight of 49 kg, representing adult humanlike proportions. The humanoid Ameca was designed by Engineered Arts with 52 degrees of freedom (DOF), allowing it to move like a human and improving its social skills. It has fluid body movements, attentive gaze-tracking behaviour, silicone skin, teeth, and lip sync of generated speech. Ameca's software now uses the EMAH robotic system developed at RRLab, which entails a sophisticated dialogue management system to perform interactions. The robotic system operates on Finroc, a robust C++/Java robot framework that facilitates real-time interaction with a structure that supports the robot's various functionalities, including speech recognition, perception system, and generation of robot gestures and poses [42]. The dialogue management system of EMAH was designed as a finite state machine powered by XML templates supported by a Finroc port-based message-passing system. The robot can perceive complex visuals and audio with a stereo vision camera and microphone. Although the appearance of the robot has been designed to portray a non-gendered look, the choice of robot text-to-speech (TTS) was set as female for English and German from Acapela group (https://www.acapela-group.com/ (accessed on 4 July 2024)) (British English: Lucy; German: Julia). Figure 2 gives an all-around view of EMAH during this study regarding location and waiting to be approached.

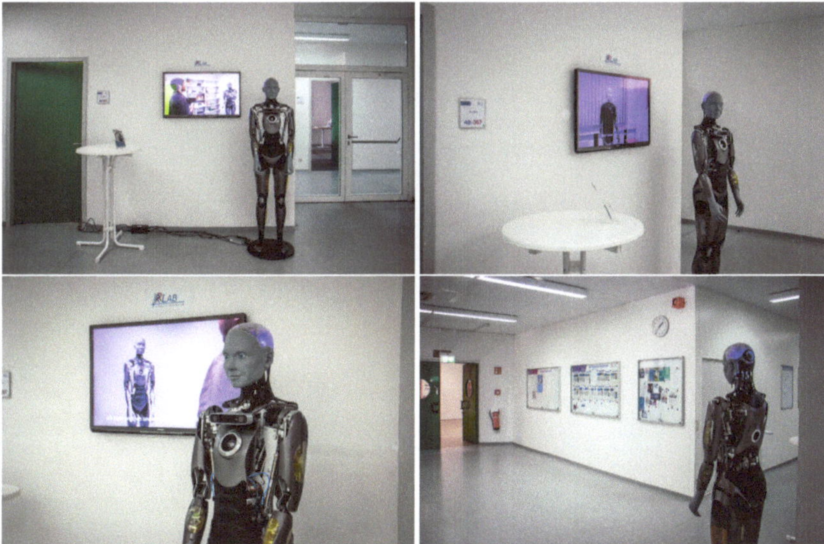

Figure 2. EMAH in the university hallway.

2.4.2. Online Study

Four 360-degree photos and one 50-s video (https://youtu.be/xCoHLO7N0cE (accessed on 26 June 2024)) of approaching robot in front view were shown to participants. The pictures were taken during the field study, as shown in Figure 2.

3. Study 2: Humanlikeness of Huamnoid Robots

Ekman's facial action coding system (FACS) has been instrumental in advancing the field of emotional robotics [43]. In our research in HRI here at RRLab with ROMAN, ROBIN, and EMAH, we implemented Ekman's inspired emotion to create realistic facial expressions, enhancing the robot's ability to engage in emotionally rich interactions with human interlocutors. As technology continues to evolve, researchers strive to build machines that recognize and respond appropriately to human emotions [44].

3.1. Research Question

R1: How does a robot's ability to depict emotions via facial expressions affect the human-likeness of the robot?

R2: How does a robot's ability to make gestures depicting movements affect the human-likeness of the robot?

3.2. Experimental Design

To answer our research questions, a within-subject online study was conducted at the RPTU Kaiserslautern using involve.me platform. A mixed-methods approach for the online survey included a 6-part questionnaire series rated on a Likert scale. The study began with the consent form and was followed by the participant demographics (age, gender, residence history, languages known, field of study, prior contact with SHRs at university). The second part included explanatory constructs of Attitudes Towards Robots and Trust Disposition Towards Robots [45]. The third instruction, supplemented with three named robot images, was to arrange the robots by their names based on low, medium, and high human likenesses (separated by a comma). The sections hereafter were repeated for all three robots in a randomized order. The fifth part included a short description (see Section 3.4) of the robot followed by the task to rate seven emotions (Anger, Disgust, Fear, Happy, Neutral, Sad, Surprise) of humanoid robot images "How accurately do you believe *robot* displayed a *emotion* face above?". The 5-point Likert scale used in this section was 1 = Not accurate at all, 2 = Slightly accurate, 3 = Moderately accurate, 4 = Very accurate, 5 = Extremely accurate. Next, participants viewed a short video featuring the robot interacting with a male human interlocutor. Unfortunately, compared to EMAH, due to robot malfunctions, old videos of ROMAN and ROBIN were used, thereby adding another variable—the content and scenario of interaction. A manipulation check was installed by surveying noticeable robot emotions in the video. Here, participants were also prompted "Describe what you think about how *robot* looks like." [12]. Finally, Godspeed constructs of anthropomorphism and likeability were measured using a 5-point Likert scale, coupled with three items of mechanical vs. humanlike (humanlikeness), strange vs. familiar (familiarity), and slightly eerie vs. extremely eerie (eeriness) [17,46].

3.3. Participants

This experiment focused on participants fluent in English, aged 18 to 35, accessing the internet via personal devices. Recruitment was effectively managed through direct emails and instant messaging platforms, targeting a broad range of university students. Our diverse group consisted of 67 participants, including 36 males, 29 females, and two whose gender was not specified. They represented various academic disciplines, predominantly from the fields of computer science and engineering, followed by other sciences and the humanities. Participants displayed considerable linguistic diversity, speaking 27 different languages, with many being multilingual in significant languages such as English, German, and various Asian languages. The geographic background of our sample included residents from multiple countries, predominantly Germany and India, the largest nationalities at RPTU Kaiserslautern, which provided a comprehensive cultural representation of students at our university. The completion times for our experiments varied widely, reflecting different engagement levels with the tasks assigned, with an average completion time of 30.17 min (approx.). Each participant consented in writing to use their anonymized data to advance this research study, ensuring ethical compliance and data integrity.

3.4. Stimuli

In accordance with state-of-the-art strategies to measure reduced robot uncanniness, this study utilized two aspects of humanoids, emotions and movement, which were considered stimuli types. We implemented seven emotions (Anger, Disgust, Fear, Happy, Neutral, Sad, and Surprise) on each robot (ROMAN, ROBIN, EMAH).

3.4.1. EMAH

The robot specifications are the same as described in Section 2.4.1, and as shown in Figure 1. Based on the available 13 facial action units (AUs) on EMAH as shown in Table 1 (left), we created the target facial expressions as shown in Figure 3. The short description used in the emotion classification task was, "EMAH, short for Empathic Mechanized Anthropomorphic huamnoid system, is developed to interact naturally with humans focusing on empathetic, personalized interactions.". An example of values used to create the Anger emotion is also shown in Table 1 (right). All emotions except for the Sad emotion have gaze set to default state due to the inability to lower the corners of the mouth (mouth movement limited to open-close); gaze target theta was set to the minimum value of -62. The video stimuli used contained the robot within an HRI with movements signified with depicting emotional expressions, hand gestures, gaze behaviour, and speech delivery as shown https://youtu.be/m8gzJQy-WXw (accessed on 15 April 2024).

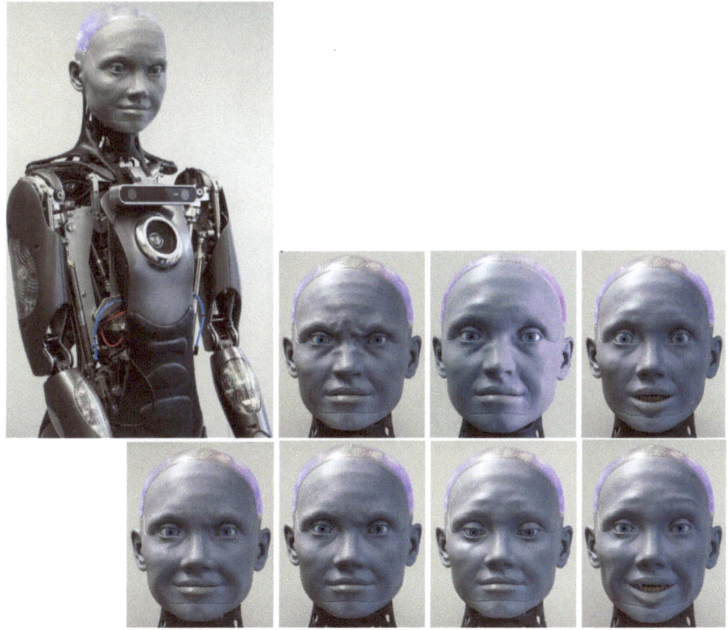

Figure 3. Emotions of EMAH along with the image shown for humanlikeness ranking.

Table 1. EMAH: Existing facial degrees of freedom available from Ameca motor control system (left), example values of Anger facial expression (right).

Action Unit	Default	Min	Max	Example: Anger	Value
Smile L	0	0	1	Smile L	0
Smile R	0	0	1	Smile R	0
Mouth open	0	0	2	Mouth open	0
Jaw slew	0.5	0	1	Jaw slew	0.5
Top eyelid L	1	−1	2	Top eyelid L	1
Top eyelid R	1	−1	2	Top eyelid R	1
Bottom eyelid L	1	−1	2	Bottom eyelid L	0
Bottom eyelid R	1	−1	2	Bottom eyelid R	0
Brow L	0.5	0	1	Brow L	0
Brow R	0.5	0	1	Brow R	0
Outer brow L	0	0	1	Outer brow L	0
Outer brow R	0	0	1	Outer brow R	0
Nose wrinkle	0	0	1	Nose wrinkle	1
Gaze target phi	0	−130	130	Gaze target phi	0
Gaze target theta	0	−62	62	Gaze target theta	0

3.4.2. ROBIN

The robot specifications are described in Figure 4. Based on the available 20 facial action units (AUs) on ROBIN, the existing facial expressions were chosen [47]. The emotions are shown in Figure 5. The short description used in the emotion classification task was "ROBIN, short for robot–human interaction, focuses on creating a seamless interaction by bridging the gap between robotic functionality and human interaction through advanced recognition of both non-verbal and verbal cues.". The video stimuli used for ROBIN with visible movements can be seen https://youtu.be/bVoVtfm9kBo (accessed on 15 April 2024).

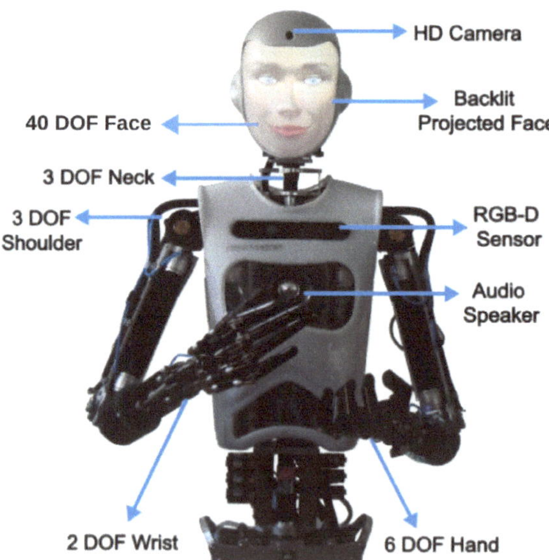

Figure 4. Robot–human interaction (ROBIN) [7] (p. 1).

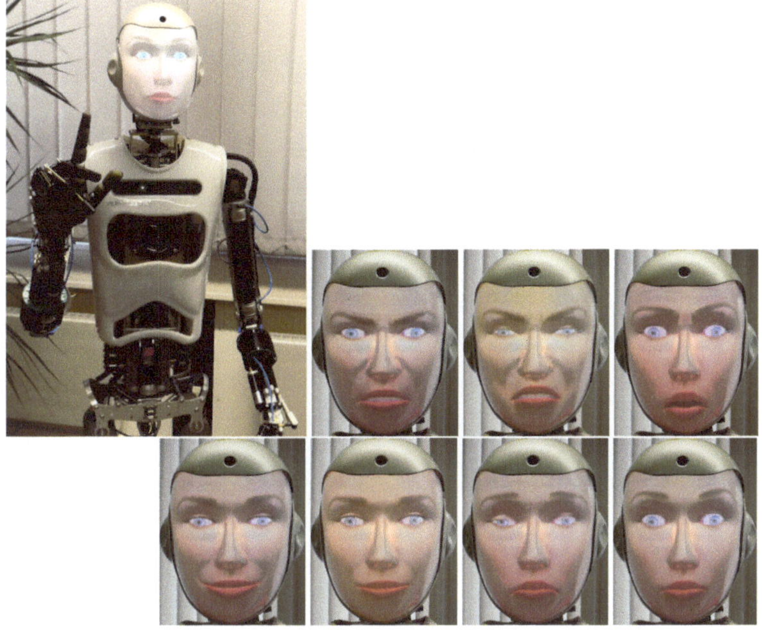

Figure 5. Emotions of ROBIN along with the image shown for ordering for humanlikeness.

3.4.3. ROMAN

The robot specifications are as described in Figure 6. Based on the available 40 facial action units (AUs) on ROMAN, the existing facial expressions were chosen [48]. The emotions are as shown in Figure 7. The short description used in the emotion classification task was, "ROMAN, short for robot–human interaction machine aimed to facilitate natural interactions beyond traditional input devices through dynamic modeling of human behaviour.". The video stimuli used to depict interaction with ROMAN can be viewed https://youtu.be/mzeE4YiggGA (accessed on 15 April 2024) [49].

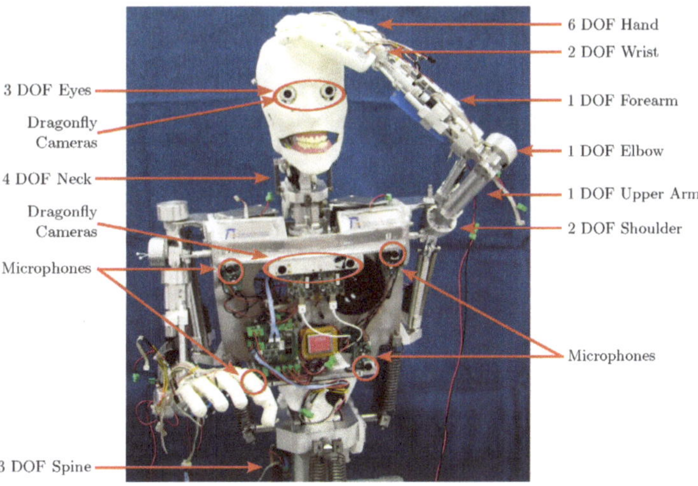

Figure 6. Robot–human interaction machine (ROMAN) [50] (p.1).

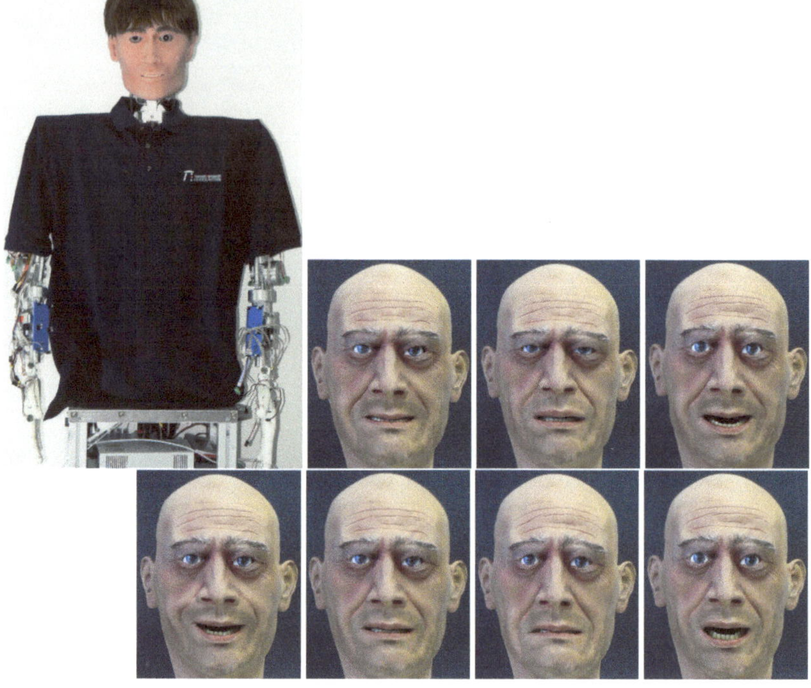

Figure 7. Emotions of ROMAN along with the image shown for ordering for humanlikeness.

4. Results and Analysis

The analyses were carried out using JASP 0.18.3, University of Amsterdam, Amsterdam, The Netherlands (https://jasp-stats.org/ (accessed 31 May 2024)).

4.1. Hallway Study

R1: What robot features are important for hallway HRI?

4.1.1. In-Person

Analyzing human behaviour during this study, the typical reactions from participants ranged from laughter and smiles to completely disregarding the robot's presence. Several participants eagerly approached EMAH, showing a keen interest in further interaction and offering positive feedback. Surprisingly, a group of participants interacted extensively, capturing selfies, making specific requests or commands, and documenting their experience with the robot through videos and photos. Several participants shared their emotions in response to EMAH's questions, ranging from fear to curiosity. Several participants interacted with EMAH by requesting information or expressing their task preferences. Despite the various reactions from participants, a few individuals opted not to provide a response. The findings highlight the complex range of human reactions to robots in real-world environments, providing insights into both common and distinct responses to humanoid robots such as EMAH.

Exploring the responses to the three questions put forth by EMAH revealed human perceptions of the robots at university. The participants exhibited various emotional responses to surveyed feelings towards the robot. Some expressed "Fear", and others noted the robot as "Interesting", with a few reporting "Nothing" to indicate indifference. When questioned about functional tasks for the robot, responses varied from "Everything" to specific requests such as "Mensa information" and "Cooking skill". Despite some inaudible responses and participants abruptly leaving amidst the experiment, the question of willingness to interact again received a recurring affirmation to "Yes" and "Yes of course". Findings, therefore, suggest a general openness to future engagements with the robot despite awkward first impressions.

4.1.2. Online

A. Robot Importance

On analyzing items from Appendix A.1, importance for robot learning (Mean = 4.120, SD = 0.881) and response rate (Mean = 4.000, SD = 0.707) was highlighted. Interpretation of non-verbal cues by the robot was also rated highly (Mean = 3.920, SD = 0.954), underscoring the value of nuanced communication. Lower scores were observed for human trust in the robot performing human tasks (Mean = 2.840, SD = 0.746) and neutral levels of comfort with the robot in the hallway (Mean = 3.000, SD = 1.000). These findings suggest that although students at university display enthusiasm for advanced robots, there remains hesitation for trust and comfort within HRI.

B. Interaction Preferences

On the analysis of Appendix A.2 yes–no questions, the findings suggest that a significant majority, 88%, favoured guidance within the university premises, while 80% and 76% supported EMAH providing Cafeteria options and course information, respectively. Interestingly, interaction initiation by EMAH was favoured by 68% of the respondents, with 60% of students agreeing to initiate interaction themselves. Similarly, 72% agreed on the importance of EMAH's ability to detect and appropriately respond to users' moods. Preferences over EMAH's personality were split, with a slight majority favouring the robot to have its own distinct personality (N = 18) over mimicking the user's mood (N = 7). Ratings for the data privacy item: The majority expressed concerns about the robot recording or storing camera data (72%) and dialogue transcripts (52%) during interactions.

The preferences for robot stance were equally divided between mobile and stationary options. Regarding gender assignment, a notable preference was shown for a neutral gender, although some responses indicated selections for either female or male gender, reflecting varied participant comfort levels with gendered interfaces. The graphical representation of the findings can be seen in Figure 8.

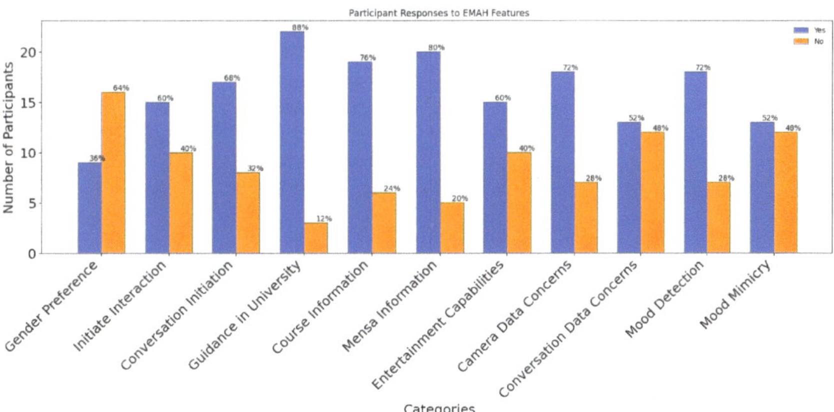

Figure 8. Results of human preferences of interaction with EMAH.

C. Open-ended

Participants stated a range of skills expected of EMAH, with one emphasizing the need for "empathetic talks" and another suggesting for "navigating, speaking different languages, and chatting" as essential. Responses also reveal the expectation for context-aware interactions beyond being an information agent. Befitting the robot in the university hallway, expected tasks centred around academic and administrative support, with one participant requesting "guidance regarding courses and locations", and another "answering questions about university procedures such as enrollment". This indicates EMAH usage as a navigational aid and a procedural guide within the university setting. Privacy concerns were also highlighted, and participants advocated for strict user data protection. One participant explicitly mentioned "just do not record my face or voice", reflecting concerns for personal data security in HRI. Regarding recommendations for robot behaviour, participants wanted EMAH to "focus on one person at a time" and employ precise turn-taking mechanisms. This can be inferred as mirroring human-to-human conversation dynamics. Regarding robot engagement during interactions, participants suggested that EMAH could use gestures such as "nodding as listening and putting a hand on a head as thinking". Suggestions for EMAH to improve daily university experiences included "being someone cool to talk to", "Make university more interesting" and providing "directions and answering different questions". The open-ended responses collectively reflect a small portion of the university community open to technological integration with clear expectations of privacy and interaction quality.

4.2. Humanlikeness Study

R1: How does a robot's ability to depict emotions via facial expressions and movement affect the humanlikeness of a robot?

4.2.1. Rankings Before-After

In a comparative analysis using the Wilcoxon signed-rank test, the study aimed to explore changes in the perception of human likeness for three social robots—ROBIN, ROMAN, and EMAH-before and after exposure to robot emotions and HRI videos. Rankings of ROBIN exhibited no significant change in perceived human likeness, suggesting it is perceived consistently, possibly indicating medium human likeness ($p = 0.935$). In contrast, ROMAN and EMAH showed significant changes in perception, suggesting a shift in how they are viewed post-experiment ($p = 0.015$ * and $p = 0.0326$ *, respectively). The results demonstrated varied changes in rankings, with significant alterations observed for ROMAN and EMAH but not for ROBIN. At the beginning of the experiment, ROBIN and ROMAN were frequently perceived as the least humanlike (N = 26 and N = 25, respectively), while EMAH (N = 45) was most commonly seen as the most humanlike. After the experiment, perceptions shifted notably, with ROMAN being the least humanlike (N = 40), suggesting a decreased perceived humanlikeness. In contrast, EMAH's position as the most humanlike decreased (N = 32), yet it remained the highest among the robots. ROBIN's perception as least humanlike decreased (N = 17) in the post-experiment rankings, and an increase in being perceived as more humanlike (N = 21).

4.2.2. Robot Emotion Classification

In analyzing the emotion recognition across the three humanoid robots, findings suggest distinctive emotional expressiveness attributed to each robot, as shown in Table 2. ROMAN displayed enhanced expressiveness in emotions such as Sad (SD = 1.152) and Anger (SD = 1.130), with a significant variability suggesting a more dramatic emotional range. Additionally, ROBIN was found to be expressive in Happy (SD = 1.132) and Neutral emotions (SD = 1.089). Overall, ROBIN maintained moderate scores across emotions, suggesting a balanced but less intense emotional expression. Notably, EMAH was rated consistently high for Neutral (Mean = 4.090, SD = 0.965), with the low variance indicating a stable and subtle emotional display. The results highlight the appropriate capability of humanoids with facial attributes to exhibit specific emotions, facilitating emotional engagement within HRI.

Table 2. Descriptive Statistics for Humanoid Emotions.

	ROMAN			ROBIN			EMAH		
Emotion	Mean	SD	SE	Mean	SD	SE	Mean	SD	SE
Sad	3.716	1.152	0.141	3.254	1.283	0.157	3.119	1.094	0.134
Surprise	3.567	1.144	0.140	3.567	1.076	0.131	2.582	1.208	0.148
Anger	2.104	1.130	0.138	3.746	1.172	0.143	3.687	1.003	0.123
Disgust	2.343	1.136	0.139	3.642	1.164	0.142	2.373	1.126	0.138
Fear	2.910	1.300	0.159	3.403	1.155	0.141	3.403	1.060	0.129
Happy	3.090	1.276	0.156	3.552	1.132	0.138	3.642	1.111	0.136
Neutral	2.209	1.162	0.142	3.582	1.089	0.133	4.090	0.965	0.118

4.2.3. Robot Movement in HRI

A. Emotion check

In the video stimuli, ROMAN's facial expression was intended to display "Neutral" and was successfully perceived as Neutral (N = 42), with notable counts of Happy (N = 17) and Surprise (N = 7). ROBIN was intended to show Happy and Other (Thinking, Skeptical, Sarcasm), but was reported showing Neutral (N = 52), followed by Happy (N = 26) and Surprise (N = 19). EMAH was intended to be "Neutral" and was correctly perceived as such (N = 56), with some perceiving Happy (N = 19).

B. Description of Robot Appearance

The analysis of participants' descriptions of the robots' appearances employed the natural language toolkit (NLTK) library, leveraging stopwords and WordNetLemmatizer for preprocessing. Negative word lists consisted of weird, scary, and uncanny, whereas positive word lists consisted of humanlike, cool, and nice. Participants frequently described ROMAN with negative terms like uncanny (N = 3), weird (N = 1), and scary (N = 1). ROBIN was also negatively perceived, with descriptors such as weird (N = 1), uncanny (N = 2), and scary (N = 1). EMAH received similar negative feedback with uncanny (N = 2) and scary (N = 1). Positive descriptions were seldom, with ROMAN termed cool once, and ROBIN and EMAH both labeled humanlike sparingly, highlighting existing challenges for humanoid robots.

4.2.4. Robot Impression

The study explored the impact of robot emotion, assessed through images of emotional robot faces and movements in HRI videos, on the perception of uncanniness.

A. Anthropomorphism and Likeability

A repeated measures ANOVA was carried out to analyze the effects of humanoid type (EMAH, ROBIN, ROMAN), the two constructs of the Godspeed questionnaire (Anthropomorphism and Likeability), and their interaction on participant responses. The results showed a significant impact for "Humanoid Type" (F = 11.870, p < 0.001), indicating distinct perceptions based on different humanoid types. The "Godspeed" items also depicted a significant effect (F = 18.278, p < 0.001), suggesting notable variations in robot perception. Furthermore, the interaction between "Humanoid Type*Godspeed" revealed significant differences (F = 21.687, p < 0.001), emphasizing the combined effect of these factors. Note: Analysis results with and without the Greenhouse-Geisser sphericity correction remained consistent across these adjustments.

A post hoc Tukey test [51] revealed the following results. For the Fake–Natural dimension, EMAH was perceived as more natural than ROMAN (MD = −1.284, 95% CI: −1.918 to −0.649, *p* < 0.001). In the Machinelike–humanlike dimension, significant differences were observed: EMAH was perceived as more humanlike than ROMAN (MD = 1.433, 95% CI: 0.815 to 2.051, *p* < 0.001). Regarding the Unfriendly–Friendly and Unkind–Kind dimensions, ROMAN was consistently perceived as less friendly and kind compared to EMAH (MD = −1.030, 95% CI: −1.576 to −0.484, *p* < 0.001) and a lesser extent compared to ROBIN (MD = −0.955, 95% CI: −1.501 to −0.410, $p < 0.001$). The Unpleasant–Pleasant and Awful–Nice dimensions showed that ROMAN was perceived more negatively compared to both ROBIN and EMAH. Additionally, in the Artificial–Lifelike dimension, ROMAN was rated significantly more artificial than both ROBIN (MD = −0.746, 95% CI: −1.381 to −0.111, $p = 0.002$) and EMAH (MD = −0.448, 95% CI: −1.083 to 0.187, $p = 0.603$), underscoring its artificial appearance. Here, MD refers to Mean Difference. The descriptive measures are shown in Table 3.

Table 3. Descriptive Statistics for Anthropomorphism and Likeability.

	ROMAN		ROBIN		EMAH	
Godspeed	**Mean**	**SD**	**Mean**	**SD**	**Mean**	**SD**
Fake–Natural	2.090	1.055	2.851	1.104	3.030	1.073
Machinelike–humanlike	2.075	1.005	2.597	1.060	2.791	1.023
Unconcious-Concious	2.478	1.146	3.373	1.071	2.776	1.098
Artificial-Lifelike	2.104	0.956	2.493	1.092	2.582	0.987
Rigidly Elegantly	2.075	0.942	2.537	1.133	3.149	0.989
Dislike-Like	2.701	1.115	3.552	0.113	3.433	0.857
Unfriendly–Friendly	3.119	1.122	3.582	0.109	3.672	0.660
Unkind–Kind	3.045	0.960	3.373	0.097	3.672	0.683
Unpleasant–Pleasant	2.925	1.078	3.522	0.113	3.537	0.927
Awful–Nice	3.015	1.037	3.582	0.095	3.567	0.857

B.   Humanlikeness, Familiarity, Eeriness

The findings indicate that the robots ROMAN, ROBIN, and EMAH elicited varied impressions of humanlikeness, familiarity, and eeriness as shown in Figure 9 and Table 4. ROMAN was perceived as least humanlike (Mean = 2.090, SD = 0.866) and moderately eerie (Mean = 2.866, SD = 1.100), whereas EMAH was perceived as relatively more humanlike (Mean = 2.761, SD = 0.889) and less eerie (Mean = 2.537, SD = 0.959). ROBIN, similar to findings of emotion classification task Section 4.2.2, scored moderately across all attributes. These variations reflect differing degrees of uncanniness, directly influenced by the combination of robotic emotional expressions and robot movement presented.

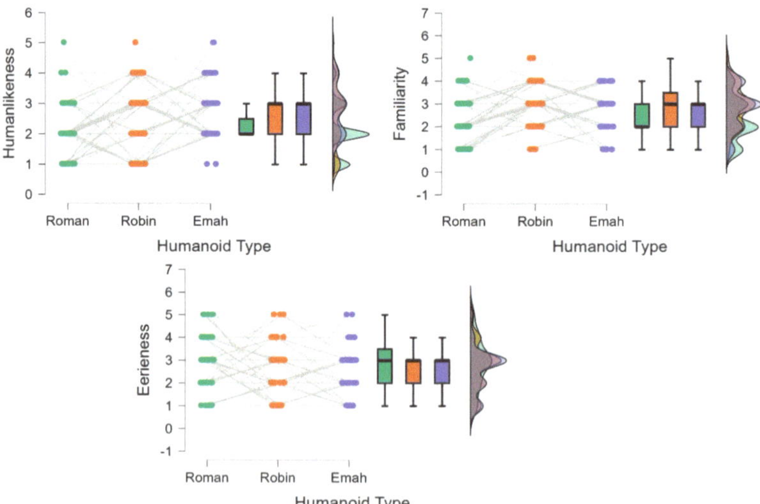

Figure 9. Raincloud plot of humanoid robots rated on bi-variate ratings of humanlikeness, familiarity, and eeriness.

Table 4. Descriptive statistics for humanlikeness, familiarity, eeriness measurements.

	ROMAN		ROBIN		EMAH	
Attribute	**Mean**	**Std. Dev.**	**Mean**	**Std. Dev.**	**Mean**	**Std. Dev.**
humanlikeness	2.090	0.866	2.552	1.063	2.761	0.889
Familiarity	2.313	0.972	2.925	0.893	2.821	0.920
Eeriness	2.866	1.100	2.716	1.042	2.537	0.959

5. Discussion and Conclusions

The exploration of humanoid robots in human–robot interaction (HRI) contexts, particularly within university settings, reveals interesting dynamics related to the uncanny valley phenomenon and the effectiveness of social humanoid robots (SHRs) like Ameca. This study summarises the findings from two critical studies conducted to investigate the impacts of robot appearance, emotional expression, and interaction skills on human perception of SHR Ameca. The uncanny valley phenomenon was evident in our studies, particularly in the varied responses to SHRs like EMAH and ROMAN, indicating a complex interaction between human expectations, robot design, and robot interactive capabilities. In Study 1, conducted at a university hallway setting, which focused on EMAH's interactions, participants displayed a range of reactions, from curiosity and engagement to wariness and indifference. EMAH's design, with its anthropomorphic features and expressive capabilities, aimed to facilitate natural interactions. The findings highlighted that while some participants were open to EMAH's presence and engaged in a positive manner, a few students remained disinterested, depicting the challenge of integrating SHRs seamlessly into public environments. The online survey complementing Study 1 revealed a strong preference for robots capable of advanced communication and emotional expression, indicating that these two features impact perceived humanlikeness and interaction quality. However, concerns about data privacy and the robot's ability to perform human tasks affected some participant's acceptance.

Study 2 aimed to understand further the depths to which robot humanlikeness affects human expectation across three types of humanoid robots—ROMAN, ROBIN, and EMAH—using emotional facial expressions and one-on-one interaction scenarios. The findings indicated that exposure to emotional stimuli significantly influenced participants' perceptions of humanlikeness, particularly in cases of ROMAN and EMAH. These robots demonstrated varied degrees of emotion regulation through facial action units and interaction capabilities, influencing how participants evaluated their humanlikeness when compared to human interlocutors. This re-emphasizes that the ability to express emotions enhances robots' acceptance and integration in social settings, diminishing aspects of the uncanny valley. Designing SHRs that balance humanlikeness with functional robot behaviour leads to user acceptance and remains an active challenge. With the integration of emotional expression and adaptive interaction skills, we aim to improve robot usability, thereby increasing acceptance.

In conclusion, this research advances the discourse on the uncanny valley phenomenon by interpreting how different aspects of humanlike humanoid robots influence human perception of robot appearance and interaction dynamics. It sets a foundation for future studies to explore how subtle changes in robot hardware design and behaviour can strategically navigate the uncanny valley, enhancing the usage and acceptability of SHRs in everyday settings. As assistance robots become increasingly prevalent in social settings, understanding and addressing the human–robot dynamics via robot appearance will be crucial for the successful integration of these technologies into human societies.

Author Contributions: Conceptualization, K.B. and A.A.; methodology, A.A.; software, K.B.; validation, A.A.; formal analysis, A.A.; investigation, A.A.; resources, K.B.; data curation, A.A.; writing—original draft preparation, A.A.; writing—review and editing, A.A.; visualization, A.A.; supervision, K.B.; project administration, K.B. All authors have read and agreed to the published version of the manuscript.

Funding: This research received no external funding.

Data Availability Statement: The data presented in this study are available on request from the corresponding author due to ethical reasons.

Acknowledgments: We express our sincere gratitude to our master students, Akil Alam Chowdhury and Golshan Shakeebaee, for their valuable contributions to this research. Akil was instrumental in conducting the online hallway study, and Golshan expertly created emotional expressions for EMAH.

Conflicts of Interest: The authors declare no conflicts of interest.

Notations and Abbreviations

HRI	Human–Robot Interaction
SHR	Social Humanoid Robot
UVE	Uncanny Valley Effect
RRLab	Robotics Research Lab
GIF	Graphics Interchange Format
DOF	Degree Of Freedom
ROMAN	Robot–human interaction machine
iB2C	integrated Behaviour-Based Control
EMAH	Empathic Mechanized Anthropomorphic Humanoid system
XML	Extensible Markup Language
TTS	Text To Speech
FACS	Facial Action Coding System
ROBIN	Robot–human interaction
AU	Action Unit
NLTK	Natural Language Toolkit

Appendix A. Hallway Study

Appendix A.1. 5-Point Likert

Here we would like to learn about your attitudes and feelings toward robots. Please read each statement carefully and provide your honest response based on your personal opinions and feelings.

1. How important do you think is a humanlike appearance for EMAH?
2. How important is it for EMAH to interpret and respond to non-verbal cues like gestures or facial expressions?
3. How important is it for EMAH to remember previous interactions with the same individuals?
4. What level of personalization in interactions do you expect from EMAH?
5. How significant is it for EMAH to display and recognize emotions?
6. How important is the response speed of EMAH during interactions?
7. How much would you trust EMAH to perform certain tasks on your behalf?
8. How important is it for EMAH to learn and improve from its interactions over time?
9. How critical is it for EMAH to manage interruptions gracefully?
10. How comfortable would you feel having a conversation with EMAH in a hallway?
11. How comfortable would you feel having a conversation with EMAH in a hallway knowing it is a stationary robot (it cannot move its legs)?

Appendix A.2. Yes–No

Please imagine yourself in the scenario where you witness EMAH in the hallway at your university and respond to these questions.

1. Would you prefer EMAH to be assigned a gender?
2. Which gender do you prefer EMAH to be assigned? (male vs. female vs. nonbinary)
3. Would you be inclined to initiate interaction with EMAH if you saw it in a hallway?
4. Should EMAH be programmed to initiate conversations first?
5. Would you prefer EMAH to guide people verbally about specific locations within the university?

6. Which robot stance do you prefer for EMAH? (Stationary (Only upper torso movement) vs. Mobile (able to navigate body))
7. Should EMAH provide information about courses offered at our university?
8. Should EMAH provide information about Mensa options for the day at our university?
9. Do you think EMAH should have entertainment capabilities, like telling jokes, singing or dancing?
10. Would you be concerned about EMAH recording or storing camera data during interaction for learning purposes?
11. Would you be concerned about EMAH recording or storing conversational transcript/dialogues for learning purposes?
12. Do you think EMAH should be able to detect and appropriately respond to a person's mood (e.g., happy, stressed)?
13. Do you think EMAH should mimic facial expressions and gestures in response to a person's mood?
14. Do you think EMAH should have it's own personality or mimic a person's mood? (it's own vs. match person's)

Appendix A.3. Open–Ended

Please imagine yourself in the scenario where you witness EMAH in the hallway at your university and respond to these questions.

1. What skills do you think the robot is able to perform?
2. What specific tasks or information would you find most useful from EMAH?
3. What topics would you like EMAH to be knowledgeable about or programmed to discuss?
4. Express your privacy concerns, if any, regarding the data EMAH might collect during interactions.
5. Share your safety concerns, if any, regarding interactions with EMAH.
6. How should EMAH manage interactions with multiple people simultaneously?
7. Describe how you would prefer EMAH to handle misunderstandings or errors during interactions.
8. Share your thoughts on how EMAH should signal that it's 'listening' or 'thinking' during an interaction.
9. How do you feel about interacting with humanoid robots like EMAH in a university setting? (if you feel negative or nothing, please suggest how EMAH could fix this)
10. In what ways could EMAH improve your daily experience in the university corridors?

References

1. Mori, M.; MacDorman, K.F.; Kageki, N. The uncanny valley [from the field]. *IEEE Robot. Autom. Mag.* **2012**, *19*, 98–100. [CrossRef]
2. Strait, M.; Vujovic, L.; Floerke, V.; Scheutz, M.; Urry, H. Too much humanness for human-robot interaction: exposure to highly humanlike robots elicits aversive responding in observers. In Proceedings of the 33rd Annual ACM Conference on Human Factors in Computing Systems, Seoul, Republic of Korea, 18–23 April 2015 ; pp. 3593–3602.
3. Seyama, J.; Nagayama, R.S. The uncanny valley: Effect of realism on the impression of artificial human faces. *Presence* **2007**, *16*, 337–351. [CrossRef]
4. Hoorn, J.F.; Huang, I.S. The media inequality, uncanny mountain, and the singularity is far from near: Iwaa and Sophia robot versus a real human being. *Int. J. Hum.-Comput. Stud.* **2024**, *181*, 103142. [CrossRef]
5. Dennler, N.S.; Kian, M.; Nikolaidis, S.; Matarić, M. Designing Robot Identity: The Role of Voice, Clothing, and Task on Robot Gender Perception. *arXiv* **2024**, arXiv:2404.00494.
6. Mishra, C.; Skantze, G. Knowing where to look: A planning-based architecture to automate the gaze behavior of social robots. In Proceedings of the 2022 31st IEEE International Conference on Robot and Human Interactive Communication (RO-MAN), Naples, Italy, 29 August–2 September 2022; IEEE: Piscataway, NJ, USA, 2022; pp. 1201–1208.
7. Ashok, A.; Paplu, S.; Berns, K. Social Robot Dressing Style: An evaluation of interlocutor preference for University Setting. In Proceedings of the 2023 32nd IEEE International Conference on Robot and Human Interactive Communication (RO-MAN), Busan, Republic of Korea, 28–31 August 2023; IEEE: Piscataway, NJ, USA, 2023; pp. 2499–2504.
8. Friedman, N.; Love, K.; LC, R.; Sabin, J.E.; Hoffman, G.; Ju, W. What robots need from clothing. In Proceedings of the 2021 ACM Designing Interactive Systems Conference, Virtual, 28 June–2 July 2021; pp. 1345–1355.

9.   Zafar, Z.; Ashok, A.; Berns, K. Personality Traits Assessment using PAD Emotional Space in Human-robot Interaction. In Proceedings of the VISIGRAPP (2: HUCAPP), Virtual, 8–10 February 2021; pp. 111–118.

10.   Oliveira, E.; Sarmento, L. Emotional valence-based mechanisms and agent personality. In *Proceedings of the Brazilian Symposium on Artificial Intelligence, Porto de Galinhas/Recife, Brazil, 11–14 November 2002*; Lecture Notes in Computer Science; Springer: Berlin/Heidelberg, Germany, 2002; pp. 152–162.

11.   Broadbent, E.; Kumar, V.; Li, X.; Sollers 3rd, J.; Stafford, R.Q.; MacDonald, B.A.; Wegner, D.M. Robots with display screens: a robot with a more humanlike face display is perceived to have more mind and a better personality. *PLoS ONE* **2013**, *8*, e72589. [CrossRef]

12.   Paetzel-Prüsmann, M.; Perugia, G.; Castellano, G. The influence of robot personality on the development of uncanny feelings. *Comput. Hum. Behav.* **2021**, *120*, 106756. [CrossRef]

13.   Kirandziska, V.; Ackovska, N. Human-robot interaction based on human emotions extracted from speech. In Proceedings of the 2012 20th Telecommunications Forum (TELFOR), Belgrade, Serbia, 20–22 November 2012, IEEE: Piscataway, NJ, USA, 2012; pp. 1381–1384.

14.   Breazeal, C.; Aryananda, L. Recognition of affective communicative intent in robot-directed speech. *Auton. Robot.* **2002**, *12*, 83–104. [CrossRef]

15.   Laakasuo, M.; Palomäki, J.; Köbis, N. Moral uncanny valley: A robot's appearance moderates how its decisions are judged. *Int. J. Soc. Robot.* **2021**, *13*, 1679–1688. [CrossRef]

16.   Clark, H.H.; Fischer, K. Social robots as depictions of social agents. *Behav. Brain Sci.* **2023**, *46*, e21. [CrossRef]

17.   MacDorman, K.F. Subjective Ratings of Robot Video Clips for Human Likeness, Familiarity, and Eeriness: An Exploration of the Uncanny Valley. In Proceedings of the ICCS/CogSci-2006 Long Symposium: Toward Social Mechanisms of Android Science, 2006, Volume 4. Available online: http://www.macdorman.com/kfm/writings/pubs/MacDorman2006SubjectiveRatings.pdf (accessed on 5 January 2024)

18.   Reich-Stiebert, N.; Eyssel, F.; Hohnemann, C. Exploring university students' preferences for educational robot design by means of a user-centered design approach. *Int. J. Soc. Robot.* **2020**, *12*, 227–237. [CrossRef]

19.   Perugia, G.; Boor, L.; van der Bij, L.; Rikmenspoel, O.; Foppen, R.; Guidi, S. Models of (often) ambivalent robot stereotypes: content, structure, and predictors of robots' age and gender stereotypes. In Proceedings of the 2023 ACM/IEEE International Conference on Human-Robot Interaction, Stockholm, Sweden, 13–16 1March 2023; pp. 428–436.

20.   Złotowski, J.; Proudfoot, D.; Yogeeswaran, K.; Bartneck, C. Anthropomorphism: opportunities and challenges in human–robot interaction. *Int. J. Soc. Robot.* **2015**, *7*, 347–360. [CrossRef]

21.   Hover, Q.R.; Velner, E.; Beelen, T.; Boon, M.; Truong, K.P. Uncanny, sexy, and threatening robots: The online community's attitude to and perceptions of robots varying in humanlikeness and gender. In Proceedings of the 2021 ACM/IEEE International Conference on Human-Robot Interaction, Boulder, CO, USA, 9–11 March 2021; pp. 119–128.

22.   Zargham, N.; Alexandrovsky, D.; Mildner, T.; Porzel, R.; Malaka, R. "Let's Face It": Investigating User Preferences for Virtual Humanoid Home Assistants. In Proceedings of the 11th International Conference on Human-Agent Interaction, Gothenburg, Sweden, 4–7 December 2023; pp. 246–256.

23.   Premathilake, G.W.; Li, H. Users' responses to humanoid social robots: A social response view. *Telemat. Inform.* **2024**, *91*, 102146. [CrossRef]

24.   Randall, N.; Sabanovic, S. A picture might be worth a thousand words, but it's not always enough to evaluate robots. In Proceedings of the 2023 ACM/IEEE International Conference on Human-Robot Interaction, Stockholm, Sweden, 13–16 March 2023; pp. 437–445.

25.   Shum, C.; Kim, H.J.; Calhoun, J.R.; Putra, E.D. "I was so scared I quit": Uncanny valley effects of robots' human-likeness on employee fear and industry turnover intentions. *Int. J. Hosp. Manag.* **2024**, *120*, 103762. [CrossRef]

26.   Li, Y.; Zhu, L.; Zhang, Z.; Guo, M.; Li, Z.; Li, Y.; Hashimoto, M. Humanoid robot heads for human-robot interaction: A review. *Sci. China Technol. Sci.* **2024**, *67*, 357–379. [CrossRef]

27.   Hirth, J. Towards socially interactive robots-designing an emotion-based control architecture. *Int. J. Soc. Robot.* **2011**, *3*, 273–290. [CrossRef]

28.   Schmitz, N. *Dynamic Modeling of Communication Partners for Socially Interactive Humanoid Robots.* Ph.D. Thesis, University of Kaiserslautern, Kaiserslautern, Germany, 2011.

29.   Ropertz, T.; Wolf, P.; Berns, K. Quality-Based Behavior-Based Control for Autonomous Robots in Rough Environments. In Proceedings of the ICINCO (1), Madrid, Spain, 26–28 July 2017; pp. 513–524.

30.   Proetzsch, M.; Luksch, T.; Berns, K. Development of complex robotic systems using the behavior-based control architecture iB2C. *Robot. Auton. Syst.* **2010**, *58*, 46–67. [CrossRef]

31.   Berns, K.; Hirth, J. Control of facial expressions of the humanoid robot head ROMAN. In Proceedings of the 2006 IEEE/RSJ international conference on intelligent robots and systems. IEEE: Piscataway, NJ, USA, 2006; pp. 3119–3124.

32.   Al-Darraji, S. *Perception of Nonverbal Cues for Human-Robot Interaction.* Ph.D. Thesis, University of Kaiserslautern, Kaiserslautern, Germany, 2016.

33.   Zafar, Z. Mutlimodal Fusion of Human Behavioural Traits: A Step towards Emotionally Intelligent Human-Robot Interaction. Ph.D. Thesis, Technische Universität Kaiserslautern, Kaiserslautern, Germany, 2020.

34. Paplu, S. *Personalized Human-Robot Interaction Based on Multimodal Perceptual Cues*. Ph.D. Thesis, Technische Universität Kaiserslautern, Kaiserslautern, Germany,

35. Schreck, J.L.; Newton, O.B.; Song, J.; Fiore, S.M. Reading the mind in robots: How theory of mind ability alters mental state attributions during human-robot interactions. In *Proceedings of the Human Factors and Ergonomics Society Annual Meeting, Seattle, WA, USA, 28 October–1 November 2019*; SAGE Publications Sage CA: Los Angeles, CA, USA, 2019; Volume 63, pp. 1550–1554.

36. Sarrica, M.; Brondi, S.; Fortunati, L. How many facets does a "social robot" have? A review of scientific and popular definitions online. *Inf. Technol. People* **2020**, *33*, 1–21. [CrossRef]

37. Hong, A.; Lunscher, N.; Hu, T.; Tsuboi, Y.; Zhang, X.; dos Reis Alves, S.F.; Nejat, G.; Benhabib, B. A multimodal emotional human–robot interaction architecture for social robots engaged in bidirectional communication. *IEEE Trans. Cybern.* **2020**, *51*, 5954–5968. [CrossRef] [PubMed]

38. Fong, T.; Nourbakhsh, I.; Dautenhahn, K. A survey of socially interactive robots. *Robot. Auton. Syst.* **2003**, *42*, 143–166. [CrossRef]

39. Dahlbäck, N.; Jönsson, A.; Ahrenberg, L. Wizard of Oz studies: why and how. In Proceedings of the 1st International Conference on Intelligent User Interfaces, Orlando, FL, USA, 4–7 January 1993; pp. 193–200.

40. Brown, J.D. Likert items and scales of measurement. *Statistics* **2011**, *15*, 10–14.

41. Schrum, M.L.; Johnson, M.; Ghuy, M.; Gombolay, M.C. Four years in review: Statistical practices of likert scales in human-robot interaction studies. In Proceedings of the Companion of the 2020 ACM/IEEE International Conference on Human-Robot Interaction, Cambridge,UK, 23–26 March 2020; pp. 43–52.

42. Reichardt, M.; Föhst, T.; Berns, K. Introducing finroc: A convenient real-time framework for robotics based on a systematic design approach. Robotics Research Lab, Department of Computer Science, University of Kaiserslautern, Kaiserslautern, Germany, Technical Report 2012. Available online: https://citeseerx.ist.psu.edu/document?repid=rep1&type=pdf&doi=aec8924f4d30b850 9cf6bb287a0644a50cb25fca (accessed on 4 February 2024).

43. Ekman, P.; Friesen, W.V. Facial action coding system. Environmental Psychology & Nonverbal Behavior 1978. Available online: https://psycnet.apa.org/doiLanding?doi=10.1037%2Ft27734-000 (accessed on 20 January 2024).

44. Mishra, C.; Verdonschot, R.; Hagoort, P.; Skantze, G. Real-time emotion generation in human-robot dialogue using large language models. *Front. Robot. AI* **2023**, *10*, 1271610. [CrossRef] [PubMed]

45. Biermann, H.; Brauner, P.; Ziefle, M. How context and design shape human-robot trust and attributions. *Paladyn J. Behav. Robot.* **2020**, *12*, 74–86. [CrossRef]

46. Bartneck, C.; Kulić, D.; Croft, E.; Zoghbi, S. Measurement instruments for the anthropomorphism, animacy, likeability, perceived intelligence, and perceived safety of robots. *Int. J. Soc. Robot.* **2009**, *1*, 71–81. [CrossRef]

47. Paplu, S.H.; Mishra, C.; Berns, K. Pseudo-randomization in automating robot behaviour during human-robot interaction. In Proceedings of the 2020 Joint IEEE 10th International Conference on Development and Learning and Epigenetic Robotics (ICDL-EpiRob), Valparaiso, Chile, 26–30 October 2020; IEEE: Piscataway, NJ, USA, 2020; pp. 1–6.

48. Hirth, J.; Berns, K. Motives as intrinsic activation for human-robot interaction. In Proceedings of the 2008 IEEE/RSJ International Conference on Intelligent Robots and Systems, Nice, France, 22–26 September 2008; IEEE: Piscataway, NJ, USA, 2008; pp. 773–778.

49. Hirth, J.; Schmitz, N.; Berns, K. Playing tangram with a humanoid robot. In Proceedings of the ROBOTIK 2012; 7th German Conference on Robotics. VDE, Munich, Germany, 21–22 May 2012; pp. 1–6.

50. Berns, K.; Zafar, Z. Emotion based human-robot interaction. In Proceedings of the MATEC Web of Conferences. EDP Sciences, St. Petersburg, Russia, 18–21 April 2018; Volume 161, p. 01001.

51. Pereira, D.G.; Afonso, A.; Medeiros, F.M. Overview of Friedman's test and post-hoc analysis. *Commun. Stat.-Simul. Comput.* **2015**, *44*, 2636–2653. [CrossRef]

 *actuators*

Article

Development of a Small-Sized Urban Cable Conduit Inspection Robot

Yiqiang You [1], **Yichen Zheng** [1], **Kangle Huang** [1], **Yuling He** [2,*], **Zhiqing Huang** [1] and **Lulin Zhan** [3]

[1] Wenzhou Power Supply Company, State Grid Zhejiang Electric Power Co., Ltd., Wenzhou 325000, China; 13587771858@139.com (Y.Y.); zyc20204@163.com (Y.Z.); 13857785722@163.com (K.H.); 13634278610@139.com (Z.H.)
[2] Hebei Engineering Research Center for Advanced Manufacturing & Intelligent Operation and Maintenance of Electric Power Machinery, North China Electric Power University, Baoding 071003, China
[3] Wenzhou Power Construction Co., Ltd., Wenzhou 325000, China; f283111@163.com
* Correspondence: heyuling1@163.com

Abstract: Cable conduits are crucial for urban power transmission and distribution systems. However, current conduit robots are often large and susceptible to tilting issues, which hampers the effective and intelligent inspection of these conduits. Therefore, there is an urgent need to develop a smaller-sized conduit inspection robot to address these challenges. Based on an in-depth analysis of the characteristics of the cable conduit working environment and the associated functional requirements, this study successfully developed a small-scale urban cable conduit inspection robot prototype. This development was grounded in relevant design theories, simulation analyses, and experimental tests. The test results demonstrate that the robot's bracing module effectively prevents tilting within the conduit. Additionally, the detection module enables comprehensive 360-degree conduit inspections, and the vacuuming module meets the negative pressure requirements for efficient absorption of dust and foreign matter. The robot has met the expected design goals, effectively enhanced the automation of the cable conduit construction process, and improved the quality control of cable laying.

Keywords: cable conduit; robot; inspection; bracing module; vacuuming module

Citation: You, Y.; Zheng, Y.; Huang, K.; He, Y.; Huang, Z.; Zhan, L. Development of a Small-Sized Urban Cable Conduit Inspection Robot. *Actuators* **2024**, *13*, 349. https://doi.org/10.3390/act13090349

Academic Editor: Ioan Doroftei

Received: 13 August 2024
Revised: 6 September 2024
Accepted: 9 September 2024
Published: 10 September 2024

1. Introduction

With the rapid development of cities, the importance of infrastructure, such as electricity, has significantly increased. Cables play a crucial role in this context, and their safe and stable operation is essential for ensuring the smooth functioning of urban areas. Typically, cables are laid in underground ducts, but due to construction activities, these ducts are prone to misalignment and the accumulation of debris, such as sand and dust. This not only affects cable operation but also poses potential risks to the power grid's operation. Given the narrow, inconvenient, and complex environment of cable ducts, developing a small-sized intelligent inspection robot capable of reliably detecting duct defects and taking corrective actions is highly necessary [1–3].

Researchers have developed a variety of conduit robots, which provide foundational cases for the research presented in this paper. Traditional conduit robots, utilizing wheels or tracks for movement, offer high stability and practical utility. Elankavi R S et al. comprehensively summarized traditional conduit robots equipped with different types of wheels, including magnetic wheels, standard wheels, and mechanical wheels [4]. Thung-Od, K et al. introduced a train-type conduit inspection robot equipped with two sets of vertical omnidirectional wheels capable of longitudinal and lateral movements to ensure the robot navigates the conduit path effectively [5]. M. Cardona et al. invented a single-degree-of-freedom wheeled conduit inspection robot featuring adjustable detection height, strong driving force, and significant engineering practicality [6]. To enhance obstacle avoidance, Tang, S. et al. developed a tracked conduit inspection robot with two independently driven side track motion mechanisms and posture adjustment systems, capable of performing well on various terrains and overcoming obstacles [7]. However, traditional wheeled and tracked robots generally lack support mechanisms, making them susceptible to tilting or even overturning within the conduit.

To address the issue of tilting within conduits, bionic conduit inspection robots have been developed. Wang, J. et al. proposed a small six-legged conduit robot equipped with a support mechanism, providing a structural foundation for subsequent obstacle avoidance robot designs [8]. Additionally, Manhong Li et al. developed a soft crawling robot for conduit detection [9], and Jingwei Liu et al. designed a snake-like conduit robot [10]. These robots mimic the characteristics of mollusks, allowing them to better adapt to conduits through creeping excavation, thus enabling stable movement and preventing tilting. Fang et al. highlighted that the design inspiration for bio-inspired robots is drawn from the wall-climbing abilities of insects and animals, offering considerable advantages in specific operational environments. This has advanced the development of bio-inspired climbing robots, making them suitable for applications on conduit walls as well [11]. However, the current application of bionic robots remains limited to ideal laboratory conditions, and their capabilities for practical engineering applications require further development.

To achieve both engineering practicality and anti-tilt capability in conduit robots, we can refer to external conduit inspection robots. For instance, Wang Z. et al. developed an external cable pipe wall-walking inspection robot that employs a non-enclosed four-bar clamping mechanism, ensuring stable operation [12]. Li J. et al. invented a flexible obstacle-crossing conduit robot, which significantly enhanced stability and obstacle-crossing capability through a rotating joint mechanism and an elastic shock-absorbing suspension system [13]. Additionally, Z. Tang et al. drew inspiration from the conduit outer diameter inspection robot developed by Z. Zheng et al. [14] and designed an adaptive diameter conduit inspection robot. This robot utilizes a support structure and a lifting structure arranged at 120°, with the outer diameter of the grinding wheel system adjusted through a screw adjustment module to accommodate different pipe diameters and prevent tilting issues [15]. Furthermore, to prevent the robot from tipping over within the conduit, Yan H. introduced a spiral conduit robot that reduces slippage by increasing the spiral angle, thereby providing an effective driving force for the adaptive conduit robot [16]. Concurrently, Zhang L. et al. conducted an in-depth study on the stability of conduit robots, providing a theoretical foundation for maintaining the stability of the robot's posture [17]. Subsequently, Y. Zhang et al. simulated and optimized the support and drive mechanisms of the adaptive diameter conduit robot, further advancing research into anti-roll solutions for conduit robots [18].

In addition to the detection function, the robot must be equipped with specialized devices to address various conduit defects. For instance, Y. Chen et al. designed an adaptive conduit robot drilling device capable of removing stones and debris from the conduit [19]. Moreover, Fanghua Liu et al. enhanced the conduit inspection robot by incorporating a rotating device to grind internal conduit defects coupled with a brush for cleaning purposes [20]. The robot developed in this paper is designed to detect internal defects of the conduit and clean the inside of the conduit at the same time, especially to transport sand and soil inside the conduit after construction.

In this context, this paper presents the development of a small-scale cable conduit inspection robot. This wheeled, single-degree-of-freedom inspection robot features a uniform four-wheel drive to prevent slipping, a bracing mechanism to avoid tilting, and a 360° all-round inspection device for accurate feedback on conduit defects. Additionally, it is equipped with a vacuuming device to collect dust and sand, thereby preventing any adverse impact on the cable insulation layer. By integrating these functionalities, the robot can complete inspection tasks more efficiently, providing a model for the development of future wheeled inspection robots.

2. Functional Requirements and Technical Indicators

The small-volume urban cable conduit inspection robot focuses on detecting and addressing internal defects in urban cable conduits. As shown in Figure 1, such defects include foreign matter such as dust, sand, and stones left in the conduit due to construction activities, as well as interface misalignment defects formed during the conduit connection process. All the defects significantly impact the laying and discharge of cables.

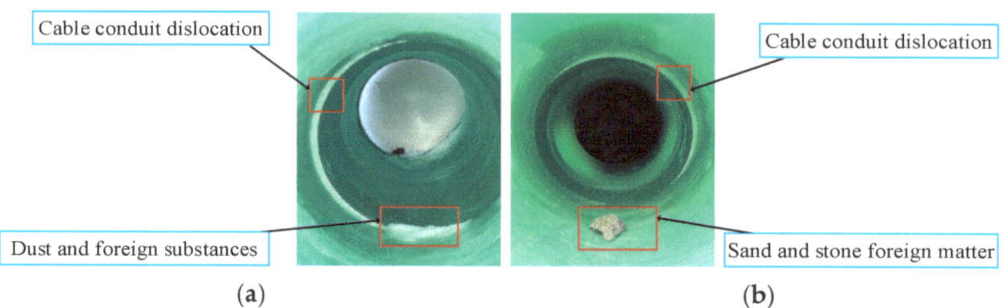

Figure 1. Different defect diagrams of cable conduit. (**a**,**b**) are conduit defect images.

Therefore, the following points are particularly important for small-sized urban cable conduit inspection robots to effectively complete their tasks in the complex environment of cable conduits:

1. Small Size: To ensure that the robot can fit into the confined space of the cable duct, it must be designed to be compact and small.
2. Reasonable Drive: When the robot is moving in a circular cross-section cable duct, it is necessary to ensure that each driving force has sufficient power to avoid slipping.
3. Comprehensive Inspection: Given the differences in conduit diameters, the inspection device needs to be appropriately adjustable to ensure effective defect detection.
4. Stable Operation: When a wheeled robot maneuvers within a circular cross-section cable conduit, the center of gravity often shifts away from the conduit's center, leading to tilting. In severe instances, this can result in the robot rolling over. Hence, strict control of the tilt angle is crucial to prevent it from becoming excessive.

In addition to these requirements, the robot should also have the capability to clean foreign objects (vacuum) to meet the practical needs of engineering and production.

According to project indicators and functional analysis, the main quantitative indicators of the small-scale urban cable conduit inspection robot are presented in Table 1.

Table 1. Quantitative indicators of robot.

Robot Mass m/kg	Conduit Diameter Range D/mm	Number of Driving Wheels	Detection Angle α_0/($^\circ$)	Inclination Angle θ/($^\circ$)	Vacuuming Negative Pressure p/kPa
$\leq$12	225–275	4	360	$\leq$8	$\geq$3

3. Robot Working Principle and Structural Design

3.1. Robot Overall Structure Design

The structural design of the small-volume urban cable conduit inspection robot (hereinafter referred to as the robot) includes a main module, a bracing module, a detection module, and a vacuuming module. The overall structure is illustrated in Figure 2.

The camera detection module is positioned at the front end of the robot's main module, equipped with components such as an industrial camera, fill light, servo, and rotary motor drive. A bracing module is located in the middle section of the robot, comprising a push-rod motor and bearing rollers. The push-rod motor is adjustable for different conduit diameters with a specific stroke. At the rear end, there is a vacuuming module designed for collecting dust and foreign matter.

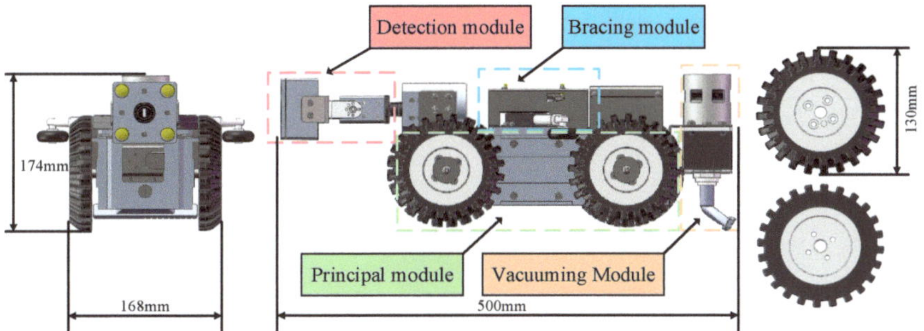

Figure 2. Three-dimensional diagram of the overall structure of the robot.

3.2. Robot Body and Bracing Module Design

As illustrated in Figure 3, power transmission within the robot's main body module occurs through a plane cross-axis gear to an odd-parallel-axis gear set, which subsequently transfers power to the front and rear output shafts. This transmission method ensures even distribution of power to the four sets of non-slip rubber drive wheels, thereby ensuring uniform power output.

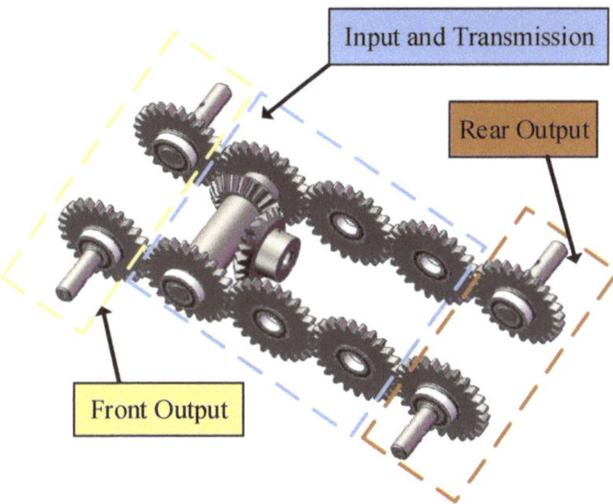

Figure 3. Three-dimensional diagram of the dynamic transmission of the robot.

The selection of the drive motor significantly influences the operational performance of the robot. Parameters such as the robot's mass, drive wheel radius, driving speed, and friction coefficient directly affect the motor's torque and power requirements. To simplify calculations, it is assumed that all four sets of drive wheels maintain non-slipping contact with the conduit wall.

T_1 is the torque required to overcome the rotational inertia:

$$\begin{cases} T_1 = J\alpha \\ \alpha = \frac{\omega}{t} \end{cases} \tag{1}$$

In the equation, J represents the equivalent rotational inertia of the drive motor itself, the transmission system, and the four sets of drive wheels. The variable α denotes the angular acceleration of the drive motor, ω represents the angular velocity of the motor, and t is the motor startup time.

The torque T_2 required by the wheel to overcome friction:

$$T_2 = \mu m g R \tag{2}$$

where μ represents the maximum static friction coefficient between the driving wheel and the conduit, m is the weight of the robot, and g denotes the acceleration due to gravity.

Thus, the power P_1 required by the motor to overcome the load and the power P_2 required to overcome friction can be determined.

$$\begin{cases} P_1 = \frac{T_1 n}{9550} \\ P_2 = \frac{T_2 n}{9550} \end{cases} \tag{3}$$

where n represents the motor speed.

From the above, the required safety torque T and safety power P for the motor can be determined:

$$\begin{cases} T = (T_1 + T_2)N \\ P = (P_1 + P_2)N \end{cases} \tag{4}$$

where N represents the safety factor.

Dynamic analysis reveals the necessary driving torque and power:

$$\begin{aligned} T &= \left(\frac{J\omega}{t} + \mu m g \right) N \\ P &= \left(\frac{J\omega n}{9550 t} + \frac{\mu m g n}{19100} \right) N \end{aligned} \tag{5}$$

Based on the calculated torque and power requirements, the selected motor dimensions, as shown in Figure 4, fulfill both the design specifications and spatial constraints. This motor is produced by DJI Technology Co., Ltd., Shenzhen, China.

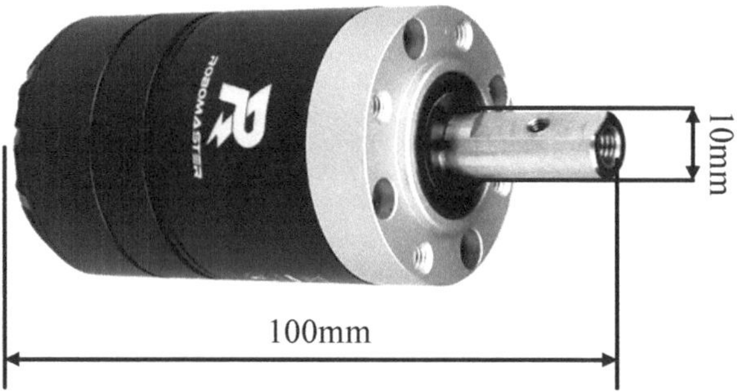

Figure 4. Drive motor dimensions and structural diagram.

To ensure stable operation and prevent tilting or tipping due to poor conduit conditions, a high center of gravity, or excessive speed, it is crucial to carefully analyze tilt phenomena and implement corresponding measures. Figure 5a illustrates the analysis of the robot without anti-tilt measures.

In Figure 5a, G and G_1 denote the positions of the center of gravity before and after tilting, respectively; θ represents the tilt angle; H is the distance between the center of gravity and the bottom of the robot; $2W$ denotes the length of the bottom of the robot; F represents the weight of the robot; F_x is the component of gravity parallel to the wheel axis; F_y is the component of gravity perpendicular to the wheel axis. Given the robot's relatively slow forward speed, only static tipping is considered in tipping scenarios. Under tilting conditions, gravity generates a tipping moment, M_1, and a tipping-preventing moment, M_2, expressed as follows:

$$M_1 = F_x \times H$$
$$M_2 = F_y \times W \tag{6}$$

The component expressions of gravity are:

$$F_x = F \times \sin\theta$$
$$F_y = F \times \cos\theta \tag{7}$$

If the tipping moment exceeds the anti-tipping moment, the robot will tilt over. To prevent tilting, the tilt angle must be maintained below a certain critical value, denoted as:

$$\theta \leq \arctan\left(\frac{W}{H}\right) \tag{8}$$

In actual conduit inspection processes, without implementing anti-tilt measures, it becomes impractical to maintain the tilt angle within a controllable range, thereby making it difficult to prevent tipping from occurring reliably.

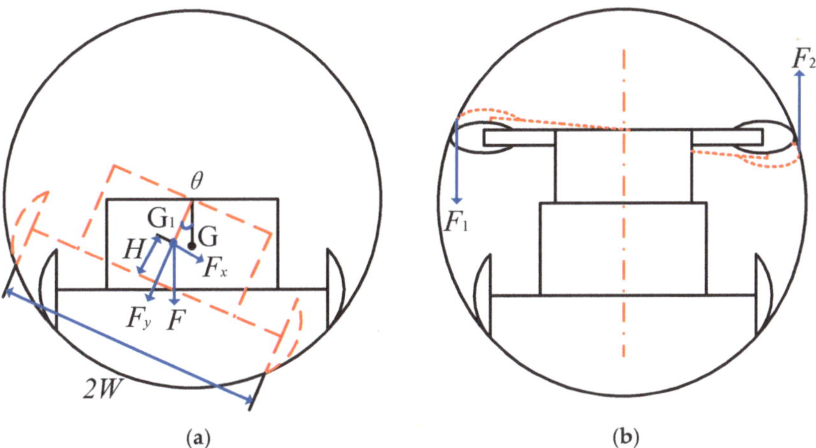

Figure 5. Simplified model of the robot's conduit traversal: (**a**) two-dimensional simplified model of the robot's inclination and rollover; (**b**) force analysis diagram of the robot's bracing device.

Therefore, it is essential to implement anti-tilt measures. As illustrated in Figure 6, bracing bearing rollers and push-rod motors are utilized to stabilize the robot and prevent tilting and rotation. The push-rod motor provides a specific amount of force and stroke to act as rigid bracing, adapting to the actual conditions within the conduit. The thrust on the duct wall is maintained by a passive system. First, the 3D model of the robot is placed within a 225–275 mm cable conduit, and the stroke of the push-rod motor is determined based on its position in the 3D space. The cable conduit is produced by Hebei Zhongming Environmental Protection Engineering Co., Ltd. in Cangzhou City, Hebei Province, China. The appropriate voltage and thrust (self-locking force) are then selected according to the power supply requirements. The selected push-rod motor operates at 12 V, with a 50 mm stroke and a thrust of 150 N.

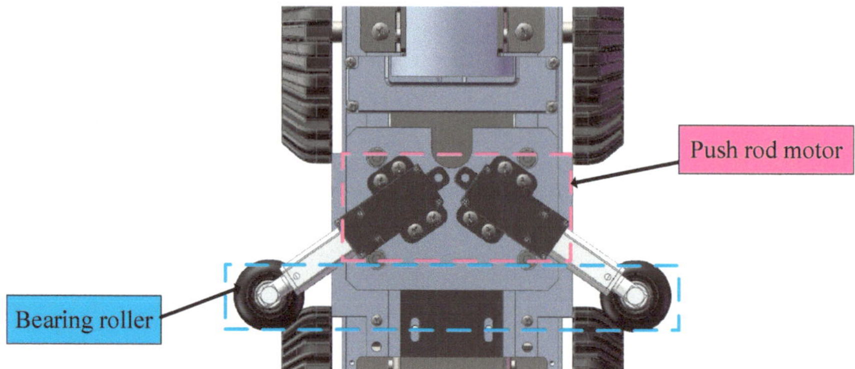

Figure 6. Three-dimensional schematic diagram of the robot's bracing device.

An MX471 current sensor module is added to the main control board, which extends the push-rod motor through the main control board. The MX471 current sensor module is produced by Shenzhen Nuojing Technology Co., Ltd., China. When an increase in current is detected, it indicates that the roller has contacted the cable conduit wall, prompting the motor to stop extending. At this point, the push-rod motor generates a self-locking force to hold against the duct wall, preventing the robot from tilting.

Additionally, the use of bracing bearing rollers reduces friction, transforming sliding friction between the bracing and the conduit wall into rolling friction, thereby minimizing the robot's travel resistance. During conduit inspections with this bracing mechanism, depicted in Figure 5b, if a clockwise tilt tendency occurs, the left bracing bearing roller generates a vertical downward force F_1, while the right side produces a vertical upward force F_2. These forces collectively create a torque that prevents the robot from tilting, ensuring smooth operation during conduit inspections.

3.3. Robot Detection and Vacuuming Module Design

Currently, most wide-angle camera modules used for inspections feature structures with adjustable relative heights, necessitating adjustments based on conduit diameter. As depicted in Figure 7a, when the camera axis deviates from the conduit centerline, the quadrilateral wide-angle area ABCD formed becomes asymmetric relative to the axis. This asymmetry results in variations between upper and lower observation areas, which can affect the assessment of conduit defects. Rotating the camera to view the asymmetric area can lead to image distortion if the rotation angle is excessive, significantly impacting the detection of conduit misalignment defects.

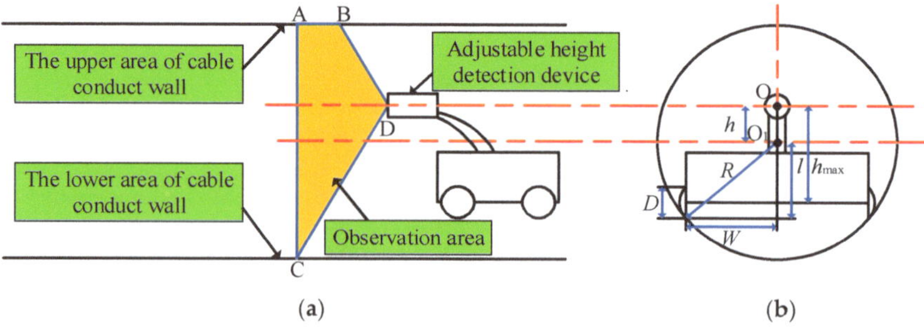

Figure 7. Analysis diagram of the optimal height of the robot's detection device: (**a**) conduit radial section diagram; (**b**) conduit axial view.

Therefore, the height of the adjustable detection device should be adjusted to align the two axes as closely as possible. In Figure 7b, where W is half of the robot width, D is the wheel diameter, R is the conduit radius, l is the vertical distance from the center of the conduit to the bottom of the wheel, and $h_{\max}$ is the maximum adjustable height of the camera relative to the bottom of the vehicle. Using these parameters, the approximate adjustable height h can be calculated as:

$$h = h_{\max} - \sqrt{R^2 - W^2} - D/2 \tag{9}$$

Therefore, the range of the conduit diameter should determine the corresponding adjustable height range, ensuring that $h_{\max}$ exceeds this adjustable height range. It is important to note that in Equation (9), the values of R and W should account for the thickness of the wheel.

Figure 8 illustrates the adjustable height range calculation, which incorporates a single-axis servo and rotary motor to compensate for the camera height in the detection module of the robot.

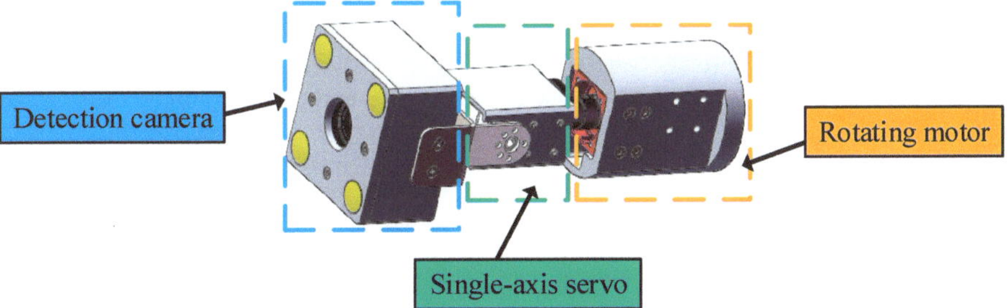

Figure 8. Three-dimensional schematic diagram of the robot's detection device.

When the detection module identifies a defect in the cable duct, targeted treatment of the defect is necessary. The small-volume urban cable duct detection and vacuuming robot utilizes a vacuuming module to absorb and collect dust and foreign matter. This prevents degradation of cable insulation performance, obstruction of heat dissipation, and corrosion damage caused by dust.

The dust collection module structure consists of a cyclone motor, a dust collection box, an air conduit joint, and a dust collection head, as illustrated in Figure 9. The cyclone motor serves as the core component, generating airflow and negative pressure to suction dust and dirt. The air conduit joint connects to the external dust collection head, directing airflow and dust entry. During operation, the cyclone motor creates negative pressure, efficiently drawing in dust and dirt from the conduit bottom. The dust collection head acts as the inlet, while the air conduit joint guides airflow and particulate matter into the dust collection box. To safeguard the motor from large particles, a filter is installed between them, and the dust collection box is sealed to securely store and collect dust and dirt.

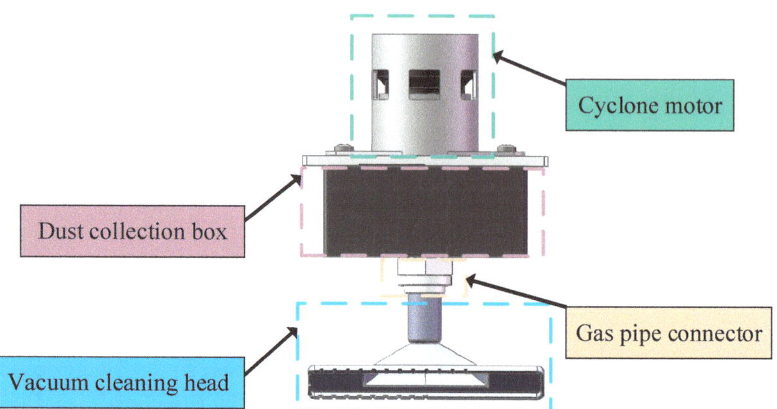

Figure 9. Three-dimensional schematic diagram of the robot's vacuuming device.

4. Robot Simulation Analysis

4.1. Simulation Analysis of Robot Body and Bracing Module

Based on the above analysis, it is evident that without anti-tilt measures, the robot moving in the conduit may tilt or even overturn. However, after incorporating the bracing module, the robot's tilt issue is mitigated. To further investigate the influence of the robot's speed within the conduit on tilting, dynamic analysis was conducted using the Motion plug-in of SolidWorks2021 software. This analysis considered different speeds and the presence or absence of bracing devices for the robot. The abbreviations for different simulation conditions are detailed in Table 2 below.

Table 2. Abbreviation for simulation of center of mass position under different conditions.

Abbreviation	Direction of Center of Mass	Bracing Mechanism Presence	Operating Speed $v/$(m/s)
Case 1	X	Without bracing	
Case 2	X	With bracing	
Case 3	Y	Without bracing	0.1
Case 4	Y	With bracing	
Case 5	X	Without bracing	
Case 6	X	With bracing	
Case 7	Y	Without bracing	0.2
Case 8	Y	With bracing	
Case 9	X	Without bracing	
Case 10	X	With bracing	
Case 11	Y	Without bracing	0.3
Case 12	Y	With bracing	

For robots equipped with bracing mechanisms, a detailed analysis of the contact forces and friction forces generated during their interaction with cable conduit walls is required. Additionally, the forces exerted by the bracing mechanisms on the duct walls at different speeds should be compared. Abbreviations for the various simulation conditions are presented in Table 3.

Table 3. Abbreviation for simulation of dynamic analysis under different conditions.

Abbreviation	Types of Forces	Operating Speed v/(m/s)
Case 13	Contact force	0.1
Case 14	Rolling friction force	
Case 15	Contact force	0.2
Case 16	Rolling friction force	
Case 17	Contact force	0.3
Case 18	Rolling friction force	

The selected parameters were validated by observing the relative position changes between the robot's center of mass and the origin, as shown in Figure 10. When the robot moves along the axial direction of the cable conduit, only the changes in the center of mass in the X and Y directions relative to the origin need to be measured to reflect the robot's inclination during motion, while changes in the Z direction can be disregarded.

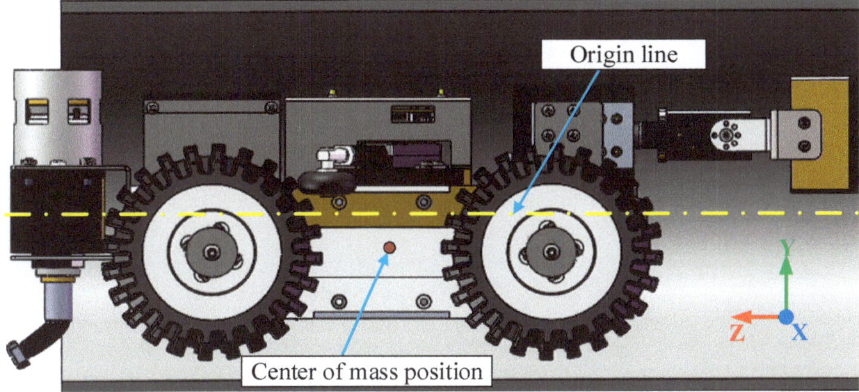

Figure 10. Schematic diagram of the simulation measurement object of the robot.

In Figure 10, the observed parameters are the distances between the robot's center of mass and the origin along the X and Y directions. Using the Motion plug-in, gravity, contact, material, and other parameters are set, providing the robot with a specific rotational speed (given speed), and the calculation time is set to 8 s. Subsequently, using the drawing plug-in integrated with SolidWorks, the vertical distance between these points is selected for data export and further processing.

Firstly, for the robot without a bracing device, vertical gravity and contact between the robot and the conduit are configured. The tire material is set to rubber, and the cable conduit material is set to glass-reinforced steel. The wheel speeds are set at 0.1 m/s, 0.2 m/s, and 0.3 m/s, respectively, allowing it to travel in a theoretically infinite cable conduit for 8 s.

Next, the same boundary conditions are applied to the robot with a bracing device, with the addition of contact settings between the auxiliary roller and the conduit. The tire material remains rubber, and the cable conduit material is glass-reinforced steel. Wheel speeds are set at 0.1 m/s, 0.2 m/s, and 0.3 m/s, allowing travel in a theoretically infinite cable conduit for 8 s.

After completing the simulations, a comparative analysis of the changes in the robot's center of mass in the X and Y directions under different conditions was conducted, as shown in Figure 11. Figure 11a illustrates the position changes of the center of mass in the X direction relative to the origin for robots with and without bracing mechanisms at a speed of 0.1 m/s. Figure 11b shows the corresponding changes at a speed of 0.2 m/s, and Figure 11c at 0.3 m/s. Similarly, Figure 11d–f display the changes in the Y direction relative to the origin at speeds of 0.1 m/s, 0.2 m/s, and 0.3 m/s, respectively, for both bracing and without bracing robots.

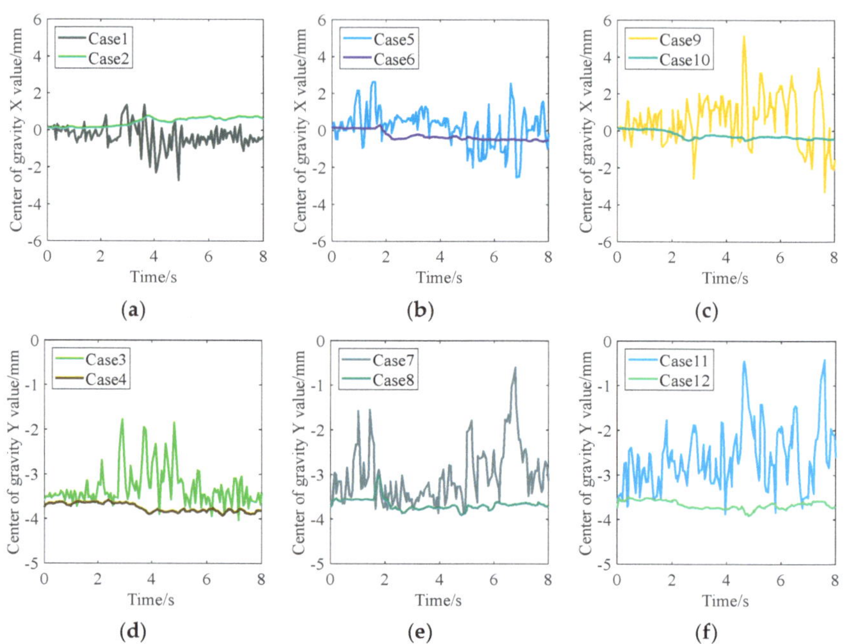

Figure 11. Comparison of displacement of centroid in X and Y directions at different velocities: (**a**) 0.1 m/s, X; (**b**) 0.2 m/s, X; (**c**) 0.3 m/s, X; (**d**) 0.1 m/s, Y; (**e**) 0.2 m/s, Y; (**f**) 0.3 m/s, Y.

A comparison of the six plots in Figure 11 reveals that robots without bracing mechanisms exhibit significant fluctuations in the center of mass in both the X and Y directions, indicating displacement in these directions and resulting in an overall inclination of the robot. In contrast, robots equipped with bracing mechanisms show minimal fluctuations in the center of mass in both directions, demonstrating the effectiveness of the bracing mechanism in mitigating the robot's inclination.

After completing the robot motion analysis, further investigation into its dynamics is required, focusing specifically on robots with bracing mechanisms. The analysis will primarily examine the relationship between the contact force and rolling friction force between the bracing mechanism and the inner wall of the cable conduit. As shown in Figure 12, Figure 12a,d present the contact force and rolling friction force at a speed of 0.1 m/s, respectively. Figure 12b,e display these forces at 0.2 m/s, while Figure 12c,f show the corresponding forces at 0.3 m/s.

Figure 12a–c show that the maximum contact forces at different speeds are approximately 7.5 N at 0.1 m/s, 12 N at 0.2 m/s, and 20 N at 0.3 m/s, respectively. This indicates that the magnitude of the contact force between the bracing mechanism and the cable conduit wall increases with the robot's speed. The fluctuations in contact force suggest that the bracing mechanism operates in a state of dynamic equilibrium, passively adjusting to mitigate inclination. Observing Figure 12d–f, it is noted that the maximum rolling friction forces are approximately 0.75 N at 0.1 m/s, 1 N at 0.2 m/s, and 1.5 N at 0.3 m/s. This demonstrates that, like the contact force, the rolling friction force also increases with the robot's speed. However, since the rolling friction force remains relatively small, its impact on the robot's forward motion is minimal.

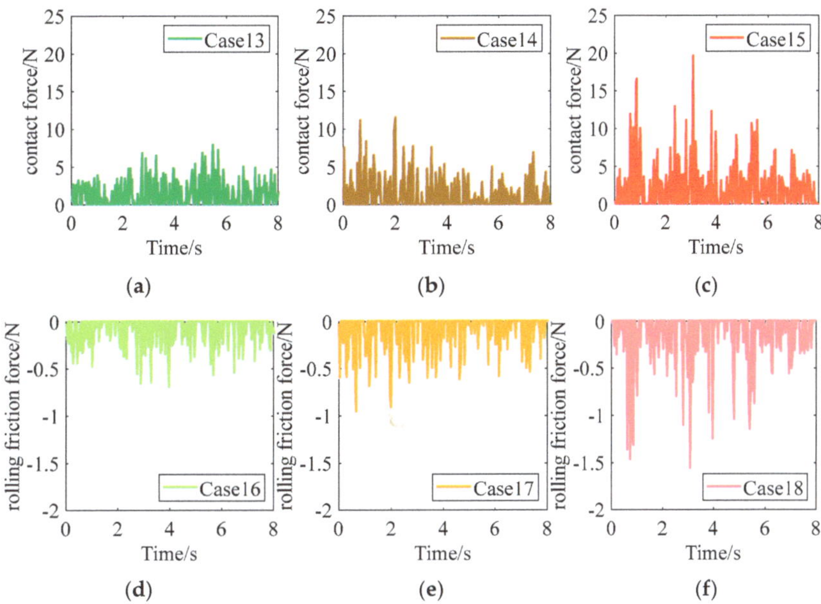

Figure 12. Variation in the contact force and rolling friction force between the robot's bracing mechanism and the cable conduit wall at different speeds: (**a**) contact force, 0.1 m/s; (**b**) contact force, 0.2 m/s; (**c**) contact force, 0.3 m/s; (**d**) rolling friction force, 0.1 m/s; (**e**) rolling friction force, 0.2 m/s; (**f**) rolling friction force, 0.3 m/s.

4.2. Simulation Analysis of Robot Detection and Dust Collection Module

In the ADAMS2019 motion simulation software, connections were added to the visual inspection module. A fixed pair was applied to the motor rack located at the right end of the module. Additionally, rotation pair 1 was established between the right-end motor shaft and the middle servo rack, while rotation pair 2 was set between the middle servo and the left-end camera rack. The configured visual inspection module setup is illustrated in Figure 13.

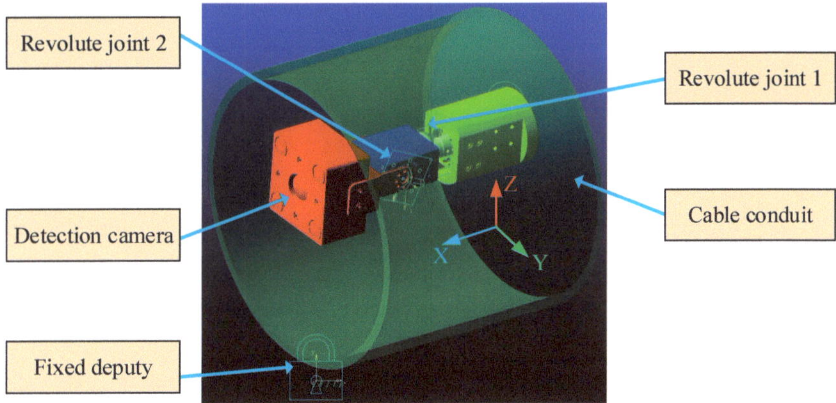

Figure 13. The interface diagram of completion of ADAMS motion settings.

The revolute pair drive is configured to simulate actual motor operation in the ADAMS motion simulation software. Initially, during setup, the visual inspection module remains static. Subsequently, revolute pair 2 rotates, orienting the camera's horizontal symmetry plane at a 30° angle relative to the horizontal plane, simulating the camera's adjustment to observe or capture road and conduit conditions ahead, crucial for transmitting conduit conditions for image recognition. Following this, revolute pair 1 initiates rotation, causing the camera and servo to rotate, mimicking a 360° observation (adjustable as per actual requirements) of the conduit's current position. Finally, revolute pair 2 rotates again to conclude the camera's capture session, returning it to its initial position. This sequence is illustrated in Figure 14.

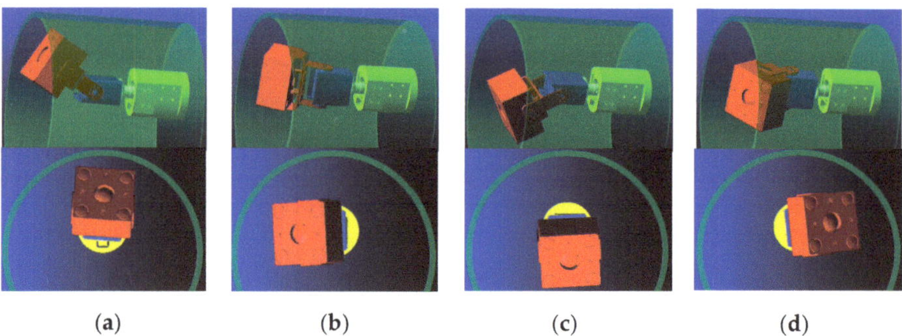

(a)   (b)   (c)   (d)

Figure 14. All orientation diagrams of the detection device in the simulation environment: (**a**) upper position, (**b**) left position, (**c**) lower position, (**d**) right position.

To limit the angle of the servo motor and prevent collisions or interference between the camera and the conduit wall, a comprehensive kinematic simulation of the servo's position within the cable conduit was conducted at its maximum operating angle, as shown in Figure 14. The results indicate that the camera does not collide or interfere with the cable conduit wall at any of the four orientations tested, confirming that the servo's maximum operating angle of 30° is a reasonable design choice.

After detecting defects such as dust and foreign matter, the detection module activates the suction module in collaboration with the main module to perform suction operations. To ensure sufficient negative pressure for effective dust and foreign matter removal, an internal fluid simulation of the suction pressure is conducted.

Using the Flow Simulation plug-in in SolidWorks, the dust collection motor was selected, and the volume flow rate at the outlet of the collection box was calculated to be 0.005 m^3/s based on its parameters. The inlet ambient pressure was set to the standard external ambient pressure, boundary conditions were established, and grids were divided prior to solving. As depicted in Figure 15, the minimum static pressure recorded was 97.24844 kPa, resulting in a negative pressure of 4.07656 kPa relative to the ambient pressure. Typically, dust collection requires a negative pressure range of 1 kPa to 3 kPa, and the maximum negative pressure generated by the dust collection motor at the collection port is 4.07656 kPa, meeting operational requirements.

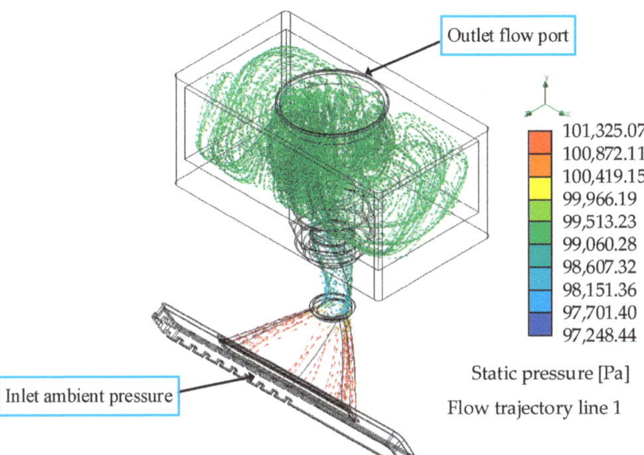

Figure 15. Simulation streamlines diagram of the internal fluid pressure of the vacuuming device.

5. Robot Experiment Verification

The designed small-volume urban cable conduit inspection robot was processed and manufactured, and the robot and its host computer terminal interface were obtained, as shown in Figure 16. Its key parameters are shown in Table 4.

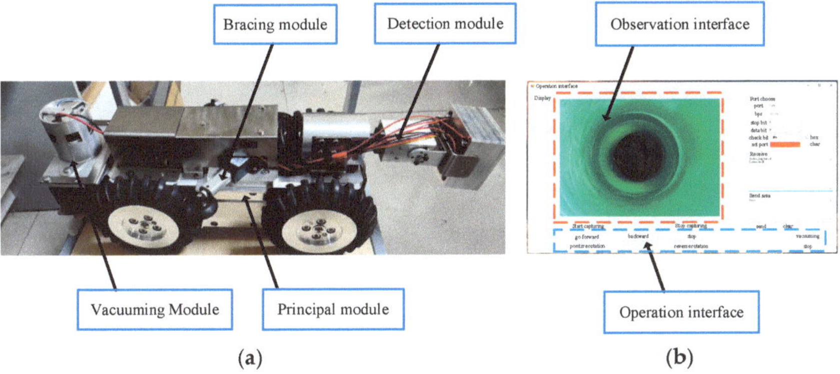

(**a**)   (**b**)

Figure 16. The physical picture of the robot: (**a**) the structural prototype of the robot; (**b**) the control interface of the robot's host computer terminal.

Table 4. The key parameter table of the robot prototype.

Key Parameter	Numerical Value
Robot mass m/kg	10
The traveling speed of the robot v/(km/h)	0.5–15
Driving motor power P/W	14
Adaptable conduit diameter D/mm	225–275
Battery endurance of the robot T/h	4

The robot can be directly connected to the power supply through the line. To enable emergency work, the robot is also equipped with an internal battery that can provide about 4 h of working time.

The robot's drive motor, suction motor, camera rotation motor, and servos communicate with the main controller via CAN communication, facilitated by the TJA1050 CAN transceiver chip (Zhidashunfa Electronics Co., Ltd., Shenzhen City, China) embedded in the main control board. The image transmission module communicates with the main controller through an Ethernet connection, as shown in Figure 17.

(a)   (b)

Figure 17. Photograph of the communication module between the robot and the main controller: (**a**) main control board; (**b**) image transmission module.

5.1. Experimental Verification of Robot Body and Bracing Module

Through the theoretical analysis and simulation verification above, it is evident that during actual conduit inspections, a robot without anti-tilt measures tilts after a period of operation. Conversely, integrating an anti-tilt device effectively mitigates this tilt.

The following is an experimental verification conducted by the robot prototype. As shown in Figure 18, in order to restore the real cable conduit scene more realistically, the prototype robot entered a conduit consisting of two sections of 225 mm diameter cable conduits with the same length and a misaligned interface in the middle at a speed of 0.2 m/s in both the bracing and unbracing states.

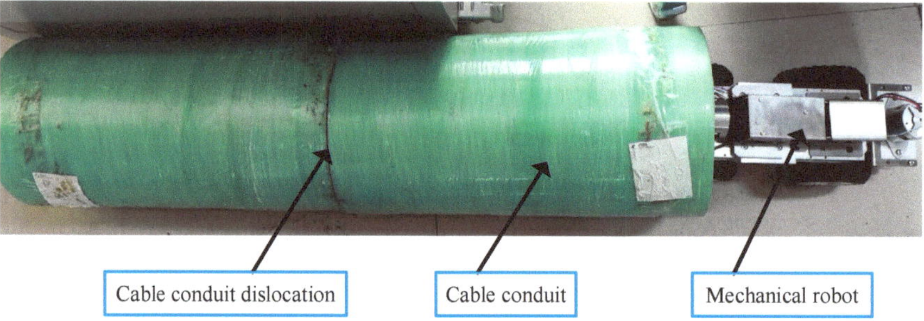

Figure 18. The experimental verification scene diagram of the robot.

During the same distance of travel, observations of the robot equipped with and without a bracing device from the entrance of the conduit are depicted in Figure 18. Figure 19a reveals that the robot equipped with a bracing device exhibits less tilt. The close proximity of the bracing wheel to the inner wall is evident through reflection, demonstrating its effective role in mitigating tilt and keeping the robot within acceptable tilt angles. Conversely, Figure 19b shows a noticeable tilt of the robot without a bracing device relative to the center of the cable conduit. This experimental validation aligns with theoretical predictions and simulations, confirming the beneficial anti-tilt effect of the bracing device during the robot's traversal in the cable conduit.

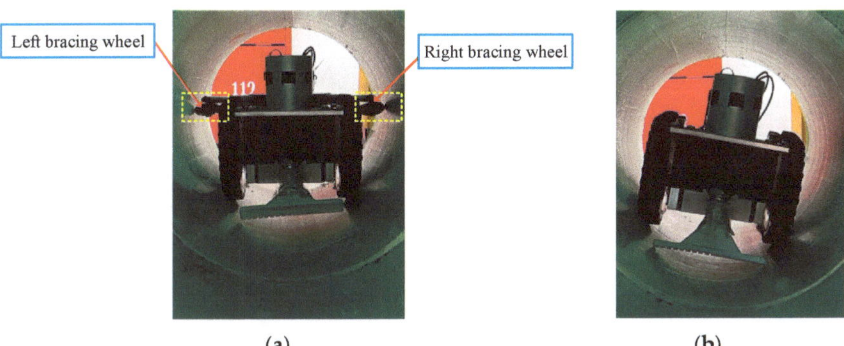

(a)   (b)

Figure 19. The experimental effect diagram of the main body and bracing devices: (**a**) the effect diagram of the bracing devices; (**b**) the effect diagram of the unbracing devices.

5.2. Experimental Verification of Robot Detection and Vacuuming Module

The robot, under coordinated action of the main module and bracing module, enters the 225 mm cable conduit. Utilizing commands from the host computer terminal, the robot's detection module proceeds to survey the conduit in all directions. As depicted in Figure 20, the experiment validates the detection device's capability to reach four positions: upper, lower, left, and right. Adjustment of the detection device enables a comprehensive 360° detection coverage of the cable conduit.

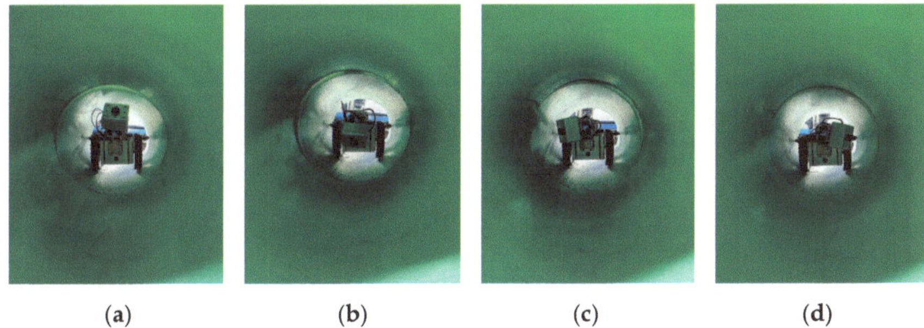

(a)   (b)   (c)   (d)

Figure 20. All orientation diagrams of the detection device in the experimental environment: (**a**) upper position, (**b**) lower position, (**c**) left position, (**d**) right position.

The simulation of the dust collection module demonstrates that the negative pressure generated by the dust collection motor, operating through the dust collection box and related components, achieves 4.07656 kPa, surpassing the required negative pressure for effective dust collection. As depicted in Figure 21, dust and other foreign particles are strategically placed within the conduit to simulate real-world conditions. Upon activation of the dust collection device, a noticeable reduction in dust content is observed before and after collection. These results affirm the device's efficacy in effectively suctioning and collecting dust and foreign matter within the conduit, thereby mitigating their potential impact on cable insulation and ensuring cable protection.

35

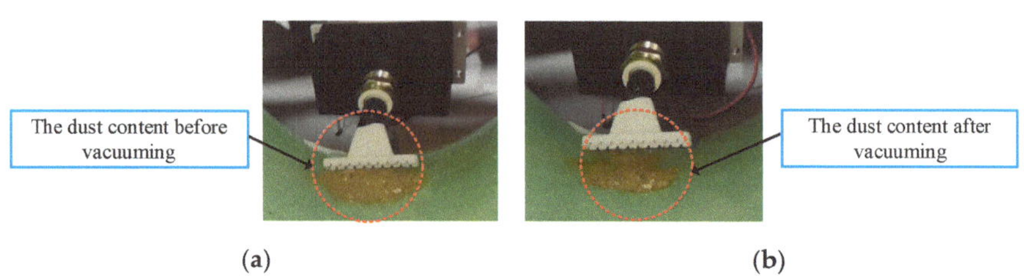

(a)　　　　　　　　　　　　　　(b)

Figure 21. Comparison chart of dust content before and after vacuuming: (**a**) before vacuuming; (**b**) after vacuuming.

6. Conclusions

Addressing the requirements for cable conduit inspection robots across various tasks, this study successfully developed a wheeled, small-volume urban cable conduit inspection robot prototype. This development was achieved through theoretical analysis, simulations, and experimental verification. The main conclusions are as follows:

To ensure smooth movement within the conduit, a four-wheel drive system and an anti-tilt and rollover bracing device based on gear transmission were designed for the robot. The feasibility of this device was validated through theoretical modeling and simulation analysis. Experimental observations at equivalent speeds and conduit distances demonstrate that the deflection angle of the robot equipped with the bracing device is smaller compared with the one without it. The results indicate that the support device effectively resolves the issues of tilting and tipping of the robot within the cable duct.

To meet the requirements of conduit inspection and cleaning, an inspection and vacuuming module for the robot was designed, followed by simulation and experimental verification. The results demonstrate that the detection device can rotate to four positions within the cable conduit without interference, thereby fulfilling the requirements for 360-degree conduit inspection. Furthermore, the dust collection module achieves a maximum negative pressure of 4.07656 kPa, effectively cleaning dust and other foreign matter from the conduit.

The research findings present an example for advancing the study of cable and conduit robots, contributing to the secure and stable operation of urban power grids.

Author Contributions: Conceptualization, Y.Y. and L.Z.; methodology, Y.Y. and K.H.; software, Y.Y. and Z.H.; validation, Y.Y., K.H. and Y.H.; formal analysis, Y.Y. and K.H.; investigation, Y.Y. and L.Z.; resources, Y.Y.; data curation, Y.Y. and Y.H.; writing—original draft preparation, Y.Y., Y.Z., K.H. and Z.H.; writing—review and editing, Y.Y., Y.Z. and Z.H.; visualization, Y.Y. and K.H.; supervision, Y.Y., Z.H. and Y.H.; project administration, Y.Y. and L.Z.; funding acquisition, Y.Y. and Y.H. All authors have read and agreed to the published version of the manuscript.

Funding: This research was funded by the Wenzhou Tusheng Holding Group Co., Ltd. Science and Technology project (CF058807002022007).

Data Availability Statement: Data are contained within the article.

Conflicts of Interest: Authors Yiqiang You, Yichen Zheng, Kangle Huang and Zhiqing Huang are employed by the company Wenzhou Power Supply Company, State Grid Zhejiang Electric Power Co., Ltd. Author Lulin Zhan is employed by the company Wenzhou Power Construction Co., Ltd. The remaining authors declare that the research was conducted in the absence of any commercial or financial relationships that could be construed as a potential conflict of interest.

References

1. Rusu, C.; Tatar, M.O. Adapting Mechanisms for In-Conduit Inspection Robots: A Review. *Appl. Sci.* **2022**, *12*, 6191. [CrossRef]
2. Verma, A.; Kaiwart, A.; Dubey, N.D.; Naseer, F.; Pradhan, S. A review on various types of in-conduit inspection robot. *Mater. Today Proc.* **2022**, *50*, 1425–1434. [CrossRef]
3. Elankavi, R.S.; Dinakaran, D.; Jose, J. Developments in conduit inspection robot: A review. *J. Mech. Contin. Math. Sci.* **2020**, *15*, 238–248.

4.   Elankavi, R.S.; Dinakaran, D.; Chetty, R.K.; Ramya, M.M.; Samuel, D.H. A review on wheeled type in-conduit inspection robot. *Int. J. Mech. Eng. Robot. Res.* **2022**, *11*, 745–754.

5.   Thung-Od, K.; Kanjanawanishkul, K.; Maneewarn, T.; Sethaput, T.; Boonyaprapasorn, A. An In-Pipe Inspection Robot with Permanent Magnets and Omnidirectional Wheels: Design and Implementation. *Appl. Sci.* **2022**, *12*, 1226. [CrossRef]

6.   Cardona, M.; Cerrato, J.; García, E. Design and Simulation of a Mobile Robot for Conduit Inspection. In Proceedings of the 2022 IEEE Central America and Panama Student Conference (CONESCAPAN), San Salvador, El Salvador, 18–21 October 2022; pp. 1–6.

7.   Tang, S.; Chen, S.; Liu, Q.; Wang, B.; Guo, X. A Small Tracked Robot for Cable Tunnel Inspection. In *Advances in Automation and Robotics*; Lee, G., Ed.; Lecture Notes in Electrical Engineering; Springer: Berlin/Heidelberg, Germany, 2011; Volume 122, pp. 591–598.

8.   Wang, J.; Mo, Z.; Cai, Y.; Wang, S. Kinematic Analysis of a Wheeled-Leg Small Conduit Robot Turning in Curved Pipes. *Electronics* **2024**, *13*, 2170. [CrossRef]

9.   Li, M.; Wang, G.; Wang, J.; Zheng, Y.; Jiao, X. Development of an inchworm-like soft conduit robot for detection. *Int. J. Mech. Sci.* **2023**, *253*, 108392. [CrossRef]

10. Liu, J.; Li, M.; Wang, Y.; Zhao, D.; Deng, R. Multi-gait snake robot for inspecting inner wall of a conduit. *Biomim. Intell. Robot.* **2024**, *4*, 100156.

11. Fang, Y.; Wang, S.; Bi, Q.; Cui, D.; Yan, C. Design and Technical Development of Wall-Climbing Robots: A Review. *J. Bionic Eng.* **2022**, *19*, 877–901. [CrossRef]

12. Wang, Z.; Wang, Y.; Zhang, B. Development and Experiment of Clamp Type Submarine Cable Inspection Robot. *Machines* **2023**, *11*, 627. [CrossRef]

13. Li, J.; Huang, F.; Tu, C.; Tian, M.; Wang, X. Elastic Obstacle-Surmounting Conduit-Climbing Robot with Composite Wheels. *Machines* **2022**, *10*, 874. [CrossRef]

14. Zheng, Z.; Zhang, W.; Fu, X.; Hazken, S.; Hu, X.; Chen, H.; Luo, J.; Ding, N. CCRobot-IV: An Obstacle-Free Split-Type Quad-Ducted Propeller-Driven Bridge Stay Cable-Climbing Robot. *IEEE Robot. Autom. Lett.* **2022**, *7*, 11751–11758. [CrossRef]

15. Tang, Z.; Li, Z.; Ma, S.; Chen, Y.; Yang, Y. Structure Design of Adaptive Conduit Detection Robot. In Proceedings of the 2021 7th International Conference on Control, Automation and Robotics (ICCAR), Singapore, 23–26 April 2021; pp. 136–140.

16. Yan, H.; Zhao, P.; Xiao, C.; Zhang, D.; Jiao, S.; Pan, H.; Wu, X. Design and Kinematic Characteristic Analysis of a Spiral Robot for Oil and Gas Conduit Inspections. *Actuators* **2023**, *12*, 240. [CrossRef]

17. Zhang, L.; Wang, X. Stable motion analysis and verification of a radial adjustable conduit robot. In Proceedings of the 2016 IEEE International Conference on Robotics and Biomimetics (ROBIO), Qingdao, China, 3–7 December 2016; pp. 1023–1028.

18. Zhang, Y.; Chen, H.; Wang, L.; Fu, Z.; Wang, S. Design of a Novel Modular Serial Conduit Inspection Robot. In Proceedings of the 2023 IEEE International Conference on Mechatronics and Automation (ICMA), Harbin, China, 6–9 August 2023; pp. 1847–1852.

19. Chen, Y.; Liu, S.; Hao, L.; Chi, F.; Xu, H.; Li, C. Analysis of Kinematic Characteristics of Cable Ducting Robot. In Proceedings of the 2022 4th International Conference on Power and Energy Technology (ICPET), Beijing, China, 28–31 July 2022; pp. 1192–1198.

20. Liu, F.; Sun, W. Analysis and Simulation of the Passability of Adaptive Conduit Grinding Robot. *Mach. Tool Hydraul.* **2021**, *49*, 15–21.

 *actuators*

Article

Research on the Methods for Correcting Helicopter Position on Deck Using a Carrier Robot

Yuhang Zhong [1,2], Dingxuan Zhao [1,2] and Xiaolong Zhao [1,2,*]

[1] Key Laboratory of Special Carrier Equipment of Hebei Province, Yanshan University, Qinhuangdao 066004, China; yuhang_zhong@stumail.ysu.edu.cn (Y.Z.); zdx@ysu.edu.cn (D.Z.)
[2] School of Mechanical Engineering, Yanshan University, Qinhuangdao 066004, China
* Correspondence: xlzhao@ysu.edu.cn

Abstract: When the landing position of a shipborne helicopter on the deck does not meet the requirements for towing it into the hangar, its position must first be corrected before towing can proceed. This paper studied the methods for using Shipborne Rapid Carrier Robots (SRCRs) to correct helicopter positions on the deck and proposed two correction methods, the stepwise correction method and the continuous correction method, aiming to improve the efficiency of the position adjustment process. Firstly, the actual helicopter landing position deviation was divided into two components—lateral offset and fuselage yaw angle—to quantitatively assess the deviations. Then, a mathematical model of the SRCR traction system was established, and its traction motion characteristics were analyzed. The kinematic characteristics and control processes of the two proposed position correction methods were subsequently studied, revealing the coordinated control relationships between key control elements. Finally, simulations were conducted to validate the feasibility of the proposed correction methods and compare their efficiencies. The results indicated that both the stepwise and continuous correction methods effectively achieved the position correction objectives. The stepwise method was more efficient when the initial yaw angle was small, while the continuous method proved more efficient when the initial yaw angle was large and the lateral offset was minimal. The results of this study may provide a valuable reference for correcting the positions of helicopters on deck.

Keywords: shipboard helicopter; helicopter landing deviation correction; helicopter deck position correction; shipborne rapid carrier robot; stepwise correction method; continuous correction method

Citation: Zhong, Y.; Zhao, D.; Zhao, X. Research on the Methods for Correcting Helicopter Position on Deck Using a Carrier Robot. *Actuators* **2024**, *13*, 342. https://doi.org/10.3390/act13090342

Academic Editors: Zhuming Bi and Ioan Doroftei

Received: 30 April 2024
Revised: 29 July 2024
Accepted: 20 August 2024
Published: 5 September 2024

1. Introduction

A shipborne helicopter needs to carry out on-board recovery operations after completing its mission at sea. This generally includes the processes of landing assistance, securing the helicopter, correcting the helicopter's deck position, and towing the helicopter to the ship's hangar. Shipborne special carrier robots are used for securing and transferring helicopters during the recovery process. Compared to the traditional manual methods of mooring and transferring, shipborne special carrier robots greatly improve the safety and efficiency of helicopter landing operations [1–5]. However, the landing position of the helicopter is affected by the motion of the deck [6] and the ship's airwakes [7]. Helicopter deck position correction is a prerequisite for helicopter warehousing. Correcting the helicopter's position in advance is necessary. This allows the helicopter traction and warehousing operations to be carried out smoothly and quickly [8–10]. Therefore, studying helicopter deck position correction methods is of great significance for ensuring helicopter warehousing and improving the efficiency of shipborne recovery operations.

Various types of shipborne helicopter traversing systems can be broadly classified into two categories; one type is the towing cart, which only possesses a traction function and needs to be used in conjunction with a harpoon grid system or recovery assist, securing, and traversing (RAST) system [11]. The other type integrates helicopter-assisted landing and towing functions, such as the Aircraft Ship Integrated Secure and Traverse System (ASIST) [12] and the Twin Claw Aircraft Ship Integrated Secure and Traverse System (TC-ASIST) [13].

Ground support equipment (GSE) [14,15] is a helicopter towing cart introduced by Curtiss-Wright, used for towing helicopters on land or aircraft carrier decks. Baury Alexandre enhanced the safety of the GSE towing system by adding a guide rail [16] and vacuum discs [17]. For traversing helicopters secured by the RAST system, Baekken Asbjorn et al. [18] designed the RAST helicopter towing system. Pesando Mario [19] designed a towing device for helicopters using both harpoon grid landing systems and RAST landing systems and used the Sea Lynx helicopter and the Sea King helicopter as examples to illustrate the towing process. However, the towing systems for harpoon grids and RAST require crew members to collaborate on the deck during landing aid and towing processes, resulting in high operational difficulty and low efficiency. To achieve automated operations for carrier-based helicopter landing and towing, Pesando Mario et al. [20] designed the ASIST system. This system completes securing and traversing operations by capturing and towing a rod on the helicopter underbelly. Zhang Z et al. [21] proposed the EASIST system with an asynchronous motor, enhancing shipborne helicopter traction efficiency and combat effectiveness. Zhang Z has enhanced ASIST performance across multiple research fields, including dynamic modeling [22], energy consumption characteristics [23], traversing performance [24], and system reliability [25].

Due to the requirement for helicopters to have rods installed on their underbellies for the ASIST system, which limits its compatibility with all helicopter models for landing and towing, Curtiss-Wright introduced the Two Claw Aircraft Ship Integrated Secure and Traverse System (TC-ASIST) [13] after launching the ASIST [12]. This system features two mechanical arms capable of capturing the probes on both sides of the helicopter's tail wheel, eliminating the need for helicopter structural modification. It is suitable for a wider range of helicopter models and can tow heavier helicopters.

In terms of control strategies for helicopter towing systems, there has been considerable research on towing carts, including towing stability control and path planning methods. In the area of towing stability control, Liu H et al. [26] proposed a method using active four-wheel steering and variable structure control to improve the stability of shipborne aircraft traction. Wang Y X [27] proposed an adaptive backstepping controller to enhance deck tractor–airplane stability, compensating for ship motion and vehicle parameter uncertainties. Neng Jian W et al. [28] proposed using rear steering control with variable structure control for robust tractor–helicopter stability. In the area of path planning methods, Meng X et al. [29] proposed the KS-RRT* algorithm for safe and time-optimal path planning in autonomous aircraft towing on carrier decks. Liu J et al. [30] propose offline trajectory planning and online tracking methods for efficient and safe dispatching of towed aircraft systems.

Despite the extensive research on trajectory planning for towing carts, the complexity of ship motion control [31,32] makes it impossible for carts to be used for helicopter towing operations in high sea conditions. It is necessary to use integrated securing and traversing helicopter systems to complete the towing work. Due to the simplicity of the RAST and ASIST towing structures, research on the towing process focuses on modeling methods [33], helicopter motion characteristics, and towing system energy consumption characteristics [34]. The shipborne special carrier robot studied in this paper, similar to the TC-ASIST, uses two claws for towing helicopters. It is suitable for a wider range of helicopter models and can tow heavier helicopters. However, the unique structural design and traction motion mode limit the flexibility of helicopter traction motion and increase the difficulty of helicopter deck position correction. There is limited research on rapid helicopter deck position correction methods.

This paper aims to add an automatic helicopter deck position correction process into the TC-ASIST system to improve the efficiency and accuracy of helicopter hangar operations. This paper proposes a stepwise correction method and a continuous correction method to correct the landing position during the transferring of helicopters. In Section 2, the composition and calibration principles of the shipborne rapid carrier robots (SRCRs) are introduced, along with an analysis of the traction motion characteristics. Sections 3 and 4, respectively, propose the stepwise calibration method and the continuous calibration method, while deriving the velocity-coordinated control equations. In Section 5, simulation studies are conducted on the correction motion trajectory and velocity characteristics. Under various traction conditions (different fuselage yaw angle θ_0 and lateral offset d_0), simulation test comparisons of the stepwise and continuous methods for correction are conducted to validate the feasibility of the proposed correction methods and analyze their simulation performance across different scenarios. The conclusion is presented in Section 6. The Nomenclature is shown in Table 1.

Table 1. Nomenclature.

Symbol	Description
θ	The helicopter fuselage yaw angle
d	The helicopter lateral offset
v_1	The traction velocity of the left winch
v_2	The traction velocity of the right winch
V_1	The traction speed provided by the left winch
V_2	The traction speed provided by the right winch
v_a	The velocity of the left traction point a of the SRCR
v_b	The velocity of the right traction point b of the SRCR
ω	The rotational angular velocity of the SRCR
L	The distance between the left and right traction tracks
Δv_a	The traction velocity difference between the left and right winches
Δv_{a1}	The velocity of the left traction point (shaft a) moving within the sliding groove
Δv_{a2}	The rotational velocity of the left traction point a on the SRCR when it performs rigid body motion around the right traction point b
v_z	The capturing point velocity of the right robotic arm
v_s	The lateral correction velocity of the right robotic arm
r_z	The distance from the right robotic arm capturing point c_R to the right traction point b
γ	The steering angle of the helicopter steerable nose wheel
x_s	The sliding displacement of the right robotic arm
L_0	The distance from the initial position of the right robotic arm (i.e., axis y_1) to the center of the helicopter's main wheel
L_1	The distance between the right traction point of the SRCR (i.e., shaft b) and the initial position of the right robotic arm
L_2	The distance between the helicopter main wheel shaft and the line connecting the left and right traction points (i.e., line a–b)
v_B	The rotational velocity of the right robotic arm's capture point around the right traction point of the SRCR (i.e., shaft b)
δ	The central angle of the continuous correction trajectory
r	The arc radius of the continuous correction trajectory
L_3	The distance from the center of the helicopter's main wheel shaft to the center of the helicopter steerable nose wheel
s	The length of the continuous correction trajectory

2. Mathematical Model and Kinematic Characteristic Analysis

2.1. The Structure and Correction Principle of Helicopter Deck Traction System

The research object in this paper is the SRCR, focusing on the helicopter deck towing operation. The simplified system principle is shown in Figures 1 and 2. Figure 1 is a schematic of the three-dimensional layout of the traction system. Figure 2 shows the schematic diagram of the towing system principle, and Figure 3 shows the three-dimensional structure of the SRCR.

As shown in Figures 1 and 2, the deck traction system includes left and right traction winches, the winch hydraulic power system, the traction cable pretightening device, the shipborne rapid carrier robot, and the left and right tracks. The winch hydraulic power system controls the hydraulic motor, thereby driving the winch. The track tractor and the SRCR are connected, as shown in Figure 3. In Figure 3, it can be seen that there are two types of motion pairs between the left track tractor and the SRCR; one is a rotational pair around shaft a, and the other is a sliding pair along the groove. There is only one rotational pair around shaft b between the right track tractor and the secure device. The left and right robotic arms slide along lateral slide rails at the front end of SRCR and are controlled by the winch hydraulic power system. The left and right robotic arms capture and hold the helicopter by clamping the securing rods of the helicopter's main wheels.

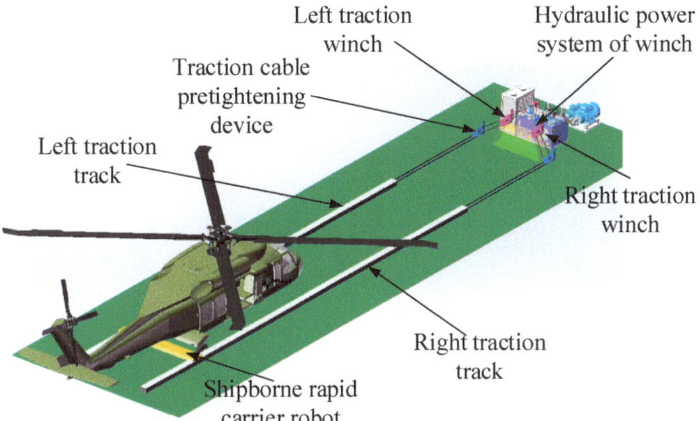

Figure 1. Three-dimensional layout schematic of the traction system.

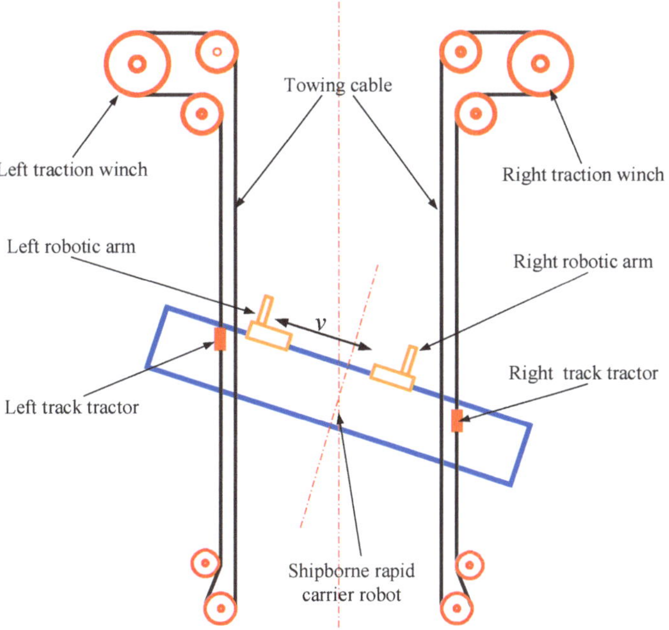

Figure 2. Schematic diagram of the towing system principle.

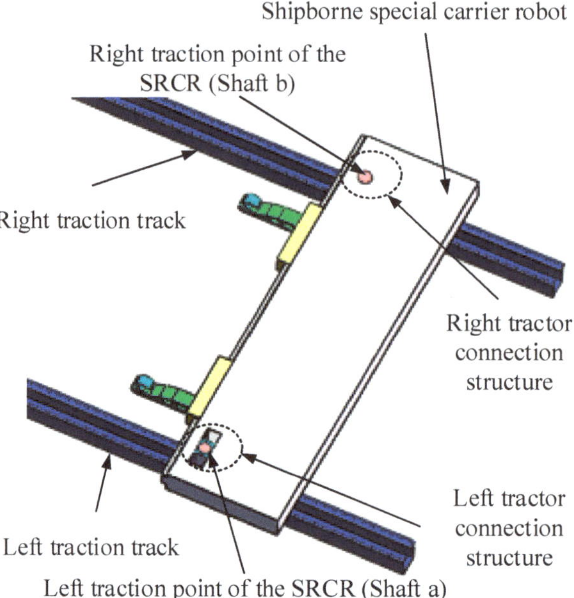

Figure 3. Three-dimensional structure of the SRCR.

The traction principle of the SRCR can be summarized as follows. The left and right traction winches, respectively, tow the SRCR for translational or rotational movement, and then SRCR robotic arms tow the helicopter for translational or rotational movement.

The ideal landing position of the helicopter is characterized by two key factors. The center point of the helicopter's main axle should be located on the center line of the two traction rails, and the fuselage should be parallel to the traction rails, as shown in Figure 4. However, the actual landing position is often uncertain due to factors such as the unpredictable motion of the ship, airwake disturbances, and the pilot's control accuracy.

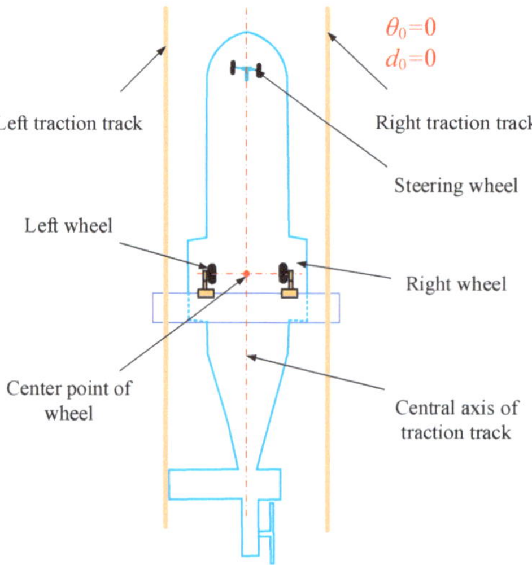

Figure 4. Position characteristics of an ideal helicopter landing.

To quantitatively evaluate the deviation between the actual and ideal landing positions, this paper decomposes the offset of the actual landing position relative to the ideal landing position. The offset can be categorized into two types: lateral position offset and fuselage direction angle offset. These are referred to as lateral offset d_0 and fuselage yaw angle θ_0, as shown in Figure 5. Therefore, the objective of the helicopter deck position is to eliminate the lateral offset and the fuselage yaw angle. The SRCR is equipped with angle sensors and displacement sensors, allowing for real-time monitoring of the helicopter's fuselage yaw angle θ and lateral offset d.

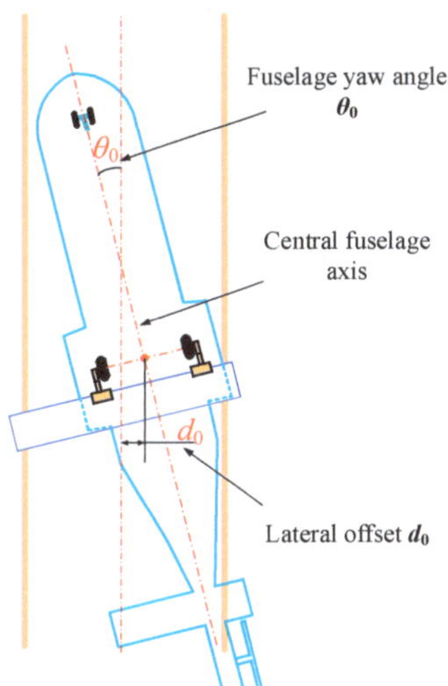

Figure 5. Deviation decomposition of a random actual landing position.

2.2. Analysis of Traction Motion Characteristics of SRCR

Figure 6 is a schematic of the SRCR structure, simplifying the body structure and the complex internal dynamic system for traction motion analysis.

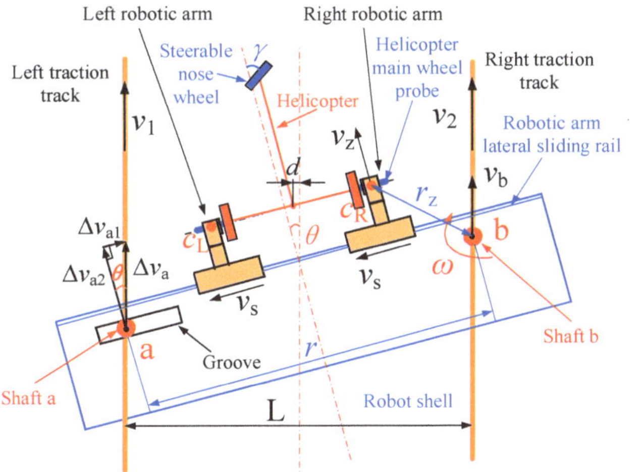

Figure 6. Schematic of SRCR traction motion analysis.

According to the connection relation between SRCR and traction tracks, the SRCR can only produce relative rotation around shaft b. In relation to the connecting shaft a, it can produce not only relative rotation around shaft a but also relative sliding along the groove. Therefore, the traction motion of the SRCR can be simplified to the planar motion of a rigid body. If the SRCR is considered a rigid body, its traction motion can be decomposed into the translational motion along the direction of the right traction track and rotational motion around shaft b.

Through the kinematic analysis of the SRCR, the schematic is shown in Figure 6. The motion trajectory of the SRCR is determined by the traction velocities of the left and right winches, denoted as v_1 and v_2, respectively. The traction speeds are assumed to be positive in the direction shown in Figure 6. The left and right winches transmit power through shafts a and b on the SRCR, with the velocities of a and b denoted as v_a and v_b, respectively. Shaft b is the base point of the SRCR's planar motion, its velocity v_b is equal to the right traction velocity v_2, and v_a is equal to the left traction velocity v_1. The relationship between the velocities v_a and v_b can be expressed as follows:

$$\vec{v_a} = \vec{v_b} + \vec{\omega} \cdot \vec{r},\tag{1}$$

In Formula (1), ω represents the rotational angular velocity of the SRCR around shaft b, and r is the distance between shaft a and shaft b.

As shown in Figure 6, r changes with the rotation angle θ and can be expressed as:

$$r = \frac{L}{\cos\theta},\tag{2}$$

In Formula (2), L is the distance between the left and right traction tracks, and θ is the fuselage yaw angle of the SRCR. Through dynamic analysis of the robot, it is known that the rotational torque of the robot comes from the difference in left and right traction velocities, Δv_a. According to the velocity decomposition theorem, Δv_a can be decomposed into velocity components Δv_{a1} and Δv_{a2}, as shown in Figure 6. Δv_{a1} is the relative motion velocity of point a in the groove direction relative to the base point b, and Δv_{a2} is the linear velocity of point a rotating around shaft b. The expressions for the traction velocity difference Δv_a and velocity component Δv_{a2} are shown in Formulas (3) and (4).

$$\Delta v_a = v_2 - v_1,\tag{3}$$

$$\Delta v_{a2} = \omega \cdot r,\tag{4}$$

The relationship between the velocity component Δv_{a2} and velocity difference Δv_a can be expressed as:

$$\Delta v_{a2} = \Delta v_a \cos\theta,\tag{5}$$

According to Formulas (2) through (5), the angular velocity of the SRCR, ω, can be expressed as:

$$\omega = \frac{(v_2 - v_1)\cos^2\theta}{L},\tag{6}$$

According to Formula (6), the rotation angular velocity ω is related to v_1, v_2 and the yaw angle θ. When the angular velocity ω is positive, it indicates counterclockwise rotation. Conversely, when ω is negative, it indicates clockwise rotation.

After analyzing the motion of the SRCR, it is also necessary to analyze the motion of the left and right robotic arms along the SRCR's lateral sliding rail. This analysis will determine the motion trajectory of the helicopter as captured by the left and right robotic arms. The left and right robotic arms move synchronously along the lateral sliding rail during the helicopter's movement. For example, consider the capture point c_R on the right robotic arm, which secures the right main wheel of the helicopter. Assume that the sliding velocity of the robotic arm along the lateral sliding rail is v_s (i.e., the lateral velocity of the right robotic arm). According to the velocity synthesis theorem, the vector expression of the capturing point velocity v_z of the right robotic arm is as follows:

$$\vec{v_z} = \vec{v_b} + \vec{\omega} \cdot \vec{r_z} + \vec{v_s}, \tag{7}$$

In Equation (7), r_z is the distance from the right robotic arm capturing point c_R to base point b. Analysis of Equations (1)–(7) reveals that the capturing point velocity, v_z, is related to the left winch traction velocity (v_1), right winch traction velocity (v_2), and the right robotic arm lateral velocity (v_s).

During the helicopter traversing process, the motion of the helicopter on deck completely depends on the traction applied by the robotic arm to the main wheel capturing probe. The fuselage yaw angle (θ) and lateral offset (d) for helicopter deck position correction must be managed through the coordinated velocity control of left and right traction winches. These factors significantly increase the difficulty of the path planning for helicopter deck position correction. Therefore, the path planning of the helicopter correction must fully consider the traction motion characteristics of SRCR.

3. The Stepwise Method of Helicopter Deck Position Correction

Based on the analysis of the SRCR traction motion characteristics, this paper proposes a stepwise correction method for helicopter deck position and investigates the cooperative control relationship between the winch traction speeds (v_1, v_2) and right robotic arm lateral velocity (v_s).

3.1. The Principle of Stepwise Correction Method

The steering angle of the helicopter's steerable nose wheel, denoted as γ, determines the helicopter's motion trajectory. There are two special steering angles: $0°$ and $90°$. When the steering angle is $0°$, the helicopter will translate along the fuselage direction. When the steering angle is $90°$, the helicopter will rotate around the center point of the main wheel shaft. The stepwise correction method proposed in this paper leverages the particularity of the two steering angles, allowing the helicopter to alternate between translation and rotation. This method aims to eliminate the helicopter's lateral offset and fuselage yaw angle. Due to its step-by-step nature, it is referred to as the stepwise correction method.

Assuming that the helicopter's lateral offset and fuselage yaw angle at the initial landing position are not zero, the stepwise correction method can be summarized as follows: first, stepwise translation correction, then stepwise rotation correction.

Firstly, the translation motion of the helicopter is shown in Figure 7a. The angle of the steerable nose wheel is adjusted to $0°$. Then, the left and right winch traction speeds and the correction speeds of the robotic arm are cooperatively controlled to move the helicopter along the direction of the steerable nose wheel. When the helicopter's lateral offset d becomes 0, it marks the completion of the stepwise translation correction. This means that the center point O of the helicopter's main axle has reached the centerline of the left and right traction tracks, indicated as point O′ in Figure 7a.

Next, the rotating motion of the helicopter is shown in Figure 7b. After adjusting the angle of the steerable nose wheel γ to $90°$, the left and right winch traction speeds and the correction speed of the robotic arm are cooperatively controlled to rotate the helicopter around the center point O′ of the main wheel shaft. When the direction of the helicopter fuselage is parallel to the tracks' centerline, the fuselage yaw angle is reduced to zero, indicating the completion of the stepwise rotational correction.

After completing these stepwise correction operations, the helicopter's steerable nose wheel angle can be set to $0°$ to tow the helicopter into the hangar. However, the qualitative motion analysis of the stepwise translational and rotational corrections does not reveal the coordinated control relationship between the left and right winch traction speeds and the robotic arm correction speed during the correction process. Section 3.2 will provide a detailed explanation of this aspect.

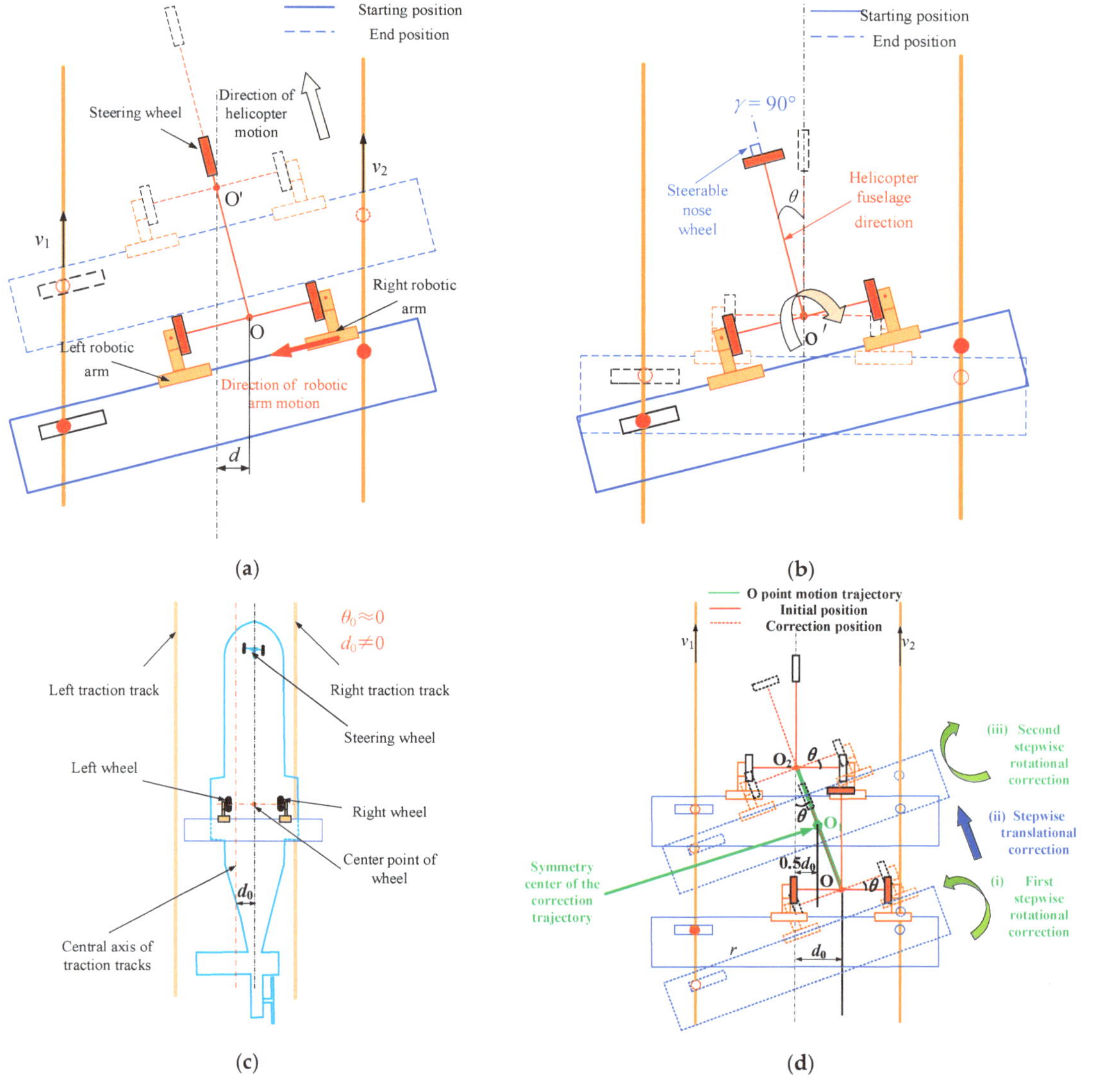

Figure 7. Schematic diagram of the stepwise correction method. (**a**) Translational motion of helicopter; (**b**) rotational motion of helicopter; (**c**) helicopter parallel landing position; (**d**) the stepwise correction method for helicopter parallel landing position.

If the lateral offset of the helicopter is zero but the yaw angle of the fuselage is not, only the stepwise rotational correction is implemented, as shown in Figure 7b.

If the yaw angle of the helicopter's initial landing position is zero and the lateral offset is not, the fuselage direction will be parallel to the track direction. This paper defines this condition as the helicopter parallel landing position, as shown in Figure 7c. During the stepwise correction for a parallel landing position, the helicopter first undergoes rotational correction by rotating around point O by a certain fuselage yaw angle θ. Next, the SRCR sequentially completes the stepwise translational correction and the stepwise rotational correction, as shown in Figure 7d. Since the correction angles for the two stepwise rotational corrections are equal in magnitude but opposite in direction, the time required for each stepwise rotational correction is the same. Therefore, the stepwise correction for a parallel landing position with a lateral offset of d_0 can be considered as two stepwise corrections for cross landing positions, each with a lateral offset of $0.5d_0$ and a fuselage yaw angle of θ. Different from cross landing positions, in the stepwise correction process for parallel landing positions, the fuselage yaw angle θ is set by the crew. Therefore, under different lateral offsets d_0, it is necessary to find an optimal θ to achieve the best correction performance.

3.2. The Relationship of Speed Cooperative Control in the Stepwise Correction Method

Based on the above analysis of the traction motion characteristics, there is a cooperative matching relationship between the left and right traction speeds (v_1, v_2) and the correction speed of the right robotic arm (v_s). Therefore, establishing a cooperative control relationship is a necessary prerequisite for achieving helicopter traction correction.

Based on the traction motion analysis of SRCR (as shown in Figure 7), the system kinematics of the helicopter during translational motion is analyzed, as shown in Figure 8. The reference coordinate system on the deck is assumed to be x_0–y_0, with the transverse axis of the deck as the x_0 axis, the longitudinal axis as the y_0 axis, and the centerline of the left and right traction tracks as the zero position of the x_0 axis. Additionally, the robotic arm's lateral sliding rail is taken as the x_1 axis, and the initial position of the right robotic arm is taken as the y_1 axis, with the origin located at the rightmost end of the robotic arm's lateral sliding rail.

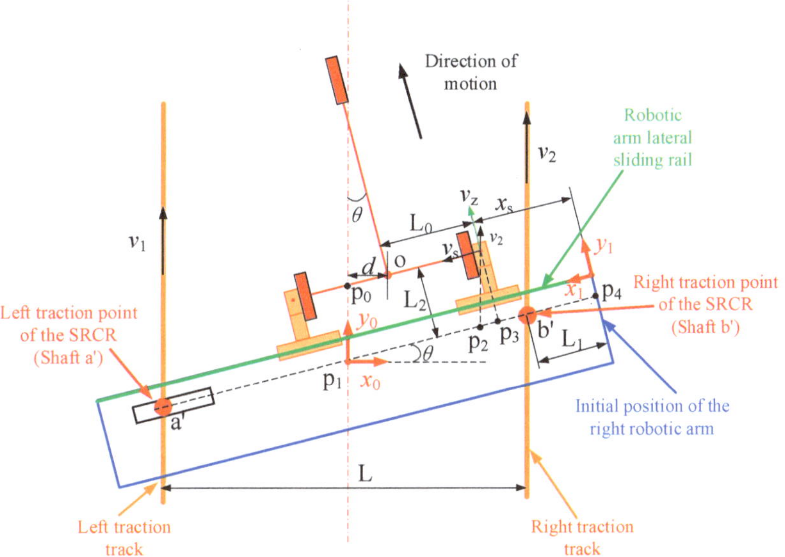

Figure 8. The system kinematics of the helicopter in translation motion.

The initial positions of the left and right robotic arms are at the left and right ends of the robotic arm's lateral sliding rail, respectively. The physical meanings of the marked parameters in Figure 8 are as follows. x_s represents the distance between the current position and the initial position of the right robotic arm (the sliding displacement of right robotic arm); L_0 represents the distance from the initial position of the right robotic arm (i.e., axis y_1) to the center point O of the helicopter's main wheel; L_1 represents the distance between the right traction point of the SRCR (i.e., shaft b') and the initial position of the right robotic arm; L_2 is the distance between the main wheel shaft and the line connecting the left and right traction points (i.e., a′–b′); points p_0~p_4 are auxiliary points, and the dotted line is the auxiliary line. It is assumed that θ counterclockwise is positive and clockwise is negative.

The length of $\overline{P_1P_4}$ is expressed as:

$$\overline{P_1P_4} = \frac{L}{2\cos\theta} + L_1, \tag{8}$$

The auxiliary line segments defined by points p_1 through p_4 satisfy the following relation:

$$\overline{P_1P_2} = \overline{P_1P_4} - \overline{P_3P_4} - \overline{P_2P_3}, \tag{9}$$

The length expression of the auxiliary line segment $\overline{P_2P_3}$ is:

$$\overline{P_2P_3} = L_2\tan\theta, \tag{10}$$

$$\overline{P_3P_4} = x_s, \tag{11}$$

After connecting Equations (8), (10) and (11) with Equation (9):

$$\overline{P_1P_2} = \frac{L}{2\cos\theta} + L_1 - x_s - L_2\tan\theta, \tag{12}$$

Th e length of $\overline{op_0}$ is expressed as:

$$\overline{op_0} = \overline{P_1P_2} - L_0, \tag{13}$$

The formula for calculating the lateral offset d is:

$$d = \overline{op_0}\cos\theta, \tag{14}$$

$$d = \frac{L}{2} + (L_1 - x_s - L_0)\cos\theta - L_2\sin\theta, \tag{15}$$

If d is positive, it means that the center point of the main wheel is located on the right side of the central axis of the traction track; if d is negative, it means that the center point is located on the left side.

During the helicopter correction process, the direction of the SRCR right traction velocity v_2 is related to lateral offset d and fuselage yaw angle θ. Define the speed value provided by the right winch system as V_2, then the SRCR right traction velocity v_2 can be expressed as:

$$v_2 = \frac{\theta d}{|\theta d|}V_2, \tag{16}$$

Equation (16) indicates that the direction of right traction velocity v_2 is determined by yaw angle θ and lateral offset d. During the traction process, the SRCR and the helicopter share the same rotational angular velocity. Therefore, when the helicopter is in translational motion, the SRCR must also move translationally. Based on the previous analysis of SRCR traction motion characteristics, when the SRCR moves translationally, the difference between the left and right traction speeds (v_1 and v_2) is zero. Consequently, v_1 should equal v_2 during the helicopter's translational traction.

Since the synthesis of the translational velocity v_2 and the correction velocity v_s of the right robotic arm results in the translational velocity v_z along the fuselage direction, the correction velocity v_s of the right robotic arm can be expressed according to the velocity synthesis theorem as:

$$v_s = \frac{\theta d}{|\theta d|}V_2\sin\theta, \tag{17}$$

Equation (17) indicates that when the helicopter is in translational motion, the correction velocity v_s is determined by yaw angle θ, lateral offset d, and the absolute value of the translational traction velocity V_2. The correction velocity v_s is directly proportional to the traction translational velocity V_2.

Next, the cooperative speed control relationship during helicopter rotational correction is analyzed. Figure 9 illustrates the system kinematic analysis for the helicopter's rotating motion. The angle β, marked in Figure 9, represents the angle between the line (connecting the right robotic arm capture point c_R and rotation base point b′) and fuselage direction. According to the previous analysis of SRCR traction motion characteristics, the velocity v_z at the right robotic arm capture point can be decomposed into three velocity components: right winch traction velocity v_2, rotation velocity v_B around base point b′, and correction velocity v_s of the right robotic arm.

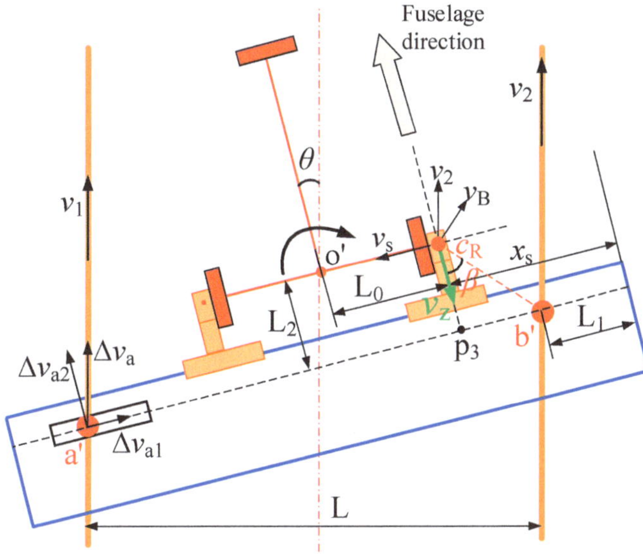

Figure 9. Schematic diagram of system kinematics analysis during helicopter rotational motion.

The rotation velocity v_B of the right robotic arm's capturing point c_R around b′ is:

$$v_B = \frac{L_2\omega}{\cos\beta}, \tag{18}$$

Since the velocity of the capturing point v_z is perpendicular to the direction of the main wheel shaft, the sum of the projections of the three velocity components (v_2, v_B, v_s) in the direction of the main wheel shaft must be zero. Therefore, the corrected velocity v_s of the right robotic arm can be expressed as:

$$v_s = v_B\cos\beta + v_2\sin\theta, \tag{19}$$

$$v_z = v_2\cos\theta - v_B\sin\beta, \tag{20}$$

The capturing point c_R of the right robotic arm can also be considered as undergoing rotational motion around the midpoint O′ of the helicopter main wheel shaft. Therefore, the capturing point speed v_z can be expressed as:

$$v_z = \omega \cdot L_0, \tag{21}$$

The relationship between the left and right traction velocities v_1 and v_2 and the relationship between the right robotic arm correction velocity v_s and traction velocity v_2 are shown in Equations (22) and (23), respectively.

$$v_1 = \left(1 - \frac{1}{(k_1 + k_2 \tan \beta) \cos \theta}\right) v_2, \tag{22}$$

$$v_s = \left(\sin \theta - \frac{k_2 \cos \theta}{k_1 + k_2 \tan \beta}\right) v_2, \tag{23}$$

In the formula, $k_1 = \frac{L_0}{L}, k_2 = \frac{L_2}{L}, \tan \beta = \frac{x_s - L_1}{L_2}$.

Equations (22) and (23) comprehensively describe the cooperative control relationship between the traction velocities (v_1, v_2) and the correction velocity (v_s) of the right robotic arm when the helicopter rotates around the center point of the main wheel shaft. If the right traction speed v_2 is taken as the reference speed, the ratios of the left traction speed v_1 and the correction speed of the right robotic arm v_s to the reference speed v_2 are nonlinear functions of the sliding displacement of the right robotic arm x_s and the helicopter fuselage yaw angle θ.

In summary, the speed cooperative control relationship for helicopter traction correction has the following characteristics. During translational correction, the left and right traction velocities (v_1, v_2) are equal, and the correction speed (v_s) of the right robotic arm is directly proportional to the right traction velocity (v_2). During rotational correction, the left traction velocity (v_1) and the correction speed (v_s) of the right robotic arm are time-varying nonlinear functions.

4. The Continuous Method of Helicopter Deck Position Correction

Drawing inspiration from the circular arc trajectory method commonly used in auto-parking systems, this paper designs a continuous correction method for SRCR. The trajectory is a single circular arc, allowing the helicopter correction process to proceed without stepwise operations.

4.1. The Principle of Continuous Correction Method

Based on whether fuselage yaw angle is zero, the helicopter landing position can be classified into parallel landing position and cross landing position. For parallel landing positions, where the fuselage direction is parallel to the track direction, two tangential arcs can be used as the correction track. For cross landing positions, where the fuselage direction is at an angle to the track direction, a single arc track can be used for correction. The following will detail the correction principles for both landing positions.

4.1.1. Parallel Landing Position

The correction principle for parallel landing position is illustrated in Figure 10. The green thick solid line curve in the figure represents the correction trajectory, which is composed of two identical and tangent arcs, $\overset{\frown}{OO_1}$ and $\overset{\frown}{O_1O_2}$. R is the center of correction trajectory, δ is the center angle of correction trajectory, and γ is the helicopter steerable nose wheel angle. The specific expression is as follows:

$$\gamma = \frac{d_0}{|d_0|}|\gamma|, \tag{24}$$

The operation process of continuous correction method under parallel landing position is as follows. Firstly, based on the transverse offset d_0 and Formula (24), the pilot sets the helicopter's steerable nose wheel angle γ. Then, the helicopter moves along the first arc $\overset{\frown}{OO_1}$ in Figure 10 under the traction of SRCR. When the current lateral offset d is reduced to half of the initial lateral offset d_0, it indicates that the center point O of the helicopter's main wheel shaft has reached the symmetry center point O_1 of the two correction trajectories.

At this time, the pilot changes the steerable nose wheel angle to the same magnitude in the opposite direction, and the helicopter moves along the second arc $\overset{\frown}{O_1O_2}$. Finally, when current lateral offset d returns to 0, the center point O of the helicopter reaches O_2, completing the correction operation.

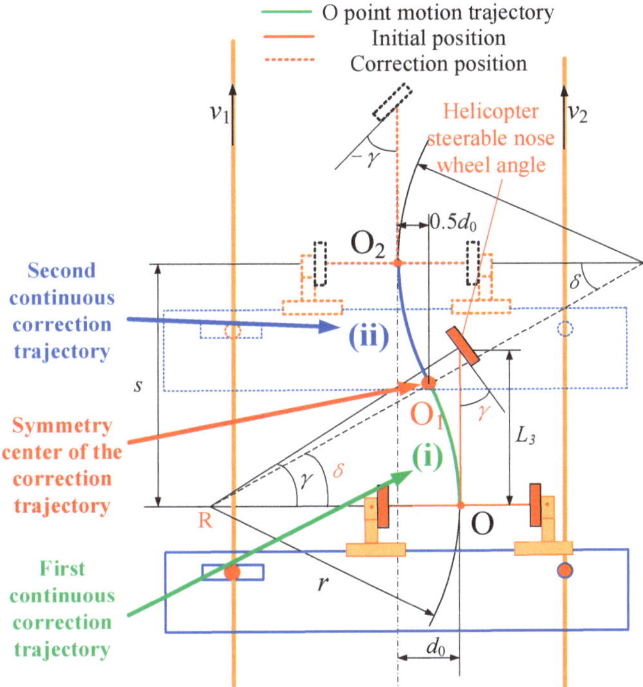

Figure 10. Schematic diagram of the two arc trajectories for parallel landing position.

Based on the geometric relationship in Figure 10, the correction arc radius r can be deduced as follows:

$$r = \frac{L_3}{|\tan\gamma|},\tag{25}$$

L_3 is the distance from the center point of the main wheel shaft to the center of the helicopter's steerable nose wheel.

The central angle δ corresponding to the arc of the correction trajectory is:

$$\delta = \arccos\frac{2L_3 - d_0\tan\gamma}{2L_3},\tag{26}$$

The correction length s in the continuous correction method is:

$$s = 2r\sin\delta,\tag{27}$$

By Formulas (25)–(27), it can be determined:

$$s = \sqrt{\frac{4L_3d_0}{\tan\gamma} - d_0^2},\tag{28}$$

As can be seen from Formula (28), the correction length s is determined by the helicopter steerable nose wheel angle γ.

The two continuous correction trajectories for a parallel landing position are symmetric about the trajectory's center point, meaning that the time required for each correction segment is equal. The correction trajectory along arc $\overset{\frown}{O_1O_2}$ is the same as a single continuous correction trajectory for a cross landing position. The following section will provide a detailed explanation of the continuous correction method for the cross landing position.

4.1.2. Cross Landing Position

The continuous correction method for the cross landing position is similar to that for the parallel landing position, but it requires only one arc. The principle diagram is shown in Figure 11. In Figure 11, the green arc represents the motion trajectory of the center point of the main wheel shaft, γ' is the helicopter steerable nose wheel angle, and r' is the correction arc radius.

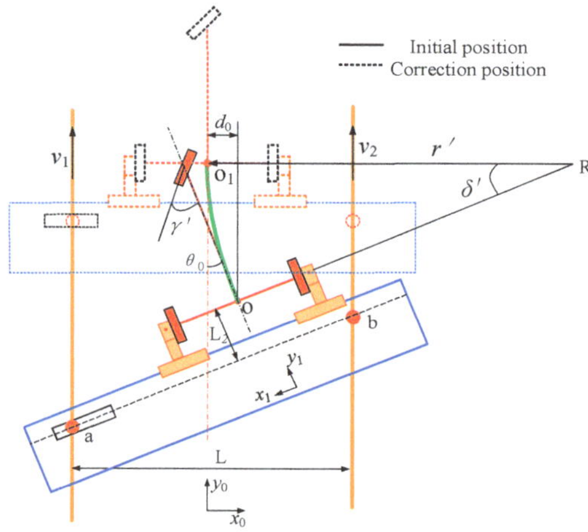

Figure 11. Schematic diagram of the arc trajectory corresponding to the cross landing position.

Unlike the parallel landing position, the helicopter's steerable nose wheel angle for a cross landing position cannot be arbitrarily set; it must be calculated based on the lateral offset and yaw angle of the helicopter's initial landing position. By analyzing the geometric relationship in Figure 11, it can be seen that the radius of the arc r', the lateral offset d_0, and the center angle δ' of the circle satisfy the following relationships:

$$r' \cos \delta' + d_0 = r', \tag{29}$$

The radius r' of the arc trajectory is:

$$r' = \frac{L_3}{|\tan \gamma'|}, \tag{30}$$

The central angle δ' of the arc is exactly equal to the initial yaw angle θ_0.

$$\delta' = \theta_0, \tag{31}$$

Using Formulas (25)–(27), the angle of the helicopter's steerable nose wheel γ' can be determined:

$$\gamma' = \arctan \frac{L_3(1 - \cos \theta_0)}{d_0} \cdot \frac{-d_0}{|d_0|}, \tag{32}$$

According to Equation (32), γ' is related to θ_0 and d_0.

To summarize, when the continuous correction method is adopted, the helicopter performs only rotating motion. In the parallel landing position, the two correction trajectories are solely related to helicopter steerable nose wheel angle, γ. This angle is set by the pilot, and the correction trajectory will vary accordingly with different angles.

4.2. The Relationship of Speed Cooperative Control in the Continuous Correction Method

When using the continuous correction method, assuming that the tires do not slip sideways, the helicopter engages only in rotational motion. The instantaneous center of rotation is located at the intersection of the vertical lines in the direction of each wheel's speed, as shown in point R in Figure 12. The angle of the helicopter's steerable nose wheel is γ, the angular speed of the helicopter's rotating motion is ω', and the fuselage yaw angle is θ.

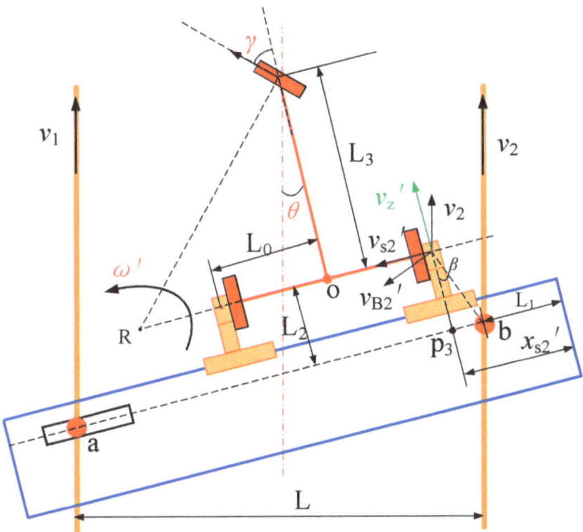

Figure 12. Schematic diagram of the system kinematics analysis corresponding to the arc straightening track.

Similar to the speed analysis of the stepwise correction method, the speed of the robotic arm capturing point v_z' can be decomposed into three speed components: the traction translational speed v_2, the rotation speed v_{B2}' around the base point b, and the robotic arm correction speed v_{s2}', as shown in Figure 12. According to the analysis of the traction motion characteristics of SRCR in the above section, the rotation angular velocity of SRCR is ω'.

$$\omega' = \frac{(v_2 - v_1)\cos^2\theta}{L}, \tag{33}$$

$$v'_{B2} = \frac{L_2\omega'}{\cos\beta}, \tag{34}$$

The sum of the projections of the three velocity components (v_2, v_{B2}', v_{s2}') in the vertical direction is zero.

$$v'_{s2} = v_2\sin\theta - v'_{B2}\cos\beta, \tag{35}$$

The helicopter is only undergoing rotational motion, so the speed of robotic arm capturing point v_z' is expressed as:

$$v'_z = (L_3\cot\gamma + L_0)\omega', \tag{36}$$

According to the velocity synthesis theorem, the right robotic arm capturing point velocity v_z' can also be expressed as:

$$v'_z = v_2\cos\theta - v'_{B2}\sin\beta, \tag{37}$$

According to the above equations, the cooperative control relationship between the two traction speeds v_1 and v_2, as well as the cooperative control relationship between the robotic arm correction speed v_{s2}' and the right traction speed v_2, can be determined, as shown in Equations (38) and (39).

$$v_1 = \left(1 - \frac{1}{(k_1 + k_2 \tan \beta + k_3 \cot \gamma) \cos \theta}\right) v_2, \tag{38}$$

$$v_{s2}' = \left(\sin \theta - \frac{k_2 \cos \theta}{k_1 + k_2 \tan \beta + k_3 \cot \gamma}\right) v_2, \tag{39}$$

In the formula, $k_1 = \frac{L_0}{L}, k_2 = \frac{L_2}{L}, k_3 = \frac{L_3}{L}, \tan \beta = \frac{x'_{s2} - L_1}{L_2}$.

Equations (38) and (39) show that if the right traction velocity v_2 is fixed, the left traction velocity v_1 and the robotic arm correction speed v_{s2}' are nonlinear functions related to the robotic arm correction displacement x_{s2}' and the fuselage yaw angle θ.

5. Simulation Verification of the Correction Method

5.1. Simulation of the Stepwise Correction Method

To verify the feasibility of the stepwise correction method, this paper utilizes Adams 2020 software [35] and MATLAB 2019b [36] to simulate and analyze the corrected motion trajectory of the helicopter. The simplified three-dimensional model of the traction system is shown in Figure 13. This study takes the controllable steering wheel helicopter as an example for simulation research. Table 2 presents the parameters of several medium and heavy helicopters that are widely used and produced in large quantities worldwide. To ensure that the SRCR can tow all types of helicopters listed in Table 2, the relevant parameters of the simulation model are shown in Table 3.

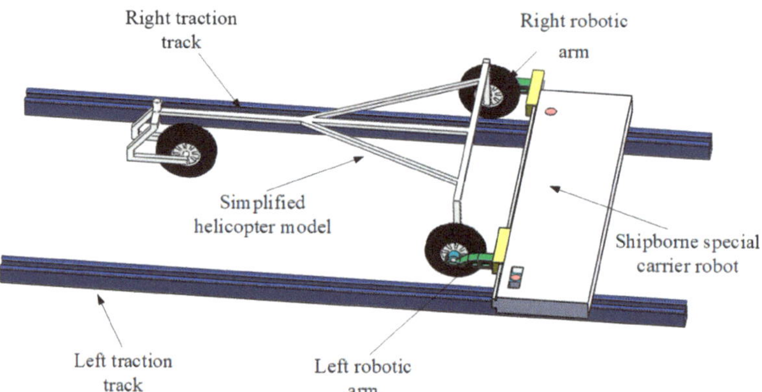

Figure 13. Simplified three-dimensional model of the helicopter traction system.

Table 2. Parameters of several medium and heavy helicopters.

Helicopter Type	Width	Rear Wheelbase	Weight
CH-53K	5.33 m	3.52 m	25,000 kg
Sea hawk (SH-60B)	2.36 m	2.8 m	10,400 kg
Sea king (SH-3)	4.98 m	4.70 m	9707 kg
MI-171helicopter	2.50 m	4.28 m	11,000 kg
SH-90	3.8 m	2.7 m	10,600 kg

Table 3. Main simulation parameters of the stepwise correction method.

Parameter	Value
The length of track spacing for left and right traction tracks L	5000 mm
The length of capturing point to fuselage center line L_0	2000 mm
The distance between the right traction point of the SRCR and the initial position of the right robotic arm L_1	150 mm
The distance between the helicopter main wheel shaft and the line connecting the left and right traction points L_2	300 mm
Initial helicopter yaw angle θ_0	10°
Initial helicopter lateral offset d_0	200 mm
Right traction speed v_2	10 mm/s

Firstly, the translation stepwise correction is simulated. With the helicopter steerable nose wheel deflection angle set to 0° and the right winch traction speed set to 10 mm/s, the main wheel motion trajectory obtained by simulation during the translational step is shown in Figure 14. This figure displays the motion trajectories of the left and right main wheel and the center point of the main wheel shaft. It can be seen that these trajectories are three parallel straight lines, and after the translational motion, the center point O of the main wheel shaft reaches the center line of the track. This indicates that the helicopter has achieved the correction target of the translation motion step. Figure 15 shows the velocity characteristic curves of the SRCR during the translational motion process. It also presents the mathematical velocity characteristic curves based on the above collaborative velocity governing equations and the simulation velocity characteristic curves based on the Adams dynamics model. By comparison, it can be seen that Figure 15a,b are in good agreement; the left and right traction velocities are equal, and the correction velocity of the right robotic arm is proportional to right traction velocity. This demonstrates mutual verification of the stepwise correction velocity cooperative control theory and the Adams dynamic simulation model at the kinematic level.

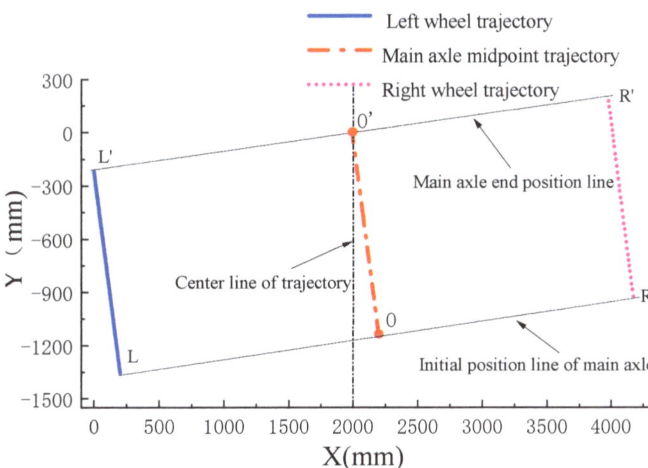

Figure 14. Trajectories of the helicopter's main wheels during the translational phase.

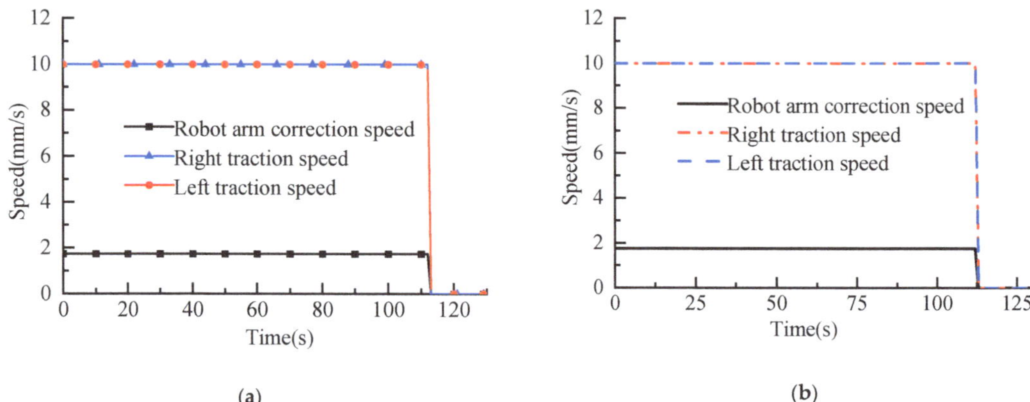

Figure 15. Velocity characteristic curves of the SRCR during the translational motion of the helicopter. (**a**) Velocity characteristic curves obtained from mathematical model; (**b**) velocity characteristic curves based on Adams simulation.

Simulation analysis of the rotating motion was conducted. The deflection angle of the helicopter's steerable nose wheel was set to 90°, and the right traction speed was set to −10 mm/s. The resulting movement trajectory of the main wheel under stepwise rotational motion of the helicopter is shown in Figure 16. The motion trajectories of the left and right wheels form two symmetrical arcs. After rotating motion of 10°, the shaft of the main wheel aligns horizontally, indicating that the helicopter's fuselage direction is now parallel to the traction track. This confirms that the rotational motion has successfully achieved the correction target. Figure 17 presents the velocity characteristic curves during the rotational motion. Comparison of Figure 17a,b reveals that the mathematical velocity characteristic curve derived from the cooperative velocity governing equation is in good agreement with the curve obtained from the Adams simulation model, thus validating the accuracy of both the simulation model and the mathematical model.

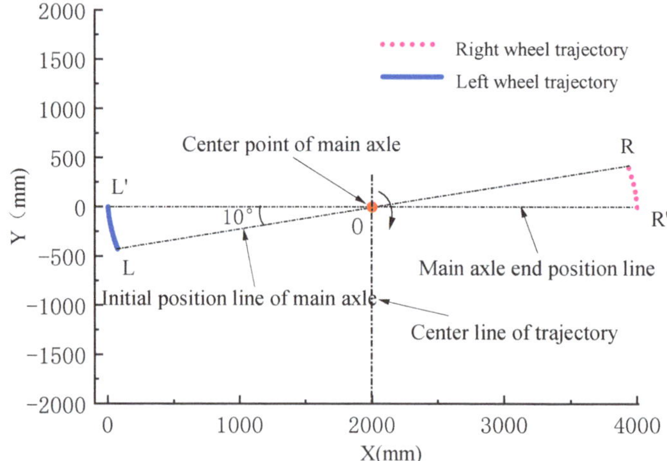

Figure 16. Trajectories of the helicopter's main wheels during the rotational phase.

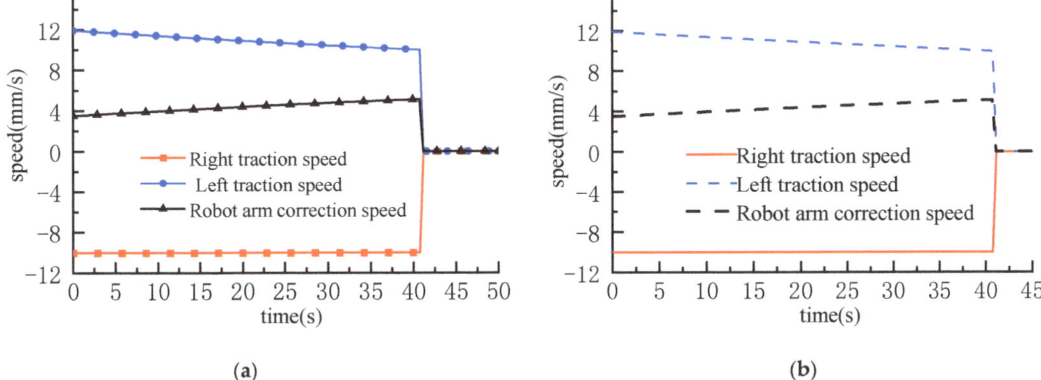

Figure 17. Velocity characteristic curves of the helicopter rotating around point O. (**a**) Velocity characteristic curves obtained from mathematical model; (**b**) velocity characteristic curves based on Adams simulation.

Through the above analysis, the correctness of the simulation model is verified by comparing the simulation results of the velocity characteristic curve with the mathematical solution results. This comparison also illustrates the correctness of the mathematical model of cooperative control relationship Equations (22) and (23). The simulation results of the helicopter trajectory based on the stepwise correction method demonstrate that the proposed method can quickly achieve the correction target, verifying its feasibility.

5.2. Simulation of the Continuous Correction Method

To verify the feasibility of the continuous correction method, Adams software is used in this paper to simulate and analyze the corrected motion trajectory of the helicopter. To ensure that SRCR can tow all types of helicopters listed in Table 2, the relevant parameters of the simulation model are shown in Table 4.

Table 4. Main simulation parameters of the continuous correction method.

Parameter	Value
The length of track spacing for left and right traction L	5000 mm
The length of grab point to fuselage center line L_0	2000 mm
The distance between the right traction point of the SRCR and the initial position of the right robotic arm L_1	150 mm
The distance between the helicopter main wheel shaft and the line connecting the left and right traction points L_2	300 mm
The distance between helicopter control wheel and the main wheel shaft L_3	8000 mm
Initial helicopter yaw angle θ_0	$-20°$
Initial helicopter lateral offset d_0	-200 mm
Right traction speed v_2	30 mm/s
Helicopter control wheel angle γ	$67.48°$

The helicopter's arc correction track simulated in Adams is shown in Figure 18. As can be seen, the tracks of the left and right wheels and the main wheel center point are three concentric arcs. After a 20° rotation, the center point O of the main wheel shaft aligns with the centerline of the traction track. Additionally, the main wheel shaft is perpendicular to the centerline of the trajectory, indicating that the helicopter has successfully completed the landing position correction task.

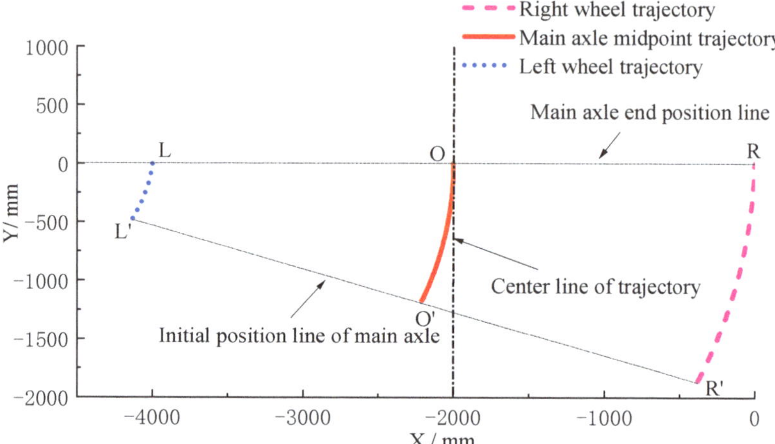

Figure 18. Trajectories of the helicopter's main wheels using the continuous correction method.

The speed characteristic curves of SRCR are shown in Figure 19. Figure 19a represents the velocity characteristic curves based on the mathematical model, while Figure 19b shows the velocity characteristic curves based on Adams simulation. The SRCR angular velocity and yaw angle curves during helicopter position correction are illustrated in Figure 20. It can be observed that the rotational angular velocity is not constant but follows a curved shape, reflecting the complex speed matching relationship between the left winch traction velocity, right winch traction velocity, and right robotic arm lateral correction velocity (v_1, v_2, v_s) during rotational motion. By comparing Figure 20a,b, it is evident that the mathematical velocity characteristic curve obtained from the collaborative velocity governing equation aligns well with the curve derived from the Adams simulation model, further validating the accuracy of both the simulation model and mathematical model.

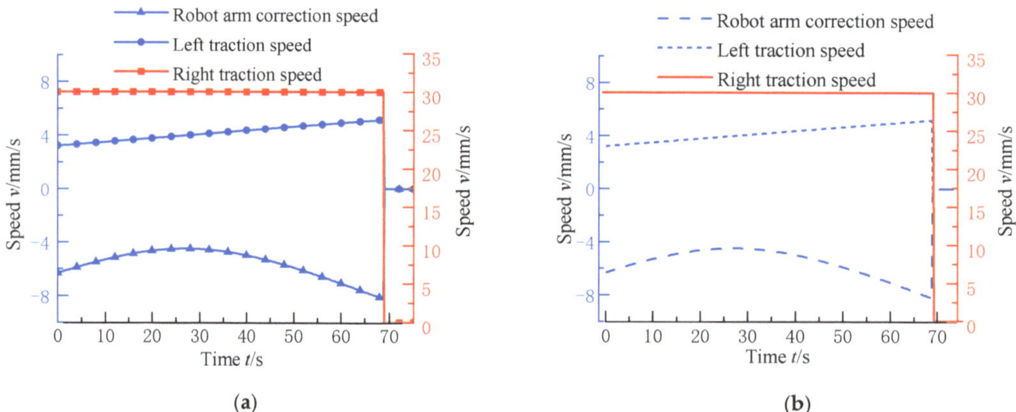

Figure 19. Velocity characteristic curves of SRCR during the continuous correction phase. (**a**) Velocity characteristic curves obtained from mathematical model; (**b**) velocity characteristic curves based on Adams simulation.

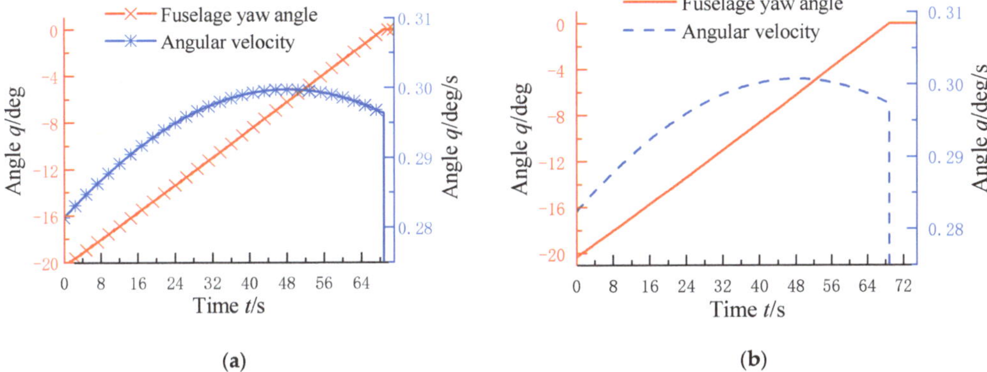

Figure 20. Angular velocity and yaw angle curves during the helicopter's continuous correction phase. (**a**) Velocity characteristic curves obtained from mathematical model; (**b**) velocity characteristic curves based on Adams simulation.

5.3. Comparative Analysis of the Efficiency of the Stepwise Correction Method and the Continuous Correction Method

Through the analysis of traction motion characteristics of SRCR in Section 2 and the simulation tests in Sections 5.1 and 5.2, it is evident that the main factors affecting the traction motion characteristics are (1) lateral offset d_0; (2) initial yaw angle θ_0; (3) right traction speed v_2. To explore the influence of these variables on operational performance, two sets of tests were conducted using both the stepwise and continuous methods.

In the cross landing position, both the fuselage yaw angle θ_0 and the landing offset d_0 of the helicopter can be measured using the angle sensors and displacement sensors on the SRCR. However, in the parallel landing position, while the landing offset d_0 can be measured using the displacement sensors, the fuselage yaw angle θ must be set through the first stepwise rotational correction shown in Figure 7d (i) or the first trajectory of the continuous correction shown in Figure 10 (i). Therefore, it is necessary to find the optimal fuselage yaw angle θ for different lateral offsets d_0 to minimize the correction time.

To analyze the impact of these factors on the correction efficiency of the stepwise and continuous methods, two sets of simulations were performed for each method: one focusing on the impact of initial yaw angle θ_0 and right traction speed v_2, and the other on the impact of initial yaw angle θ_0 and lateral offset d_0. The factors affecting correction performance under different conditions were analyzed, and finally, the performance of the stepwise correction method and the continuous correction method was compared.

5.3.1. Efficiency Analysis of the Stepwise Correction Method on the Impact of Initial Yaw Angle and Traction Speed

The performance comparison test is shown in Table 5 and Figure 21. By analyzing the data of different right traction speeds v_2 at initial fuselage yaw angles of $-10°$, $-20°$, and $-30°$, the following conclusions can be drawn:

(1) By comparing serial numbers 1-2-3, 4-5-6, 7-8-9, it can be seen that increasing v_2 can effectively improve the efficiency of the correction operation and reduce the total correction time t.

(2) By comparing serial numbers 1-4-7, 2-5-8, 3-6-9, it can be seen that during the translational operation, a larger initial yaw angle $|\theta_0|$ results in a shorter translation time t_1; however, during the rotational operation, a larger initial yaw angle $|\theta_0|$ results in a longer rotation time t_2.

Table 5. Analysis results data on the impact of initial yaw angle and traction speed on efficiency of the stepwise correction method.

Serial Number	Initial Fuselage Yaw Angle θ_0/deg	Lateral Offset d_0/mm	Right Traction Speed v_2/mm/s	Translation Time t_1/s	Rotation Time t_2/s	Total Correction Time t/s
1	−10°	−200	10	108.71	43.80	152.51
2	−10°	−200	20	54.56	21.80	76.36
3	−10°	−200	30	36.42	14.60	51.02
4	−20°	−200	10	78.29	91.00	169.29
5	−20°	−200	20	39.18	45.60	84.78
6	−20°	−200	30	26.16	30.40	56.56
7	−30°	−200	10	95.56	149.20	244.76
8	−30°	−200	20	47.82	74.60	122.42
9	−30°	−200	30	31.90	49.80	81.70

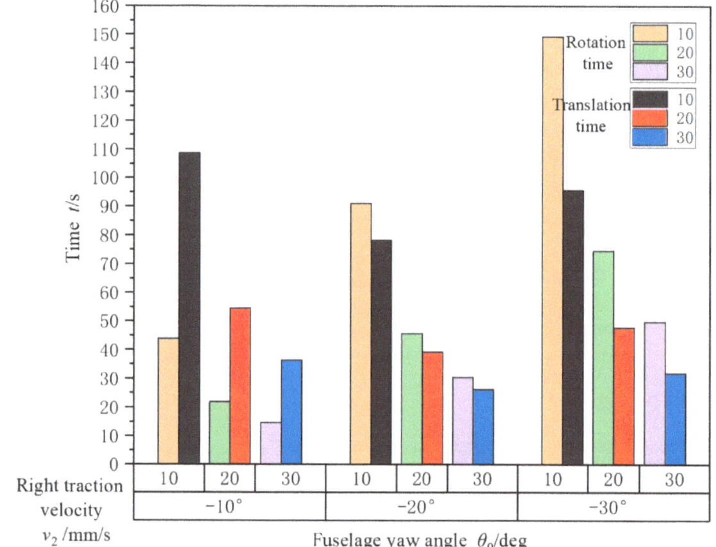

Figure 21. Bar chart of analysis results of the impact of initial yaw angle and traction speed on efficiency of the stepwise correction method.

5.3.2. Efficiency Analysis of the Stepwise Correction Method on the Impact of Initial Yaw Angle and Lateral Offsets

The performance comparison test is shown in Table 6 and Figure 22. By analyzing the data for different lateral offsets d_0 when the initial yaw angle θ_0 is −10°, −20°, and −30°, respectively, the following conclusions can be drawn:

(1) By comparing serial numbers 1-2-3-4-5-6, 7-8-9-10-11, 12-13-14-15-16 and referring to Figure 22, it can be seen that when the initial yaw angle $|\theta_0|$ is fixed, the smaller the lateral offset $|d_0|$, the shorter the translation time t_1, with no effect on rotation time t_2.

(2) According to Figure 22, in the stepwise correction for the cross landing position, when the lateral offset $d_0 = -600$ mm and -400 mm, the total correction time t is minimized when the initial yaw angle θ_0 is −20°. For lateral offset $d_0 = -200$ mm, -150 mm, and -100 mm, the total correction time t is minimized when the initial yaw angle θ_0 is −10°.

(3) According to Figure 22 and Table 6, in the correction process for the parallel landing position, when the lateral offset d_0 is -1200 mm or -800 mm, the total correction time t is minimized when the fuselage yaw angle θ for the first stepwise rotational correction is $-20°$. For lateral offsets d_0 of -400 mm, -300 mm, and -200 mm, the total correction time t is minimized when the fuselage yaw angle θ for the first stepwise rotational correction is $-10°$.

Table 6. Analysis results data on the impact of initial yaw angle and lateral offsets on efficiency of the continuous correction method.

Serial Number	Initial Fuselage Yaw Angle θ_0/deg	Lateral Offset d_0/mm	Right Traction Speed v_2/mm/s	Translation Time t_1/s	Rotation Time t_2/s	Total Correction Time t/s
1	$-10°$	-600	30	109.26	14.60	123.86
2	$-10°$	-400	30	72.84	14.60	87.44
3	$-10°$	-200	30	36.42	14.60	51.02
4	$-10°$	-150	30	27.32	14.60	41.92
5	$-10°$	-100	30	18.21	14.60	32.81
6	$-10°$	-50	30	9.11	14.60	23.71
7	$-20°$	-600	30	78.48	30.40	108.88
8	$-20°$	-400	30	52.32	30.40	82.72
9	$-20°$	-200	30	26.16	30.40	56.56
10	$-20°$	-150	30	19.62	30.40	50.02
11	$-20°$	-100	30	13.08	30.40	43.48
12	$-30°$	-600	30	95.70	49.80	145.50
13	$-30°$	-400	30	63.80	49.80	113.60
14	$-30°$	-200	30	31.90	49.80	81.70
15	$-30°$	-150	30	23.88	49.80	73.68
16	$-30°$	-100	30	15.85	49.80	65.65

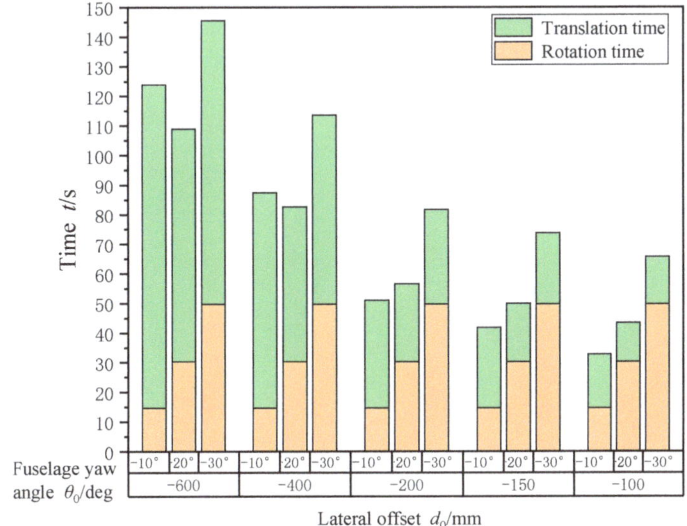

Figure 22. Bar chart of analysis results of the impact of initial yaw angle and lateral offsets on efficiency of the stepwise correction method.

The following conclusions can be drawn regarding the stepwise correction method:

(1)   In the correction process for the cross landing position, a larger initial yaw angle θ_0 results in a longer time t_2 required for stepwise rotational correction. However, for the same initial lateral offset d_0, the stepwise translational correction time t_1 will be shorter. Moreover, with the same initial yaw angle θ_0, the variation in stepwise correction time t is solely dependent on the initial lateral offset d_0.

(2)   In the correction process for the parallel landing position, according to the different lateral offset d_0, weighing the translation time t_1 and the rotation time t_2 and choosing the optimal fuselage yaw angle θ for the first stepwise rotational correction is the key to improving the efficiency of the stepwise correction method.

5.3.3. Efficiency Analysis of the Continuous Correction Method on the Impact of Initial Yaw Angle and Traction Speed

The performance comparison test is shown in Table 7 and Figure 23. The following conclusions can be drawn by analyzing the data of different right traction speeds v_2 when the initial yaw angle θ_0 is $-10°$, $-20°$, and $-30°$, respectively:

(1)   By comparing serial numbers 1-2-3, 4-5-6, 7-8-9, it can be seen that an increase in right traction speed v_2 can effectively improve correction efficiency and reduce the total correction time t.

(2)   By comparing serial numbers 1-4-7, 2-5-8, it can be seen that the larger the initial yaw angle $|\theta_0|$, the larger the helicopter control wheel angle $|\gamma|$, and the smaller the helicopter correction radius r in the continuous correction method.

Table 7. Analysis results data on the impact of initial yaw angle and traction speed on efficiency.

Serial Number	Initial Fuselage Yaw Angle θ_0/deg	Lateral Offset d_0/mm	Right Traction Speed v_2/mm/s	Helicopter Control Wheel Angle γ/deg	Helicopter Correction Radius r/mm	Total Correction Time t/s
1	$-10°$	-200	10	31.29	13162.86	271.60
2	$-10°$	-200	20	31.29	13162.86	135.80
3	$-10°$	-200	30	31.29	13162.86	90.60
4	$-20°$	-200	10	67.48	3316.98	200.60
5	$-20°$	-200	20	67.48	3316.98	100.40
6	$-20°$	-200	30	67.48	3316.98	67.20
7	$-30°$	-200	10	79.43	1492.82	207.60
8	$-30°$	-200	20	79.43	1492.82	103.80
9	$-30°$	-200	30	79.43	1492.82	69.20

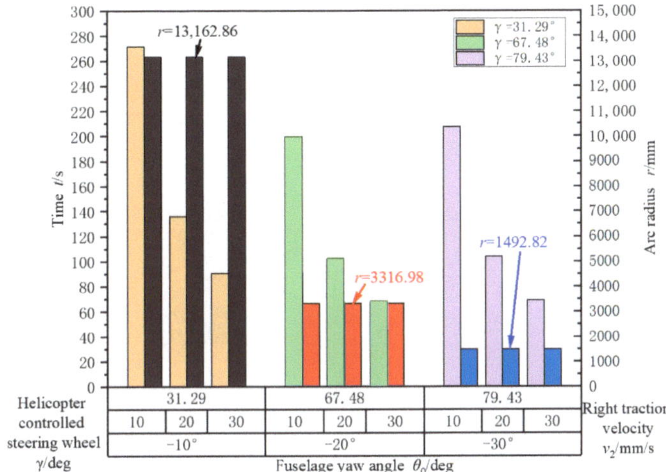

Figure 23. Bar chart of analysis results of the impact of initial yaw angle and traction speed on efficiency of the continuous correction method.

5.3.4. Efficiency Analysis of the Continuous Correction Method on the Impact of Initial Yaw Angle and Lateral Offsets

The performance comparison test is shown in Table 8 and Figure 24. By analyzing the data of different lateral offsets d_0 when the initial yaw angle θ_0 is $-10°$, $-20°$, and $-30°$, the following conclusions can be drawn:

(1) As the initial yaw angle $|\theta_0|$ increases and the lateral offset d_0 decreases, the length of the helicopter correction radius r also decreases, as indicated by the downward trend of the red line in Figure 24.

(2) Comparing serial numbers 1-2-3-4, 5-6-7, 8-9-10 and Figure 24, it is evident that with a fixed initial yaw angle $|\theta_0|$, a larger helicopter control wheel angle $|\gamma|$ results in a smaller helicopter correction radius r and a shorter correction time t.

(3) According to Figure 24, in the continuous correction for the cross landing position, although the helicopter correction radius r varies from large to small under working conditions from 1 to 10, a comparative analysis of serial numbers 1-5-8, 2-6-9, and 3-7-10 shows that with lateral offsets d_0 of -200 mm, -150 mm, and -100 mm, the shortest correction time t occurs with the initial yaw angle θ_0 of $-20°$.

(4) According to Figure 24, in the continuous correction for the parallel landing position, when the lateral offset d_0 is -400 mm, -300 mm, and -200 mm, the total correction time t is minimized when the fuselage yaw angle θ for the first trajectory of the continuous correction is $-20°$.

The following conclusions can be drawn regarding the stepwise correction method:

(1) In the correction process for the cross landing position, the continuous correction time t is negatively correlated with the initial yaw angle $|\theta_0|$ and positively correlated with the initial lateral offset $|d_0|$.

(2) In the correction process for the parallel landing position, it is crucial to balance the helicopter correction radius r and choose the optimal helicopter control wheel angle γ according to different lateral offset d_0 to improve the efficiency of the continuous correction method.

Table 8. Analysis results data on the impact of initial yaw angle and lateral offsets on efficiency.

Serial Number	Initial Fuselage Yaw Angle θ_0/deg	Lateral Offset d_0/mm	Right Traction Speed v_2/mm/s	Helicopter Control Wheel Angle γ/deg	Helicopter Correction Radius r/mm	Total Correction Time t/s
1	$-10°$	-200	30	31.29	13162.86	90.60
2	$-10°$	-150	30	39.01	9875.65	71.60
3	$-10°$	-100	30	50.55	6582.97	52.80
4	$-10°$	-50	30	67.63	3292.46	33.80
5	$-20°$	-200	30	67.48	3316.98	67.00
6	$-20°$	-150	30	72.73	2487.13	57.60
7	$-20°$	-100	30	78.29	1658.18	48.40
8	$-30°$	-200	30	79.43	1492.82	69.20
9	$-30°$	-150	30	82.03	1120.06	63.40
10	$-30°$	-100	30	84.67	746.36	57.40

5.3.5. Comparison of Performance between Stepwise and Continuous Correction Methods

Figure 25 shows the performance comparison between the stepwise correction method and the continuous correction method under different initial yaw angles θ_0 and lateral offsets d_0.

As can be seen from Figure 25, when the initial yaw angles are $\theta_0 = -10°$ and $\theta_0 = -20°$, the total correction time t for the stepwise correction method is less. However, when the initial yaw angle is $\theta_0 = -30°$, the total correction time t for the continuous correction method is less.

According to the conclusions in Sections 5.3.2 and 5.3.4, when the lateral offset d_0 is constant and the initial yaw angle $|\theta_0|$ is small, the rotation time t_2 for the stepwise correction method is shorter. In contrast, the continuous correction method takes more time because the helicopter control wheel angle $|\gamma|$ is small.

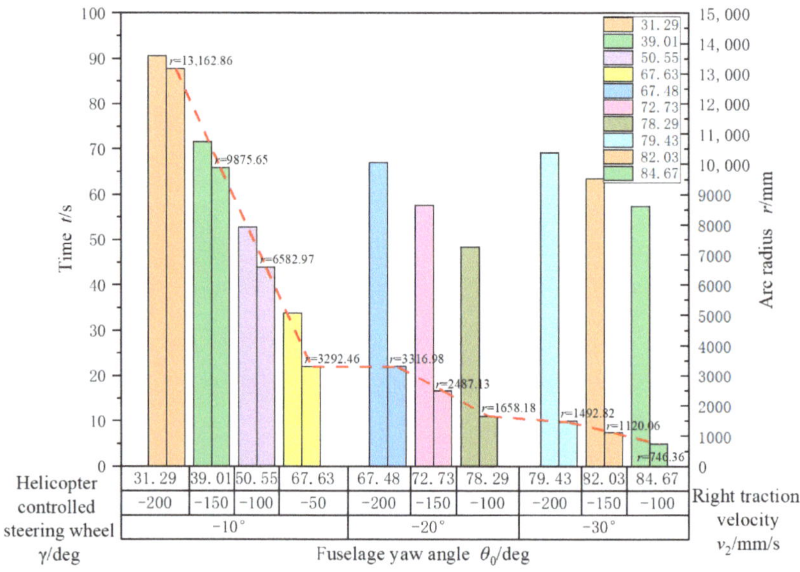

Figure 24. Bar chart of analysis results of the impact of initial yaw angle and lateral offsets on efficiency of the continuous correction method.

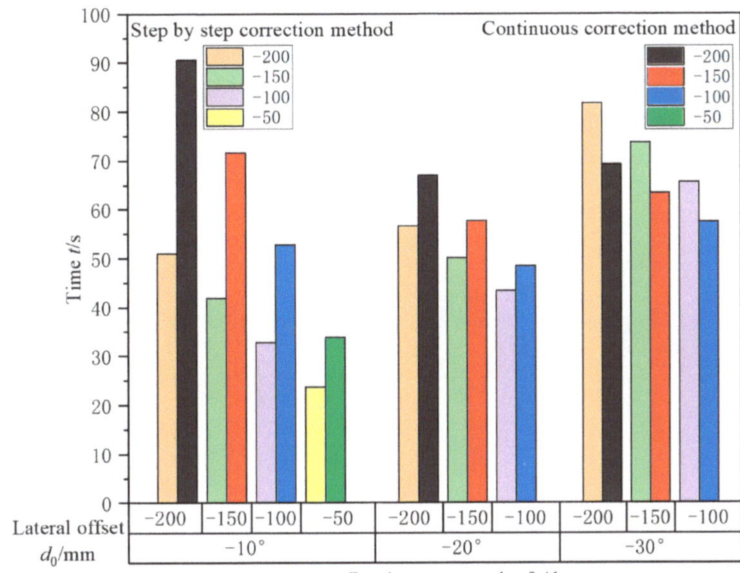

Figure 25. Bar chart of comparison results for the two methods.

However, when $|\theta_0|$ is large, the increase in rotation time for the stepwise correction method is greater than the decrease in translation operation time. In the continuous correction method, a larger helicopter control wheel angle $|\gamma|$ results in a smaller helicopter correction arc radius r. Thus, the helicopter's continuous correction path becomes smaller, reducing the time spent.

The performance comparison results indicate that when the initial yaw angle $|\theta_0|$ is small, the stepwise correction method is faster. Conversely, when the initial yaw angle $|\theta_0|$ is large and the lateral offset $|d_0|$ is small, the continuous correction method is faster.

6. Conclusions

This paper studies the methods for correcting the position of a helicopter on a deck, which is a critical operational process to ensure smooth towing into the hangar. This process is of great importance for improving the efficiency of helicopter towing operations. The paper proposes two methods for correcting a helicopter's position based on SRCR towing: the stepwise method and the continuous method. The effectiveness and correction efficiency of these methods are studied, leading to the following conclusions.

(1) To address the helicopter deck position correction of SRCR, the actual landing position deviation is decomposed into two components: lateral offset d_0 and fuselage yaw angle θ_0. This allows for a quantitative evaluation of the landing position deviation.

(2) To correct the helicopter landing deviation, this paper proposes both the stepwise correction method and the continuous correction method, based on the motion characteristic model of the SRCR. The feasibility of these two correction methods is verified through mathematical analysis and simulation research.

(3) The three key factors affecting correction efficiency are lateral offset d_0, fuselage yaw angle θ_0, and right traction speed v_2. The stepwise method and continuous method were simulated and analyzed. Results indicate that in the correction process for the parallel landing position, improving the efficiency of the stepwise correction method hinges on selecting the optimal fuselage yaw angle θ for the first stepwise rotational correction based on the different lateral offsets d_0. Similarly, enhancing the efficiency

of the continuous correction method requires choosing the optimal helicopter control wheel angle γ according to the varying lateral offsets d_0.

(4) Comparing the stepwise method with the continuous method, results show that when the initial yaw angle $|\theta_0| \leq 20°$, the stepwise correction method has a significant advantage. However, when the initial yaw angle $|\theta_0| > 20°$ and the lateral offset $|d_0| \leq 200$ mm, the efficiency of continuous correction is superior to the stepwise correction method because the helicopter's correction arc radius r is smaller.

The limitations of this study lie in it not accounting for trajectory deviations caused by environmental disturbances during pilot operations and control system processes. Future research will address these trajectory errors resulting from environmental disturbances, pilot operation accuracy, and towing speed control during real-time correction to design controllers for real-time deviation correction.

Author Contributions: Conceptualization, Y.Z. and X.Z.; methodology, Y.Z. and X.Z.; software, Y.Z. and X.Z.; formal analysis, Y.Z.; data curation, Y.Z.; writing—original draft preparation, Y.Z.; writing—review and editing, X.Z. and D.Z.; supervision, D.Z.; funding acquisition, D.Z. All authors have read and agreed to the published version of the manuscript.

Funding: The 2024 Hebei Province doctoral students innovation ability training funding project [grant number CXZZBS2024052].

Institutional Review Board Statement: Not applicable.

Informed Consent Statement: Not applicable.

Data Availability Statement: Data is contained within the article.

Conflicts of Interest: The authors declare that they have no conflicts of interest.

References

1. Wu, H.; Tan, D.; Li, Q. Comparative analysis of Comprehensive Performance of Foreign Shipborne Helicopter Landing Aid and Traction Equipment. *Ship Sci. Technol.* **2021**, *43*, 185–189. (In Chinese)
2. Feldman, A.R.; Langlois, R.G. Autonomous straightening and traversing of shipboard helicopters. *J. Field Robot.* **2006**, *23*, 123–139. [CrossRef]
3. Crocco, J. *US Coast Guard ASIST Probe Evaluation on a H-65 Dolphin*; Army Aviation and Missile Research Development and Engineering Center Fort Eustis VA Aviation Applied Technology Directorate: Fort Eustis, VA, USA, 2010; pp. 1122–1127.
4. Curtiss-Wright. Aircraft Ship Integrated Secure and Traverse System (ASIST). 2017. Available online: https://www.cw-ems.com/indal/products/helicopter-securing-and-traversing/default.aspx (accessed on 9 May 2024).
5. Leveille, M. *Development of a Spacial Dynamic Handling and Securing Model for Shipboard Helicopters*; Carleton University: Ottawa, ON, Canada, 2013.
6. Sharma, A.; Padthe, A.K.; Friedmann, P.P. Simulation of Helicopter Hover and Landing on a Moving Ship Deck using a Dynamic Ground Effect Model. In Proceedings of the AIAA Scitech 2020 Forum, Orlando, FL, USA, 6–10 January 2020. [CrossRef]
7. Thedin, R.; Kinzel, M.P.; Schmitz, S. Coupled simulations of atmospheric turbulence-modified ship airwakes and helicopter flight dynamics. *J. Aircr.* **2019**, *56*, 812–824. [CrossRef]
8. Tremblay, J.P. *Development and Validation of a Tire Model for a Real Time Simulation of a Helicopter Traversing and Manoeuvring on a Ship Flight Deck*; Carleton University: Ottawa, ON, Canada, 2007; pp. 10–34.
9. Cai, B.; Xu, T.; Zhang, Z. A Method of Lateral Correction of Helicopter. Patent CN10962530-6B, 17 August 2021. (In Chinese).
10. Zheng, X.; Zhou, L.; Zhou, H.; Chen, C.; Feng, Y. Method of Automatic Traction of Carrier-Based Helicopter. CN106428606B, 22 February 2019. (In Chinese).
11. Curtiss-Wright. Recovery Assist, Secure and Traverse System (RAST). CW-EMS. Available online: https://www.cw-ems.com/indal/products/helicopter-securing-and-traversing/rast/default.aspx (accessed on 9 May 2024).
12. Curtiss-Wright. Aircraft Ship Integrated Secure and Traverse System (ASIST). CW-EMS. Available online: https://www.cw-ems.com/indal/products/helicopter-securing-and-traversing/asist/default.aspx (accessed on 9 May 2024).
13. Curtiss-Wright. An Intelligent Claw Capture System for Non-Probe Installed Aircraft. CW-EMS. Available online: https://www.cw-ems.com/indal/products/helicopter-securing-and-traversing/tc-asist/default.aspx (accessed on 9 May 2024).
14. Curtiss-Wright. Aviation Ground Support Equipment (GSE). CW-EMS. Available online: https://www.cw-ems.com/indal/products/helicopter-securing-and-traversing/mantis-elp-aircraft-tug/default.aspx (accessed on 9 May 2024).
15. William, H.D. Aircraft Handler. U.S. Patent No. 9,428,283, 30 August 2016.
16. Alexandre, B.; Raspic, P.; Perillaud, J. Naval Platform Provided with a Deck Landing/Take-Off Zone and Means for Handling an Aircraft. U.S. Patent No. 10,836,510, 17 November 2020.

17. Alexandre, B.; Raspic, P.; Cellario, S. Naval Platform Provided with a Zone for the Deck Landing/Take-Off of at least One Aircraft and Dolly-Type Means for Handling Said Aircraft. U.S. Patent No. 11,260,990, 1 March 2022.
18. Asbjorn, B.; William, G.S. Helicopter Rapid Securing Device. C.A. Patent No. 781808, 2 February 1968.
19. Pesando, M. Traverser and Components Therefor. C.A. Patent No.1,116,150, 12 January 1982.
20. Pesando, M.; Boris, V. Helicopter Rapid Securing & Traversing System. U.S. Patent No. 4,786,014, 22 November 1988.
21. Zhang, Z.; Liu, Q.; Zhao, D.; Wang, L.; Jia, T. Electrical Aircraft Ship Integrated Secure and Traverse System Design and Key Characteristics Analysis. *Appl. Sci.* **2022**, *12*, 2603. [CrossRef]
22. Jia, T.; Shao, T.; Liu, Q.; Yang, P.; Li, Z.; Zhang, H. Research into a Marine Helicopter Traction System and Its Dynamic Energy Consumption Characteristics. *Appl. Sci.* **2023**, *13*, 12493. [CrossRef]
23. Liu, Q.; Zhang, Z.; Jia, T.; Wang, L.; Zhao, D. Energy consumption analysis of helicopter traction device: A modeling method considering both dynamic and energy consumption characteristics of mechanical systems. *Energies* **2022**, *15*, 7772. [CrossRef]
24. Zhang, Z.; Liu, Q.; Zhao, D.; Wang, L.; Yang, P. Research on shipborne helicopter electric rapid secure device: System design, modeling, and simulation. *Sensors* **2022**, *22*, 1514. [CrossRef] [PubMed]
25. Zhang, Z.; Li, W.; Sun, H.; Wang, B.; Zhao, D.; Ni, T. Research on Reliability Growth of Shock Absorption System in Rapid Secure Device of Shipboard Helicopter. *IEEE Access* **2023**, *11*, 134652–134662. [CrossRef]
26. Liu, H.; Yang, X.; Wang, N. A Stability Control Strategy for Tractor-Aircraft System on Deck. In Proceedings of the 2021 IEEE International Conference on Real-Time Computing and Robotics (RCAR), Xining, China, 15–19 July 2021; IEEE: Piscataway, NJ, USA; pp. 997–1002.
27. Wang, Y.X. Adaptive Backstepping Controller to Improve Handling Stability of Tractor-Airplane System on Deck. *Appl. Mech. Mater.* **2013**, *389*, 389–393. [CrossRef]
28. Wang, N.J.; Zhou, L.J.; Liu, H.B. Study on stability control for tractor-helicopter system on deck. *J. Shanghaiiaotong Univ.* **2012**, *46*, 1146.
29. Meng, X.; Wang, N.; Liu, Q. Aircraft Parking Trajectory Planning in Semistructured Environment Based on Kinodynamic Safety RRT∗. *Math. Probl. Eng.* **2021**, *2021*, 3872248. [CrossRef]
30. Liu, J.; Han, W.; Peng, H.; Wang, X. Trajectory planning and tracking control for towed carrier aircraft system. *Aerosp. Sci. Technol.* **2019**, *84*, 830–838. [CrossRef]
31. Wang, L.; Wu, Q.; Liu, J.; Li, S.; Negenborn, R.R. State-of-the-art research on motion control of maritime autonomous surface ships. *J. Mar. Sci. Eng.* **2019**, *7*, 438. [CrossRef]
32. Jing, Q.; Wang, H.; Hu, B.; Liu, X.; Yin, Y. A universal simulation framework of shipborne inertial sensors based on the ship motion model and robot operating system. *J. Mar. Sci. Eng.* **2021**, *9*, 900. [CrossRef]
33. Yang, H.; Ni, T.; Wang, Z.; Wang, Z.; Zhao, D. Dynamic modeling and analysis of traction operation process for the shipboard helicopter. *Aerosp. Sci. Technol.* **2023**, *142*, 108661. [CrossRef]
34. Byeong-Woo, Y.; Park, K.P. Dynamics simulation model for the analysis of aircraft movement characteristics on an aircraft carrier deck. *Int. J. Nav. Archit. Ocean. Eng.* **2024**, *16*, 100591.
35. Wikipedia contributors. MSC Adams. In Wikipedia, The Free Encyclopedia. Available online: https://en.wikipedia.org/wiki/MSC_Adams (accessed on 23 May 2024).
36. Wikipedia Contributors. MATLAB. In Wikipedia, The Free Encyclopedia. Available online: https://en.wikipedia.org/wiki/MATLAB (accessed on 23 May 2024).

Article

Analysis and Simulation of Permanent Magnet Adsorption Performance of Wall-Climbing Robot

Haifeng Ji , **Peixing Li *** and **Zhaoqiang Wang**

School of Mechanical and Automotive Engineering, Shanghai University of Engineering Science, Shanghai 201620, China; 011619115@sues.edu.cn (H.J.); wangzhaoqiang_2008@sues.edu.cn (Z.W.)
* Correspondence: lpxsuesmail@sues.edu.cn

Abstract: In response to problems such as insufficient adhesion, difficulty in adjustment, and weak obstacle-crossing capabilities in traditional robots, an innovative design has been developed for a five-wheeled climbing robot equipped with a pendulum-style magnetic control adsorption module. This design effectively reduces the weight of the robot, and sensors on the magnetic adsorption module enable real-time monitoring of magnetic force. Intelligent control adjusts the pendulum angle to modify the magnetic force according to different wall conditions. The magnetic adsorption module, using a Halbach array, enhances the concentration effect of the magnetic field, ensuring excellent performance in high-load tasks such as building maintenance, bridge inspection, and ship cleaning. The five-wheel structural design enhances the stability and obstacle-crossing capability, making it suitable for all-terrain environments. Simulation experiments using Maxwell analyzed the effects of the magnetic gap and the angle between the adsorption module and the wall, and mechanical performance analysis confirmed the robot's ability to adhere safely and operate stably.

Keywords: wall-climbing robot; adsorption performance; simulation analysis; Halbach

Citation: Ji, H.; Li, P.; Wang, Z. Analysis and Simulation of Permanent Magnet Adsorption Performance of Wall-Climbing Robot. *Actuators* **2024**, *13*, 337. https://doi.org/10.3390/act13090337

Academic Editor: Ioan Doroftei

Received: 8 August 2024
Revised: 29 August 2024
Accepted: 31 August 2024
Published: 3 September 2024

1. Introduction

With the continuous advancement of science and technology, robotics has made tremendous progress and has shown great potential for application in various fields [1,2]. Wall-climbing robots are special robots, which are important branches in the field of robotics, which comprehensively utilize various technologies such as mobile technology, adsorption technology, and control [3]. Scholars have conducted a lot of research on wall-climbing robots, and the adsorption methods are mainly divided into three types: negative pressure adsorption, bionic adsorption, and magnetic adsorption. Originally, Professor Ryo Nishi of Japan developed a wall-climbing robot in 1966, which uses the negative pressure adsorption principle generated by the rotation of the electric fan to make it adsorb on the wall. Since then, other countries have also joined the research of wall-climbing robots. Wall-climbing robots can crawl on vertical, curved, and corner walls and carry specific equipment to complete dangerous work such as cleaning, testing, and welding of special machinery [4]. In addition, the robot adsorbed on the overhanging wall, compared with manual inspection, has the outstanding advantages of strong safety and high detection efficiency and has attracted more and more attention in the field of wall inspection [5–8], which can be completed instead of humans to ensure personal safety, and has important applications in industry and shipbuilding.

In the design of wall-climbing robots, adsorption technology is a crucial element. Permanent magnet adsorption, as a common method, offers advantages such as a simple structure and stable operation, making it widely recognized. The structural forms of magnetic adsorption mechanisms are mainly categorized into flat-plate and wheel types. The flat-plate type consists of a rectangular permanent magnet mechanism, typically embedded in the tracks. Robots using annular array magnetic wheels often experience significant magnetic leakage and low magnetic energy utilization, leading to energy waste. Although existing wall-climbing robot technologies have been widely applied in industrial inspections, they often face limitations such as insufficient adsorption force and difficulty adapting to complex surfaces. This study significantly enhances the robot's stability and adsorption efficiency on various surfaces by employing a Halbach array and a suspension-type magnetic-controlled adsorption mechanism. The study conducts an in-depth analysis and validation of the factors influencing the permanent magnet adsorption performance of wall-climbing robots, providing theoretical and practical support for improving the adsorption efficiency and stability of these robots.

Through data analysis, the influencing factors and their mechanisms are clarified, which provides a practical, theoretical basis and technical support for future design optimization and provides technical support and guidance for practical engineering applications so as to promote the further development and application of wall-climbing robot technology.

2. Wall-Climbing Robot Design

2.1. Structural Design

In order to meet the special working scene of flexible movement and steering of the wall-climbing robot on the wall, the wheeled movement mode is adopted, which not only avoids the shortcomings of large crawler volume and difficult steering but also avoids the disadvantages of foot-type structure and complex control [9,10]. Common adsorption methods are mainly divided into three categories: magnetic adsorption, negative pressure adsorption, and bionic adsorption [11]. For metal walls, magnetic adsorption has the advantages of energy saving, no heat, no noise, safety, and low cost compared with other adsorption methods [12]. In order to ensure that the wall-climbing robot can be safely adsorbed on the wall surface and at the same time meet the requirements of a simple and lightweight structure, the permanent magnet unit is composed of NdFeB rare earth permanent magnet material, which will not be dangerous due to power failure like electromagnetic structure.

As shown in Figure 1, the wall-climbing robot mainly consists of walking parts and deformed parts. This division clarifies the functions and interactions of each part, which is crucial for the overall design.

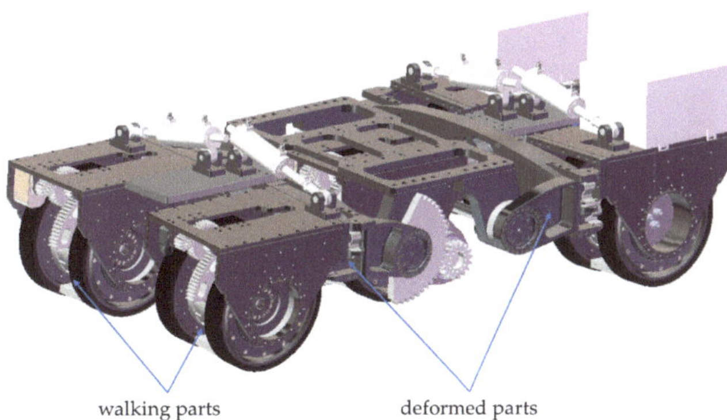

Figure 1. The overall structure of the wall-climbing robot.

The walking parts and deformed parts are symmetrically distributed. Figure 2 shows the specific structure of the deformed parts, where the design of the swing rod allows the two front walking parts to lift appropriately when encountering obstacles, thereby overcoming them. The walking parts are connected to the swing rod through twisting connectors, allowing the walking parts to rotate around the wheel set's axis to adjust the angle between the wheels and the contact surface. This design enhances the robot's ability to adapt to different terrains, allowing it to walk on vertical and curved surfaces. The mechanism parameters of the wall-climbing robot are shown in Table 1.

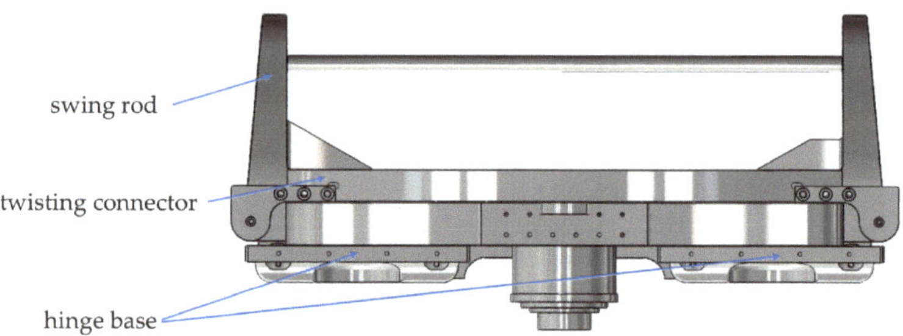

Figure 2. Structure diagram of deformed parts.

Table 1. Structural parameters of wall-climbing robots.

Items	Parameters
Length × width × height	1587 × 762 × 514
Robot weight	300
Number of wheel sets	5

Compared to traditional three- or four-wheeled climbing robots, the five-wheeled design significantly improves stability, better distributes the robot's weight, and improves maneuverability on uneven and vertical surfaces, which is a common challenge in real-world environments such as hulls and industrial complexes.

2.2. Magnetic Chuck Design

The core technology of the wall-climbing robot relies on the interaction between the magnetic suction unit and the wall, which enables the robot to closely adhere to and move on the wall by generating a strong magnetic attraction, and its performance depends on the selected magnetic material and magnetic circuit design [13]. The magnetic suction unit is composed of high-performance NdFeB permanent magnets (NdFe35), which ensures that the robot remains stable under various special working conditions. The traditional magnetic attraction method is to achieve magnetic attraction adsorption through the magnetic wheel; the permanent magnet wheel is usually composed of a whole magnet, its magnetic field is symmetrically distributed, its weight is large, and the magnetic energy utilization rate is low. Only the magnetic wheel and the wall contact part produce magnetic force, and the magnetic attraction force is very small. The designed magnetic suction unit is suspended between the two wheels by the connecting shaft between the driving wheels, which replaces the traditional permanent magnet design on the wheels, effectively reduces the overall weight, improves the magnetic mass ratio, and thus improves the utilization rate of magnetic energy.

The force sensors monitor the magnetic attraction force in real time while a motor rotates a cylindrical gear to drive the magnetic chuck to the appropriate position. This rotation adjusts the angle between the magnet and the yoke, thereby altering the magnetic attraction force between the wall-climbing robot and the wall. This ensures attachment stability under various wall conditions. Maximum magnetic attraction force is generated when the surface is parallel to the wall, and the magnetic attraction force decreases when the magnetic chuck is at an angle to the wall. The structure is shown in Figure 3.

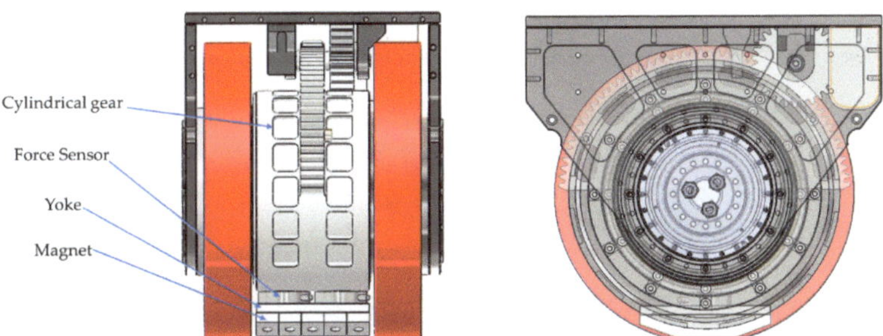

Figure 3. Schematic diagram of the magnetic adsorption unit structure.

The specific parameters of the magnetic suction unit are shown in Table 2 below.

Table 2. Parameters of magnetic suction units.

Parameters	Value (mm)
Overall length	140
Width	120
The height of the yoke	10
Radius of the lower arc	124.5

The permanent magnet and the conductive magnet are arranged in a variety of different ways to make more magnetic flux converge into the working air gap to improve the utilization rate of the permanent magnet. The Halbach-type permanent magnet array was first discovered and proposed by Klaus Halbach of the United States [14], with good uniformity and reduced self-weight.

During the design process, stability, flexibility, and magnetic attraction are mainly considered:

(1) The five-wheel design can improve the stability of the robot in vertical and inclined planes and avoid tipping over.
(2) The deformed parts allow the robot to adjust its attitude and adapt to different wall angles to enhance its adaptability.
(3) The adsorption angle is adjusted by the pendulum magnetic suction mechanism to improve the adsorption efficiency and ensure adhesion stability under different wall conditions.

The pendulum magnetron adsorption module marks a significant improvement over traditional stationary magnetic systems. By allowing the magnetic force to be dynamically adjusted according to surface conditions, this design improves the safety and efficiency of the operation of complex structures. The simulation results show that the magnetic adhesion is significantly increased, especially on irregular surfaces, where conventional systems may not be able to maintain sufficient contact.

In order to explore whether the designed magnetic chuck meets the safety requirements, the following simulation will be based on Maxwell and the influence of the distance between the magnetic chuck and the wall on the magnetic attraction and the angle between the magnetic chuck and the wall will be explored, respectively.

2.3. Magnetic Suction Simulation Based on Maxwell

2.3.1. Analysis of the Influence of Air Gap Distance on Magnetic Attraction

In order to obtain a large magnetic attraction and improve the performance under heavy load conditions, the permanent magnet is made of NdFeB material, and the yoke iron and metal wall are made of steel_1008 structural steel. The performance parameters of the selected NdFeB materials are shown in Table 3 below.

Table 3. NdFe35 neodymium iron boron performance parameters.

Model	N35
Remanence	1.17–1.22 (Br/T)
Intrinsic coercivity	$\geq$955 (Hcj/kJ·m^{-1})
Coercivity	$\geq$860 (Hcb/kA·m^{-1})
Maximum energy product	263–287 (BH/kJ·m^{-3})
Restore permeability	1.05

Compared with the traditional arrangement of magnets that are magnetized in the same direction, that is, the plate-type static magnetic field. The Halbach array has better space and material utilization. Two types of magnets with the same volume produce a much stronger magnetic field than the plate type [15]. The principle is to combine permanent magnets with different magnetization directions according to certain rules and superimpose the two magnetization methods of ordinary tangential magnetization and radial magnetization. A more ideal sinusoidal distributed magnetic field can be obtained, the structure of which is shown in Figure 4.

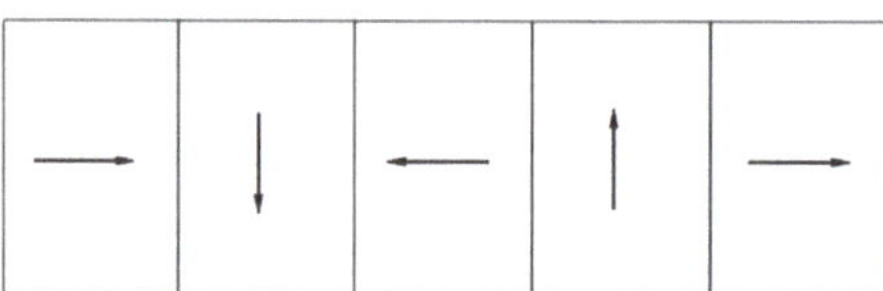

Figure 4. Magnetization schematic of stacking method.

The magnetic field of the obtained magnet has the characteristics of obvious enhancement on one side and obvious weakening on the other side [16]. The comparison of the magnetic induction intensity on both sides is shown in Figure 5.

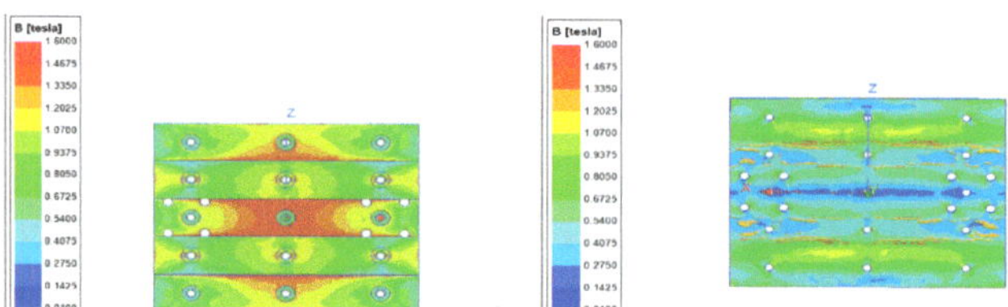

Figure 5. The magnetic induction intensity contour map of the front and back sides of the magnetic chuck.

Simulation analysis of magnetic attraction force was conducted using ANSYS-Maxwell software to calculate the magnetic attraction force of the magnet on the metal wall. Simulation analysis: Magnetic force simulation was performed using Ansys Electronics Desktop—Maxwell software, with the following simulation steps:

1. Create a model of the magnetic element and set the material properties, the yoke iron and steel plate materials are Steel_1008, and the permanent magnet materials are NdFe35.
2. The permanent magnet material is magnetized according to the Halbach arrangement.
3. Establish a closed simulation space, i.e., boundary conditions, and offset 20% outward in the XYZ direction according to the maximum volume of the model, and then mesh the yoke iron and steel plate to 10 mm.
4. Set the magnetic gap range from 1 mm to 20 mm, and the angle between the magnetic unit and the wall clamp is 0~45°.
5. Run the simulation to record the magnetic attraction at different parameters.

6. Analyze the simulation results and plot the magnetic attraction as a function of the gap and angle.

When the magnetic unit is 5.5 mm away from the wall, the magnetic attraction force of the structural simulation using the Halbach array is 1994 N, while the traditional directional consistency magnetization simulation result is 1012.6 N, and the magnetic energy utilization rate is increased by 97.2%, which shows the significant effect of the Halbach array on the magnetic field focusing. The relationship between the magnet and the wall and the magnetic attraction is shown in Figure 6.

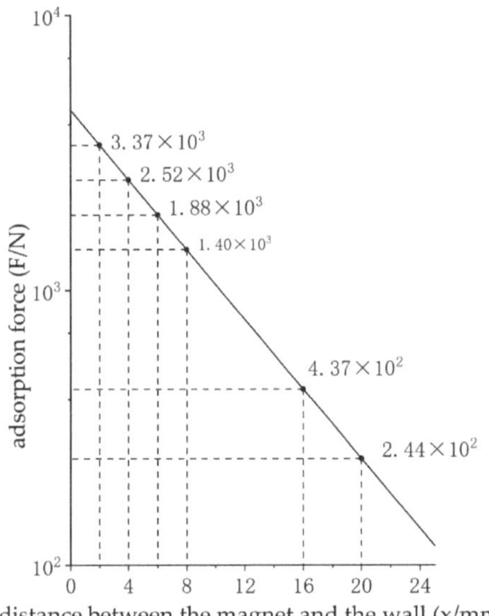

Figure 6. Relationship between magnet and wall spacing and magnetic attraction.

Fitting the above data, the relationship between the distance between the magnet and the wall and the magnetic attraction is obtained.

$$F = 4509 \times e^{-0.1458x} \tag{1}$$

with a gap of 8 mm, Maxwell simulation showed a magnetic attraction of 1400.7 N, while the magnetic attraction predicted by the exponential attenuation fitting model was 1404.5 N, with a corresponding error of only 0.271%, showing a high degree of accuracy of the model. In addition, the model had a goodness-of-fit (R^2) of 0.9986, a value that reflects the proportion of total variation explained by the model, indicating that the agreement between the fit curve and the experimental data is extremely high. This high R^2 value confirms the reliability of the model in predicting the magnetic attraction simulation results at different clearances.

Simulation analysis highlights the advantages of the pendulum magnetron adsorption module in maintaining high adhesion across a wide range of surface angles and types. This feature is critical for tasks that require high reliability, such as bridge inspections and high-rise building maintenance, as previous technologies may not provide adequate performance.

2.3.2. Effect of the Relative Wall Angle of the Magnet on the Magnetic Attraction

As a kind of special robot, the wall-climbing robot often has to adapt to various complex working conditions and wall transitions. Completing the wall chamfer transition is a big problem, and Maxwell was used to simulate and analyze the wall-climbing robot during the right-angle transition. During the rotation of the magnetic structure, the horizontal and vertical forces are constantly changing, so it is necessary to measure the magnetic attraction at all angles first. The horizontal force and the vertical force are symmetrical processes in the change of the magnetic structure from 0~90°, so only the horizontal and vertical directions in the range of 0~45° need to be measured, and the results are shown in Figure 7.

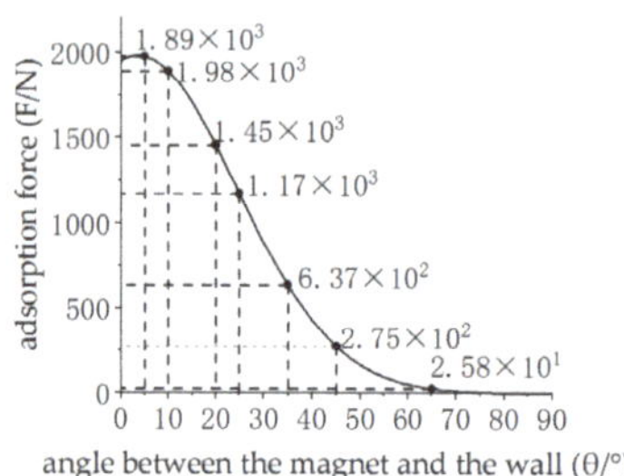

Figure 7. Vertical force variation during the angular transition.

The data fitting of the simulation results is carried out, and the relationship between the rotation angle of the magnet and the force in the vertical direction is obtained.

$$F = 5612 \times e^{\left(-\frac{\theta-96.88}{29.27}\right)^2} \tag{2}$$

The R^2 of the above fitting is 0.9998, and the fitting effect is superior. As shown in Figure 8, when the robot transitions at a right-angle wall, the magnetic induction intensity contour of the magnetic chuck shows that during the rotation process, the force in the vertical direction gradually decreases while the force in the horizontal direction gradually increases. This change has an important impact on the navigation and stability of the robot on complex wall surfaces, ensuring the adaptability and safety of the robot when performing tasks.

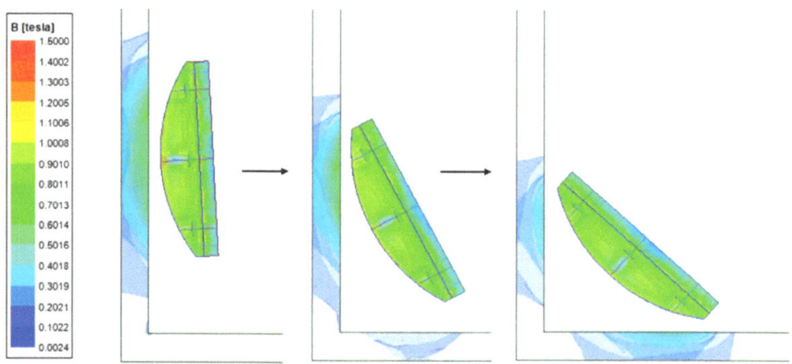

Figure 8. Contour of magnetic induction intensity.

3. Analysis of Mechanical Properties of Five-Wheeled Wall-Climbing Robot

3.1. Robot Static Analysis

This section may be divided by subheadings. It should provide a concise and precise description of the experimental results, their interpretation, as well as the experimental conclusions that can be drawn. The designed five-wheeled wall-climbing magnetic robot can achieve a horizontal plane to a vertical plane, as well as an inner fold from the vertical plane to the ceiling and an outer fold from the vertical plane to the roof. The circumferential motion process of the inner fold angle is shown in Figure 9.

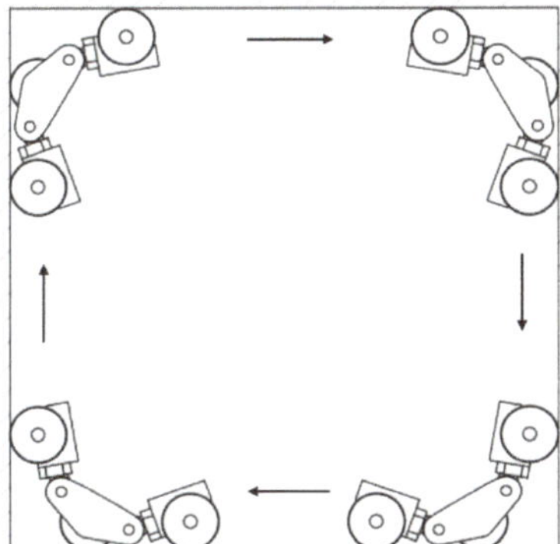

Figure 9. Circumferential motion diagram of the internal folding angle.

When the five-wheeled wall-climbing robot is adsorbed on the metal wall surface for movement, the horizontal plane movement is similar to that of ordinary wheeled robots driven by a motor, and basically, no failure will occur. If you move on a vertical wall, the direction of sliding can only be in the vertical direction, the support force of the robot will be reduced, and the friction will be weakened accordingly, so there must be enough magnetic attraction to generate enough friction to overcome the gravity of its own sliding (see Figure 10). The intermediate wheel has two magnetic chucks, so the conditions for not slipping on the vertical wall are as follows:

$$5F_M \cdot \mu \geq Mg \tag{3}$$

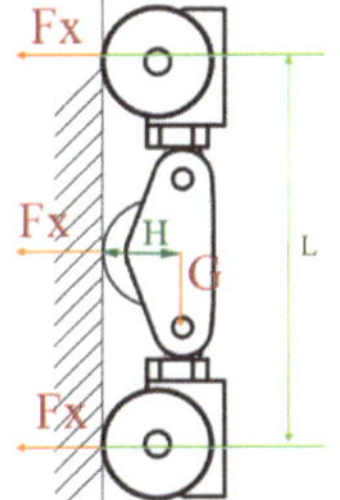

Figure 10. Schematic diagram of the force under the vertical state of the robot.

Additionally, on vertical surfaces, the condition of moment balance must be met; otherwise, longitudinal overturning along the rear wheels may occur.

Calculating the torque at the contact point between the rear wheel and the wall.

$$\frac{L}{2} \cdot F_X + L \cdot 2F_X \geq Mg \cdot H \tag{4}$$

L—the length between the two wheels of the wall-climbing robot is 910 mm;
H—the distance from the center of gravity of the robot to the wall is 150 mm;
M—The overall mass of the robot is 181.66 kg;
μ—The coefficient of static friction between the wheel and the wall is about 0.5.

Through the magnetic attraction data and calculations obtained from the simulation results, it can be seen that the robot can not slip and overturn on the vertical wall. In addition, the main forms of instability of the robot include failure in the transition from the horizontal ground to the vertical wall, the slipping and falling of the vertical wall and the dangerous working conditions caused by the magnetic force change during the vertical wall facing the ceiling. On both horizontal and vertical walls, the distance between the magnetic suction mechanism and the wall is usually constant, so there is almost no danger. In the process of the transition between the vertical wall and the ceiling, the magnetic force of the wall-climbing robot decreases sharply, and there is no ground support force to counteract gravity. The overall stress state and the minimum safe magnetic attraction under this condition are discussed below.

In this section, we will analyze the entire process from before the front wheels of the five-wheel wall-climbing robot touch the ceiling to the full adsorption to the ceiling. The process is shown in Figure 11.

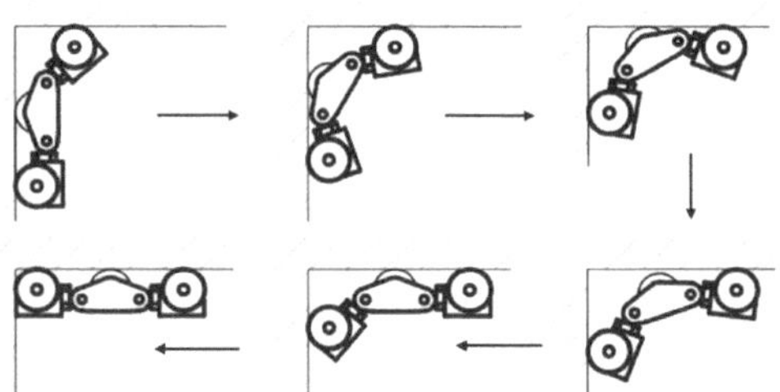

Figure 11. Diagram of transition from vertical wall to ceiling.

At the beginning of the transition of the front wheels to the ceiling, the suction of the vertical wall decreases sharply, and the suction of the ceiling to the wheels gradually increases. Let the suction force of a single wheel in the horizontal direction be F_x, and the suction force of a single wheel in the vertical direction be F_y, then it is necessary to meet the:

$$2F_y + \mu \cdot (3F_X + 2F_{X2}) \geq Mg \tag{5}$$

F_{X2} is the magnetic attraction of the wheel at the moment of transition, and its size changes from moment to moment. Maxwell's magnetic simulation results from different angles show that F_{x2} increases from 1994 N to 506.73 N and F_y rises from 1.2 N to 101.4 N when the magnetic simulation results are changed from 0 to 38 degrees. At 40 degrees, F_y is 133.12 N, so it can be seen that 0~38° and 40~90° are absolutely safe processes. The range angle of the most likely danger is 38~40°, and the known simulation data are substituted, and the five-wheeled wall-climbing robot can still maintain safe working conditions under this angle.

After the front wheel is completely adsorbed to the ceiling, the middle wheel has a suspension process in the span to ensure that the robot does not fall. The conditions are:

$$2(F_y + \mu \cdot F_X) \geq Mg \tag{6}$$

Based on the known magnetic attraction, this process does not cause failure such as the robot falling.

3.2. Robot Dynamics Analysis

The robot spanning from the vertical wall to the ceiling is one of the most dangerous stages. In the spanning stage, the middle wheel gradually moves away from the wall, the speed is small, the acceleration can be ignored, and the resultant moment is calculated at the rear wheel B point, wherein the wall support force F_{N3} for the intermediate wheel is 0 (see Figure 12), then it should be satisfied:

$$F_{d1}\, L_{BA} + F_{N1}L + GLg - (F_{M!}L + F_{M3}L_{BC} + F_{f1}L_{BA} = 0 \tag{7}$$

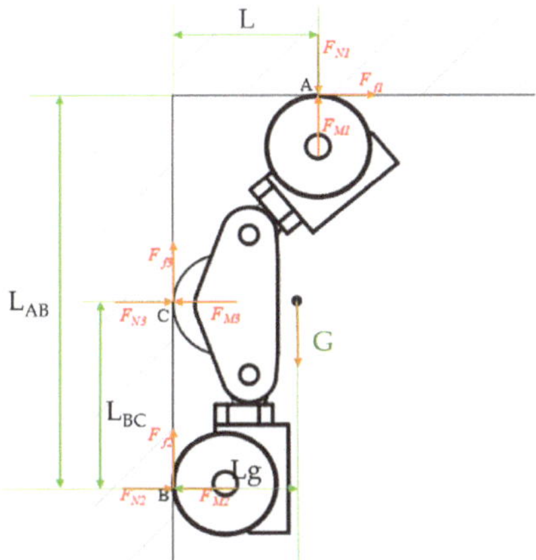

Figure 12. Force diagram of crossing the critical point.

F_{d1} and F_{f1} are the driving force and friction force of the front wheel, respectively, and F_{M1}, F_{M2}, and F_{M3} are the magnetic attraction force of the front wheel, the rear wheel, and the middle wheel on the wall, respectively. F_{N1} and F_{N2} are the normal support forces of the front and rear wheels, respectively. L and L_{BA} are the projection distances of the contact point between the front wheel and the wall surface and point B in the X and Y directions, respectively; L_{BC} is the vertical distance between the rear wheels in the robot; Lg is the projection of the center of gravity of the robot and point B in the Y direction. In the process of moving the intermediate wheel away from the wall, the magnetic attraction of the wall to the intermediate wheel is also gradually decreasing.

4. Climbing Robot Prototype Testing

Prototype testing takes place in a lab environment and is designed to simulate real-world conditions encountered by wall-climbing robots in industrial applications. Test surfaces include vertical steel walls, slopes, and curved surfaces to replicate a variety of operating environments. The robot is equipped with sensors to measure adhesion, speed, and stability. There is also anti-falling rope protection; even when the adsorption force is insufficient, there will be no safety accidents.

The experimental setup involved the following key tests: The adsorption force of the robot as it moves in horizontal, 60° inclined, and vertical planes was measured and recorded using a force transducer to verify the effectiveness of the suspended magnetron adsorption module. Obstacle crossing test: The robot was tested on a cuboid obstacle with a height of 12 cm to evaluate the obstacle crossing performance of the five-wheeled wall-climbing robot. The process is that the magnetic chuck of the front wheel is staggered with the wall surface at first, forming an angle so that the magnetic attraction of the front wheel and the wall surface is reduced so that the front wheel is more convenient to lift and cross the obstacle, and then the middle wheel and the rear wheel cross the obstacle in turn. The obstacle-crossing process is shown in Figure 13.

Wall Transition Test: Validates the robot's ability to transition from horizontal to vertical and from vertical to horizontal. The process of transition from the horizontal plane to the vertical plane is that the front wheel is lifted first and then adsorbed to the vertical wall, and the wall transition is gradually completed and the experimental process of inner and outer fold angles is shown in Figures 14 and 15.

Obstacle crossing tests have shown that the robot can successfully cross obstacles up to 12 cm without losing any stability or adhesion. During the wall transition test, when the robot transitions from a vertical surface to a horizontal surface, the front wheel gradually moves away from the wall, the adsorption decreases, and then the front wheel adsorbs to the horizontal plane, the whole process force changes within a certain range, but the overall force remains within a safe operating range without slipping or falling.

Figure 13. Robot obstacle-crossing process diagram.

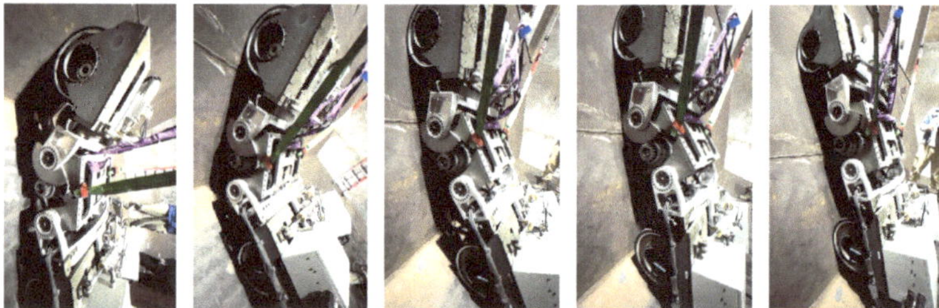

Figure 14. Diagram of the robot's inner folding angle transition.

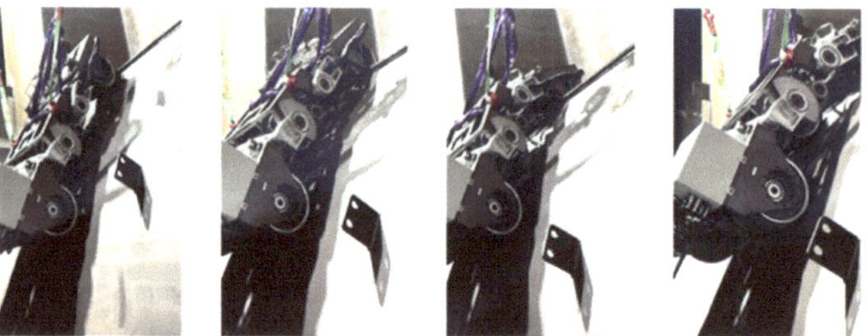

Figure 15. Diagram of the robot's outer folding angle transition.

5. Conclusions

This paper designed and developed a prototype of a wheeled magnetic climbing robot, conducted simulation analysis using Maxwell, and explored the effects of the gap and angle between the magnet and the wall on the magnetic attraction, validating its adhesion performance based on the mechanical model and ensuring a safe transition at the critical turning point. Motion tests on metal surfaces demonstrated that the five-wheel climbing robot meets the design requirements. Further research on the prototype's dynamic and operational performance will focus on validating its performance and reliability in application areas such as industrial inspection, ship cleaning, and building maintenance.

In summary, the analysis and validation of the permanent magnet adhesion performance of the climbing robot were successfully conducted, and the development of the five-wheel climbing robot prototype was completed. This research achievement has significant implications for advancing robot technology applications in engineering and other fields and provides a useful reference and model for related research.

Author Contributions: Conceptualization, P.L. and H.J.; methodology, Z.W.; software, H.J.; validation, P.L., H.J. and Z.W.; formal analysis, H.J.; writing—original draft preparation, H.J.; writing—review and editing, H.J. and P.L.; supervision, P.L.; project administration, Z.W. All authors have read and agreed to the published version of the manuscript.

Funding: This work was supported by Shanghai Jiao Tong University and Leader Harmonious Drive Systems Co., Ltd. Robot and CNC Technology Research and Development Center.

Data Availability Statement: The data that support the findings of this study are available on request from the corresponding author.

Conflicts of Interest: The authors declare no conflicts of interest.

References

1. Zheng, K.; Hu, Y.; Wu, B. Trajectory planning of multi-degree-of-freedom robot with coupling effect. *J. Mech. Sci. Technol.* **2019**, *33*, 413–421. [CrossRef]
2. Albitar, H.; Dandan, K.; Ananiev, A.; Kalaykov, I. Underwater robotics: Surface cleaning technics, adhesion and locomotion systems. *Int. J. Adv. Robot. Syst.* **2016**, *13*, 7–18. [CrossRef]
3. Zhou, J. *Research on Design and Control Method of Rolling Adsorption Wall-Climbing Robot*; Science and Technology of China: Hefei, China, 2023. [CrossRef]
4. Lee, G.; Wu, G.; Kim, J.; Seo, T.W. High-payload climbing and transitioning by compliant locomotion with magnetic adhesion. *Robot. Auton. Syst.* **2012**, *60*, 1308–1316. [CrossRef]
5. Meng, X.Y.; Dong, H.L. Research on the structure design of climbing robot and the key technology of surface magnetic force absorption. *Manuf. Autom.* **2018**, *40*, 19–22.
6. Bisht, R.; Pathak, P.; Panigrahi, S. Modelling, simulation and experimental validation of wheel and arm locomotion based wall-climbing robot. *Robotica* **2023**, *41*, 433–469. [CrossRef]
7. An, L.; Zhang, C.; Chu, S.; Zhang, J.; Yang, Y.; Liu, J.; Li, S.; Liu, J.; Wang, Z. Finite element analysis of permanent magnet attraction device based on Halbach array wall-climbing robot. *Mech. Res. Appl.* **2020**, *33*, 36–39.
8. Dian, S.; Fang, H.; Zhao, T.; Wu, Q.; Hu, Y.; Guo, R.; Li, S. Modeling and trajectory tracking control for magnetic wheeled mobile robots based on improved dual-heuristic dynamic programming. *IEEE Trans. Ind. Inform.* **2021**, *17*, 1470–1482. [CrossRef]
9. Zhang, Y. *Structural Design and Motion Performance Analysis of Permanent Magnet Adsorption Wall-Climbing Robot*; Chang'an University: Xi'an, China, 2023.
10. Sun, Z. *Research on Structural Design and Control System of Underwater Hull Defouling Robot*; Dalian University of Technology: Dalian, China, 2022. [CrossRef]
11. Chen, L. *Research on Obstacle Crossing and Wall Transition of Permanent Magnet Adsorption Wall-Climbing Robot*; China Jiliang University: Hangzhou, China, 2018.
12. Song, W.; Jiang, H.; Wang, T. Optimal design and experimental study of magnetic adsorption components for wall-climbing robots. *J. ZheJiang Univ.* **2018**, *52*, 6–13.
13. Chen, Y. *Theoretical Analysis and Experimental Study of Magnetic Adsorption Unit of Halbach Array Robot*; Nanjing University of Science and Technology: Nanjing, China, 2013.
14. Yang, B. *Design of Mechanical System of Wall-Spanning Magnetic Adsorption Wall-Climbing Robot and Analysis of Mechanical Characteristics under Typical Working Conditions*; Southeast University: Dhaka, Bangladesh, 2015.
15. Zhang, D.; Yang, P.; Huang, Z.; Sun, L.; Zhang, M. Design and optimization of pendulum magnetic adsorption mechanism for wall-climbing robot. *Chin. J. Eng. Des.* **2023**, *30*, 334–341.
16. Zhang, X. *Research on Wheeled Suspended Magnetic Adsorption Wall-Climbing Robot (Applied Research).Heilongjiang*; Harbin Institute of Technology: Harbin, China, 2012. [CrossRef]

Article

Designing a Robotic Gripper Based on the Actuating Capacity of NiTi-Based Shape Memory Wires

Adrian Petru Teodoriu [1], Bogdan Pricop [1], Nicoleta-Monica Lohan [1], Mihai Popa [1], Radu Ioachim Comăneci [1], Ioan Doroftei [1,2] and Leandru-Gheorghe Bujoreanu [1,*]

1. Faculty of Materials Science, "Gheorghe Asachi" Technical University of Iasi, Blvd. D. Mangeron 41, 700050 Iasi, Romania; nicoleta-monica.lohan@academic.tuiasi.ro (N.-M.L.)
2. Romanian Academy of Technical Sciences, 26 Dacia Blvd., 030167 Bucharest, Romania
* Correspondence: leandru-gheorghe.bujoreanu@academic.tuiasi.ro

Abstract: In the present study, the capacity of two commercial NiTi and NiTiCu shape memory alloy (SMA) wires to develop work-generating (WG) and constrained-recovery (CR) shape memory effects (SMEs), as well as the capacity of a commercial NiTiFe super-elastic wire to act as cold-shape restoring element, have been investigated. Using differential scanning calorimetry (DSC), the reversible martensitic transformation to austenite of the three NiTi-based wires under study was emphasized by means of an endothermic minimum of the heat flow variation with temperature. NiTi and NiTiCu wire fragments were further tested for both WG-SME and CR-SME developed during the heating, from room temperature (RT) to different maximum temperatures selected from the DSC thermograms. The former tests revealed the capacity to repetitively lift various loads during repetitive heating, while the latter tests disclosed the repetitive development of shrinkage stresses during the repetitive heating of elongated wires. The tensile behavior of the three NiTi-based SMA wires was analyzed by failure and loading–unloading tests. The study disclosed the actuation capacity of NiTi and NiTiCu shape memory wires, which were able to develop work while being heated, as well as the resetting capacity of NiTiFe super-elastic wires, which can restore the initial undeformed shape of shape memory wires which soften while being cooled down. These features enable the design of a robotic gripper based on the development of NiTi-based actuators with repetitive action.

Keywords: shape memory alloy; super-elastic alloy; martensitic transformation; work; constrained recovery; repetitive actuation

Citation: Teodoriu, A.P.; Pricop, B.; Lohan, N.-M.; Popa, M.; Comăneci, R.I.; Doroftei, I.; Bujoreanu, L.-G. Designing a Robotic Gripper Based on the Actuating Capacity of NiTi-Based Shape Memory Wires. *Actuators* **2024**, *13*, 319. https://doi.org/10.3390/act13080319

Academic Editor: Giulia Scalet

Received: 3 July 2024
Revised: 14 August 2024
Accepted: 18 August 2024
Published: 21 August 2024

1. Introduction

Shape memory alloys (SMAs) are able to recover a "hot shape", which was induced in the austenite high-temperature state, by simply heating an element to which a "cold shape" was induced in the martensite low-temperature state. [1] This heating-triggered phenomenon is called shape memory effect (SME) and can be observed in alloys, ceramics and polymers [2]. Depending on the constraints that the SMA has to overcome during its hot shaped recovery, SME can be: (i) with free recovery (FR) without constraints, (ii) with constrained recovery (CR) with stress development at a constant strain, or (iii) work generating (WG) with stroke development under a constant force [3].

The work generating capacity of SMAs inspired the first exposure of an application of these materials at the Brussels International Fair in 1958. It was a cycling lifting device actuated by an Au-Cd single crystal which lifted a load during electrical resistive heating and lowered it back during air-fan cooling. This was the first ever mentioned SMA-activated thermal actuator [4].

With the discovery of NiTi SMA in 1963 [5], a remarkable burst in the development of actuators with active elements made from NiTi-based SMAs was observed [6]. This progress was enhanced by the advance of ternary alloyed NiTi-based SMAs, such as Ni-Ti-Cu or Ni-Ti-Fe [7]. The former has been able to decrease transformation hysteresis, stabilize critical transformation temperature, soften the martensite, and cause a two-step shape change [8]. The latter depressed the critical temperatures of the martensitic transformation below room temperature (RT) which enabled the alloys to be austenitic at RT [9]. Some SMAs experience a reversible thermoelastic martensitic transformation which is the mechanism of a work-generating SME [10]. These commercial SMAs, based on NiTi, CuZnAl, or CuAlNi, when deformed in an austenitic state, enable the obtainment of an unstable stress-induced martensite which forms during loading and reverts to austenite during unloading. This mechanism governs super-elasticity (SE) behavior, characterized by RT unloading recovery strains that, in special cases, can be as high as 25% [11].

NiTi-based SMAs have been produced in the form of wires [12], beams [13], coil [14] or helical springs [15], thin plates [16], etc., which can provide various useful properties [17]. One of these properties is the quiet and soft operation that enabled the development of various bioinspired applications, mimicking the operation of the tendons, such as robotic arms capable of pan–tilt movements [18]. By connecting several NiTi wires in series, large values for specific strokes and work outputs were obtained at a parallel gripper [19].

NiTi-based SMAs have a remarkable storage capacity, releasing as much as 4 J/g upon heating-induced SME and 6.5 J/g during super-elastic RT unloading [20]. Based on these features, a large variety of work-generating SME-driven applications have been fabricated [21], such as mini modular mechanical devices [22], small bioinspired robots [23], self-sensing compact actuators [24], artificial limbs [25], and dynamic vibration absorbers [26]. These applications are generally characterized by a long functional fatigue life, being able to withstand up to 10^5 cycles [27].

One of the main drawbacks of all SMA-driven actuators is the requirement of "cold shape" resetting, since the material is unable to develop a perceptible amount of work during cooling [28]. A possible solution could be the application of a training thermomechanical treatment meant to obtain two-way shape memory effects (TWSME). Through TWSME, the active SMA element can instantly recover its hot shape during heating and its cold shape during cooling [29]. Various training procedures have been devised, such as the bidirectional memory effect training method, which enables maximum deflections of about 10 mm [30]. Nevertheless, due to some structural and functional fatigue phenomena, both hot and cold shapes become gradually deformed, as compared to the initial ones, and the total stroke is reduced [31].

Other constructive solutions involved the use of "dead loads" that diminish the useful stroke developed by WG-SME during heating and have the merit of resetting the cold shape, thus enhancing the cycling functioning of the actuator [32].

Several constructive variants have been adopted for "cold shape" resetting. The first variant was the use of regular steel springs which had the disadvantage of producing the maximum resetting force at the beginning of the backstroke [6]. The improved variants involved using reciprocating SMA elements [33] or a pair formed by an SMA and a super-elastic element [34]. The functioning principle of the SMA/super-elastic coupling consists in the deformation of the latter during the heating of the former (that becomes tougher during heating and softer during cooling) and its shape resetting under the effect of the super-elastic element that forces it to recover its cold shape [35].

Based on the actuation capacity of NiTi-based SMAs, several types of grippers were developed. Some representative examples include a soft robotic gripper capable of handling delicate things [36], a module with adjustable grasping stiffness [37], or a modular soft gripper driven by large wire tendons [38].

A particular type of solution was introduced by Guo et al., who designed a compliant differential SMA actuator with two antagonistic wires, coupled by a torsion steel spring, that developed an angular motion [39].

All these applications have their specific disadvantages concerning cold shape resetting. The use of bias steel springs caused an increase in the resistance force with the increase in the stroke, while the backstroke is subjected to maximum resetting force at the beginning. On the other hand, using SMA/super-elastic couplings did not involve a direct connection between the two executive elements, which caused an efficiency loss from an energetic point of view.

The present paper aims to: (i) evaluate the capacities of NiTi-based SMAs to develop WG and CR-SME as well as to accommodate tensile strains both during single and multiple cycling); (ii) to emphasize the effect of partial substitution of Ni with Cu, to enhance SM response; (iii) to analyze the functioning of an actuation setup where an SM and an SE wire are working against each other, and (iv) to introduce a three-fingers gripper with a fast grasping speed, caused by three SMA wires and a slow releasing capacity, due to three super-elastic wires which are directly connected to the SMA ones.

2. Materials and Methods

Here, 0.5 mm-diameter wires were purchased from Kellogg's Research Labs, (New Boston, NH, USA) with three chemical compositions: $Ni_{48}Ti_{52}$; $Ni_{45}Ti_{50}Cu_5$, and $Ni_{48}Ti_{50}Fe_2$. The wires were subjected to: (i) thermal analysis, (ii) WG-SME experiments, (iii) RT tensile testing, (iv) tensile CR-SME investigations, and (v) SM shrinkage biased by SE recovery tests.

Thermal analysis was performed by differential scanning calorimetry (DSC) using a NETZSCH calorimeter type DSC 200 F3 Maya, with a sensitivity below 1 μW, temperature accuracy of 0.1 K, and an enthalpy accuracy below 1%. Calibration was performed with Bi, In, Sn, and Zn standards. Heating scans were performed between RT and 200 °C with a heating rate of 10 °C/min. Calibration curves were also used to exclude any background noise. NiTi, NiTiCu, and NiTiFe wire fragments weighing less than 5 mg were cut and were investigated under an Ar-protective atmosphere. The resulting DSC thermographs, comprising heat flow variations with temperature, were evaluated with the Proteus software v.6.1.

WG-SME experiments were achieved by means of an experimental setup which was recently fully described [40]. Based on the critical temperatures determined on the DSC thermographs, the maximum heating temperatures were selected as 39 °C, 41 °C, 43 °C, and 50 °C. Both NiTi and NiTiCu wires were subjected to WG-SME experiments during which they were elongated by four different loads (2.5, 3, 3.5 and 4 kg) at RT, lifted them during heating and lowered them back during cooling. The entire heating–cooling cycle was controlled by an experimental electronic device specially built for this purpose. The device instantly switched between heating and cooling modes when the maximum pre-set temperature was reached. Experiments were performed with a single heating–cooling cycle for each of the above four maximum temperatures under the effect of each of the four applied loads. In the case of the 4 kg load, five heating–cooling cycles were applied in order to check the reproducibility of the displacement variation with temperature during WG-SME experiments. The results of these experiments consisted of plotting the variations in temperature and the wire's free-end displacement, which were recorded by an acquisition module.

RT tensile tests of NiTi, NiTiCu, and NiTiFe wires were performed on an INSTRON 3382 tensile machine (Norwood, MA, USA). The machine was equipped with a thermal chamber able to heat up to 250 °C. The tests were first performed up to the wire failure and then consisted in loading–unloading up to a maximum strain ranging between 6–8%. Both failure and loading–unloading tests were performed with a cross-head speed of 1 mm/min. To firmly fasten the 0.5 mm-diameter wires, a special type of grips was designed and manufactured. Figure 1 illustrates the details of the experimental grip assemblies which were designed to accommodate a longer wire length, such as to increase the friction force between the wire and grip and to firmly fasten the wire specimen.

Figure 1. Images of the experimental grip assemblies specially designed for wire testing: (**a**) individual images of the grip assembly in a dissembled state (up) and assembled state with the fixed wire specimen; (**b**) wire specimen failure while being fixed in the lower grip assembly fastened on the tensile testing machine.

The grips were machined from E 335 plain carbon steel. The drilled central elasticizing hole and the final central slot were created using wire spark erosion and enabled the grip body to act as a tweezer, as illustrated in the upper part of Figure 1. The wire specimen passed between the two tightening screws, and then it was tilted through the deviatory hole and wound around the fixing screw. After the final tightening of all three screws, the wire specimen was firmly fastened, as shown in the lower part of Figure 1a. Figure 1b illustrates a wire specimen that failed in tension while being fixed in the lower grip assembly fastened in the thermal chamber of the tensile testing machine.

It must be noted that the SMA wire specimens are rather stiff and hard to bend around the load reduction roller. Moreover, bending alters the structure and induces parasitic mechanical stress on the specimens. To keep the same length for each experiment, the specimens were fastened in the dedicated clamping mechanism by the two large M8 screws. The results of the RT tensile tests comprised stress–strain curves recorded either during tensile failure or during loading–unloading tests. The same grips were used to emphasize CR-SME occurrence.

Tensile CR-SME investigations were performed using the above described tensile testing machine and wire testing grips assemblies. In this case, NiTi and NiTiCu wires were fastened and subjected to two types of investigations. The former comprised emphasizing single CR-SME occurrence, by RT loading–unloading and heating to 56 or 60 °C in an elongated state. The latter consisted of (i) RT loading up to 4% strain, (ii) heating–cooling under 4% constant strain, starting from an elevated stress, (iii) RT additional loading from 4 to 5%, and (iv) heating–cooling under 5% constant strain. The results of these investigations comprised ternary stress–strain–temperature diagrams.

The experiments aiming to monitor the interaction between the SM shrinkage of pre-strained NiTiCu wire and the SE recovery of NiTiFe wire were performed on a special experimental setup designed and fabricated for this study. Figure 2 illustrates the main elements of this setup.

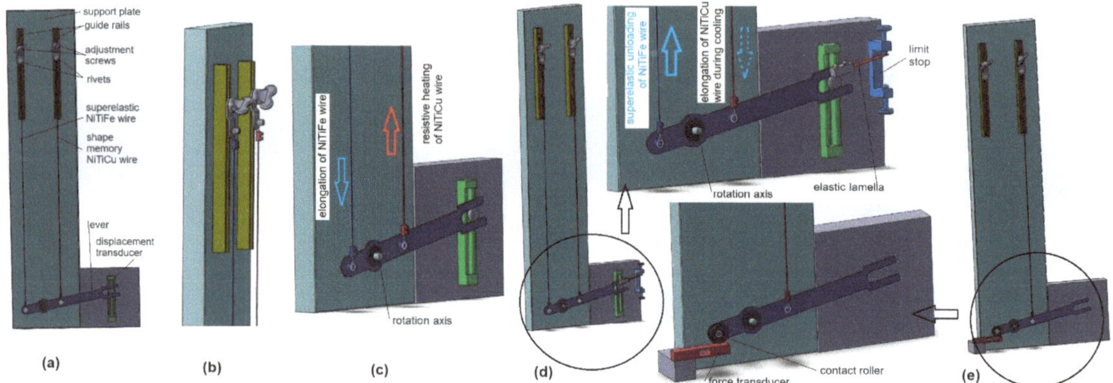

Figure 2. Schematic illustration of the experimental setup for monitoring the interaction between SM shrinkage of NiTiCu wires and SE recovery of NiTiFe wires: (**a**) general view of the experimental setup, displaying the SM (NiTiCu) and the SE (NiTiFe) wires; (**b**) fixing detail of the wire upper ends; (**c**) detail of the lever mechanism with a displacement transducer; (**d**) experimental setup equipped for thermomechanical cycling, with details of the mechanism of stroke reversion; (**e**) experimental setup equipped for force measurement.

Two guide rails are fixed on the support plate. On each rail, one rivet is fixed by means of an adjustment screw. The right-side screw fastens the SM NiTiCu wire specimen, which was pre-strained with approx. 5% force, as it will be pointed out later. The left-side screw fastens the SE NiTiFe wire specimen. The SM-NiTiCu and the SE-NiTiFe wires are fastened in parallel, as shown in Figure 2a. The upper ends of the SM and the SE wires are fixed, as shown in the detail from Figure 2b. The lower ends are connected to a 1st degree leverage, which transmits the displacement from one to the other by rotating around a rotation axis. In addition, the SM NiTiCu wire is coupled to electric connectors at its two ends. When heated, the wire will develop WG-SME and will shrink. The lever will stretch the SE NiTiFe wire, as illustrated in Figure 2c. After cutting off the electric power, the NiTiCu wire will cool down, will become martensitic, and will soften, transmitting less force to the lever, which will allow the SE wire to shrink in order to recover its initial undeformed state. Therefore, in turn, each wire will be elongated by the other, by means of the leverage: the SE NiTiFe wire will be elongated during SM NiTiCu heating which, in turn, will be elongated after becoming softer during cooling as a result of the unloading of the SE NiTiFe wire. Figure 2d illustrates the setup equipped for thermomechanical cycling. The detail shows a flexible elastic lamella fixed at one end of the leverage, which alternatively touches the upper and lower limit stops. During NiTiCu wire heating, it will shrink and rotate the leverage, which will elongate the SE NiTiFe wire, until the flexible elastic lamella touches the upper limit stop, as displayed in Figure 2d. This will stop the electric resistive heating, and the NiTiCu wire will start very quickly to cool down, thus becoming softer. When its stress would become lower than that of the NiTiFe wire, the latter would rotate the leverage clockwise, thus elongating the NiTiCu wire. Thus, the electric resistive heating will be turned on and off, thus determining the oscillating rotation of the leverage. Finally, Figure 2e exemplifies how the force developed by WG-SME of the NiTiCu wire is measured with a force transducer. The results of these experiments have the form of ternary diagrams of the wire's free-end displacement as a function of time and temperature.

3. Results and Discussion

The results of the thermal analysis, WG-SME experiments, RT tensile tests, CR-SME investigations, and SM shrinkage vs. SE recovery experiments are presented and discussed in this section.

3.1. Thermal Analysis

The DSC thermograms recorded during the first heating and second heating–cooling cycle, up to 60 °C, of fragments of the three wires, are illustrated in Figure 3. The first cooling to RT was performed without a controlled temperature variation rate and is missing from the figure.

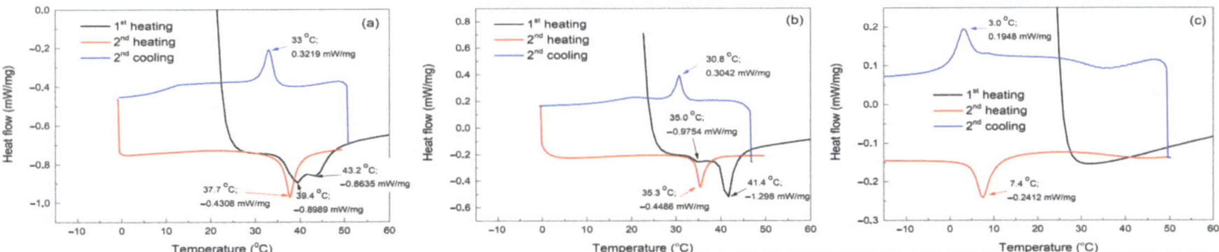

Figure 3. DSC thermograms recorded during the first heating and second heating–cooling cycle to 100 °C of different wire fragments: (**a**) NiTi; (**b**) NiTiCu, and (**c**) NiTiFe.

The presence of an endothermic minimum on the DSC charts recorded during the heating of a martensitic SMA is typically associated with martensite reversion to austenite and represents the mechanism of SME [41].

The martensitic $Ni_{48}Ti_{52}$ SMA wire, which is Ti-rich, would experience, in the first heating, a reverse martensitic transformation, which can be associated with the major endothermic peak located at 39.4 °C, and an R-phase transition [42] located, in this case, at 43.2 °C, according to Figure 3a. During the second heating–cooling cycle, martensite reversion occurred at 37.7 °C because uncontrolled cooling resulted in unstable martensite, which reversed to austenite at lower temperature [17]. When comparing the specific absorbed enthalpies that accompany the two heating-induced transitions, values between 0.86 and 0.9 mW/mg can be observed at the first heating and a value of 0.4308 mW/mg can be observed at the second heating. The second controlled cooling caused martensite formation at 33 °C.

Figure 3b, displaying the heat flow variation during the heating of a fragment of $Ni_{45}Ti_{50}Cu_{5}$ wire, illustrates two endothermic peaks, during the first heating. The smaller one, located at 35 °C, can be ascribed to the reverse transformation of B19′ monoclinic martensite into B19 orthorhombic martensite [43], while the larger one, located at 41.4 °C corresponds to the reversion of the intermediate B19 martensite [44] to B2 body-centered cubic austenite [45]. From the point of view of the specific absorbed enthalpies, Figure 3b shows that the fragment of $Ni_{45}Ti_{50}Cu_{5}$ wire required about 0.98 mW/mg for the first transition and 1.3 mW/mg for the second one. During the second heating, martensite reversion occurred in a single step at a lower temperature (35.3 °C), and martensite formation occurred at 30.8 °C.

Finally, Figure 3c displays a thermogram without any transition that corresponds to the super-elastic character of the $Ni_{48}Ti_{50}Fe_{2}$ wire, which is austenitic at RT [46].

From the above results, it can be concluded that the NiTiCu wire is more suitable for developing WG-SME since it requires a larger amount of specific energy for the reverse martensitic transformation.

3.2. Work-Generating Shape Memory Effect

To compare the WG-SME capacity of the two SM-wires, they were elongated at RT by four loads weighing 2.5, 3, 3.5, and 4 kg, and were resistively heated up to four maximum temperatures selected from the DSC thermograms from Figure 3 as 39 °C, 41 °C, 43 °C, and 50 °C, respectively. The variations in time (which was omitted for simplicity reasons) of the displacement of wire's end where the load is fastened and maximum heating temperature are illustrated in Figure 4 for the TiNi wires and in Figure 5 for the NiTiCu wires.

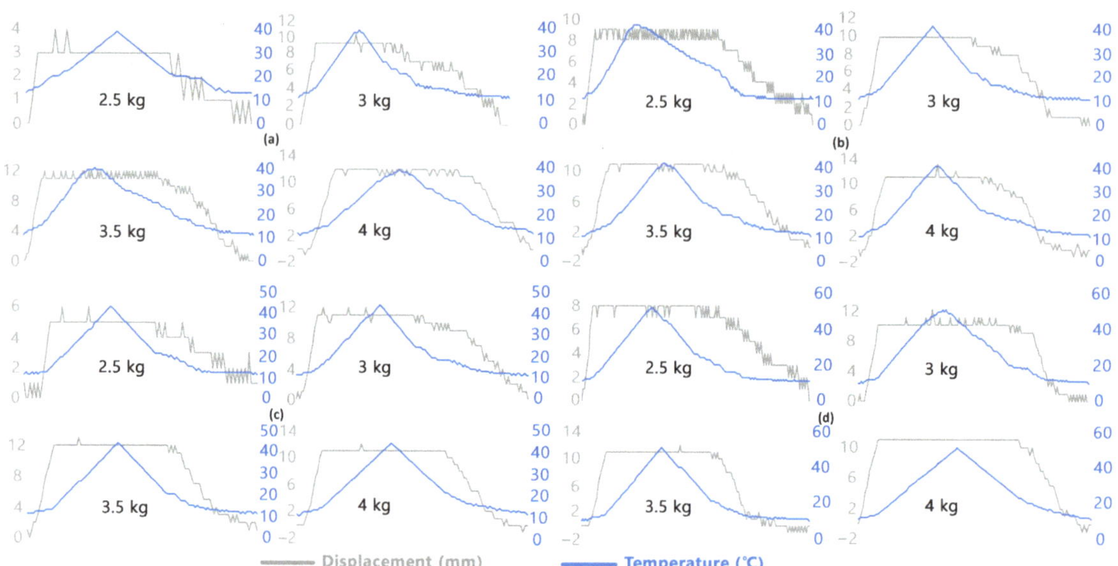

Figure 4. Highlighting the development of WG-SME by NiTi wires, by lifting four loads of 2.5, 3, 3.5 and 4 kg, respectively, during the heating from RT to different temperatures: (**a**) 39 °C; (**b**) 41 °C; (**c**) 43 °C; (**d**) 50 °C.

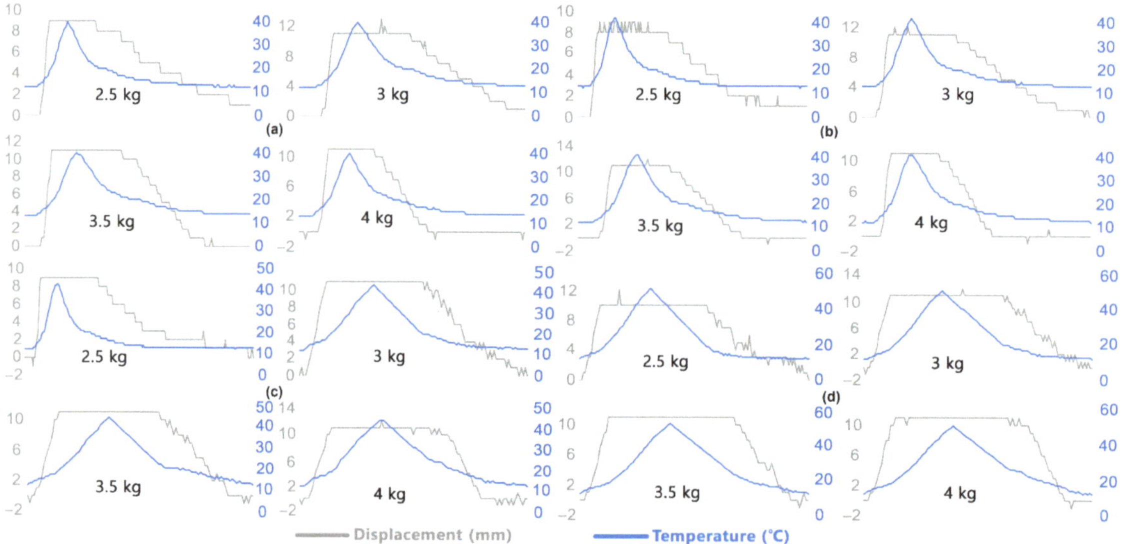

Figure 5. Highlighting the development of WG-SME by NiTiCu wires, by lifting four loads of 2.5, 3, 3.5 and 4 kg, respectively, during the heating from RT to different temperatures: (**a**) 39 °C; (**b**) 41 °C; (**c**) 43 °C; (**d**) 50 °C.

In all of the 16 diagrams from Figure 4, the wires that were elongated at RT by the applied loads shrunk during heating and lifted the loads. The smaller loads, such as 2.5 and 3 kg, could not elongate the wire to a greater extent at RT. For this reason, the displacements associated with these loads were lower. Conversely, the highest displacements were associated with the higher loads, namely 3.5 and 4 kg.

The maximum vertical displacement was 12 mm = 0.012 m and the maximum load lifted along this distance was 4 kg, as in Figure 4a. It follows that the maximum work, developed by the NiTi wire was $0.012 \text{ m} \times 9.8 \text{ ms}^{-2} \times 4 \text{ kg} \approx 0.47 \text{ J}$.

Another noticeable fact is the delay between the variations of displacement and temperature. The former varies much faster, reaches the maximum value, and remains constant, while the latter is still varying. This is due to the thermal inertia of the temperature-measuring thermistor.

Figure 5 displays the variations of the displacement of the $Ni_{45}Ti_{50}Cu_5$ wire end with ah attached load and the maximum electric resistive heating temperature.

When comparing Figures 4 and 5, two features can be noticed, as follows:

- NiTiCu wires warmed up faster than the NiTi ones;
- The displacements developed by the NiTiCu wires were slightly lower than those developed by the NiTi wires, compared to which they were smoother and had shorter maintaining times.

Aiming to compare the cycling capacities of the two SM wires, during repetitive WG-SME development, they were subjected to five cycles of resistive heating up to a maximum temperature of 50 °C, with an applied load of 4 kg. The results are illustrated in Figure 6.

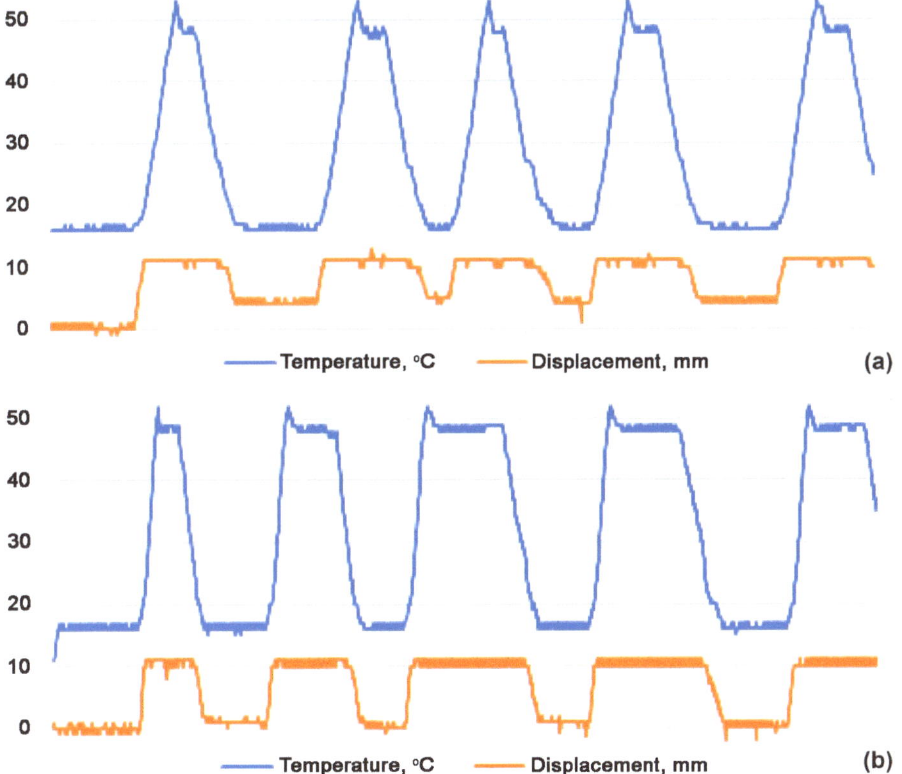

Figure 6. Highlighting the development of repetitive WG-SME during the repetitive heating from RT to 50 °C, with a hanging load of 4 kg: (**a**) NiTi; (**b**) NiTiCu.

Figure 6 shows that there are the following differences between the repetitive development of WG-SME by the two SM wires:

- From the point of view of temperature variation, the NiTiCu wire warmed up faster, cooled down faster and maintained the maximum temperature longer than the NiTi wire, though the two wires had the same diameter and length;
- From the point of view of displacement variation, it is obvious that the NiTiCu wire developed a larger stroke, about 10 mm, while the NiTi wire remained elongated after the first heating–cooling cycle.

These results demonstrate that both SM wires were able to develop repetitive WG-SME, but the NiTiCu wire was able to develop a larger stroke and a maximum work of 0.392 J.

3.3. Tensile Behavior at Room Temperature

The tensile failure curves at RT of the three wires are illustrated in Figure 7

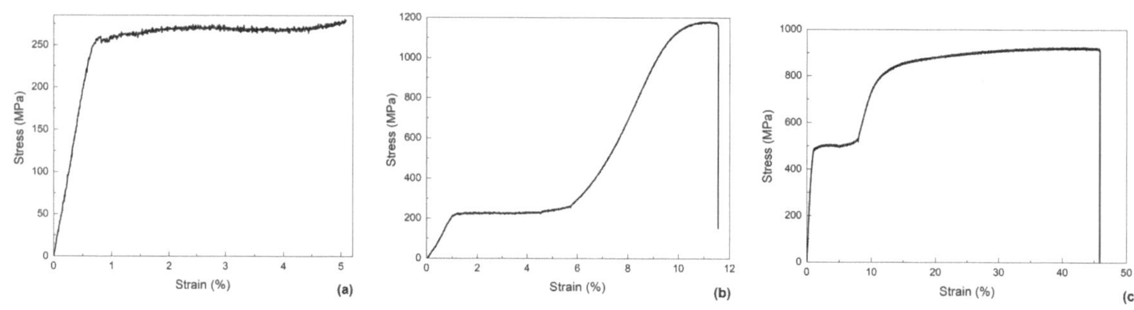

Figure 7. Tensile failure curves at RT of different wire specimens: (**a**) NiTi; (**b**) NiTiCu; (**c**) NiTiFe.

The main mechanical parameters, (E—Young's modulus; $R_{y,m}$—yield and failure strength, and $\varepsilon_{y,m}$—yield and failure strains) of the three failure curves are summarized in Table 1.

Table 1. Mechanical parameters of the tensile failure curves from Figure 7.

Wire Material	E	R_y	R_m	ε_y	ε_m
	GPa	MPa	MPa	%	%
$Ni_{48}Ti_{52}$;	38	259	279	0.8	5
$Ni_{45}Ti_{50}Cu_5$	23	220	1177	1.2–5.7	12
$Ni_{48}Ti_{50}Fe_2$	56	481	925	1–7.7	46

These results confirm the beneficial effects of ternary alloying of NiTi SMAs, if one considers that both the NiTiCu and NiTiFe wires exhibit larger values of failure stress and strain.

The next tensile tests were performed by loading–unloading, with the aim of monitoring the mechanical response of NiTi-based wires when subjected to reversible deformation. The resulting tensile loading–unloading curves are illustrated in Figure 8, for the three alloy wires under study.

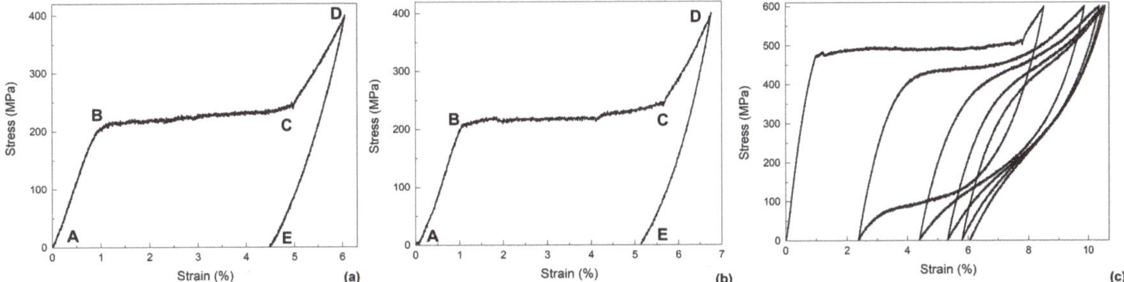

Figure 8. Tensile loading–unloading curves at RT of different wire specimens: (**a**) NiTi; (**b**) NiTiCu; (**c**) NiTiFe (please see text for explanations).

The two diagrams from Figure 8a,b are typical for the plastic deformation of the SMAs that experience a thermoelastic martensitic transformation [47]. The first quasi-linear portion (AB) corresponds to the elastic deformation of the martensitic wires (as shown in Figure 3, both $Ni_{48}Ti_{52}$ and $Ni_{45}Ti_{50}Cu_5$ wires were martensitic at RT). The stress plateaus (BC) are generally associated with the crystallographic reorientation of martensite by detwinning. Further deformation caused the elastic deformation of detwinned martensite (CD) which springs back during unloading (DE) [48]. On the other hand, the diagrams from Figure 8c, revealing an unloading stress plateau, are typical for the super-elastic response of the $Ni_{48}Ti_{50}Fe_2$ SMA wires [49]. When increasing the number of cycles, both the loading and unloading stress plateaus became shorter and more inclined, while permanent strain increased in terms of absolute value but relatively decreased at the cycle level [50]. For this reason, the SE description was reduced from an unloading stress plateau to an unloading inflection point.

3.4. Constrained Recovery Shape Memory Effect

The typical curves, displaying stress variation during RT loading–unloading followed by constant strain-heating, are illustrated in Figure 9, for the NiTi and NiTiCu wires.

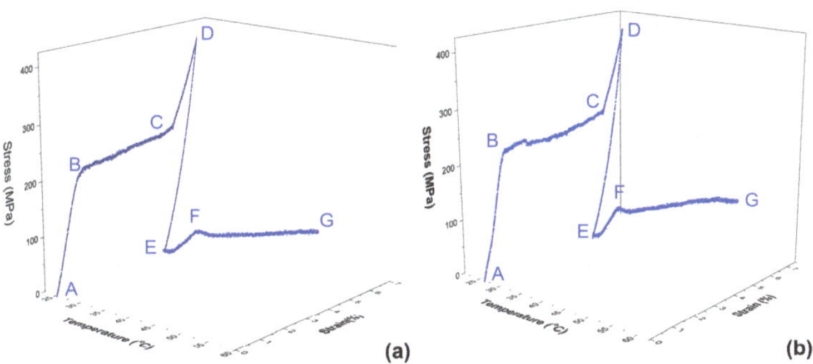

Figure 9. Emphasizing CR-SME after RT tensile loading–unloading and heating at a constant strain: (**a**) NiTi heated to 56 °C; (**b**) NiTiCu heated to 60 °C (please see text for explanations).

The characteristic stages of the RT loading–unloading can be easily identified along the A–B–C–D–E route, which coincides with the points designated in Figure 8. In point E, approximate values of 4.5% and 5% were obtained for the permanent strain in Figure 9a and in Figure 9b, respectively.

The crosshead of the tensile testing machine was blocked, so the respective permanent strains were subsequently kept constant and the heating chamber was turned on. During heating, stress increased abruptly up to point F which corresponds to the first endothermic peak of the DSC thermograms from Figure 3a,b and represents the midpoint of the reverse martensitic transformation.

To explain this initial stress increase, one must consider that, during the first half of the reverse martensitic transition, the transformation rate increases and stabilizes. Then, in the second half, it remains constant upon further heating, and then it finally decreases [51].

The transformation rate increase, between E and F, could be the cause of the abrupt stress rise at the beginning of heating, when SME-induced strain recovery counteracts thermal expansion. The former causes a stress rise, since it tends to stretch the wire specimen, while the latter tends to increase the wire's length, thus reducing the tensile stress.

The continuation of the heating process initially caused an initial stress decrease with several MPa, then a slow continuous gradual increase up to 123 MPa for the NiTi wire and 147 MPa for NiTiCu.

The initial stress decrease can be associated with the transformation rate diminution in the second half of the reverse martensitic transformation in such a way that thermal expansion becomes the most prominent phenomenon. At higher temperatures, R-phase/B19 martensite reversion to B2 austenite is completed in such a way that SME-induced shrinkage again becomes the prominent phenomenon and the tensile stress rises because B2 austenite has a smaller relative volume than both R-phase and B19 martensite [52].

Aiming to determine the behavior of the two SM-wires when undergoing repetitive CR-SME tests, according to the mechanical response from Figure 10, the specimens were (i) elongated to 4% strain at RT, 0–1 and kept in a stressed state in point 1, (ii) heated under a constant strain up to 56 or 60 °C, 1–2, (iii) cooled to RT under constant strain, 2–3, and additionally loaded from 4 to 5%, 3–4, for a second cycle of CR heating (4–5) and cooling (5–6), according to the same routine as in the first cycle.

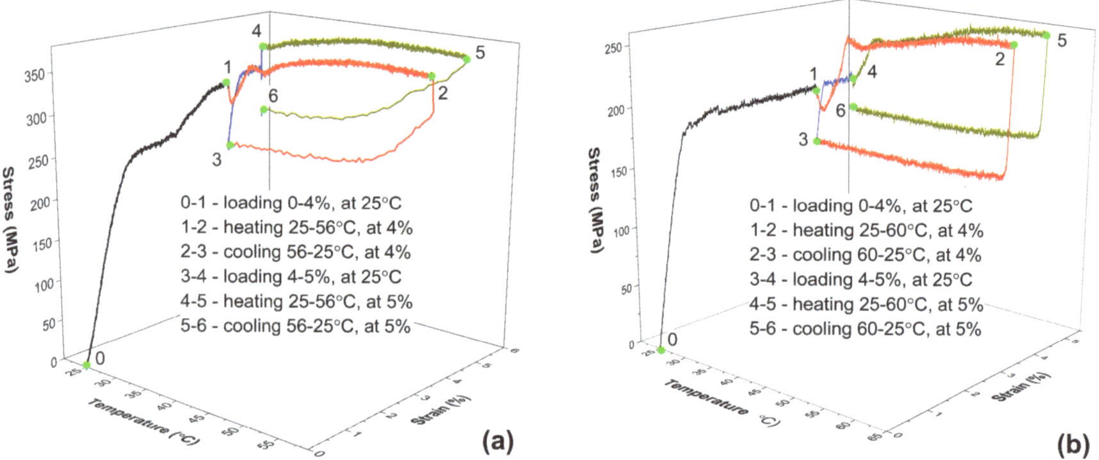

Figure 10. Emphasizing CR-SME during two cycles comprising 0–1 = RT tensile loading to 4%, 1–2 = heating at constant strain, 2–3 = cooling at constant strain, 3–4 = RT loading from 4 to 5%, 4–5 = heating at constant strain, 5–6 = cooling at constant strain: (**a**) NiTi heated up to 56 °C; (**b**) NiTiCu heated up to 60 °C.

It is obvious that even when the two wire specimens were heated up to the same maximum temperatures as in Figure 9, the stress increase was smaller in Figure 10. Thus, for the NiTi wire, stress increased by 123 MPa, when starting from 0, and with approx. 50 MPa after additional loading between two heating–cooling cycles, when starting from 300 MPa. On the other hand, the corresponding rise was 147 MPa and with approx. 70 MPa, when starting from 180 MPa, respectively.

As such, the following can be concluded:

- The NiTiCu wire experienced a larger relaxation capacity during cooling, in good accordance with the WG-SME response from Figure 6;
- Both SM-wires were able to develop repetitive CR-SME.

3.5. Actuation System Based on SM Shrinkage vs. SE Recovery

The final set of tests was performed on the experimental setup and monitored the evolution of the NiTiCu SM-wire working against that of the SE NiTiFe wire. The results are summarized in Figure 11.

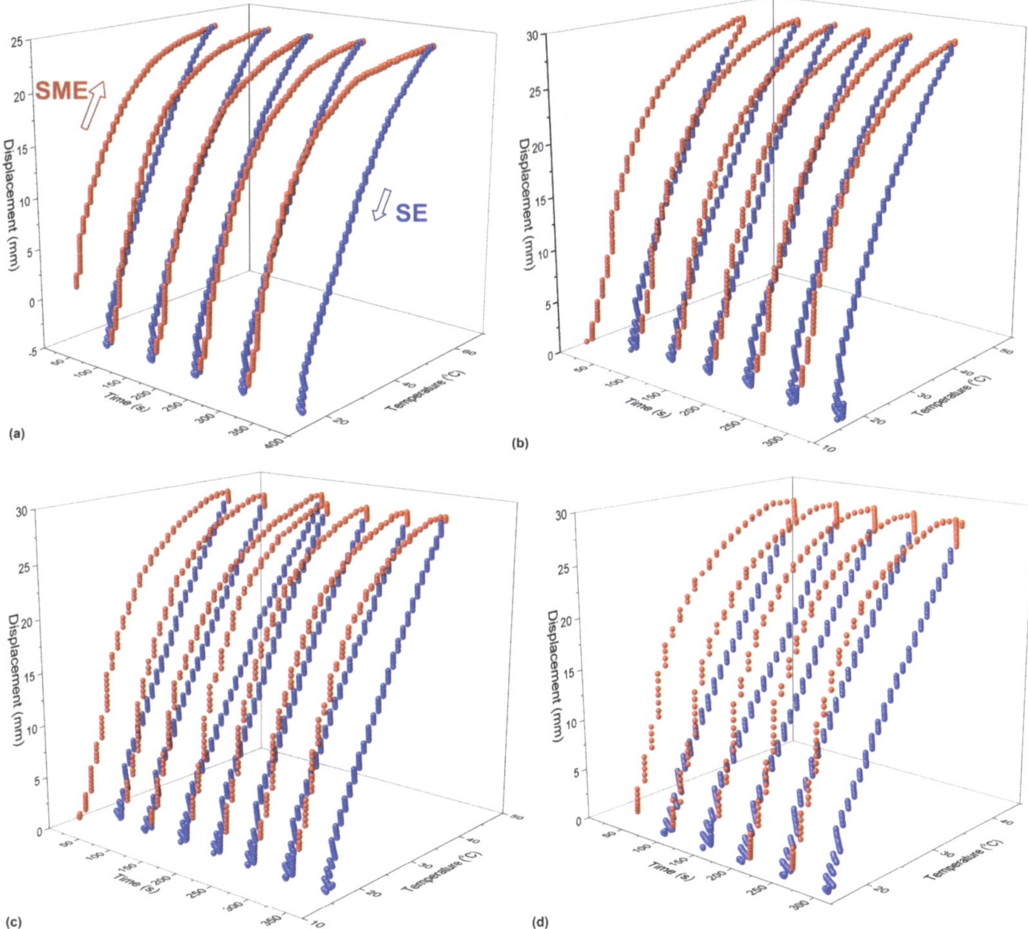

Figure 11. Emphasizing the variation in the shrinkage of the $Ni_{45}Ti_{50}Cu_5$ SM wire biased by the unloading recovery of the SE $Ni_{48}Ti_{50}Fe_2$ wire, at different maximum voltage values of the electric current used for resistive heating: (**a**) 3.5 V; (**b**) 4 V; (**c**) 4.5 V; (**d**) 5 V.

The martensitic NiTiCu wire was pre-strained at RT with 5%. For actuation, four voltages were used, as pointed out in Figure 11. During resistive heating, SME occurred and stretched the NiTiCu wire which, in turn, elongated the SE NiTiFe wire by means of the leverage, as specified in Figure 2. When the elastic lamella shown in Figure 2d touched the upper limit stop, resistive heating was interrupted and the NiTiCu wire started to cool down. During the transformation of austenite into martensite, the NiTiCu wire softened and gradually became elongated by the NiTiFe wire. As can be observed from Figure 11 and from the Supplementary Material Video S1, the displacement rate was much larger during SME development than during super-elastic recovery. Actually, SM shrinkage occurred in a few seconds while SE recovery took about 50 s. Nevertheless, the process has been reproducible for several tens of cycles. For example, Figure S1 from the Supplementary Materials displays cycles recorded during a 40 min period. Displacement variation was recorded every five cycles, with SME being caused by a 4 V voltage-heating effect. It can be noticed that the displacement loops tend to become narrower with the increase in the number of cycles. This could be a training effect that is frequently observed in NiTiCu SMAs, as it was previously reported by some of the present authors [53].

From an energetic point of view, the setup consumed energy during SME development when loading the NiTiFe wire and restored it during super-elastic unloading. So, the area under the loading curve (corresponding to displacement variation during SME development) is proportional to the amount of consumed energy, W_0, while the area under the unloading curve (corresponding to super-elastic recovery) is proportional to the amount of restored energy, W_{rec}. These area values are summarized in Table 2.

Table 2. Variation in the surface areas proportional to specific consumed (W_0) and restored (W_{rec}) energies during SM vs. SE displacement cycles (a.u.).

	1st Cycle		2nd Cycle		3rd Cycle		4th Cycle		5th Cycle	
	W_0	W_{rec}	W_0	W_{rec}	W_0	W_{rec}	W_0	W_{rec}	W_0	W_{rec}
3.5 V	640.35	393.55	687.2	452.5	644.7	455.6	663.35	455.8	762.85	504.7
4 V	630.9	418.3	680.45	458.2	664.75	465.35	644.8	443.6	637.65	468.95
4.5 V	550.45	341.55	550.5	356.2	666.8	429.8	529.7	338.85	549.6	362.8
5 V	281.7	240.5	460	260.6	506.15	282.3	502.95	291.95	509.45	302.8

By subtracting the restored energy from the consumed energy and dividing the result ($\Delta W = W_0 - W_{rec}$) by the consumed energy one, can calculate the specific energy dissipated by internal friction, as shown in Figure 12.

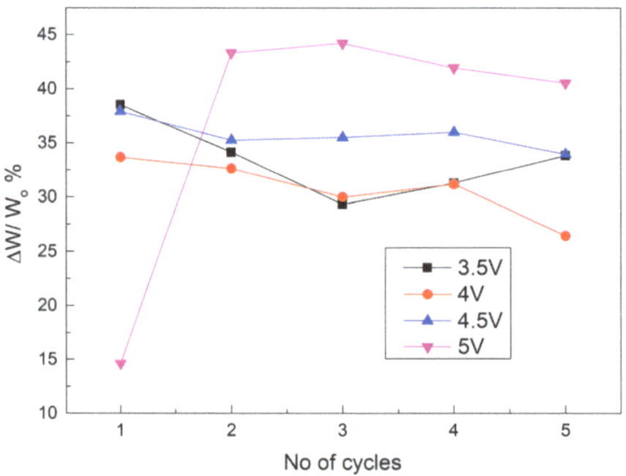

Figure 12. Variations in the specific energy, dissipated by internal friction, during consecutive cycles of resistive heating-induced SM shrinkage of the $Ni_{45}Ti_{50}Cu_5$ SM wire compensated by unloading recovery of the $Ni_{48}Ti_{50}Fe_2$ SE wire.

Figure 12 confirms the decreasing tendency of the thermal hysteresis, proportional to the area between the loading and unloading curves [54], with the increase in the number of cycles.

Based on the above observations, a robotic gripper was designed. The device was built up by adapting a NiTiCu and a NiTiFe wire, as an actuating element, on a classical gripper, as illustrated in Figure 13.

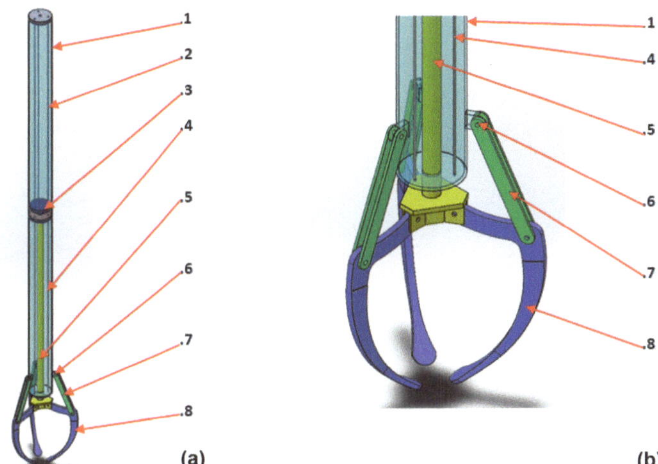

Figure 13. Design of a robotic gripper, actuated with a NiTiCu SM and NiTiFe SE wires: 1—housing; 2—NiTiFe wires; 3—piston; 4—NiTiCu wires; 5—fork; 6—pins; 7—connecting rods; 8—claws. (**a**) general view; (**b**) detail of the grasping system.

The system comprises a housing (1), three NiTiFe SE wires (2), a piston (3), three NiTiCu SM wires (4), a fork (5), nine pins (6), six connecting rods (7), and three claws (8) that act as gripping elements. The NiTiCu wires (4) are in the martensitic state at room temperature, being pre-strained and connected with one end at the housing base and with the other end to the piston, as illustrated in Figure 13b. The NiTiFe wires (2) are connected with one end at the piston and with the other end to the top of the housing. Their role is to act as a bias element. Thus, while being non-actuated, the NiTiCu wires are softer than the NiTiFe wires, being pre-strained and fixed in the assembly in an elongated martensitic state.

During resistive heating, by means of the electric connection wires shown in Figure 14a, the NiTiCu wires stretch back, since they become tougher than the NiTiFe wires, thus causing the translation of the piston (3) which moves the fork (5). After cutting the power, the NiTiCu wires start to cool, regain the martensitic state, and became softer than the NiTiFe wires, which move the piston back in the initial position.

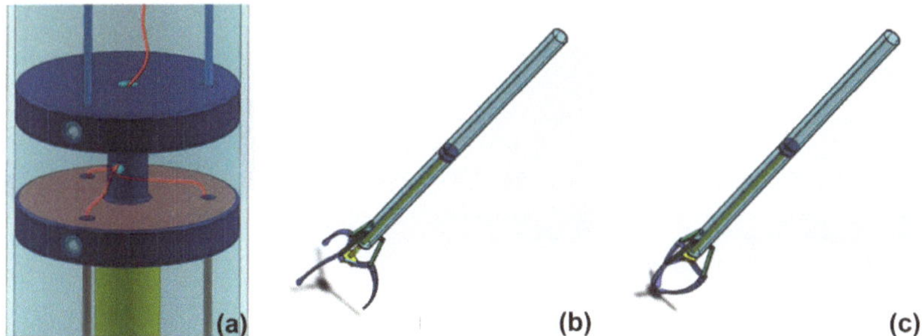

Figure 14. Details of the gripping device: (**a**) electric connection wires; (**b**) open position; (**c**) closed position.

The claws (8) are articulated at one end to the fork and at the middle to the housing, by means of the connecting rods (7). The piston translation causes a rotation movement of the claws, hence the device being able to grip and release loads, as illustrated in Figure 14b,c, respectively. The robotic gripper is able to grab various objects, with the action controlled by electric resistive heating of the NiTiCu wire, and to release them in a fixed position, with the action controlled by the super-electic recovery of the elongated NiTiFe wire.

4. Summary and Conclusions

By summarizing the above results, the following conclusions may be drawn.

1.   From the point of view of WG-SME:

- The SM response (RT displacement) increased with the applied load up to 12 mm, corresponding to a work of 0.47 J;
- Cu addition caused higher heating and cooling rates during single tests and larger displacement during cyclic tests, developing a repetitive work of 0.392 J.

2.   From the point of view of the RT tensile behavior:

- Both $Ni_{48}Ti_{52}$ and $Ni_{45}Ti_{50}Cu_5$ wires exhibited stress plateaus during loading associated with the crystallographic reorientation of martensite;
- The $Ni_{48}Ti_{50}Fe_2$ wire had stress plateaus both during loading and unloading, associated with a reversible stress induced martensitic transformation, which changed to inflection points during cycling.

3.   From the point of view of CR-SME:

- $Ni_{48}Ti_{52}$ and $Ni_{45}Ti_{50}Cu_5$ wires elongated with 4.5 and 5% developed CR stresses of 123 MPa and 147 MPa, when heated to 56 and 60 °C, respectively;
- Both $Ni_{48}Ti_{52}$ and $Ni_{45}Ti_{50}Cu_5$ wires, when heated in elongated state, developed repetitive stress–temperature variations;
- Cu addition caused a larger relaxation capacity during cooling.

4.   From the point of view of the trade-off between SM shrinkage and SE recovery:

- When heated under the effect of a voltage of 4.5–5 V, the 5% pre-strained $Ni_{45}Ti_{50}Cu_5$ wires developed a 28 mm stroke which elongated the $Ni_{48}Ti_{50}Fe_2$ wire by means of leverage;
- When air-cooled, the $Ni_{45}Ti_{50}Cu_5$ wires were elongated again by the unloading recovery of the $Ni_{48}Ti_{50}Fe_2$ wire;
- SM shrinkage lasted a few seconds, while and SE recovery lasted up to 50 s;
- SM vs. SE displacement cycles could be repeated for tens of cycles, during which time a decreasing tendency of the specific energy dissipated by internal friction was noted;
- A robotic gripper was designed based on the antagonistic deformation of $Ni_{45}Ti_{50}Cu_5$ and $Ni_{48}Ti_{50}Fe_2$ wires;
- Compared to classical solutions, with hydraulic or mechanical actuation, the proposed system is more compact, rapid, economical, and ergonomic. Compared to other SMA-driven systems, the present solution can provide the manipulated objects' fast grasping and slow release speeds.
- General conclusions:
- Here, 0.5 mm-diameter martensitic $Ni_{48}Ti_{52}$ and $Ni_{45}Ti_{50}Cu_5$ wires were able to perform both WG-SME and CR-SME by repetitively lifting loads up to 4 kg, (developing works of 0.392–0.47 J) and by generating stress of 123–147 MPa (when pre-strained with 4.5 and 5%), respectively;
- Here, 5% pre-strained martensitic $Ni_{45}Ti_{50}Cu_5$ wires and austenitic $Ni_{48}Ti_{50}Fe_2$ wires developed a 28 mm stroke by SM/SE coupling, which was repeated during tens of cycles and accompanied by a decreasing tendency of energy dissipation by internal friction.

- The simulated three-fingers gripper has the potential to develop rapid grasping forces and slow release rates due to the direct coupling of pre-strained SM and SE wires.
- After manufacturing the gripper, additional experiments have to be performed to determine its real grasping and releasing rates and to test the training effect on these rates.

Supplementary Materials: The following supporting information can be downloaded at: https://www.mdpi.com/article/10.3390/act13080319/s1, Video S1: SM vs. SE 2 cycles; Figure S1: SM vs. SE displacement cycles recorded during a 40 min period.

Author Contributions: Conceptualization, L.-G.B.; methodology, B.P.; software, A.P.T., B.P. and R.I.C.; validation, B.P.; formal analysis, N.-M.L. and M.P.; investigation, A.P.T., N.-M.L. and R.I.C.; resources, L.-G.B. and M.P.; data curation, R.I.C.; writing—original draft preparation, L.-G.B.; writing—review and editing, M.P. and I.D.; visualization I.D.; supervision, N.-M.L.; project administration, A.P.T. and L.-G.B.; funding acquisition, I.D. All authors have read and agreed to the published version of the manuscript.

Funding: This research received no external funding.

Data Availability Statement: The original contributions presented in the study are included in the article/Supplementary Material, further inquiries can be directed to the corresponding author.

Conflicts of Interest: The authors declare no conflicts of interest.

References

1. Wayman, C.M. Deformation, Mechanisms and Other Characteristics of Shape Memory Alloys. In *Shape Memory Effects in Alloys*; Perkins, J., Ed.; Plenum Press: New York, NY, USA, 1975; pp. 1–27.
2. Otsuka, K.; Wayman, C.M. (Eds.) Introduction. In *Shape Memory Materials*; Cambridge University Press: Cambridge, UK, 1998; pp. 1–26.
3. Lohan, C.; Pricop, B.; Comaneci, R.I.; Cimpoesu, N.; Bujoreanu, L.G. Variation tendencies of tensile constrained recovery behaviour and associated structural changes during thermal cycling of a Fe-Mn-Si-Cr-Ni shape memory alloy. *Optoelectron. Adv. Mater.* **2010**, *4*, 816–820.
4. Lieberman, D.S.; Schmerling, M.A.; Carz, R.W. Ferroelastic "Memory" and Mechanical Properties in AuCd. In *Shape Memory Effects in Alloys*; Perkins, J., Ed.; Plenum Press: New York, NY, USA, 1975; pp. 203–244.
5. Buehler, W.J.; Gilfrich, J.V.; Wiley, R.C. Effect of Low-Temperature Phase Changes on the Mechanical Properties of Alloys near Composition TiNi. *J. Appl. Phys.* **1963**, *34*, 1475–1477.
6. Ohkata, I.; Suzuki, Y. The design of shape memory alloy actuators and their applications. In *Shape Memory Materials*; Otsuka, K., Wayman, C.M., Eds.; Cambridge University Press: Cambridge, UK, 1998; pp. 240–266.
7. Nayan, N.; Singh, G.; Murty, S.V.S.N.; Narayana, P.R.; Mohan, M.; Venkitakrishnan, P.V.; Ramamurty, U. Effect of ternary additions of Cu and Fe on the hot deformation behavior of NiTi shape memory alloy—A study using processing maps. *Intermetallics* **2021**, *131*, 107084. [CrossRef]
8. Velmurugan, C.; Senthilkumar, V. The effect of Cu addition on the morphological, structural and mechanical characteristics of nanocrystalline NiTi shape memory alloys. *J. Alloys Compd.* **2018**, *767*, 944–954. [CrossRef]
9. Liu, X.; Li, H.; Zhang, Y.; Yang, Z.; Gu, Q.; Wang, X.; Zhang, Y.; Yang, J. Cryogenic temperature deformation behavior and shape memory mechanism of NiTiFe alloy. *Intermetallics* **2023**, *162*, 107997. [CrossRef]
10. Lotkov, A.; Grishkov, V.; Timkin, V.; Baturin, A.; Zhapova, D. Yield stress in titanium nickelide-based alloys with thermoelastic martensitic transformations. *Mater. Sci. Eng. A* **2019**, *744*, 74–78. [CrossRef]
11. Oliveira, J.P.; Zeng, Z.; Berveiller, S.; Bouscaud, D.; Fernandes, F.B.; Miranda, R.M.; Zhou, N. Laser Welding of Cu-Al-Be Shape Memory Alloys: Microstructure and Mechanical Properties. *Mater. Design* **2018**, *148*, 145–152. [CrossRef]
12. Mohan, S.; Banerjee, A. Particle filter based self sensing Shape Memory Alloy wire actuator under external cooling. *Mech. Syst. Signal Proc.* **2023**, *185*, 109779. [CrossRef]
13. Rao, Z.; Wang, X.; Leng, J.; Yan, Z.; Yan, X. Design methodology of the $Ni_{50}Ti_{50}$ shape memory alloy beam actuator: Heat treatment, training and numerical simulation. *Mater. Design* **2022**, *217*, 110615. [CrossRef]
14. Holschuh, B.; Obropta, E.; Newman, D. Low Spring Index NiTi Coil Actuators for Use in Active Compression Garments. *IEEE/ASME Trans. Mechatron.* **2015**, *20*, 1264–1277. [CrossRef]
15. Wang, J.; Huang, B.; Gu, X.; Zhu, J.; Zhang, W. Actuation performance of machined helical springs from NiTi shape memory alloy. *Int. J. Mech. Sci.* **2022**, *236*, 107744. [CrossRef]
16. Lu, Y.; Jiang, J.; Zhang, J.; Zhang, R.; Zhang, Q.; Zhou, Y.; Wang, L.; Yue, H. A dynamic stiffness improvement method for thin plate structures with laminated/embedded shape memory alloy actuators. *Thin Walled Struct.* **2022**, *175*, 109286. [CrossRef]

17. Jani, J.M.; Leary, M.; Subic, A.; Gibson, M.A. A review of shape memory alloy research, applications and opportunities. *Mater. Design* **2014**, *56*, 1078–1113. [CrossRef]
18. Yurtsever, O.; Küçük, H. Design, production and vision based analysis of a wireless operated 2-DOF SMA driven soft robotic arm. *Mater. Mater. Today Commun.* **2023**, *34*, 105176. [CrossRef]
19. Lu, Y.; Xie, Z.; Wang, J.; Yue, H.; Wu, M.; Liu, Y. A novel design of a parallel gripper actuated by a large-stroke shape memory alloy actuator. *Int. J. Mech. Sci.* **2019**, *159*, 74–80. [CrossRef]
20. Amadi, A.; Mohyaldinn, M.; Ridha, S.; Ola, V. Advancing engineering frontiers with NiTi shape memory alloys: A multifaceted review of properties, fabrication, and application potentials. *J. Alloys Compd.* **2024**, *976*, 173227. [CrossRef]
21. Nicholson, D.E.; Benafan, O.; Bigelow, G.S.; Pick, G.; Demblon, A.; Mabe, J.H.; Karaman, I.; Van Doren, B.; Forbes, D.; Sczerzenie, F.; et al. Standardization of Shape Memory Alloys from Material to Actuator. *Shape Mem. Superelasticity* **2023**, *9*, 353–363. [CrossRef]
22. Nespoli, A.; Besseghini, S.; Pittaccio, S.; Villa, E.; Viscuso, S. The high potential of shape memory alloys in developing miniature mechanical devices: A review on shape memory alloy mini-actuators. *Sens. Actuators A* **2010**, *158*, 149–160. [CrossRef]
23. Baek, H.; Khan, A.M.; Kim, Y. A bidirectional rotating actuator by using a single shape memory alloy wire in a double bend shape. *Sens. Actuators A* **2023**, *360*, 114526. [CrossRef]
24. Kennedy, S.; Shougat, M.R.E.U.; Perkins, E. Robust self-sensing shape memory alloy actuator using a machine learning approach. *Sens. Actuators A* **2023**, *354*, 114255. [CrossRef]
25. Hamid, Q.Y.; Wan Hasan, W.Z.; Azmah Hanim, M.A.; Nuraini, A.A.; Hamidon, M.N.; Ramli, H.R. Shape memory alloys actuated upper limb devices: A review. *Sens. Actuators Rep.* **2023**, *5*, 100160. [CrossRef]
26. Shukla, S.; Barjibhe, R. An experimental and numerical comparison of traditional spring DVA in parallel and shape memory alloy actuated DVA in series and parallel for fixed beam vibration control. *Mater. Today Proc.* **2023**. [CrossRef]
27. Schuleit, M.; Becher, M.; Franke, F.; Maaß, B.; Esen, C. Development of form-fit connection for NiTi shape memory wire actuators using laser processing. *Procedia CIRP* **2020**, *94*, 546–550. [CrossRef]
28. Sgambitterra, E.; Greco, C.; Rodino, S.; Niccoli, F.; Furgiuele, F.; Maletta, C. Fully coupled electric-thermo-mechanical model for predicting the response of a SMA wire activated by electrical input. *Sens. Actuators A* **2023**, *362*, 114643. [CrossRef]
29. Liu, Y.N.; Liu, Y.; Van Humbeek, J. Two-way shape memory effect developed by martensite deformation in NiTi. *Acta Mater.* **1998**, *47*, 199–209. [CrossRef]
30. Liu, R.; Zhang, C.; Ji, H.; Zhang, C.; Qiu, J. Training, Control and Application of SMA-Based Actuators with Two-Way Shape Memory Effect. *Actuators* **2023**, *12*, 25. [CrossRef]
31. Eggeler, G.; Hornbogen, E.; Yawny, A.; Heckmann, A.; Wagner, M. Structural and functional fatigue of NiTi shape memory alloys. *Mater. Sci. Eng. A* **2004**, *378*, 24–33. [CrossRef]
32. Mizzi, L.; Hoseini, S.F.; Formighieri, M.; Spaggiari, A. Development and prototyping of SMA-metamaterial biaxial composite actuators. *Smart Mater. Struct.* **2023**, *32*, 035027. [CrossRef]
33. Ben Zohra, M.; Er-Remyly, O.; Riad, A.; Alhamany, A. Finite element analysis for fatigue phenomena in mechatronic shape memory alloy actuator. *Sci. Afr.* **2023**, *20*, e01601. [CrossRef]
34. Liu, Q.; Ghodrat, S.; Jansen, K.M.B. Design and modelling of a reversible shape memory alloy torsion hinge actuator. *Mater. Design* **2024**, *237*, 112590. [CrossRef]
35. O'Toole, K.T.; McGrath, M.M.; Coyle, E. Analysis and Evaluation of the Dynamic Performance of SMA Actuators for Prosthetic Hand Design. *J. Mater. Eng. Perform.* **2009**, *18*, 781–786. [CrossRef]
36. Dinakaran, V.P.; Balasubramaniyan, M.P.; Muthusamy, S.; Panchal, H. Performa of SCARA based intelligent 3 axis robotic soft gripper for enhanced material handling. *Adv. Eng. Softw.* **2023**, *176*, 103366. [CrossRef]
37. Yang, C.; Wu, X.; Zhu, Y.; Wu, W.; Tu, Z.; Zhou, J. Recent Progress of an Underwater Robotic Avatar. In *Intelligent Robotics and Applications*; Yang, H., Liu, H., Jun, Z., Yin, Z., Liu, L., Yang, G., Ouyang, X., Wang, Z., Eds.; ICIRA 2022 Lecture Notes in Computer Science; Springer: Cham, Switzerland, 2022; Volume 13455, pp. 615–626. [CrossRef]
38. Lee, J.-H.; Chung, Y.S.; Rodrigue, H. Long Shape Memory Alloy Tendonbased Soft Robotic Actuators and Implementation as a Soft Gripper. *Sci. Rep.* **2019**, *9*, 11251. [CrossRef]
39. Guo, Z.; Pan, Y.; Wee, L.B.; Yua, H. Design and control of a novel compliant differential shape memory alloy actuator. *Sens. Actuator A-Phys.* **2015**, *225*, 71–80. [CrossRef]
40. Teodoriu, A.P.; Doroftei, I.; Pricop, B.; Bujoreanu, L.G. Study of the Load-Response Effect of NiTi-Based Thermal Actuators. Application to a Robotic Gripper. In *25th International Symposium on Measurements and Control in Robotics*; ISMCR 2023. Mechanisms and Machine Science; Doroftei, I., Kiss, B., Baudoin, Y., Taqvi, Z., Keller Fuchter, S., Eds.; Springer: Cham, Switzerland, 2024; Volume 154, pp. 237–251. [CrossRef]
41. Frenzel, J.; Wieczorek, A.; Opahle, I.; Maas, B.; Drautz, R.; Eggeler, G. On the effect of alloy composition on martensite start temperatures and latent heats in Ni–Ti-based shape memory alloys. *Acta Mater.* **2015**, *90*, 213–231.
42. Zhang, X.Y.; Sehitoglu, H. Crystallography of the B2→R→B 19′ phase transformations in NiTi. *Mater. Sci. Eng. A* **2004**, *374*, 292–302. [CrossRef]
43. Nam, T.H.; Saburi, T.; Shimizu, K. Cu-content dependence of shape memory characteristics in Ti-Ni-Cu alloys. *Mater. Trans. JIM* **1990**, *31*, 959–967. [CrossRef]

44. Wu, S.-K.; Lin, H.; Lin, T. Electrical resistivity of Ti–Ni binary and Ti–Ni–X (X = Fe, Cu) ternary shape memory alloys. *Mater. Sci. Eng. A* **2006**, *438–440*, 536–539. [CrossRef]

45. Harikrishnan, K.; Chandra, K.; Misra, P.S.; Agarwala Vinod, S. B19 orthorhombic martensitic transformations in aged TiNiCu shape memory alloys. In *ESOMAT 2009, Proceedings of the 8th European Symposium on Martensitic Transformations, Prague, Czech Republic, 7–11 September 2009*; Šittner, P., Heller, L., Paidar, V., Eds.; EDP Sciences: Les Ulis, France, 2009; p. 02020. [CrossRef]

46. Karthik, G.; Kashyap, B.; Ram Prabhu, T. Processing, properties and applications of Ni-Ti-Fe shape memory alloys. *Mater. Today Proc.* **2017**, *4*, 3581–3589. [CrossRef]

47. Kumar, A.; Palani, I.A.; Yadav, M. Comprehensive study of microstructure, phase transformations, and mechanical properties of nitinol alloys made of shape memory and superelastic wires and a novel approach to manufacture Belleville spring using wire arc additive manufacturing. *Mater. Today Commun.* **2024**, *38*, 107881. [CrossRef]

48. Li, X.C.; Su, Y. A phase-field study of the martensitic detwinning in NiTi shape memory alloys under tension or compression. *Acta Mech.* **2020**, *231*, 1539–1557. [CrossRef]

49. Isalgue, A.; Soul, H.; Yawny, A.; Auguet, C. Functional fatigue recovery of superelastic cycled NiTi wires based on near 100 °C aging treatments. In *ESOMAT 2015, Proceedings of the Tenth European Symposium on Martensitic Transformations, Antwerp, Belgium, 14–18 September 2015*; Schryvers, N., van Humbeeck, J., Eds.; EDP Sciences: Les Ulis, France, 2015; Volume 33, p. 03019. [CrossRef]

50. Delville, R.; Malard, B.; Pilch, J.; Sittner, P.; Schryvers, D. Transmission electron microscopy investigation of dislocation slip during superelastic cycling of Ni-Ti wires. *Int. J. Plasticity* **2011**, *27*, 282–297. [CrossRef]

51. Lohan, N.M.; Pricop, B.; Bujoreanu, L.-G.; Cimpoesu, N. Heating rate effects on reverse martensitic transformation in a Cu–Zn–Al shape memory alloy. *Int. J. Mater. Res.* **2011**, *102*, 1345–1351. [CrossRef]

52. Heller, L.; Sittner, P. On the habit planes between elastically distorted austenite and martensite in NiTi. *Acta Mater.* **2024**, *269*, 119828. [CrossRef]

53. Popa, M.; Lohan, N.-M.; Pricop, B.; Cimpoeşu, N.; Porcescu, M.; Comăneci, R.I.; Cazacu, M.; Borza, F.; Bujoreanu, L.-G. Structural-Functional Changes in a $Ti_{50}Ni_{45}Cu_5$ Alloy Caused by Training Procedures Based on Free-Recovery and Work-Generating Shape Memory Effect. *Nanomaterials* **2022**, *12*, 2088. [CrossRef] [PubMed]

54. Mihalache, E.; Pricop, B.; Comăneci, R.I.; Suru, M.-G.; Lohan, N.-M.; Mocanu, M.; Ozkal, B.; Bujoreanu, L.-G. Structural Effects of Thermomechanical Processing on the Static and Dynamic Responses of Powder Metallurgy Fe-Mn-Si Based Shape Memory Alloys. *Adv. Sci. Technol.* **2017**, *97*, 153–158.

Article

The Design and Application of a Vectored Thruster for a Negative Lift-Shaped AUV

Hong Zhu [1], Lunyang Lin [1,2], Chunliang Yu [1], Yuxiang Chen [1,*], Hong Xiong [1], Yiyang Xing [1] and Guodong Zheng [1]

[1] Institute of Deep-Sea Science and Engineering, Chinese Academy of Sciences, Sanya 572000, China; zhuh@idsse.ac.cn (H.Z.); linly@idsse.ac.cn (L.L.); yucl@idsse.ac.cn (C.Y.); xiongh@idsse.ac.cn (H.X.); xingyy@idsse.ac.cn (Y.X.); zhenggd@idsse.ac.cn (G.Z.)

[2] University of Chinese Academy of Sciences, Beijing 100049, China

* Correspondence: chenyux@idsse.ac.cn

Abstract: Autonomous underwater vehicles (AUVs), as primary platforms, have significantly contributed to underwater surveys in scientific and military fields. Enhancing the maneuverability of autonomous underwater vehicles is crucial to their development. This study presents a novel vectored thruster and an optimized blade design approach to meet the design requirements of a specially shaped AUV. Determining the ideal blade characteristics involves selecting a maximum diameter of 0.18 m and configuring the number of blades to be four. Furthermore, the blades of the AUV were set to rotate at a speed of 1400 revolutions per minute (RPM). The kinematics of the thrust-vectoring mechanism was theoretically analyzed. A propulsive force test of the vectored thruster with ductless and ducted propellers was performed to evaluate its performance. A ductless propeller without an annular wing had a higher propulsive efficiency with a maximum thrust of 115 N. Open-loop control was applied to an AUV in a water tank, exhibiting a maximum velocity of 0.98 m/s and a pitch angle of 53°. The maximum rate of heading angle was 14.26°/s. The test results demonstrate that the specially designed thrust-vectoring mechanism notably enhances the effectiveness of AUVs at low forward speeds. In addition, tests conducted in offshore waters for depth and heading control validated the vectored thruster's capability to fulfill the AUV's motion control requirements.

Keywords: autonomous underwater vehicles; novel vectored thruster; thrust-vectoring mechanism; specially shaped AUV

Citation: Zhu, H.; Lin, L.; Yu, C.; Chen, Y.; Xiong, H.; Xing, Y.; Zheng, G. The Design and Application of a Vectored Thruster for a Negative Lift-Shaped AUV. *Actuators* **2024**, *13*, 228. https://doi.org/10.3390/act13060228

Academic Editor: Keigo Watanabe

Received: 14 May 2024
Revised: 11 June 2024
Accepted: 16 June 2024
Published: 19 June 2024

1. Introduction

As an extremely significant tool, underwater vehicles play a significant role in various ocean-related activities, such as military applications, scientific research, port operations, and maritime safety. Underwater vehicles are primarily used for inaccessible, hazardous, and expensive tasks for divers [1]. Engineers have an essential mission to improve the capability of underwater vehicles to explore oceans while performing these tasks [2].

These missions require underwater vehicles equipped with advanced and vital technologies to expand their capabilities such as maneuvering and autonomy. Sophisticated sensors and intelligent algorithms can improve the autonomy of underwater vehicles. The propulsion system is a crucial factor determining the maneuverability of underwater vehicles; hence, the question of how to incorporate underwater propulsion technology with high navigation efficiency and excellent maneuverability has always been a research focus in the field of underwater robots [3].

Different missions and applications determine the shapes, configurations, and propulsion systems of underwater vehicles [4]. The propulsion systems of underwater vehicles can be classified into three types: classical rear propeller propulsion, bio-inspired propulsion [5,6], and vectored thrust propulsion [1]. The propulsion strategy of an underwater vehicle determines its movement-control mode. Underwater vehicles intended for high-speed cruising are designed with streamlined hulls to reduce hydrodynamic drag. A main propeller equipped at the tail cone enables these streamlined-shaped vehicles to propel, and the steering ability is achieved by changing the angle of fins and rudders [7]. However, a major disadvantage of propulsion technology is that the underwater vehicle control authority is unstable at low speeds or suspended [8]. Under such conditions, the effectiveness of fins and rudders for controlling the vehicle is diminished or rendered nearly ineffective.

However, this limitation can be overcome by adding additional thrusters to underwater vehicles [9,10]. The ineffectiveness of the control surfaces was solved using the differential propulsive forces of the thrusters. This results in an overall hull configuration of the underwater vehicle complex. The weight, manufacturing cost, hydrodynamic drag, and energy consumption of underwater vehicles have also increased [11].

Another effective method to overcome the drawbacks discussed above is vectored thrust propulsion (VT), which replaces conventional rear or multiple propeller propulsion and bio-inspired propulsion [7]. Typically, an underwater vehicle is equipped with only one vectored thruster. The vectored thrust provided by the vectored thruster can be decomposed into two components: a control force and a driving force.

In recent decades, the application of vector propulsion technology in the field of underwater robots has undergone rapid development. There are relevant studies documenting the application of vector propellers in various underwater robots. However, some vector propellers remain at the theoretical design stage or prototype verification stage. In other words, during the process of transforming vectored thruster prototypes into practical applications, a multitude of issues tend to become pronounced. The following will introduce the design or applications of typical vector propellers.

Some vector propulsion technologies have their main propulsion motor placed at the end of the propeller, and the propeller is directly driven by a waterproof motor, meaning that the motor is exposed to water. The vector mechanism changes the thrust direction by altering the orientation of the motor itself. Obviously, this kind of mechanism has a large rotational inertia at its end, and when the propeller deflects, the power required by the reversing drive motor is also relatively large. Below, we will introduce several vector propellers of this type.

The Bluefin series AUV [12] is equipped with a vector propulsion system that changes the orientation of the mobile platform through the linear movement of a push rod to alter the direction of the propeller thrust. This vector propulsion device boasts a simple mechanical structure, flexible movement, and easy sealing.

The vectored thruster designed by E. Cavallo consists of a spherical parallel mechanism, which comprises a fixed platform and a moving platform connected by three identical chains [13,14]. By rotating the three active joints on the fixed platform, it drives the rotation of the three driven joints connected to the moving platform, thereby changing the attitude of the moving platform.

Spherical underwater robots employ three or four vectored thrusters based on pumped-water jets, with each vectored thruster controlled by two servo motors for a total of 2 DOFs (degrees of freedom) [11,15]. This vector mechanism is only suitable for this kind of small underwater robot like SUR-I and SUR-III. This article refers to this design method, but there have been critical issues in practical applications, which will be discussed later in the text.

For AUVs with a torpedo shape and limited tail space, Tao Liu designed a vector mechanism based on the 3-RPS parallel mechanism. The author conducted simulations on this mechanism but did not provide any practical application cases. Compared to the method of using two servo motors to achieve two degrees of freedom, this mechanism utilizes three linear motors. The control of various kinematic and dynamic methods for this mechanism is overly complex, and there are few documented cases of its actual application in AUVs. The biggest issue is that for the three linear motors of this mechanism to achieve a certain angle of the vectored thruster, a certain response time is required, resulting in input time delays for the AUV [2,7].

In addition to the direct drive of the propeller by a waterproof motor, there are also some vector propellers that utilize a non-waterproof motor to drive the propeller through a drive shaft or magnetic coupling. The drive shaft is a flexible transmission shaft formed by connecting a ball gear and a universal joint in series. The propeller is rotated in different thrust directions by driving the flexible shaft with the main motor. The magnetic coupling drive method achieves the rotation of the propeller in different thrust directions through a magnetic coupling drive.

A study has reported the implementation of a spherical reconfigurable magnetic coupling (S-RMC) method for motor-driven propellers [16]. A servo motor is utilized to ensure that the change in thrust direction can be achieved with one degree of freedom (DOF). However, the vectored thruster remains in the design phase and has not been applied to underwater robots.

Based on reference [16], which utilized a single steering gear to achieve one degree of freedom for the vectored thruster, Yaxin Li et al. [3] proposed the use of two servo motors to form a vector steering actuator, enabling two degrees of freedom for the vectored thruster. To achieve spatial vector thrust output, the steering actuator is assembled from two intersecting arc-shaped slide rails. At the end of each slide rail, a servo motor is mounted on the support frame to change the deflection angle. The propeller's driving method still utilizes the spherical reconfigurable magnetic coupling (S-RMC) approach.

Luca Pugi and his team proposed a method similar to the one in reference [16] for designing a vectored thruster. To address the sealing issue, this vectored thruster employs two magnetic couplings. One magnetic coupling is used by the main motor to drive the propeller rotation, while the other magnetic coupling is used by the steering gear to adjust the attitude of the thruster, thus achieving one degree of freedom for this vectored thruster [8]. It has not been actually applied to underwater robots.

The magnetic coupling-based vector propulsion device scheme is conducive to achieving the sealing of the main motor, but the non-collinear torque transmission performance of the magnetic coupling shaft coupler will deteriorate when the propeller deflects.

There are also some studies that provide cases of propeller rotation driven by the combination of drive motors, spherical gears, and transmission shafts [17]. For instance, the gear-and-ring vector propulsion device developed by MIT [18] utilizes gears meshed with incomplete gear rings to control the horizontal and vertical movement of the propeller. This solution is still in the theoretical research stage.

In conclusion, a considerable amount of research is currently focused on underwater vector propellers. While there are mature design schemes for vector propellers with classic application cases, most of the research is still in the model testing phase, and practical applications are not yet mature. The main reasons for this inability to achieve practical application are poor sealing reliability, as well as corrosion issues faced by mechanisms in seawater; the non-collinearity of the propeller shaft and motor shaft during deflection, which can easily lead to low transmission efficiency; complex vector mechanisms prone to processing errors, reduced coupling between branches, and increased control difficulty; and the cumulative errors of each joint in serial vector mechanisms, which can result in low deflection accuracy at the end.

There is a comprehensive plan available as a reference for designing an underwater propeller prototype. The design of the propeller prototype typically requires considerations of the driving motor, the shape and diameter of the propeller, the number of blades, the shape of the duct or nozzle, the drive circuit, and the coupling transmission components [19,20]. After the design is completed, the thrust, friction, torque, and efficiency values of the propeller are first calculated through finite element analysis [21]. Secondly, an open-water test is conducted to identify the thruster parameters [22]. Finally, performance tests are carried out on an actual underwater vehicle to verify the performance of the propeller [23].

The main contribution of this study lies in the design of two prototype vectored thrusters based on linear actuators and steering motors, which aim to fulfill the design requirements of specially shaped AUV. Both vectored thrusters were utilized for AUVs to perform underwater missions lasting for weeks and months, and the linear actuator-based vectored thruster was selected after evaluation. Propulsion force tests were conducted to evaluate the performance of the vectored thruster, and kinematic analysis was performed on the thrust-vectoring mechanism to achieve reliable and precise control. To verify the design principles, an AUV equipped with this vectored thruster was tested in a water tank. Different thrust forces for propelling the AUV were tested to investigate the vehicle's open-loop response at various vector angles. Additionally, the closed-loop control performance of the vectored thruster was tested in a real marine environment. Experimental results demonstrate that the thrust-vectoring mechanism can adjust the vector thrust in a directionally, and the vectored thruster can achieve a sufficient level to meet the control requirements of the AUV.

This study focuses on the design and validation of a novel vectored thruster that exhibits reliable performance and precise control. Successfully applied to the motion control of AUV, it provides a new solution for enhancing the design and performance of AUV. Section 2 presents the design requirements of the vectored thruster, and the general design of the vectored thruster is analyzed. In Section 3, the kinematics of the thrust-vectoring mechanism are discussed and the propulsive force test of the vectored thruster is described. Section 4 conducts water tank tests and offshore depth and heading control tests using an actual AUV to verify the performance of the developed prototype. Finally, Section 5 summarizes the study and discusses future work.

2. The General Design of the Vectored Thruster

2.1. Specifications of the Solar-Powered AUV

As illustrated in Figure 1, an underwater vehicle preparing to install the designed vectored thrust is a specially shaped AUV. The required design parameters for the AUV are listed in Tables 1 and 2.

Figure 1. A specially shaped AUV equipped with a vectored thruster.

Table 1. Main parameters of AUV.

Parameter	Value
Weight	205 kg
Dimension	2391 mm × 2130 mm × 519 mm
Max. speed	2.5 kn
Designed positive buoyancy	1 kg

Table 2. The relationship between power and speed of AUV.

Parameter	Value			
V (knots)	1	1.5	2	2.5
P_E (W)	8.12	24.76	56.93	109.74
P_E (hp)	0.0107	0.0328	0.0753	0.1452
R (N)	15.79	32.09	55.33	85.33

The designed vectored thruster was installed below the stern of the AUV. The AUV was not equipped with traditional fins or rudders. The driving and control forces generated by the vectored thruster combine to form the resultant force for controlling the motion of the AUV. The inclination of the thruster is a critical variable that governs the operational efficacy of the AUV. The deflection angle of the vectored thruster determines the direction and magnitude of the driving and control force.

The design of a vectored thruster consists of two parts. The first part involves the design of the thruster itself, including the selection of the motor and the design of the propeller. The second part is the vector mechanism that enables the spatial vector output of the thruster, typically with two degrees of freedom. This study first designed the corresponding propeller based on the motor specifications, and then designed and fabricated two types of vector mechanisms using steering motors and linear motors. Finally, open-water tests of the thruster and control tests of the autonomous underwater vehicle (AUV) were conducted to verify the practical application performance of the vectored thruster.

2.2. Design of Propeller Blades

The design of underwater propeller blades requires comprehensive consideration of various factors, including blade shape, angle, pitch, rotational speed, noise and vibration, load, and strength, as well as material and manufacturing processes. These factors interact with each other and jointly determine the performance and service life of the propeller. In this study, the underwater D5085 motor was selected, and the motor parameters are presented in Table 3. To ensure that the propeller and motor can work efficiently and stably together to achieve the best propulsion effect, the matching relationship between the underwater propeller blade and the motor must be fully considered during the design process.

Table 3. The parameters of the motor.

Parameter	Specification
Product Model	D5085
Voltage	22.2 V
No-load Current	0.8 A
No-Load Speed	3100 rpm
Load Current	29.5 A
Power	650 W
Weight	680 g

To meet the design requirements of an AUV in Table 1, an analysis of the propeller design was conducted, including variations in the number of blades and propeller diameters. This investigation was conducted within the framework of an optimum blade design technique. The present study considered two, three, and four blades. The diameter values for the propeller were in the range of 0.15~0.18 m. The other relevant parameters for the propeller are listed in Table 4.

Table 4. The parameters of the propeller.

Parameter	Specification
Airfoil Profile	NACA
Number of Blades	2–4
Number of Propeller	Single
Propeller Rotation Direction	Clockwise
Diameter of Blade	0.15~0.18 m
Pivot Diameter	0.058 m
Rotate Speed of the Propeller	400 r/min~1600 r/min

The propulsion efficiency was calculated for different rotational speeds, diameters, and number of blades to obtain the optimum combination. The values of the rotation speeds were in the range of 400–1600 rpm. Two, three, or four blades were used. The calculation results are shown below.

As shown in Figure 2, the larger the diameter of the blade, the higher the efficiency. When the maximum diameter of the blade was 0.18 m, the propulsion efficiency of the propellers with different numbers of blades did not differ significantly. Therefore, the number of blades of the propeller was set to four. When the rotation speed was 1400 rpm, the propulsion efficiency was the highest, and in the range of 1000–1600 rpm, the propulsion efficiency was also relatively high. Considering the motor's efficiency, the propeller speed was selected as 800–1600 rpm. A preliminary determination of the propeller parameters was obtained, which is shown in Table 5.

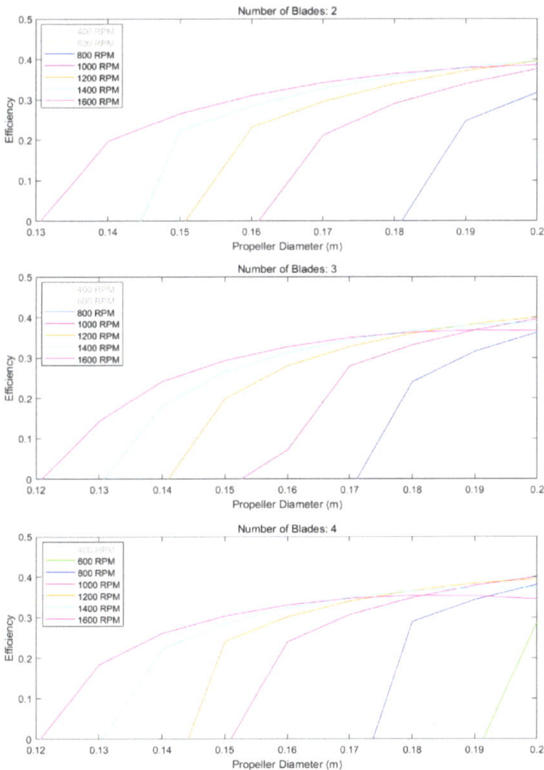

Figure 2. Optimization results of rotation speed, blade number, and diameter.

Table 5. The preliminary parameters of the propeller.

Parameter	Specification
Airfoil Profile	NACA
Number of Blades	4
Diameter of Blade	0.18 m
Rotate Speed of the Propeller	800 r/min~1600 r/min

For a rotational speed that meets the thrust requirements, the Lifting Line Theory is used to optimize the chord length of the propeller and the torque, M_{Prop}, generated by the optimized chord length of the propeller at a specific speed. We assumed that the torque generated by the motor was M_{motor}. Considering the torque loss due to sealing, friction, and other reasons for the motor, the efficiency of the motor was set to $M_{motor} = 0.8 \cdot M_{Prop}$. The blade design employed the foil NACA65A010 thickness distribution, with a modified mean line of NACA a = 0.8 modified mean line.

Load coefficient, σ_{T0}, is defined as

$$\sigma_{T0} = \frac{T}{\frac{1}{2}\rho\pi R^2 V_a^2} \tag{1}$$

The efficiency formula, η_{disk}, for an ideal actuating disc is

$$\eta_{disk} = \frac{2}{1 + \sqrt{1 + \sigma_T}} \tag{2}$$

Classical propeller open-water characteristics were used as the performance metrics in this study. The metrics were the thrust coefficients K_T, torque coefficient K_Q, propeller efficiency J, efficiency η, propeller speed n, and diameter of blade D.

$$K_T = \frac{T}{\rho n^2 D^4} \tag{3}$$

$$K_Q = \frac{Q}{\rho n^2 D^5} \tag{4}$$

$$J = \frac{V_a}{nD} \tag{5}$$

$$\eta = \frac{K_T}{K_Q}\frac{J}{} \tag{6}$$

The propulsion coefficients K_{T0} and K_{T1} were used to represent the load coefficients σ_{T0} and σ_{T1}. The advance ratios were taken across the range of 0.14 to 0.34. Under the condition of satisfying the thrust coefficient, the propeller design was optimized for different advance ratios. Finally, the relationship between K_{T0} and K_{T1}, as well as the open-water performance curve K_T and the efficiency of the propeller under different design inlet ratios, were obtained, as shown in Figure 3.

The open-water characteristic curves of K_T, K_Q, and η based on the designed blade are shown in Figure 4.

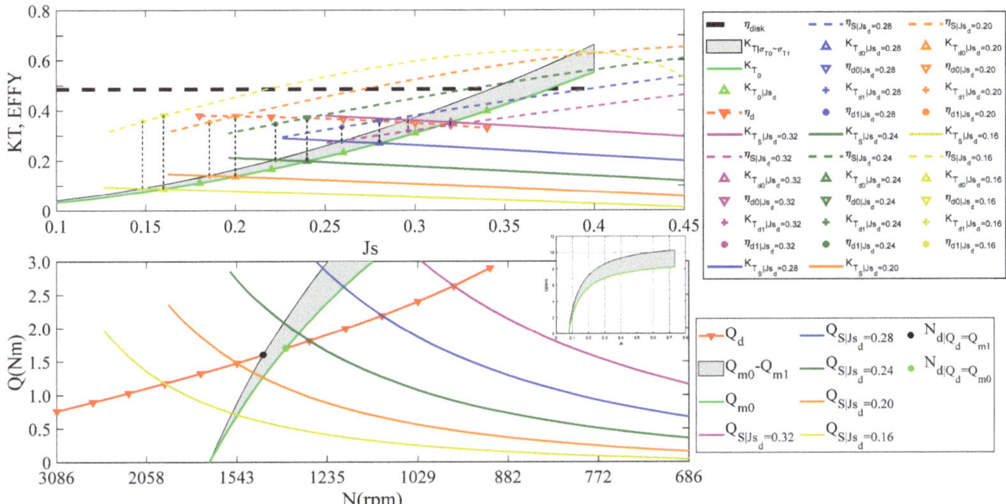

Figure 3. Thrust-torque balance propeller design points and performance curves.

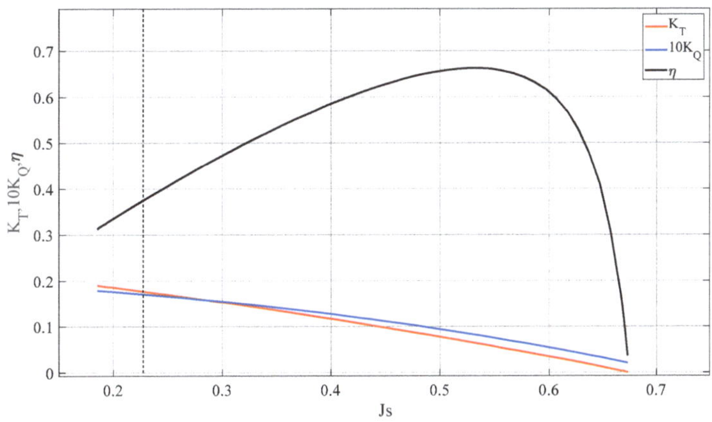

Figure 4. Open-water characteristic curves.

2.3. Performance Testing of the Thrust-Vectoring Mechanism

After summarizing the relevant literature on the design of vectored thrusters, it can be seen that there are two types of vector mechanisms used to achieve the spatial movement of thrusters: steering motors and linear motors. In this study, two types of vectored thrusters were designed and manufactured, as shown in Figure 5.

The vectored thruster in Figure 5a was designed with reference to the vectored thrusters used in the small underwater robots SUR-I and SUR-III, which utilize two steering motors to achieve two degrees of freedom of rotation of the vectored thruster. Steering motor 1 enables the rotation of the thruster around the x-axis with a range of $[-180°, 180°]$, while steering motor 2 achieves rotation around the y-axis within a range of $[-20°, 20°]$. The significant advantage of this approach lies in its simple structural design, and the steering motors' faster response speed compared to linear actuators. Standard steering motors typically take from 0.11 s to 0.21 s to rotate $60°$, which can be considered almost instantaneous for the motion control of AUVs with slower movement speeds.

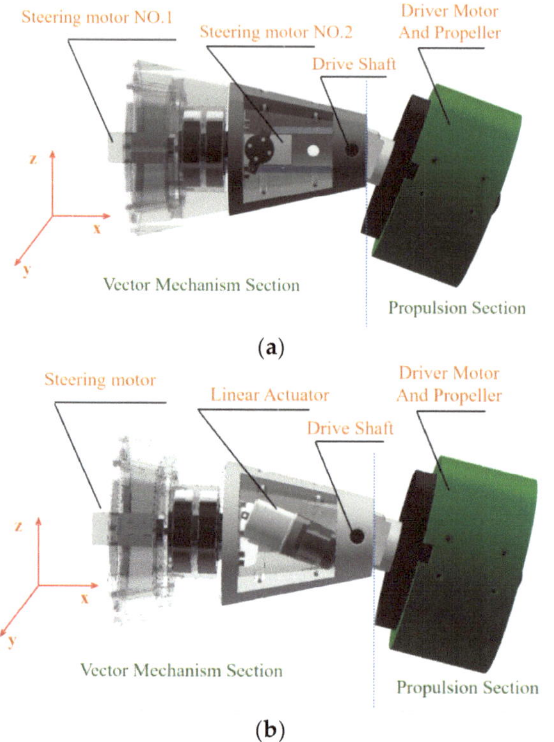

Figure 5. Schematic of vectored thruster configuration. (**a**) Two steering motors; (**b**) a steering motor and a linear actuator.

When this vector mechanism is applied to an AUV, the circular duct of the thruster functions as a rudder surface. As steering motor 2 rotates around the y-axis up and down, the greater rotation speed generates higher resistance. Figure 6 shows the vector diagram of the surface pressure field of the AUV operating at speeds of 0.5 kn, 1 kn, and 2 kn, with a detailed vector diagram of the thruster enclosed in the red box. The surface pressure distribution of the AUV remained almost the same at these three speeds, and the regions of maximum and minimum pressure distribution were identical. The key point to note is that when the speed was 0.5 kn, 1 kn, and 2 kn, the maximum pressure at the front of the annular duct of the AUV's vectored thruster was approximately 30 Pa, 130 Pa, and 520 Pa, respectively, while the rear of the duct was under negative pressure. The pressure at the front of the circular duct was significantly greater than that at the rear, indicating that the circular duct encountered significant hydrodynamic resistance. And, the hydrodynamic resistance increased exponentially with the increase in speed, and steering motor 2 needs to provide torque equal to the hydrodynamic force to maintain the attitude of the vectored thruster. If steering motor 2 sustains a large torque for a prolonged period, it can lead to damage. When this solution was applied to an AUV for sea trials, the lifespan of steering motor 2 often lasted for only a few hours. From this, it can be concluded that steering motor-based vectored thrusters are not suitable for underwater robots.

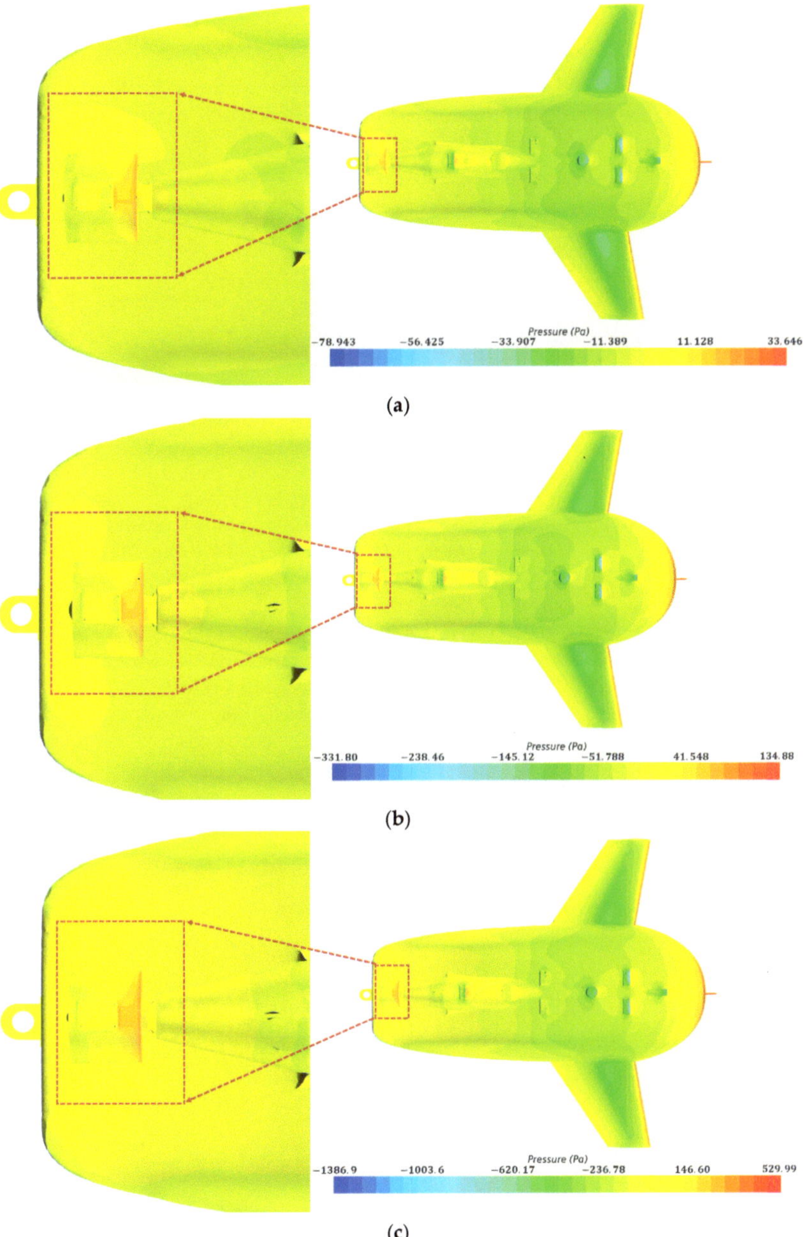

Figure 6. Velocity and pressure cloud maps of an AUV. (**a**) 0.5 kn; (**b**) 1 kn; and (**c**) 2 kn.

In Section 2.2, the design of the blade of the vectored thruster is described. This section discusses the vector mechanism of the vectored thruster. As shown in Figure 5, the vectored thruster designed in this work was composed of a linear actuator and a steering motor, a stepping motor, and a steering motor, which achieved a change in the tilt angle. The tilt angle combines two angles: elevation α and azimuth β. Referring to Figure 5, when the vectored thruster rotates around the y-axis, this was defined as a change in the elevation angle α of the vectored thruster, and the red arrow indicates that the rudder is pointing downward at this moment, as shown in Figure 7a. Similarly, when the vectored thruster rotates around the x-axis, this was defined as a change in the azimuth angle β of the vectored thruster, and the red arrow indicates that the rudder is rotating clockwise at this moment, as illustrated in Figure 7c.

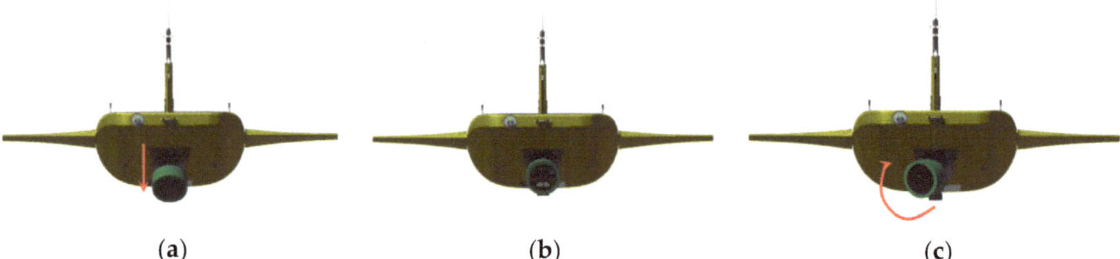

(a)   (b)   (c)

Figure 7. Tilt angle adjustment of vectored thruster relative to AUV: (**a**) elevation angle α; (**b**) initial position; (**c**) azimuth angle β.

Due to the frequent damage to steering motor 2, this study proposed an improved design for the vectored thruster. This design replaces steering motor 2 with a linear actuator. The vectored thruster comprises two parts: a propeller and a thrust-vectoring mechanism. The ductless propeller and ducted propeller are shown in Figure 8a,b, respectively, and their physical structure and local design details are shown in Figure 8c. A ducted propeller combines a screw propeller and annular wing. The annular wing also mitigates the threat posed by marine animals and seaweeds to screw propellers. However, a ductless propeller without an annular wing has higher propulsive efficiency, as shown in Figure 9. Underwater vehicles can quickly accelerate from zero or low speed to a predetermined speed.

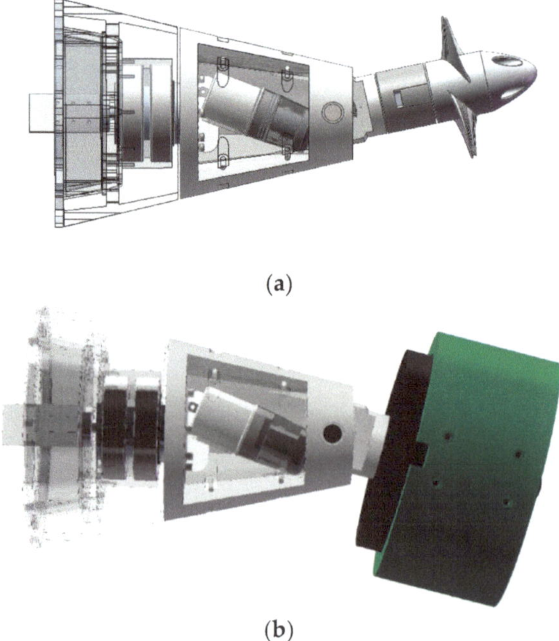

(a)

(b)

Figure 8. *Cont.*

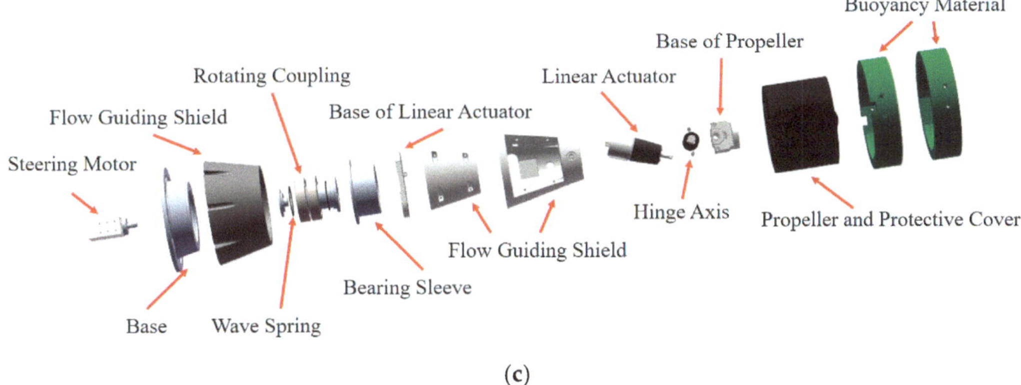

Figure 8. Schematic of vectored thruster configuration. (**a**) The ductless propeller; (**b**) ducted propeller; and (**c**) design details.

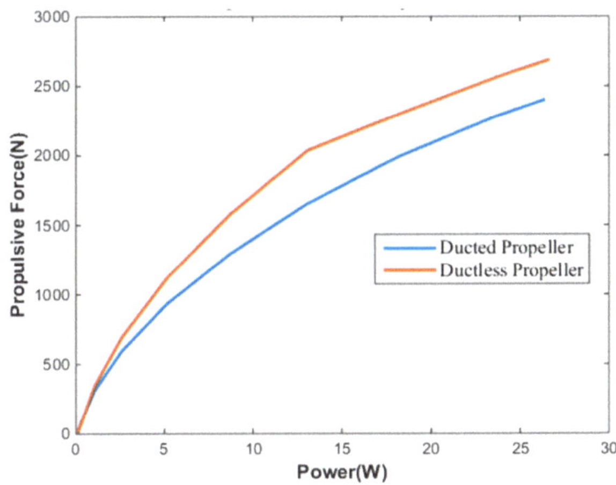

Figure 9. Thrust testing of two types of vectored thrusters.

3. Analysis and Test of the Thrust-Vectoring Mechanism

3.1. Thrust-Vectoring Mechanism Design

As an executing actuator, the vectored thruster is a critical component in the overall system of the AUV. Depending on the collocation requirements and application circumstances, there are many design schemes for the mechanical structure of thrust-vectoring mechanisms. Implementing thrust-vectoring mechanism functions for AUVs relies entirely on the chosen mechanism structure. Based on the application background of AUVs, the new design concept presented in this article for a thrust-vectoring mechanism for AUVs is shown in Figure 10.

The thrust-vectoring mechanism has two rotational degrees of freedom, which are achieved through four rotational axes. These two rotational degrees determine the spatial posture of the vectored thruster. Rotating coupling, namely rotary axis 1, joins the steering motor to a platform in a circular truncated cone, as shown in Figure 10. The steering motor can achieve platform rotation, thereby changing the azimuth angle of the vectored thruster. The platform can be considered as a fixed base relative to the ducted propeller and linear actuator.

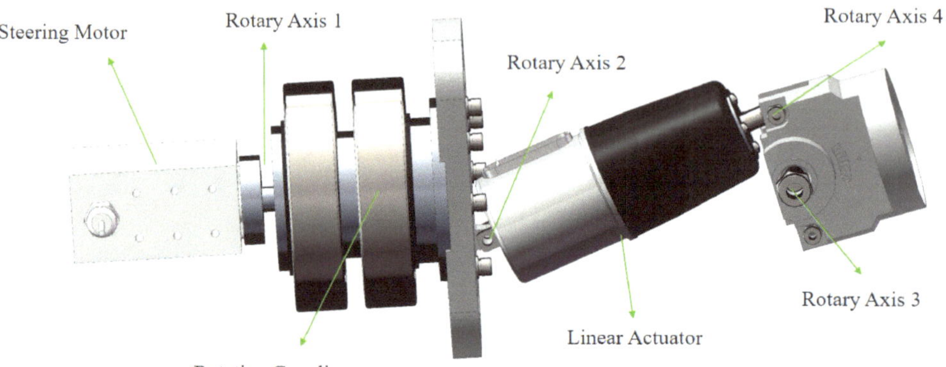

Figure 10. Thrust-vectoring mechanism.

The linear actuator assembled on a fixed platform with two joints was an actuating prismatic joint. One of the joints, rotary axis 2, was attached to the platform by a rotational joint, which had only one rotational degree of freedom. Another joint, rotary axis 4, was connected to the ducted propeller by a passive rotational joint. The linear actuator consisted of a ball screw assembly, magnet, Hall sensor, and stepping motor. The Hall sensor was installed in a fixed position inside the linear actuator, and a magnet was installed on the ball screw assembly. Thus, the feedback signal from the Hall sensor can be used to determine the operation length range of the linear actuator during working when the Hall sensor approaches the magnet. Similarly, the ductless propeller was installed on a fixed platform by a rotational joint, rotary axis 3, which had only one rotational degree of freedom. The length variation of the linear actuator achieved a rotational degree of freedom.

3.2. Kinematic Analysis of the Thrust-Vectoring Mechanism

The tilt angle of the thrust-vectoring mechanism is essential. Accurate control of the tilt angle ensures the safety and reliability of AUV operations. An accurate tilt angle measurement can be realized by mounting an attitude transducer on a fixed platform. However, this practical and straightforward approach has not been adopted because of the limited space available for the vectored thrusters. A kinematic analysis of the thrust-vectoring mechanism was performed to achieve reliable and accurate control of the tilt angle of the thrust-vectoring mechanism. This study's kinematic model of the thrust-vectoring mechanism is more accessible than the 3SPS-S parallel manipulator-based vectored thruster. There is only a linear actuator and a steering motor.

The linear actuator connected the ductless propeller and stationary platform via revolute joints. The two connecting joints are identified as points A and B. The ductless propeller was connected to the linear actuator and stationary platform by two designated connection points, points R and B, where the distance between locations A and B exhibited variability. It should be noted that points A, B, and C correspond to rotary axis 2, rotary axis 3, and rotary axis 4 in Figure 10 respectively. The distance between the RB and RA remained constant, with RB measuring 2 cm and RA measuring 10 cm, as shown in Figure 11.

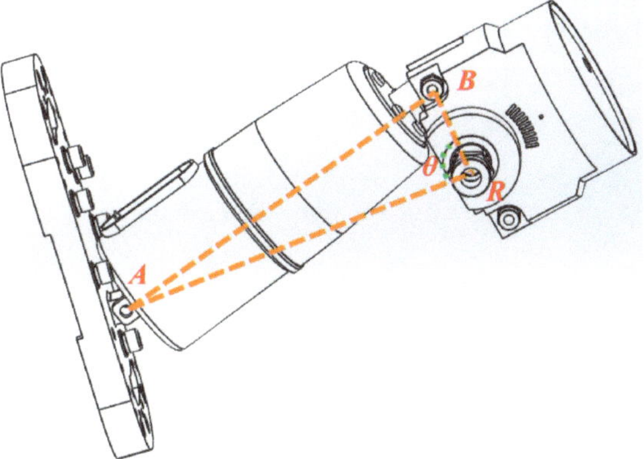

Figure 11. The schematic representation of revolute and spherical joints.

The cosine theorem describes the length of the RB, AB, and RB line segments and the angles between the neighboring line segments. The angle between RB and RA is denoted by α.

$$AB^2 = RB^2 + RA^2 + 2 \times RB \times RA \times cos(\alpha) \tag{7}$$

The relationship between the change in length of AB and α is

$$\alpha = arccos\left(\frac{RB^2 + RA^2 - AB^2}{2 \times RB \times RA}\right) \tag{8}$$

Because $RB = 2$ and $RA = 10$, the range of length changes of linear actuators is limited to

$$8 \leq AB \leq 12$$

Due to the structural limitations, the range of changes in α is limited to

$$-20° \leq \alpha \leq 20°$$

As shown in Figure 12, when the ductless propeller moved regularly according to the input of elevation angle α, the length changes of the linear actuator present a regular trend that echoes the inputs of elevation angle α.

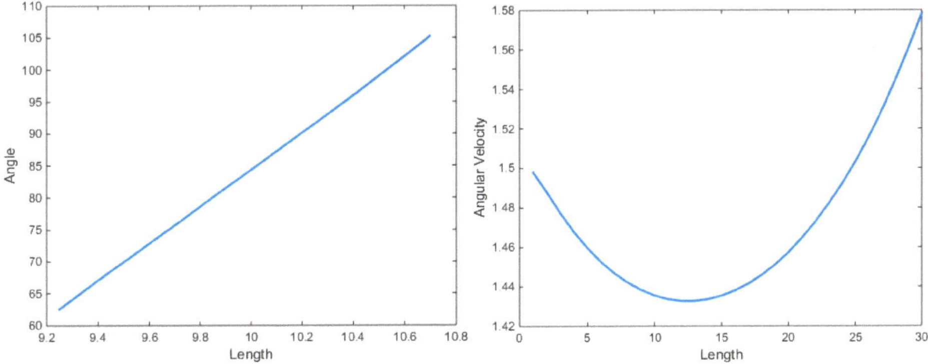

Figure 12. Changes in α, when the length of the linear actuator changes.

Thrust T in the two-dimensional plane can be rotated by adjusting the tilt angle α. The components of T are expressed as follows:

$$\begin{bmatrix} F_x \\ F_{yz} \end{bmatrix} = \begin{bmatrix} T\cos\alpha \\ T\sin\alpha \end{bmatrix} \tag{9}$$

β is achieved by a rotating motor, with a variation range of $[-180° \ 180°]$. At this time, the three components of T in three-dimensional space can be represented by the following equation:

$$\begin{bmatrix} F_x \\ F_y \\ F_z \end{bmatrix} = \begin{bmatrix} F_x \\ F_{yz}\cdot\sin\beta \\ F_{yz}\cdot\cos\beta \end{bmatrix} = \begin{bmatrix} T\cos\alpha \\ T\sin\alpha\sin\beta \\ T\sin\alpha\cos\beta \end{bmatrix} \tag{10}$$

Assuming T is a fixed value of 112 N, the spatial motion range of the output terminal can be obtained when the sliding range of the linear actuator and steering motor are given, as shown in Figure 13.

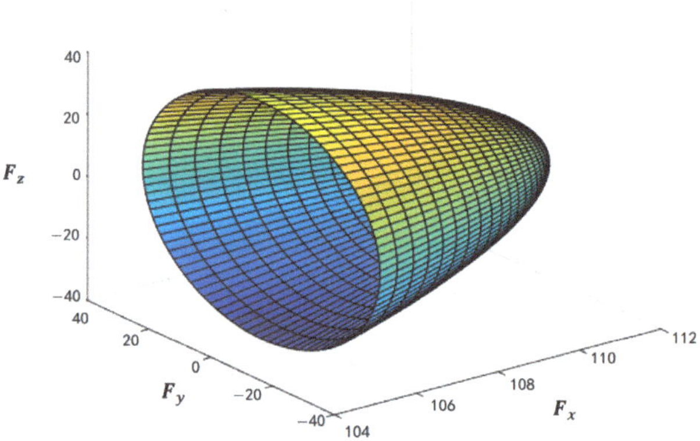

Figure 13. Propulsive surfaces of vectored thruster.

The AUV was capable of conducting six-degree-of-freedom (6DOF) movements in water, but the vectored thrusters can only control three of them, namely surge, pitch, and yaw. In this study, the component forces $\begin{bmatrix} F_x & F_y & F_z \end{bmatrix}$ for controlling the three-degree-of-freedom movement of the AUV were the outputs of PID (Proportional-Integral-Derivative) controllers. That is, given the output of the vectored thruster as $\begin{bmatrix} F_x & F_y & F_z \end{bmatrix}$, the input $\begin{bmatrix} T & \alpha & \beta \end{bmatrix}$ to the vectored thruster can be expressed using the following equation:

$$\begin{bmatrix} T \\ \alpha \\ \beta \end{bmatrix} = \begin{bmatrix} F_x{}^2 + F_y{}^2 + F_z{}^2 \\ \arctan\left(\left(F_y{}^2 + F_z{}^2\right)^{1/2} / F_x\right) \\ \arctan\left(F_y / F_z\right) \end{bmatrix} \tag{11}$$

3.3. Control of Thrust-Vectoring Mechanism

The hardware and software implementation strategies are provided in detail based on the thrust-vectoring mechanism of the AUV. The main control board used to control the vectored thruster in Figure 14a was designed by our team. The MUC is an STM32F103C8T6 chip, which is responsible for generating PWM (Pulse Width Modulation) signals to control the steering and thruster. The PWM was set at 50 Hz, with a pulse width ranging from 1 ms to 2 ms. The propeller power was the smallest when the pulse width equaled 1 ms. When the pulse width was 2 ms, the propeller power was at its maximum. The control of the linear actuator was achieved through a driver board and a position sensor. The position sensor was a Hall sensor.

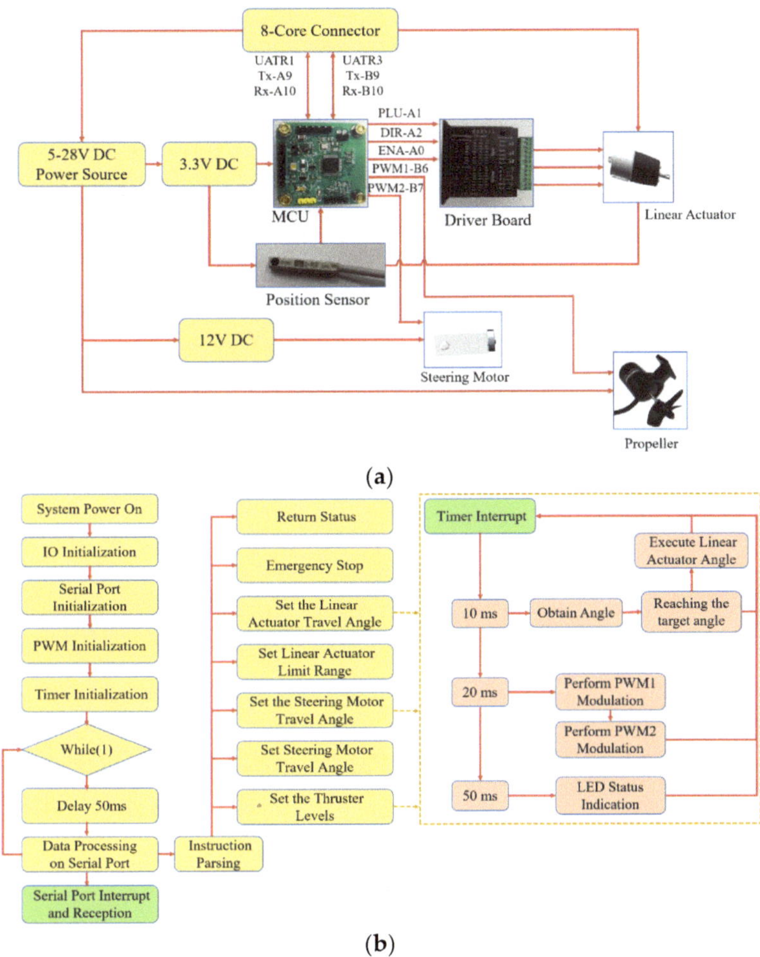

Figure 14. The (**a**) hardware and (**b**) software implementation strategies for vectored thruster.

The propeller and steering motor were controlled using PWM frequency, which was set at 50 Hz, with a pulse width ranging from 1 ms to 2 ms. The propeller power was the smallest when the pulse width equaled 1 ms. When the pulse width was 2 ms, the propeller power was at its maximum.

The relationship diagram between the propulsive force and current was obtained based on the thrust test in Section 2.3, and is shown in Figure 15a. To facilitate control of the upper computer, it was necessary to map the pulse width to the control input. The control inputs on the upper computer were set between 0 and 5. The mapping of the pulse width of 1 ms–2 ms to the control input was 0–5. The propulsive force and control inputs were fitted to obtain two curves, Fitted Curve 1 and Fitted Curve 2, as shown in Figure 15b. The control input was selected to be Level 0–4 to avoid excessive current, and the program used Fitted Curve 2 as the mapping relationship. When the pulse width was 1 ms, the control input was 0, and when the pulse width was 2 ms, the control input was 4.

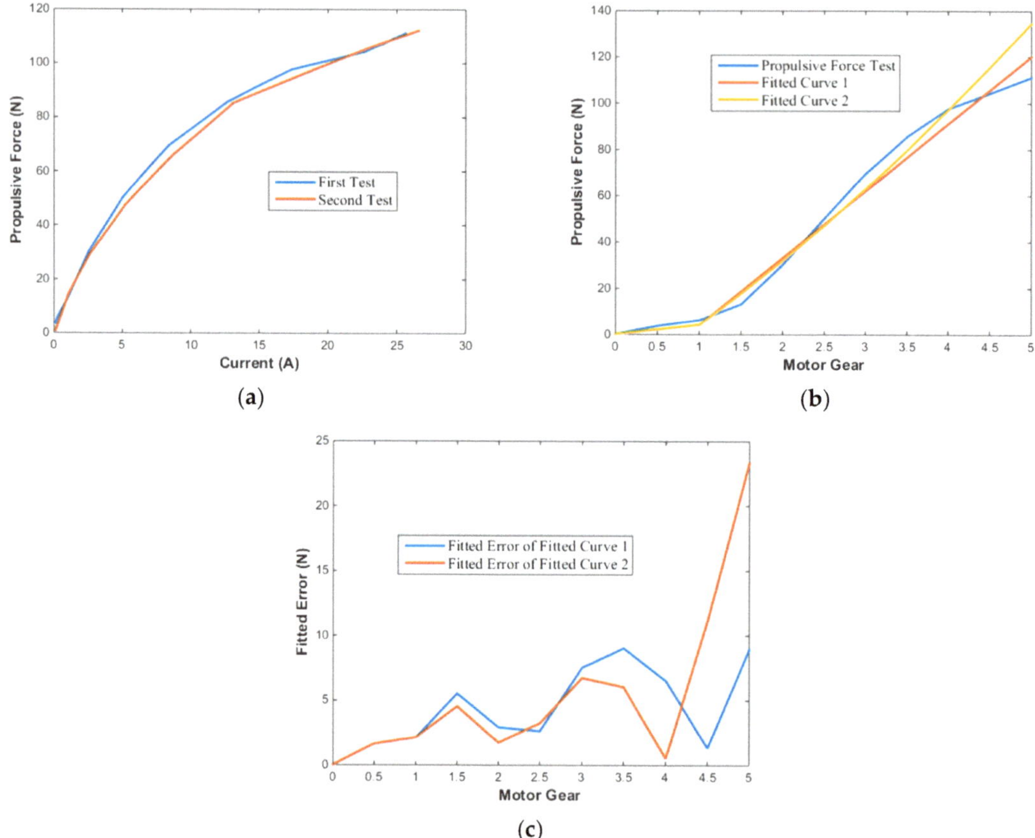

Figure 15. (**a**) Vectored thruster thrust testing; (**b**) propulsive force fitted curve; (**c**) fitting error of fitted curve.

The two curves in Figure 15c represent two types of fitted error curves. As the control input was selected as Level 0–4, the error of Fitted Curve 2 was smaller. Finally, Fitted Curve2 was selected as the mapping relationship between the pulse width and control input.

4. Field Tests and Results

Assuming that the thrust generated by the vectored thruster is T, the thrust T acting on the AUV can be illustrated as shown in Figure 16 below. Here, coordinate system $O_{NED} - X_nY_nZ_n$ is defined as the reference inertial coordinate system, while coordinate system $O_{Body} - X_bY_bZ_b$ is defined as the body-fixed coordinate system attached to the AUV. The thrust and its direction generated by the vectored thruster are indicated by the red arrow in the figure.

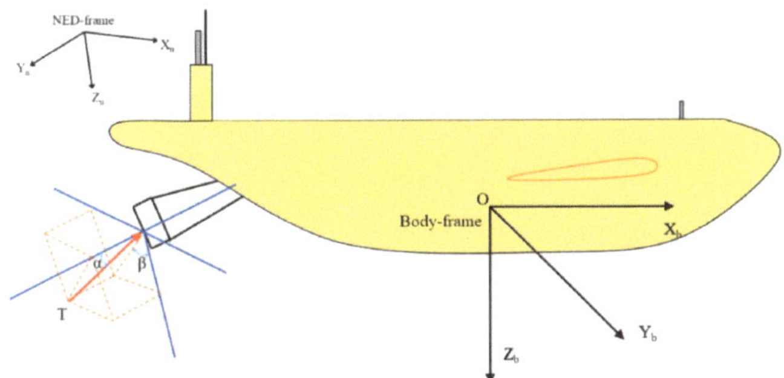

Figure 16. Coordinate frame of the AUV.

4.1. Water Tank Tests

Underwater tests of an actual AUV are necessary to determine the propulsive effect of the designed vectored thruster. A validation test was conducted in a water tank. The test environment was a water tank that was 22 m in length, 10 m in width, and 20 m in depth. The AUV was placed in an underwater environment, as shown in Figure 17, and the detailed venue and methods of the water tank tests can be found in the field test videos in the Supplementary Materials. Open-loop control for the horizontal and vertical motion of a specially shaped AUV is an essential and efficient method for evaluating the propulsive efficiency of a newly developed vectored thruster. The horizontal and vertical motions include three basic motions: surge, pitch, and yaw. The durations of the three basic motions were restricted to relatively small ranges. Surge and yaw motions were conducted in the horizontal plane, and pitch motions were conducted in the vertical plane. Owing to the asymmetric shape in the XY-plane and YZ-plane of the AUV, it was evident that the motion and hydrodynamic characteristics of the three basic motions may be distinct.

Figure 17. The AUV was tested in a water tank.

4.1.1. Surge Motion

In the surge motion, the AUV moved forward along the X-axis, and the rudder angle α and elevator angle β of the vectored thruster were equal to 0. The tests conducted were assigned to three groups, where the thrust outputs were Level 2, Level 3, and Level 4. Each test was repeated three times.

The DVL sensor measured the speed. Figure 18a–c shows the velocity–time relationship during the surge stage of the AUV. The reproducibility of the three test results was satisfactory. The maximum velocity corresponding to the three levels of propulsion inputs was 0.41 m/s, 0.65 m/s, and 0.98 m/s, respectively. Owing to the limited space in the test pool, the AUV did not accelerate to its maximum speed. The maximum speed measured did not reach the designed speed of 2.5 knots.

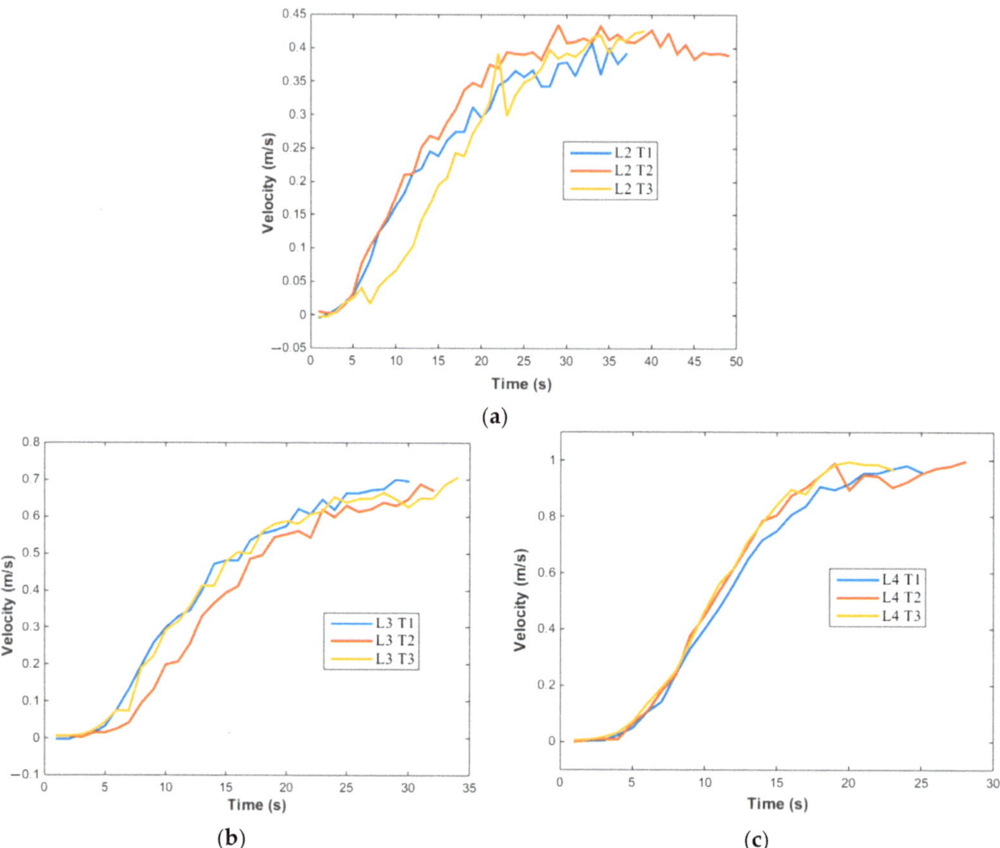

Figure 18. (a–c) Velocity–time relationship during surge motion.

4.1.2. Pitch Motion

An AUV is an underactuated underwater robot, whose depth and descent speed are related to the pitch angle. The heave motion was achieved using pitch motion. The maximum designed working depth of the AUV was 300 m. The maximum depth of the water tank was 20 m; tests of the heave motion could only be carried out in shallow water. The pitch motion was performed to verify the maximum diving speed of the AUV. To obtain the AUV's maximum pitch angle, the elevator angle α of the vectored thruster was set to a maximum angle of $20°$, and the azimuth β was $0°$. Simultaneously, the control input to the thrusters was fixed at 4. The test was repeated three times.

The IMU sensor was responsible for measuring the rotational angle. The depth sensor measured the depth. As Figure 19a–c indicate, the tests were repeated three times and the results matched each other reasonably well. The maximum errors appeared in the three pitch curves; the pitch curve L4 T3 represented the fastest descent speed during the tests.

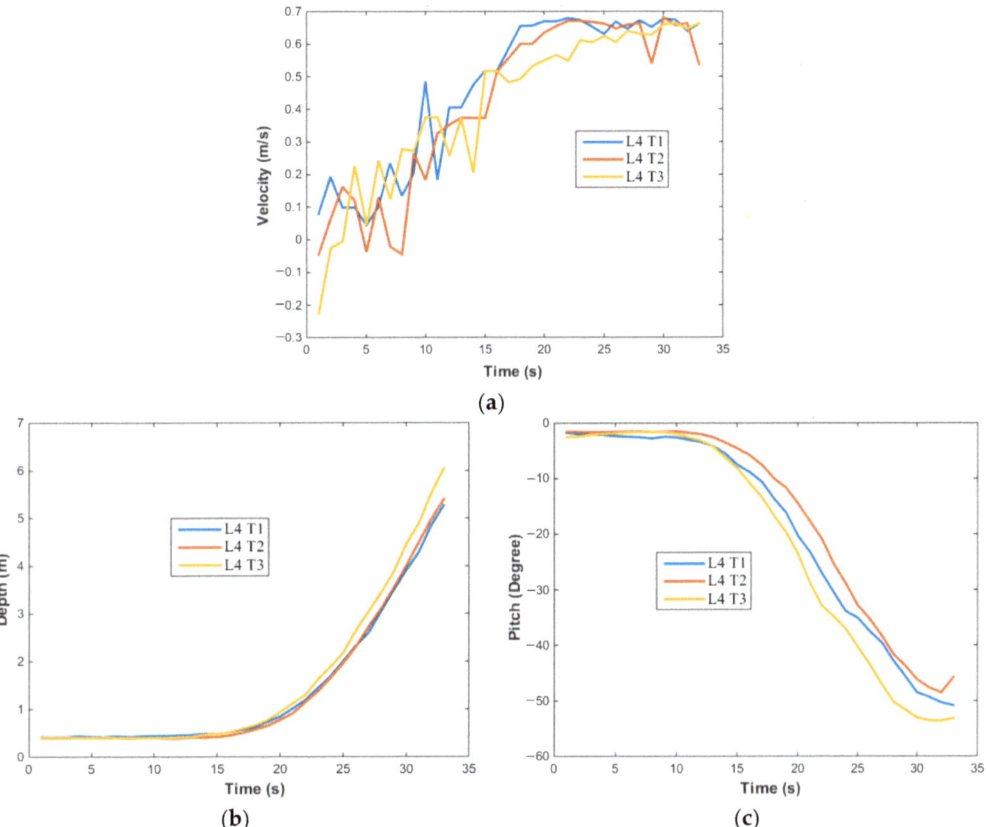

Figure 19. Heave motion of AUV. (**a**) Velocity; (**b**) depth; and (**c**) pitch.

In Figure 19a, the maximum speed along the X-axis was reduced from 0.98 m/s to 0.67 m/s. Figure 19b indicates that, in the initial stage of motion, the depth of the AUV remained unchanged from 0 to 15 s; the depth gradually increased after 15 s, and the descent speed remained stable after 25 s. As shown in Figure 19c, the pitch angle of the AUV remained unchanged from 0 to 12 s, gradually increased after 15 s, and remained stable after 30 s. The maximum pitch angle of the AUV was 53°, and the fastest descent speed of the AUV was 0.45 m/s.

4.1.3. Yaw Motion

The yaw motion of an AUV refers to its rotational movement around the z-axis. The azimuth angle of the vectored thruster was configured to −90 degrees to achieve the highest rate of change in the AUV heading angle. For comparison, the elevator angle was set to its maximum value of 20°. This configuration was intended to induce a counterclockwise rotation in the AUV. Because of the symmetrical shape of the shell of the AUV in the XZ plane, the hydrodynamic characteristics in the counterclockwise and clockwise directions were the same. Simultaneously, the control input to the thrusters was fixed at 4. The test was repeated three times. As shown in Figure 20, the results from the repeated tests matched each other reasonably well. The heading angle of the AUV remained unchanged from 0 s to 12 s and gradually increased after 12 s. The maximum rate of the heading angle was 14.26°/s, and the turning radius was less than 1 m.

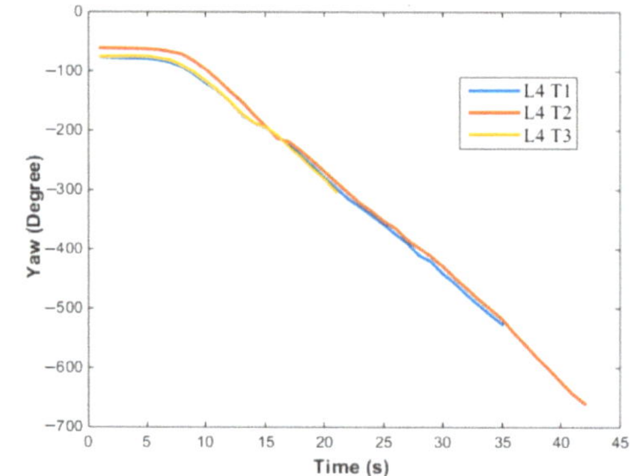

Figure 20. Yaw motion of AUV.

4.2. Offshore Tests

The open-loop test of the vector propeller applied to an AUV was completed in a safe and controllable still-water environment, and the basic performance parameters of the propeller were obtained. Next, a closed-loop test of the AUV's depth and heading control were conducted in an actual marine environment. The test was carried out in Nanshan Port, Sanya City, as shown in Figure 21, and the detailed testing results of offshore depth and heading control can be found in the field test videos provided in the Supplementary Materials.

Figure 21. Offshore testing area of AUV.

The AUV only utilized one vectored thruster, making it an underactuated system. This vectored thruster could only control three degrees of freedom of the AUV, namely speed, depth, and heading direction. The combined force T of the vectored thruster and its three component forces Fx, Fy, and Fz are defined in Equation (10). Specifically, Fx serves as the driving force to control the speed (u) of the AUV, Fy is the steering force for the AUV's heading direction, and Fz regulates the depth (h) of the AUV. The control scheme of the AUV is shown in Figure 22 below, where all three controllers utilize PID control.

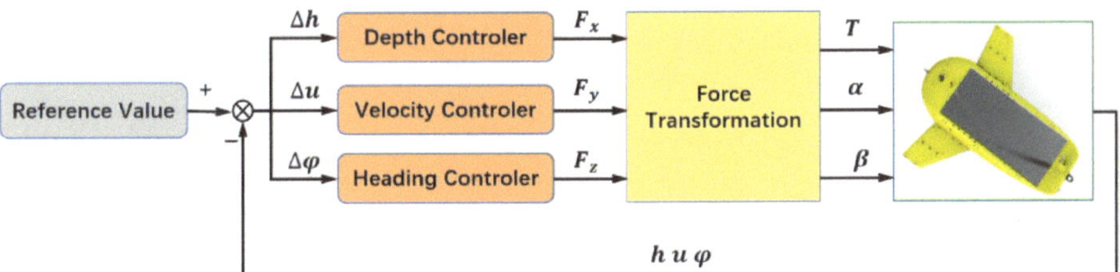

Figure 22. Motion control of AUV.

From Figure 13, we can see that the variation ranges of the three components *Fx*, *Fy*, and *Fz* of the thrust *T* generated by the vectored thruster were [0 112], [−38 38] and [−38 38]. Within these ranges, the control input for the AUV can be calculated using Equation (11) based on the three thrust components $\begin{bmatrix} T & \alpha & \beta \end{bmatrix}$.

Generally, the three parameters K_P, K_i, and K_d wee adjusted and set by trial and error. Finally, a set of relatively ideal adjustment parameters were obtained, as shown in Table 6.

Table 6. Coefficients of PID controller.

Parameter	Heading	Depth	Pitch	Velocity
K_P	0.2	1.5	1	1.3
K_i	0.01	0.1	0.1	0.15
K_d	3	5	5	3

During the test, the rotational angular velocity of the linear actuator was set at 1.6°/s, and the rotational angular velocity of the steering motor was set at 25°/s. The target AUV was set with a reference depth of $h_r = 6$ m, a reference speed of $u_r = 0.9$ m/s, and a reference heading of $\varphi_r = 50°$, and multiple tests were conducted. Figure 23 below shows the results of two of these tests.

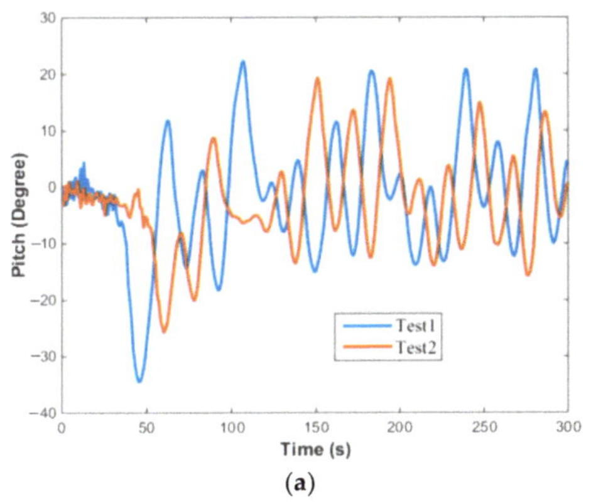

(a)

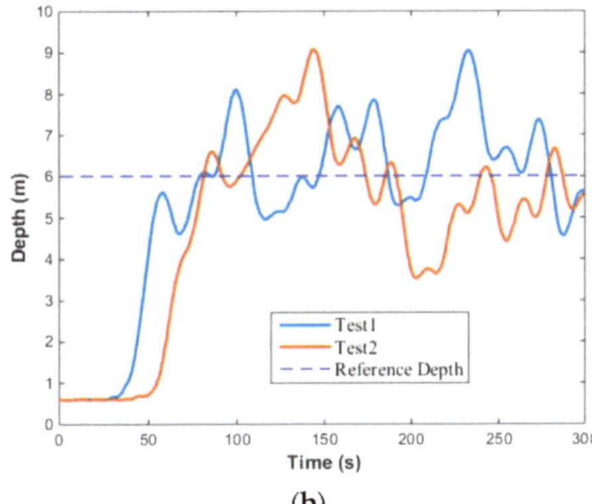

(b)

Figure 23. *Cont.*

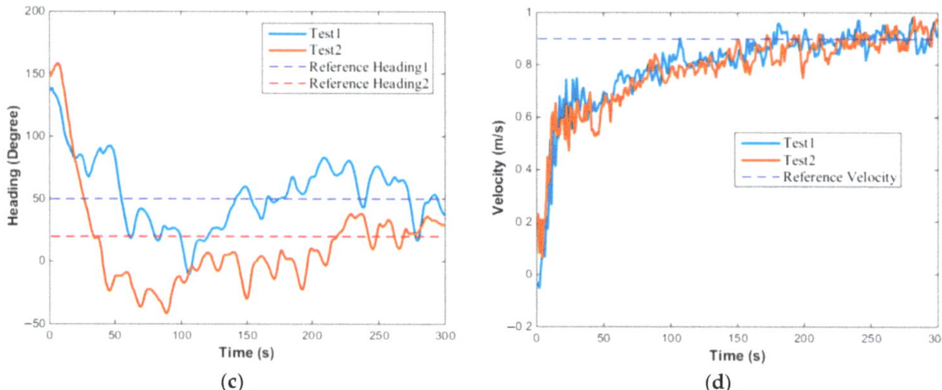

Figure 23. Results of motion of AUV (the maximum angular velocity of linear actuator is set at $1.6°/s$). (**a**) Pitch; (**b**) depth; (**c**) yaw; and (**d**) velocity.

As can be seen from the diagram, the depth control and heading effect of the AUV based on this vectored thruster were not satisfactory, as the AUV oscillated near the reference values without converging to the target depth and heading. Further discussion is needed on the response of the AUV's steering motor and linear actuator, as illustrated in Figure 24 below.

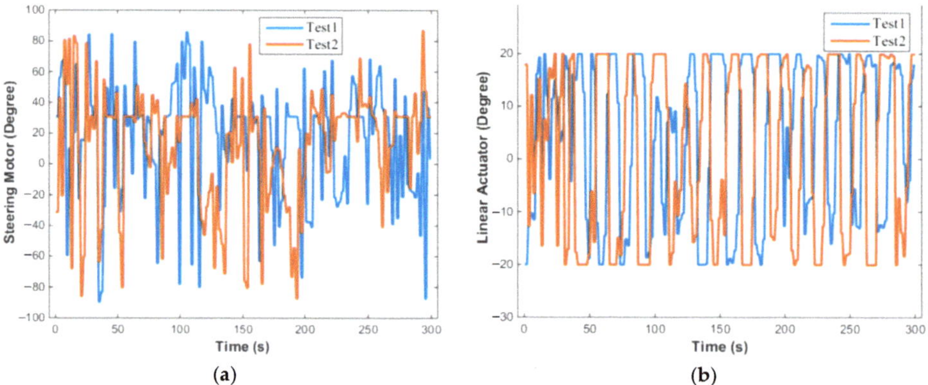

Figure 24. The target angle output by the PID. (**a**) Steering motor; (**b**) linear actuator.

Because the depth and heading control did not converged, resulting in repeated oscillations around the reference values, the two actuators needed to continuously adjust the attitude of the vectored thruster based on the PID output. Figure 24 shows the target angle output by the PID, and the linear actuator and the steering motor needed to track this target angle. As seen in Figure 24, the linear actuator would have to adjust from $20°$ to $-20°$ within 1 s, but with a maximum adjustment angular velocity of $1.6°/s$, it was unable to track the target angle output by the PID. Similarly, the steering motor would also need to adjust from $80°$ to $-80°$ within 1 s, but it was also unable to track the target angle output by the PID. There was a relatively long input delay for the AUV, preventing it from accurately tracking the target depth and heading.

The output angular velocity of the linear actuator needed to be adjusted, and the rotational angular velocity of the linear actuator was reset to $3.2°/s$, while the rotational angular velocity of the steering motor was set to $25°/s$. With the PID control parameters remaining unchanged, multiple tests were conducted. Figure 25 shows the results of two tests, where the reference depth of the target AUV was set as $h_r = 6$ m, the reference velocity was $u_r = 1$ m/s, and the reference headings were $\varphi_r = 140°$ and $\varphi_r = 110°$.

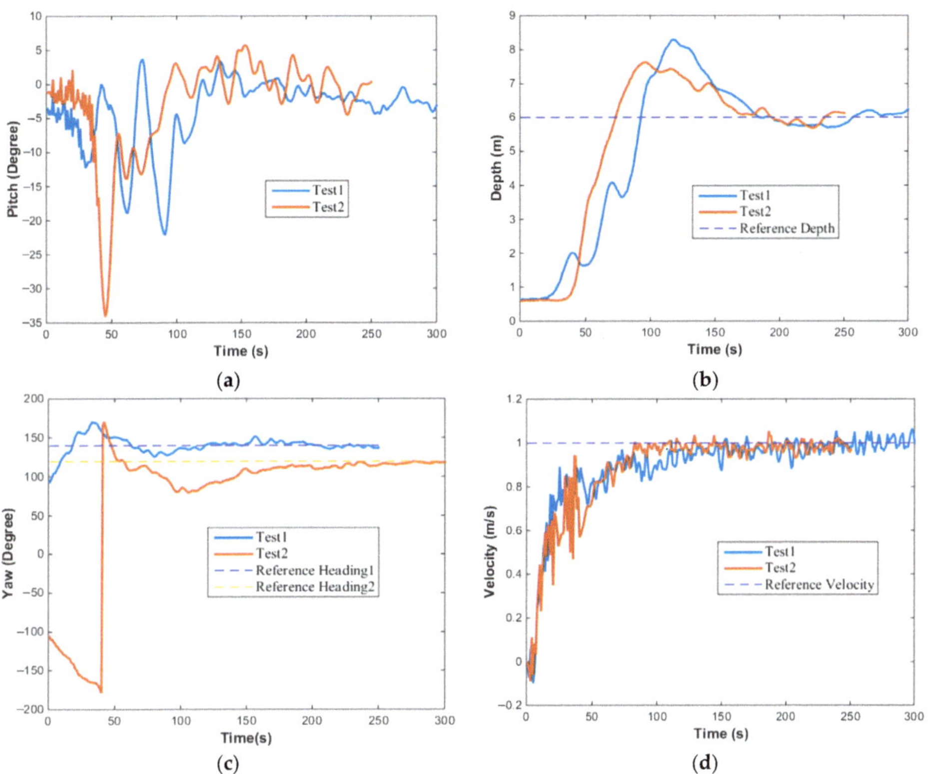

Figure 25. Results of motion of AUV (the maximum angular velocity of linear actuator is set at 3.2°/s). (**a**) Pitch; (**b**) depth; (**c**) yaw; and (**d**) velocity.

From Figure 25, it can be seen that the depth control and heading of the AUV based on this vectored thruster achieved the expected results, both converging to the reference values. The response of the steering motor and linear actuator is also presented in Figure 26.

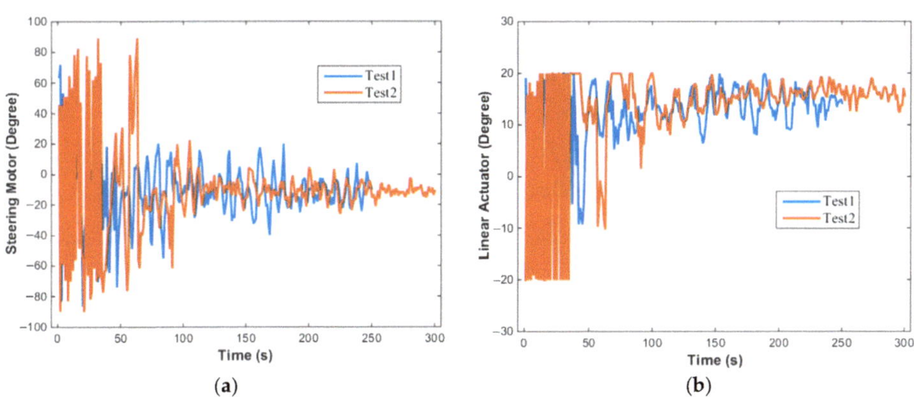

Figure 26. Yaw motion of AUV. (**a**) Steering motor; (**b**) linear actuator.

As seen in Figure 26, during the first 40 s, there was a significant error between the actual pitch angle and heading angle when compared to the reference values. The range of target angle variations output by the PID was relatively wide, making it difficult for the linear actuator and the steering motor to track the target angle. However, after 40 s, the error between the actual pitch and heading angles and their reference values gradually decreased, allowing the linear actuator and the steering motor to track the target angle output by the PID. As the movement of the AUV stabilized, the adjustment range of the linear actuator and the actuator's output gradually narrowed, eventually reaching a stable state. This indicates that when the rotational angular velocity of the linear actuator is set to $3.2°/s$ and the rotational angular velocity of the steering motor is set to $25°/s$, the configured vectored thruster can meet the AUV's motion control requirements. In future research, a more suitable controller can be designed for this vectored thruster to further enhance the stability of the AUV's motion control.

5. Conclusions

In this study, we designed a vectored thruster based on a linear actuator and steering motor to satisfy the design requirements of a specially shaped AUV. It employs a ductless propeller as a thrust supply to produce a vectored thrust for driving the AUV. Propulsive force tests were performed to evaluate its performance. Kinematic analyses of the thrust-vectoring mechanism were performed to achieve reliable and accurate control of the thrust-vectoring mechanism.

An AUV equipped with the designed vectored thruster was built to verify the design principles. The AUV was tested in a water tank. The thrust used for propelling and controlling the AUV was selected to study the vehicle's response at various vectoring angles.

Three basic motions (surge, pitch, and yaw) of an AUV are essential and efficient methods for evaluating the propulsive efficiency of a newly developed vectored thruster. The results showed that the thrust-vectoring mechanism directionally regulates vectored thrust. The vectored thruster can reach a sufficient level for AUV control.

The performance of the vectored thrusters in actual marine environments was tested, indicating that the vectored thrusters can meet the motion control requirements of AUVs. In the future, long-term sea trials will be conducted on the AUV to verify the durability and reliability of the vectored thruster. Considering the characteristics of the vectored thruster from the perspective of control algorithms, such as the rate and range of variation in the elevation angle α, the motion control of the AUV can be more efficient. In addition, in subsequent tests, improvements and upgrades will be made to the vectored thruster based on the problems encountered.

Supplementary Materials: The following supporting information can be downloaded at: https://www.mdpi.com/article/10.3390/act13060228/s1. This section provides videos of field tests, including the open-loop control test video in the water tank, and the offshore heading control test video and depth control test video to demonstrate the effectiveness and mechanism of the work.

Author Contributions: We confirm that the authors mentioned in the manuscript have been approved as the sole contributors and authors. We confirm that the manuscript has been read and approved by the authors and that there are no other persons who satisfied the criteria for authorship. All authors agree to be accountable for all aspects of the work. The main contributions of the authors are as follows: H.Z. designed the mechanical structure of the vectored thruster. L.L. drafted the manuscript and reviewed it critically. C.Y. provided the software implementation strategy for the vectored thruster. Y.C. agreed to be accountable for all aspects of the work in ensuring that questions related to the accuracy or integrity of any part of the work are appropriately investigated and resolved. H.X. completed the design and optimization of the blades. Y.X. completed the experiments mentioned in the manuscript. G.Z. provided the hardware implementation strategy for the vectored thruster. All authors have read and agreed to the published version of the manuscript.

Funding: The research was supported by the Hainan Provincial Joint Project of Sanya Yazhou Bay Science and Technology City (No. 2021CXLH0003) as well as Sanya Science, Technology, Industry and Information Technology Bureau (No. 2022KJCX65).

Data Availability Statement: The data that support the findings of this study are available from the corresponding author, Chen Yuxiang, upon reasonable request.

Acknowledgments: We wish to express our heartfelt appreciation to our research institutes. Their unwavering support, intelligent recommendations, and vital resources were critical to completing this research on the design and application of a novel vectored thruster for a specially shaped AUV. We are excited to investigate the possibilities of our unique vectored thruster for specially shaped autonomous underwater vehicles and to contribute to the growth of maritime technology.

Conflicts of Interest: The authors declare no conflict of interest.

References

1. Fagundes Gasparoto, H.; Chocron, O.; Benbouzid, M.; Siqueira Meirelles, P. Advances in Reconfigurable Vectorial Thrusters for Adaptive Underwater Robots. *J. Mar. Sci. Eng.* **2021**, *9*, 170. [CrossRef]
2. Liu, T.; Hu, Y.; Xu, H.; Wang, Q.; Du, W. A Novel Vectored Thruster Based on 3-RPS Parallel Manipulator for Autonomous Underwater Vehicles. *Mech. Mach. Theory* **2019**, *133*, 646–672. [CrossRef]
3. Li, Y.; Tian, C.; Wang, Y.; Gao, P.; Ma, X. Design and Simulation of a Collaborative Propulsion System for the Underwater Robot. *Int. J. Robot. Eng.* **2019**, *4*, 1–19.
4. Li, Y.; Guo, S.; Wang, Y. Design and Characteristics Evaluation of a Novel Spherical Underwater Robot. *Robot. Auton. Syst.* **2017**, *94*, 61–74. [CrossRef]
5. Kadiyam, J.; Mohan, S. Conceptual Design of a Hybrid Propulsion Underwater Robotic Vehicle with Different Propulsion Systems for Ocean Observations. *Ocean Eng.* **2019**, *182*, 112–125. [CrossRef]
6. Hu, H.; Wang, Y. Vertical Force Generation of a Vectorial Thruster That Employs a Rigid Flapping Panel. *Phys. Fluids* **2021**, *33*, 061906. [CrossRef]
7. Liu, T.; Hu, Y.; Xu, H.; Zhang, Z.; Li, H. Investigation of the Vectored Thruster AUVs Based on 3SPS-S Parallel Manipulator. *Appl. Ocean Res.* **2019**, *85*, 151–161. [CrossRef]
8. Pugi, L.; Allotta, B.; Pagliai, M. Redundant and Reconfigurable Propulsion Systems to Improve Motion Capability of Underwater Vehicles. *Ocean Eng.* **2018**, *148*, 376–385. [CrossRef]
9. Kadiyam, J.; Mohan, S.; Deshmukh, D. Control of a Vectorial Propulsion Underwater Vehicle Considering Thruster Hydrodynamics Constraints and Actuator Saturation. In Proceedings of the Global Oceans 2020: Singapore–US Gulf Coast, Biloxi, MS, USA, 5–30 October 2020; pp. 1–10.
10. Kadiyam, J.; Mohan, S.; Deshmukh, D.; Seo, T. Simulation-Based Semi-Empirical Comparative Study of Fixed and Vectored Thruster Configurations for an Underwater Vehicle. *Ocean Eng.* **2021**, *234*, 109231. [CrossRef]
11. Lin, X.; Guo, S. Development of a Spherical Underwater Robot Equipped with Multiple Vectored Water-Jet-Based Thrusters. *J. Intell. Robot. Syst.* **2012**, *67*, 307–321. [CrossRef]
12. Kopman, V.; Cavaliere, N.; Porfiri, M. A Thrust-Vectored Submersible for Animal Behavior Research: Design and Proof of Concept. In Proceedings of the ASME 2010 Dynamic Systems and Control Conference, DSCC2010, Cambridge, MA, USA, 12–15 September 2010; Volume 1.
13. Cavallo, E.; Michelini, R.C.; Filaretov, V.F. Conceptual Design of an AUV Equipped with a Three Degrees of Freedom Vectored Thruster. *J. Intell. Robot. Syst.* **2004**, *39*, 365–391. [CrossRef]
14. Cavallo, E.; Michelini, R.C. A Robotic Equipment for the Guidance of a Vectored Thurstor AUV. In Proceedings of the 35th International Symposium on Robotics ISR, Paris, France, 23–26 March 2004.
15. Guo, S.; Lin, X.; Tanaka, K.; Hata, S. Modeling of Water-Jet Propeller for Underwater Vehicles. In Proceedings of the 2010 IEEE International Conference on Automation and Logistics, Hong Kong, China, 16–20 August 2010; pp. 92–97.
16. Chocron, O.; Prieur, U.; Pino, L. A Validated Feasibility Prototype for AUV Reconfigurable Magnetic Coupling Thruster. *IEEE/ASME Trans. Mechatron.* **2013**, *19*, 642–650. [CrossRef]
17. Gao, F.; Han, Y.; Wang, H.; Ji, G. Innovative Design and Motion Mechanism Analysis for a Multi-Moving State Autonomous Underwater Vehicles. *J. Cent. South Univ.* **2017**, *24*, 1133–1143. [CrossRef]
18. Papachristos, C.; Alexis, K.; Tzes, A. Efficient Force Exertion for Aerial Robotic Manipulation: Exploiting the Thrust-Vectoring Authority of a Tri-Tiltrotor UAV. In Proceedings of the 2014 IEEE International Conference on Robotics and Automation (ICRA), Hong Kong, China, 31 May–7 June 2014.
19. Gucer, Ç.A.; Onur, A.; Kantarcioglu, B.; Cenk, U. Thruster Design for Unmanned Underwater Vehicles. *Selçuk-Tek. Derg.* **2020**, *19*, 196–208.
20. Darmawan, S.; Raynaldo, K.; Halim, A. Investigation of Thruster Design to Obtain the Optimum Thrust for Rov (Remotely Operated Vehicle) Using CFD. *Evergreen* **2022**, *9*, 115–125. [CrossRef]
21. Atalı, G. Prototyping of a Novel Thruster for Underwater ROVs. *Int. J. Appl. Math. Electron. Comput.* **2022**, *10*, 11–14. [CrossRef]

22. Laidani, A.; Bouhamida, M.; Benghanem, M.; Sammut, K.; Clement, B. A Low-Cost Test Bench for Underwater Thruster Identification. *IFAC-PapersOnLine* **2019**, *52*, 254–259. [CrossRef]
23. Li, C.; Guo, S.; Guo, J. Performance Evaluation of a Hybrid Thruster for Spherical Underwater Robots. *IEEE Trans. Instrum. Meas.* **2022**, *71*, 7503110. [CrossRef]

Article

Research on the Deviation Correction Control of a Tracked Drilling and Anchoring Robot in a Tunnel Environment

Chuanwei Wang [1,2,*], Hongwei Ma [1,2], Xusheng Xue [1,2], Qinghua Mao [1,2], Jinquan Song [3], Rongquan Wang [3] and Qi Liu [4]

1   School of Mechanical Engineering, Xi'an University of Science and Technology, Xi'an 710054, China; mahw@xust.edu.cn (H.M.); xuexsh@xust.edu.cn (X.X.); maoqh@xust.edu.cn (Q.M.)

2   Shaanxi Key Laboratory of Mine Electromechanical Equipment Intelligent Detection and Control, Xi'an University of Science and Technology, Xi'an 710054, China

3   Xi'an Heavy Industry Hancheng Coal Mining Machinery Co., Ltd., Hancheng 715400, China; 17382631379@163.com (J.S.); wangrongquan1@163.com (R.W.)

4   Langfang Jinglong Heavy Equipment Co., Ltd., Langfang 065300, China; l15137591890@163.com

*   Correspondence: wangchuanwei228@xust.edu.cn

Abstract: In response to the challenges of multiple personnel, heavy support tasks, and high labor intensity in coal mine tunnel drilling and anchoring operations, this study proposes a novel tracked drilling and anchoring robot. The robot is required to maintain alignment with the centerline of the tunnel during operation. However, owing to the effects of skidding and slipping between the track mechanism and the floor, the precise control of a drilling and anchoring robot in tunnel environments is difficult to achieve. Through an analysis of the body and track mechanisms of the drilling and anchoring robot, a kinematic model reflecting the pose, steering radius, steering curvature, and angular velocity of the drive wheel of the drilling and anchoring robot was established. This facilitated the determination of speed control requirements for the track mechanism under varying driving conditions. Mathematical models were developed to describe the relationships between a tracked drilling and anchoring robot and several key factors in tunnel environments, including the minimum steering space required by the robot, the minimum relative steering radius, the steering angle, and the lateral distance to the sidewalls. Based on these models, deviation-correction control strategies were formulated for the robot, and deviation-correction path planning was completed. In addition, a PID motion controller was developed for the robot, and trajectory-tracking control simulation experiments were conducted. The experimental results indicate that the tracked drilling and anchoring robot achieves precise control of trajectory tracking, with a tracking error of less than 0.004 m in the x-direction from the tunnel centerline and less than 0.001 m in the y-direction. Considering the influence of skidding, the deviation correction control performance test experiments of the tracked drilling and anchoring robot at $dy = 0.5$ m away from the tunnel centerline were completed. In the experiments, the tracked drilling and anchoring robot exhibited a significant difference in speed between the two sides of the tracks with a track skid rate of 0.22. Although the real-time tracking maximum error in the y-direction from the tunnel centerline was 0.13 m, the final error was 0.003 m, meeting the requirements for position deviation control of the drilling and anchoring robot in tunnel environments. These research findings provide a theoretical basis and technical support for the intelligent control of tracked mobile devices in coal mine tunnels, with significant theoretical and engineering implications.

Keywords: drilling and anchoring robot; track mechanism; skidding and slipping; deviation correction control; path tracking

Citation: Wang, C.; Ma, H.; Xue, X.; Mao, Q.; Song, J.; Wang, R.; Liu, Q. Research on the Deviation Correction Control of a Tracked Drilling and Anchoring Robot in a Tunnel Environment. *Actuators* **2024**, *13*, 221. https://doi.org/10.3390/act13060221

Academic Editor: Ioan Doroftei

Received: 4 May 2024
Revised: 2 June 2024
Accepted: 12 June 2024
Published: 13 June 2024

1. Introduction

During the coal mine tunneling process, the permanent support time accounts for 60% of the total time [1,2]. There are issues such as excessive personnel, heavy support tasks, and high labor intensity [3–5]. This paper proposes a novel tracked drilling and anchoring robot that is required to maintain alignment with the centerline of the tunnel during operation. Owing to the complex nonlinear interaction between the track and the floor, accompanied by high-speed track skidding and low-speed track slipping [6], the drilling and anchoring robot is prone to collisions with the sidewalls of the tunnel in tunnel environments.

In a study on tracked chassis steering, Jia W. [7] determined the track force of tracked vehicles during steering based on a shear stress model, established a dynamic model with high computational accuracy, and investigated the steering stability of tracked vehicles in sandy road environments under different steering conditions. Xiong H. [8] provided a dynamic model of the tracked mechanism and a method for motion control in a new underwater environment with nonholonomic constraints and used the Lyapunov principle to verify its safety. Qin H.W. [9] presented a comprehensive overview of recent advancements and breakthroughs in the field of path planning for mobile robots while conducting an in-depth examination and comparison of various path-planning algorithms. Sabiha A.D. [10] optimized the backstepping controller as a kinematic controller and verified the stability analysis of the entire system based on Lyapunov theory. Saglia J. [11] studied the establishment of dynamic models and the adjustment of control gains. The research optimizes the control parameters by analyzing and reducing the execution conflicts and tracking errors. Simulations and experiments have been conducted to implement the analysis and control strategies in mechanisms. Fang Y. [12] established a dynamic model of a negative-pressure suction-tracked wall-climbing robot based on the discrete method of track force load and analyzed the influence of design parameters on the motion performance of tracked robots. Ishikawa T. [13] developed a trajectory tracking control system for dump trucks tracked on both hard and soft surfaces, enabling path tracking under varying ground conditions. Zhang M.J. [14] addressed the slippage and tunnel gradient issues. They established a neural network PID-based motion control algorithm for a boom-type roadheader, achieving real-time correction control of the roadheader. Qu Y.Y. [15] achieved deviation correction for roadheader body walking based on pose deviation path tracking control. Zhang X.H. [16] proposed an automatic directional excavation control method for boom-type roadheaders based on visual navigation, thereby realizing the automatic directional excavation function of the boom-type roadheader. The motion control accuracy is within ±20 mm. Mao Q.H. [17] proposed a deviation correction path planning method based on an Improved Particle Swarm Optimization (I-PSO) algorithm for a full-width horizontal axis roadheader, achieving path planning for the EJM340/4-2 type full-width horizontal axis roadheader.

In summary, the interaction between the track mechanism and the roadway surface is complex, particularly in the confined spaces of underground coal mine tunnels. The track mechanism cannot freely change direction, necessitating a thorough analysis of the track mechanism and steering performance. Therefore, this paper will study the following aspects. First, the article introduces the overall structure and key parameters of the drilling and anchoring robot, laying the foundation for its subsequent analysis. Secondly, the paper analyzes the steering kinematics of the drilling and anchoring robot, investigating the fundamental principles behind its maneuverability. Next, the article discusses the deviation correction control strategy for the drilling and anchoring robot, which is crucial for ensuring the accuracy and reliability of its drilling and anchoring operations. Furthermore, the paper analyzes the path tracking control testing conducted on the drilling and anchoring robot, evaluating its ability to accurately follow a specified trajectory. Finally, the article tests and analyzes the lane deviation control of the drilling and anchoring robot in tunnel environments, assessing its overall applicability in such challenging applications.

2. Structure and Parameters of the Drilling and Anchoring Robot

Drilling and anchoring robots can be used in conjunction with traditional roadheaders, robotic roadheaders, bolter miners, and other equipment to complete support and drilling tasks, reduce the labor intensity of workers, and improve drilling efficiency. The drilling and anchoring robot mainly consists of a main frame, left-track mechanism, right-track mechanism, and other components, as shown in Figure 1. Depending on the research requirements, the tunnel coordinate system O_h and the drilling and anchoring robot coordinate system O_r were established where θ represents the yaw angle, α represents the pitch angle, and φ represents the roll angle.

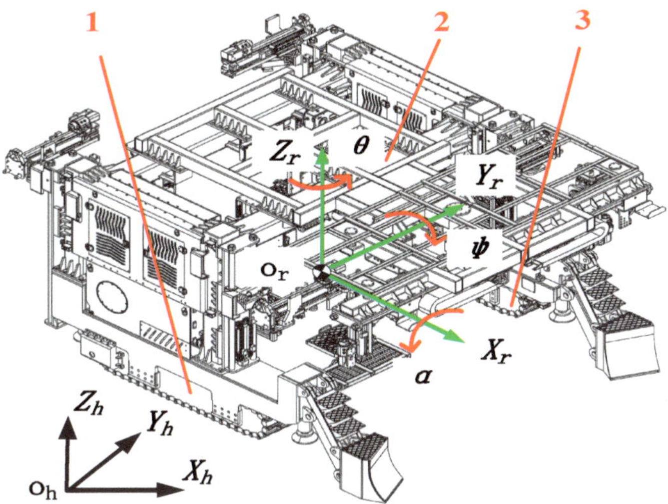

Figure 1. Composition of drilling and anchoring robot. 1. Left track mechanism, 2. main frame, 3. right track mechanism.

A 3D model of the track mechanism of the drilling and anchoring robot is shown in Figure 2, where B_0 represents the track width, r_1 is the radius of the driving wheel, r_2 is the radius of the driven wheel, L is the length of the track contact segment with the ground, D is the overall length of the track, d is the track pitch, and h is the track thickness. Owing to the thickness of the track, the influence of track thickness cannot be ignored when determining the traveling speed of the robot.

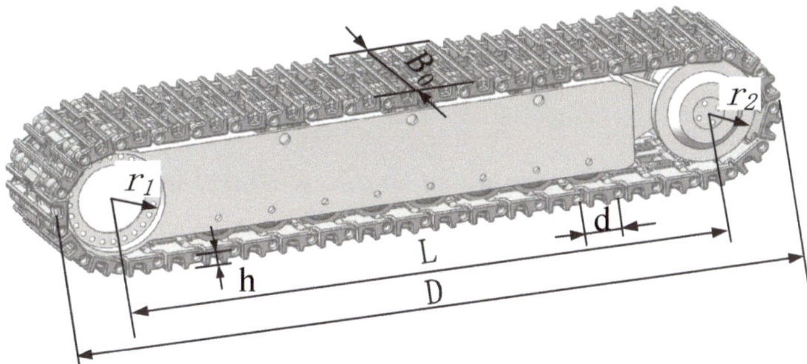

Figure 2. The 3D model of the track mechanism.

The drilling and anchoring robot uses typical differential steering, where the speed and direction are mainly controlled by adjusting the rotation speed of the driving wheels on the left and right tracks to achieve precise tracking control of the predetermined trajectory. The planar steering model of the drilling and anchoring robot is illustrated in Figure 3, where O represents the symmetrical center of the two track axes of the drilling and anchoring robot, CM denotes the center of mass of the robot, a is the distance from O to CM, and B is the distance between the centerlines of the left and right tracks. The primary parameters of the drilling and anchoring robots are listed in Table 1.

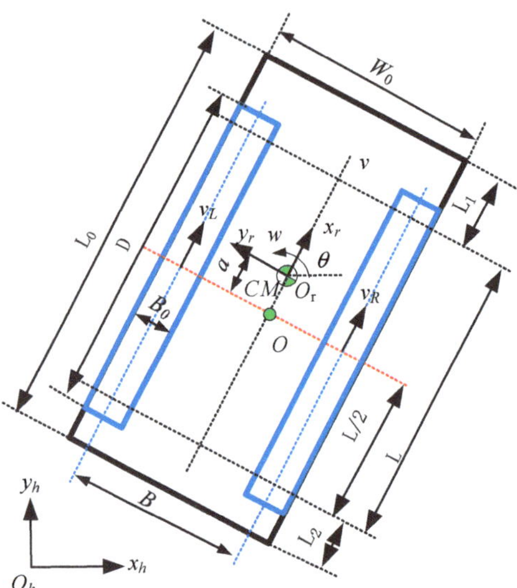

Figure 3. The geometric configuration of the tracked drilling and anchoring robot's differential drive system in the xy plane.

Table 1. Main parameters of the drilling and anchoring robot.

Parameters	Value
Dimensions ($L_0 \times W_0 \times H$)	$4.95 \times 4.28 \times 3.1$ m
Track distance (B)	3.9 m
Distance from front axle to front end (L_1)	2.01 m
Distance from rear axle to rear end (L_2)	0.83 m
The distance of CM offset (a)	0.1 m
Travel speed (v)	0~30 m/min
Track length (D)	2.67 m
Track width (B_0)	0.38 m
Track thickness (h)	0.04 m
Length of track contact with ground (L)	2.11 m

3. Steering Kinematics Analysis of the Drilling and Anchoring Robot

3.1. Kinematic Analysis

The definition of the state variables of the drilling and anchoring robot in the global coordinate system of the tunnel is denoted by $P = (x, y, \theta)^T$, whereas the state variables in the local coordinate system of the drilling and anchoring robot are denoted by $P_r = (x_r, y_r, \theta_r)^T$. The orthogonal rotation matrix $R(\theta)$ mapping of the local coordinate system of the drilling and anchoring robot to the global coordinate system is as follows [18]:

$$R(\theta) = \begin{bmatrix} \cos\theta & -\sin\theta & 0 \\ \sin\theta & \cos\theta & 0 \\ 0 & 0 & 1 \end{bmatrix} \tag{1}$$

The relationship between P and P_r can be expressed as $P = R(\theta)P_r$, $P_r = R^{-1}(\theta)P$.

Without considering the skidding factor, based on the principles of theoretical mechanics, the kinematic equations [19] of the drilling and anchoring robot in the tunnel coordinate system can be established as follows.

$$\begin{cases} v_L = \omega_L(r_1 + h) \\ v_R = \omega_R(r_1 + h) \\ v = (v_L + v_R)/2 \\ \omega = (v_R - v_L)/B \\ v_r = \omega a \end{cases} \tag{2}$$

where v_L and v_R represent the linear velocities of the left and right tracks; ω_L and ω_R represent the angular velocities of the left and right track drive wheels; v is the linear velocity of the robot at point o; ω represents the angular velocity of the robot at point o; and v_r represents the linear velocity of the drilling and anchoring robot's CM. The equation for the motion relationship of the CM of the drilling and anchoring robot is given as follows [20]:

$$\dot{P} = \begin{bmatrix} \dot{x} \\ \dot{y} \\ \dot{\theta} \end{bmatrix} = \begin{bmatrix} \frac{(r_1+h)\cos\theta}{2} + \frac{dr\sin\theta}{B} & \frac{(r_1+h)\cos\theta}{2} - \frac{dr\sin\theta}{B} \\ \frac{(r_1+h)\sin\theta}{2} - \frac{dr\cos\theta}{B} & \frac{(r_1+h)\sin\theta}{2} + \frac{dr\cos\theta}{B} \\ -\frac{(r_1+h)}{B} & \frac{(r_1+h)}{B} \end{bmatrix} \begin{bmatrix} \omega_L \\ \omega_R \end{bmatrix} \tag{3}$$

As can be seen from the above equation, the drilling and anchoring robot consists of three variables, whereas the control inputs have only two, making it a typical nonholonomic system.

3.2. Steering Curvature Analysis of the Drilling and Anchoring Robot

To eliminate the influence of the drilling and anchoring robot's own width on the steering performance, the concept of relative steering radius is introduced, defined as

$$\rho = R/B \tag{4}$$

where R represents the track steering radius, and B represents the track center distance.

When ω_L and ω_R are constant, the steering curvature κ of the robot is given by:

$$\kappa \le \frac{\dot{\theta}}{v} = \frac{2(v_R - v_L)}{B(v_R + v_L)} = \frac{2(\omega_R - \omega_L)}{B(\omega_R + \omega_L)} \tag{5}$$

Because the drilling and anchoring robot's travel speed ranged from 0 to 30 m/min, four different values of v_L (0, 10, 20, and 30 m/min) were chosen. Simultaneously, the v_R range was set to [−30, 30] m/min. A simulation analysis was conducted to examine the changes in the steering curvature and travel speed of the drilling and anchoring robots under these four scenarios, as illustrated in the following Figure 4.

From the simulation results, it can be concluded that the greater the difference between the speeds of the left and right track wheels, the larger is the steering curvature, indicating a smaller turning radius for the drilling and anchoring robot. Conversely, when the speeds of the left and right track wheels were closer, the steering curvature was smaller and the steering of the drilling and anchoring robot was smoother. When the drilling and anchoring robot traveled at the maximum speed ($v = v_{\max}$), the steering curvature of the robot was 0. At this point, the drilling and anchoring robot could only move along a straight line.

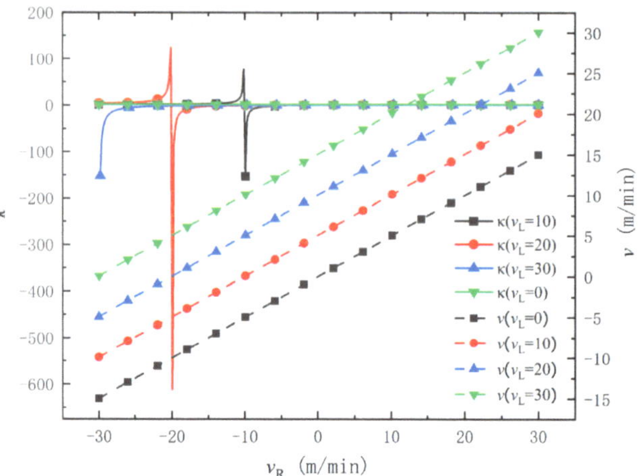

Figure 4. The relationship curve between the robot's steering curvature κ, velocity v, and the speeds of the left track v_L and right track v_R ($v_L = 0, 10, 20, 30$).

3.3. Analysis of Skid Steering in Drilling and Anchoring Robots

The walking mechanism of the drilling and anchoring robot is a tracked structure that is always accompanied by skidding and slipping phenomena during movement. When describing the characteristics of a robot's track skidding and slipping, it is assumed that the tracks at all points cannot be stretched, meaning that no deformation occurs. The motion of the drilling and anchoring robot was analyzed based on the ground mechanics model developed by Wong [21]. i represents the state of track skidding and slipping, i_s represents the skid rate, and i_t represents the slip rate. The following relationship can be obtained:

$$i = \begin{cases} \frac{\omega_R(r_1+h)-v}{\omega_R(r_1+h)} = (1 - \frac{r_e}{r_1+h}) \times 100\% = i_s, if\, r_1 \geq r_e \\ \frac{v-\omega_R(r_1+h)}{v} = (1 - \frac{r_1+h}{r_e}) \times 100\% = i_t, if\, r_1 \leq r_e \end{cases} \tag{6}$$

where r_e represents the effective rolling radius of the drive wheel of the track.

In this equation, the skidding and slipping ratios can be used for the entire motion process of different drilling and anchoring robots, including acceleration or braking. Because the drilling and anchoring robot travels at a slow speed in tunnel environments, only the slip situation of the track is considered in this study ($i = i_s$). Therefore, the actual traveling speed and relative steering radius of the drilling and anchoring robots can be determined using the following equations:

$$\begin{cases} v = \frac{1}{2}[\omega_R(r_1 + h)(1 - i_R) + \omega_L(r_1 + h)(1 - i_L)] \\ \rho = \frac{\omega_R(1-i_R)+\omega_L(1-i_L)}{\omega_R(1-i_R)-\omega_L(1-i_L)} \end{cases} \tag{7}$$

where i_L, i_R represent the slip rates of the left and right tracks, respectively.

From Equation (7), it is evident that when the left and right track speeds of the robot are limited, the robot achieves its maximum speed when $i_R = i_L = 0$. As the speed of the drilling and anchoring robot varied from 0 to 30 m/min, the analysis was conducted for cases where the linear speed during the steering process was set to 0, 5, 15, and 20 m/min. The relationship between the drive wheel speed and turning radius of the drilling and anchoring robot is shown in Figure 5 for both scenarios: without considering ($i = 0$) and considering ($i = 0.3$) the track slip factor.

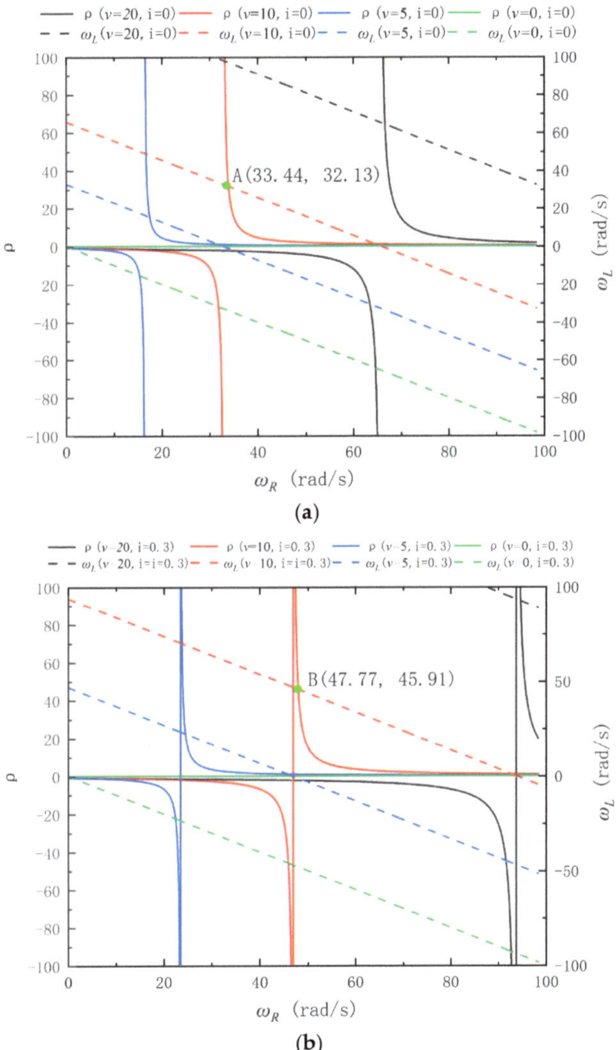

Figure 5. The relationship graph between the robot's turning radius ρ, driving speed v, slippage ratio i, and the speeds of the left track v_L and right track v_R: (**a**) $i = 0$, (**b**) $i = 0.3$.

From Figure 5, it can be observed that considering or not considering the effects of track skidding and slipping causes significant differences in the control requirements for the speed of the drilling and anchoring robots. By controlling the speed difference between the robot's left and right tracks, the robot can achieve a specified turning radius during its steering motion, based on the state of track skidding and slipping and the required driving speed. When the robot's speed remains constant and completes the same radius turning motion, the speeds of the tracks on both sides of the robot are much higher when considering the track's slipping than when not considering it. For instance, at points A and B in the graph, where the robot's speed is 10 m/min and it completes a steering motion with $\rho = 50$, without considering track skidding and slipping ($i = 0$), the drive-wheel speeds of the drilling and anchoring robot are A(33.44, 32.13) rad/s, whereas when considering track skidding and slipping ($i = 0.3$), the drive-wheel speeds of the drilling and anchoring robot are B(47.77, 45.91) rad/s.

4. Analysis of the Deviation Correction Control Strategy for the Drilling and Anchoring Robot

4.1. Analysis of the Steering Space of the Robot

An analysis of the steering motion of the drilling and anchoring robots was conducted to understand the relationship between the driving motion of the drilling and anchoring robots and their spatial environment. The principle diagram of the steering motion space of the drilling and anchoring robots is shown in Figure 6. ICR represents the instantaneous center of rotation of the steering motion of the robot. R_{max} denotes the radius of the largest arc space occupied by the robot during steering and R_{min} represents the radius of the smallest arc space occupied by the robot during steering. During the steering motion of the drilling and anchoring robot, the minimum travel space required was the area enclosed by segments R_{max} and R_{min}.

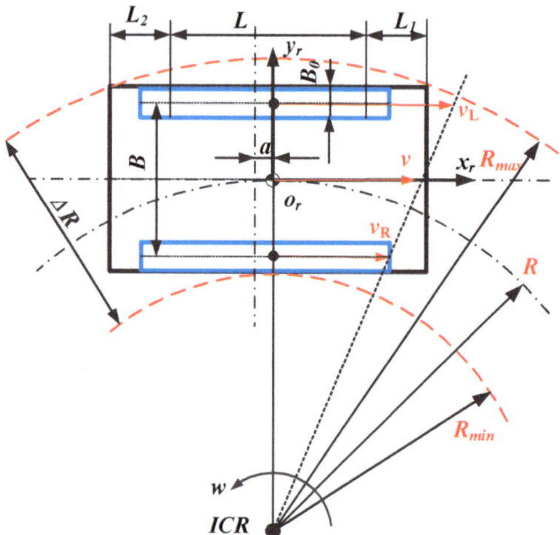

Figure 6. Analysis of the steering space drilling and anchoring robot.

The formulas for calculating the minimum and maximum steering radii of the robot are as follows.

$$\begin{cases} R_s = \frac{B}{k-1} - \frac{B_0}{2} \\ R_m = \sqrt{\left(\frac{B}{k-1} + \frac{B_0}{2} + B\right)^2 + (L/2 + L_1 + a)^2} \end{cases} \tag{8}$$

When $R_s = 0$, the minimum relative steering radius required for the drilling and anchoring robot is

$$\rho_0 = \frac{B_0/2 + B/2}{B} = \frac{1}{2}\left(\frac{B_0}{B} + 1\right) \tag{9}$$

4.2. Relationship between the Steering Angle of the Robot and the Distance to the Sidewall

Influenced by factors such as the width of the tunnel, equipment, and distance from the sidewall, the steering capability of the drilling and anchoring robot is closely related not only to its intrinsic structural parameters and driving parameters, but also to the distance from the tunnel sidewall. Therefore, when traveling in a tunnel, drilling and anchoring robots cannot steer freely. Based on an analysis of the structural dimensions of the drilling and anchoring robot and the tunnel parameters, as shown in Figure 7a, the relationship between the steering angle of the robot and its distance from the sidewall coal seam can be obtained.

$$
\begin{cases}
R_1 = \sqrt{(L + 2L_2 + 2a)^2 + (B + B_0)^2}/2 \\
Y_R = R_1 \cos(\alpha - \theta) \\
\tan \alpha = \frac{L + 2L_2 + 2a}{B + B_0}
\end{cases}
\tag{10}
$$

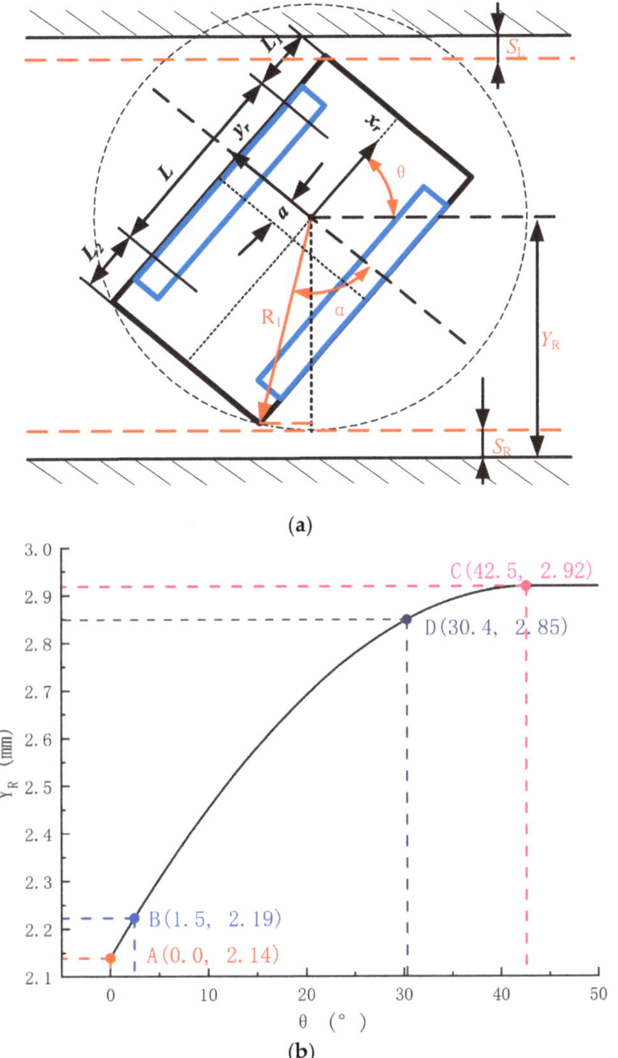

Figure 7. Relationship between the maximum steering angle of the robot and the distance to the sidewall: (**a**) analysis graph of robot steering capability, (**b**) robot steering angle θ vs. sidewall distance Y_R.

Substituting the parameters related to Table 1 into the above equation, the parameters determining the structure of the drilling and anchoring robot are $\alpha = 42.5°$ and $R_1 = 2.92$ m, respectively. When the robot was at a distance $Y_R = 2.14$ m from the right sidewall, at this point, $\theta = 0°$, the robot could not turn and could only move forward, as indicated by point A in Figure 7b. When considering the robot's safe distance, at this point where Y_R = 2.19 m, the robot's steering angle is $\theta = 1.5°$, as shown by point B in Figure 7b. During $|Y_R| > 2.92$ m, the robot can turn freely without being limited by obstacles, as shown by point C in Figure 7b. Assuming that the centerline of the drilling and anchoring robot coincides with the centerline of the tunnel, with a tunnel width of 5.8 m, the maximum distance between the robot and the sidewall will be $Y_R = 2.9$ m < = 2.92 m, and the robot will not be able to achieve a 360° pivot turn. Considering a safety distance of $S_R = 0.05$ m between the robot and the tunnel wall during travel, the drilling and anchoring robot will be able to adapt to a tunnel width of 5.7 m, and the maximum steering angle for the drilling and anchoring robot is $\theta = 30.4°$, as indicated by point D in Figure 7b.

4.3. Correction and Steering Control Strategy for the Drilling and Anchoring Robot

To ensure safe and smooth operation of the drilling and anchoring robot in the tunnel, the structure of the drilling and anchoring robot must not collide with the sidewalls of the tunnel. For ease of analysis, let A, B, C, and D represent the vertices of the drilling and anchoring robots, respectively. The coordinates of these four vertices relative to the robot's body coordinate system can be obtained using geometric relationships [22–24]. If the coordinates of the robot's center of mass in the tunnel coordinate system are $P = (x,y,z)$ and the coordinates of point A in the robot's coordinate system are $P_{Ar} = (\frac{L}{2} + \frac{L_{01}}{2}, -\frac{B}{2} - \frac{B_0}{2}, \theta)^T$, then the coordinates of point A in the xy-plane coordinate system of the tunnel can be calculated as follows:

$$P_A = \begin{cases} x + (\frac{L}{2} + \frac{L_{01}}{2} - a)\cos\theta + (\frac{B}{2} + \frac{B_0}{2})\sin\theta \\ y + (\frac{L}{2} + \frac{L_{01}}{2} - a)\sin\theta - (\frac{B}{2} + \frac{B_0}{2})\cos\theta \end{cases} \tag{11}$$

The coordinates of points B, C, and D on the drilling and anchoring robot can be obtained in the same manner.

$$P_B = \begin{cases} x + (\frac{L}{2} + \frac{L_{01}}{2} - a)\cos\theta - (\frac{B}{2} + \frac{B_0}{2})\sin\theta \\ y + (\frac{L}{2} + \frac{L_{01}}{2} - a)\sin\theta + (\frac{B}{2} + \frac{B_0}{2})\cos\theta \end{cases}$$

$$P_C = \begin{cases} x + -(\frac{L}{2} + \frac{L_{01}}{2} - a)\cos\theta - (\frac{B}{2} + \frac{B_0}{2})\sin\theta \\ y - (\frac{L}{2} + \frac{L_{01}}{2} - a)\sin\theta + (\frac{B}{2} + \frac{B_0}{2})\cos\theta \end{cases}$$

$$P_D = \begin{cases} -(\frac{L}{2} + \frac{L_{01}}{2} + a)\cos\theta + (\frac{B}{2} + \frac{B_0}{2})\sin\theta \\ -(\frac{L}{2} + \frac{L_{01}}{2} + a)\sin\theta - (\frac{B}{2} + \frac{B_0}{2})\cos\theta \end{cases}$$

To ensure that the drilling and anchoring robot does not collide with the sidewalls of the tunnel, it is necessary to restrict the positions of the four vertices of the drilling and anchoring robot. Let the width of the tunnel be W, the safe distance of the robot from the left side be S_L, and the safe distance of the drilling and anchoring robot from the right side be S_R, as illustrated in Figure 8. Then, the vertical coordinate distances of points A, B, C, and D from the centerline have the following constraints.

$$\begin{cases} S_{Rmin} = \min(|y_A|, |y_D|) < \frac{W}{2} - S_R \\ S_{Lmin} = \min(|y_B|, |y_D|) < \frac{W}{2} - S_L \end{cases} \tag{12}$$

Based on the above relationships, the following collision avoidance driving control strategy can be formulated for drilling and anchoring robots:

1. During $|S_{Rmin} - S_{Lmin}| < s_a$, the drilling and anchoring robot maintains its current state and does not require correction. s_a represents the permissible error for robot navigation, which must be set based on the tunnel environment, typically defaulting to 0.01 m.

2. When $|S_{R\min} - S_{L\min}| > s_a$ and $S_{R\min} < S_{L\min}$, the drilling and anchoring robot deviates towards the right side of the tunnel, requiring leftward correction.

3. When $|S_{R\min} - S_{L\min}| > s_a$ and $S_{R\min} > S_{L\min}$, the drilling and anchoring robot deviates towards the left side of the tunnel, requiring rightward correction.

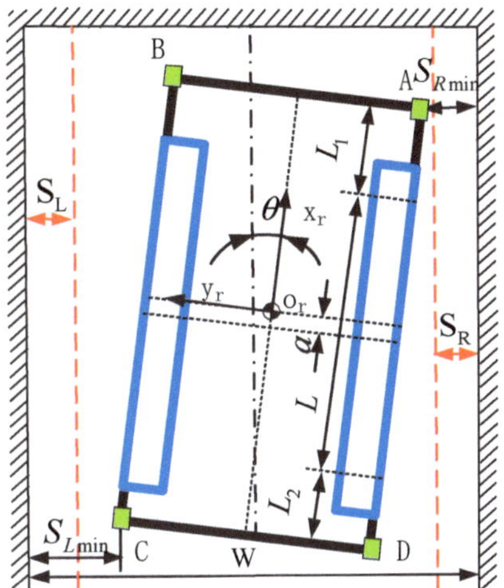

Figure 8. Relationship between robot steering angle and distances from points A, B, C, and D to the sidewall.

5. Drilling and Anchoring Robot Path Tracking Control Test Analysis

5.1. Anchoring Robot Deviation Correction Path Planning in a Tunnel Environment

To verify the effectiveness of the deviation correction path planning of the anchoring robot in a tunnel environment, we formulated the following requirements for steering and driving.

The steering angle of the robot during turning must be limited within a reasonable range, $\theta < 30.4°$.

The starting and ending angles should align with the direction of the tunnel, $\theta = 0°$.

Based on the above requirements, we devised a smooth path denoted as a-b-o-c-d, as shown in Figure 9. In this path, segment a-b is a straight line, b-o represents the left-turn path with a turning radius of R_1, and o-c represents the right-turn path with a turning radius of R_2. When the CM of the drilling and anchoring robot coincides with the centerline of the tunnel, the robot travels in a straight line, as illustrated by curves c-d in the diagram. The entire adjustment curve b–c represents the displacement segment for robot adjustment. Assuming that the driving speed of the drilling and anchoring robot is v, and the steering angular velocity is ω, the equations for the path curves of the robot during each stage of travel are as follows:

a-b:

$$\begin{cases} x = vt \\ y = 0 \end{cases}, t < t_1$$

b-o:

$$\begin{cases} x = x_{AB} + R_1 \cos(-\frac{\pi}{2} + \omega(t - t_1)) \\ y = R_1 + R_1 \sin(-\frac{\pi}{2} + \omega(t - t_1)) \end{cases}, t < t_2$$

o-c:

$$\begin{cases} x = x_{AB} + x_{BO} + R_1 \cos(\frac{\pi}{2} + \omega(t_2 - t_1) + \omega(t - t_2)) - R_1 \cos(\frac{\pi}{2} + \omega(t_2 - t_1)) \\ y = y_{AB} + y_{BO} + R_1 \sin(\frac{\pi}{2} + \omega(t_2 - t_1) + \omega(t - t_2)) - R_1 \sin(\frac{\pi}{2} + \omega(t_2 - t_1)) \end{cases}, t_2 \leq t < t_3$$

c-d:

$$\begin{cases} x = x_{AB} + x_{BO} + x_{OC} - R_2 \cos(\frac{\pi}{2} + \omega(t_3 - t_2)) \\ y = y_{AB} + y_{BO} + y_{OC} - R_1 \sin(\frac{\pi}{2} + \omega(t_3 - t_2)) + R_2 \end{cases} , \; t \geq t_3$$

When the drilling and anchoring robots are traveling in a tunnel environment, with distances Y_R and Y_L representing the distances to the right and left sidewalls, respectively, the steering angle is θ. During $R_1 = R_2$, the steering parameters of the robot can be determined as follows:

$$\begin{cases} Y_R = Y_L \in [2.19, 2.8] \\ d_y = \frac{W}{2} - Y_R \in [0, 0.56] \\ R_1 = R_2 = \frac{180B}{\pi\theta} \end{cases} \tag{13}$$

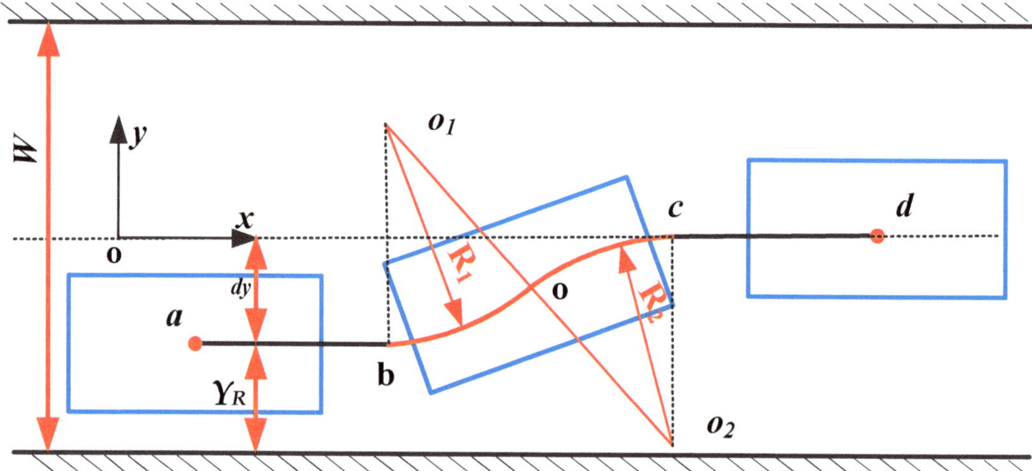

Figure 9. Path planning for robot correction in a tunnel environment.

5.2. Design of the Robot Kinematic Controller

In the tunnel coordinate system, the formula for calculating the pose error of the drilling and anchoring robots is as follows:

$$\zeta_e^h = \left\{ \begin{array}{c} x_e^h \\ y_e^h \\ \theta_e^h \end{array} \right\} = \left\{ \begin{array}{c} x_r - x \\ y_r - y \\ \theta_r - \theta \end{array} \right\} \tag{14}$$

The formula for transforming the pose error of the drilling and anchoring robot from the global coordinate system ζ_e^h to the local coordinate system ζ_e is as follows.

$$\zeta_e = R^{-1}(\theta)\zeta_e^h = \begin{bmatrix} \cos\theta & -\sin\theta & 0 \\ \sin\theta & \cos\theta & 0 \\ 0 & 0 & 1 \end{bmatrix} \left\{ \begin{array}{c} x_r - x \\ y_r - y \\ \theta_r - \theta \end{array} \right\} \tag{15}$$

The differential equation for the pose error of the center of mass of the drilling and anchoring robot is

$$\dot{\zeta}_e \doteq \left\{ \begin{array}{c} \dot{x}_e \\ \dot{y}_e \\ \dot{\theta}_e \end{array} \right\} = \begin{bmatrix} \omega y_e - v + v_r \cos\theta_e - \omega_r d \sin\theta_e \\ -\omega y_e - \omega d + v_r \sin\theta_e - \omega_r d \cos\theta_e \\ \omega_r - \omega \end{bmatrix} \tag{16}$$

Due to the elongated nature of the coal mine's tunneling roadway, this paper designs a PID controller using the difference in the y-direction (*ey*) between the drilling and bolting robot and the roadway centerline. The PID controller is as follows:

$$u_y = K_p y_e + K_i \int y_e dt + K_d \frac{dy_e}{dt} \tag{17}$$

The control inputs are the linear velocity v and angular velocity ω. To relate u_y to these inputs, we need to find a control law for v and ω that minimizes y_e. One approach is to assume that v is constant and adjust ω to minimize y_e.

Choose a Lyapunov function:

$$V = \frac{1}{2} y_e^2 > 0 \tag{18}$$

Compute the time derivative of the Lyapunov function:

$$\dot{V} = y_e \dot{y}_e \tag{19}$$

Substitute $\dot{y}_e$ into the equation:

$$\dot{y}_e = -\omega y_e - \omega d + v_r \sin \theta_e - \omega_r d \cos \theta_e \tag{20}$$

Considering the PID controller input,

$$\omega = K_{p\omega} y_e + K_{i\omega} \int y_e dt + K_{d\omega} \frac{dy_e}{dt} \tag{21}$$

Assuming the linear velocity v is constant and equal to v_r, the error dynamics simplifies to

$$\dot{y}_e = -\omega y_e - \omega d + v \sin \theta_e - \omega_r d \cos \theta_e \tag{22}$$

Approximate for small errors:

$$\dot{y}_e \approx -(K_{p\omega} y_e + K_{i\omega} \int y_e dt + K_{d\omega} \frac{dy_e}{dt}) y_e \tag{23}$$

Calculate the time derivative of the Lyapunov function:

$$\dot{V} = -K_{p\omega} y_e^2 + K_{i\omega} y_e \int y_e dt + K_{d\omega} y_e \frac{dy_e}{dt} \tag{24}$$

Since $K_{p\omega}$, $K_{i\omega}$, and $K_{d\omega}$ are positive constants,

$$\dot{V} \leq -K_{p\omega} y_e^2 \leq 0 \tag{25}$$

Through the analysis of the Lyapunov function $V = \frac{1}{2} y_e^2$ and its time derivative, it is proven that when $K_{p\omega}$, $K_{i\omega}$, and $K_{d\omega}$ are appropriately chosen, the derivative of the Lyapunov function $\dot{V}$ is always negative or zero. This indicates that the system is stable, meaning that the Y-direction error y_e gradually decreases over time and eventually approaches zero. This proves the stability of the PID controller in controlling the Y-direction error of the drilling and anchoring robot.

In this study, a control system model of the robot is constructed in MATLAB/Simulink, and a PID kinematic control model is designed, as shown in Figure 10.

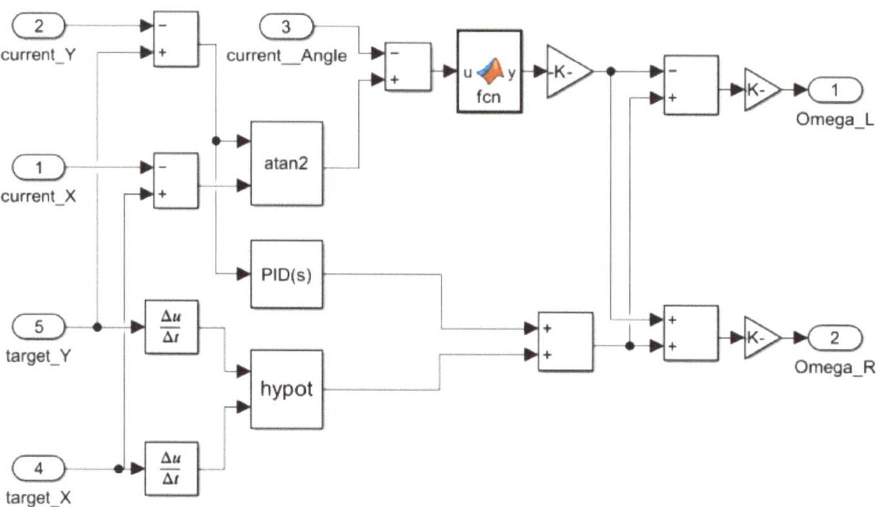

Figure 10. The robot path tracking PID kinematic controller.

5.3. Construction of the Robot Path Tracking Control System

Utilizing the kinematic control system of the drilling and anchoring robot, a simulation model of the robot path-tracking control system was built to validate the feasibility of the method. Two scenarios were considered in the path-planning $d_y = \{0.1, 0.5\}$ m. The straight-line distance traveled on the driving curve was 1 m, and the turning radius was $R = \{0.57, 6.6\}$ m. In the kinematic model, the maximum driving speed of the robot was v = 30 m/min, and the rotation speeds of the left and right wheels of the drilling and anchoring robot were restricted to $[-1.64, 1.64]$ rad/s. The block diagram of the drilling and anchoring robot is shown in Figure 11 and the simulink model of the robot path tracking control system is shown in Figure 12.

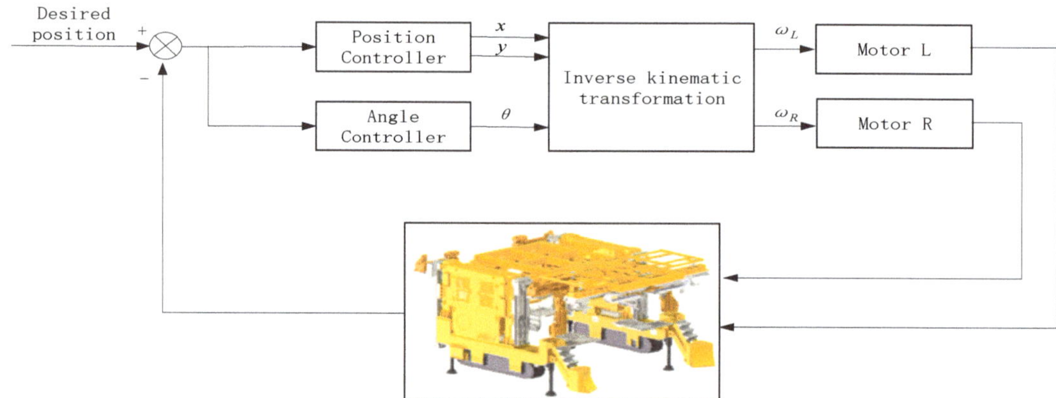

Figure 11. Block diagram of the robot control system.

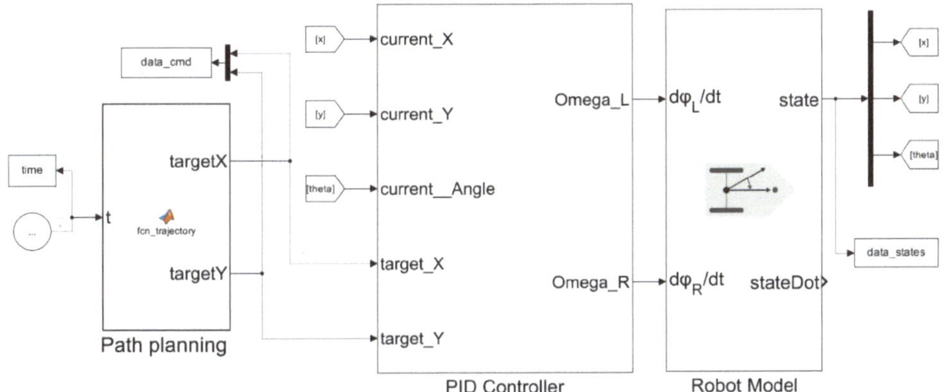

Figure 12. Simulink simulation model of the robot path tracking control system.

5.4. Analysis of Path Tracking Control Errors

From the simulation results shown in Figure 13, the condition of the robot being 0.1 m away from the center line of the roadway was first set. By simulation, the PID parameters were adjusted and set to {50, 10, 1} and {10, 1, 0.1}, with a simulation time of 60 s. The simulation results show that under these two PID parameter settings, the drilling and anchoring robots can accurately track the planned trajectory. It can also be found that when the PID controller parameters are smaller, such as {10, 1, 0.1}, the tracking accuracy of the robot is higher.

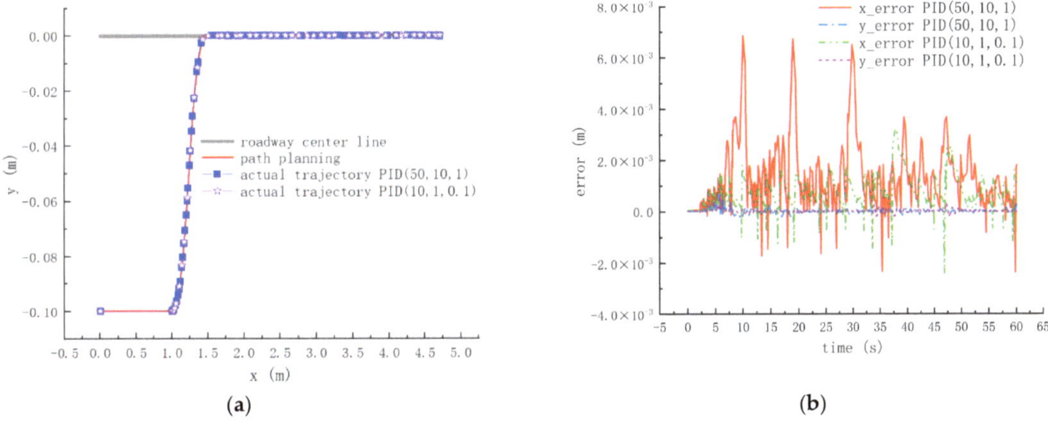

Figure 13. Robot path tracking control simulation under different PID parameters: (**a**) path tracking simulation under different PID parameters; (**b**) x-y path tracking error under different PID parameters.

To verify the path tracking control performance of the drilling and anchoring robot in the roadway space, the robot was set at distances of 0.1 m and 0.5 m from the center line of the roadway. After multiple simulation tests, the PID controller parameters were set to {0.1, 0.01, 0}. Additionally, since the deviation between the robot and the center line of the roadway became very small after 30 s of simulation, the simulation time was set to 30 s in this study. From the simulation results shown in Figure 14, the tracking errors in the x-direction from the tunnel centerline were less than 0.005 m, and those in the y-direction were less than 0.001 m, satisfying the precise control requirements of the robot in the tunnel environment. This indicated that the motion controller design was reasonable.

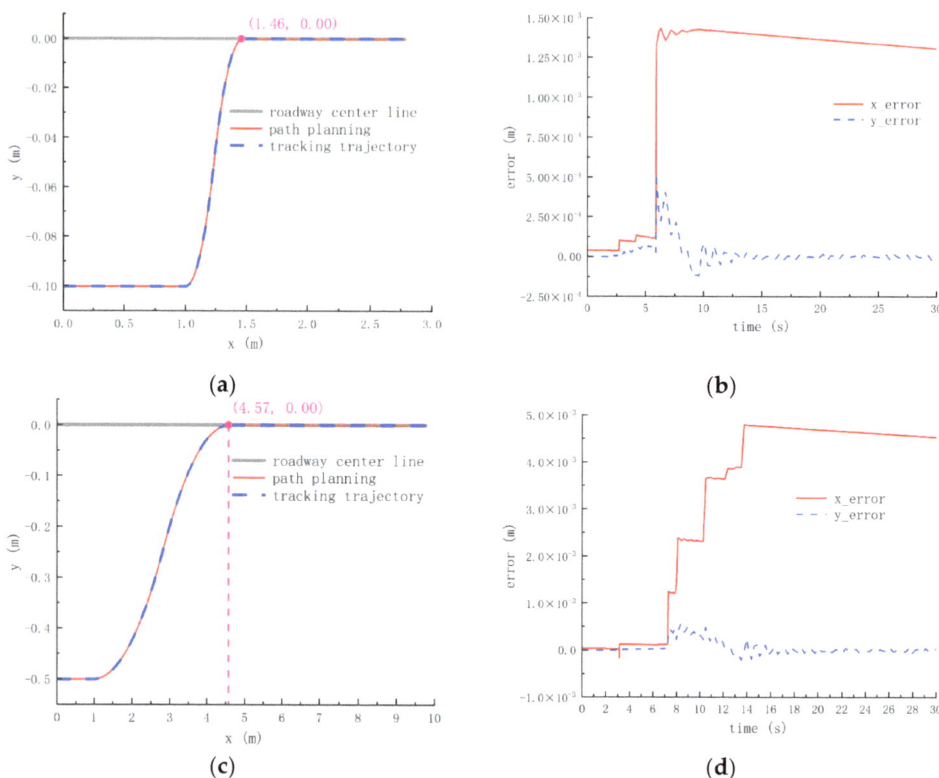

Figure 14. Path tracking control simulation of the robot PID {0.1, 0.01, 0}: (**a**) path tracking simulation ($dy = 0.1$), (**b**) x-y path tracking error ($dy = 0.1$), (**c**) path tracking simulation (($dy = 0.5$), (**d**) x-y path tracking error (($dy = 0.5$).

At the same time, it can be observed that the drilling and anchoring robot can only reach the centerline of the tunnel at distances of 1.46 m and 4.57 m, respectively. This suggests that in a constrained tunnel environment, the robot requires a longer distance to travel to the centerline when it is closer to the sidewall. Therefore, in tunnel environments, it is preferable for the robot to be closer to the centerline of the tunnel to ensure a safe and smooth operation.

6. Testing and Analysis of Lane Deviation Control in Tunnel Environments

An experimental platform was set up to verify the lane deviation performance of the drilling and anchoring robot, as shown in Figure 15. The left and right sidewalls are movable components that can be adjusted according to the experimental requirements to control their distance from the drilling and anchoring robots. In this experiment, the path shown in Figure 9 was followed. The width of the experimental site was adjusted to $W = 5.8$ m, with the robot's distance from the centerline of the lane set to $dy = 0.5$ m and $S_R = 0.05$ m. Lane deviation performance tests were conducted on the drilling and anchoring robot both with and without considering skidding and slipping.

The performance of the drilling and anchoring robot's steering control was tested by considering the skidding and slipping of the robot tracks ($i = 0$ and $i = 0.22$). The speed of the robot's driving wheel is limited to $[-1.64, 1.64]$ rad/s. The drive curves of the left and right track motors of the drilling and anchoring robots are shown in Figure 16a. As can be seen from Figure 16a, when skidding and slipping are not considered ($i = 0$), it takes

13.9 s for the robot to move to the center position of the roadway, while when skidding and slipping are considered ($i = 0.22$), it takes 64.4 s for the robot to move to the center position of the roadway. The change trend of the result is consistent with the analysis results of Equation (7) and Figure 5, indicating that the theoretical analysis is correct. The displacement curves of the drilling and anchoring robots are shown in Figure 16b. When considering track skidding and slipping, the speed difference between the two sides of the drilling and anchoring robot tracks was significant. At the same time, the real-time tracking error in the y direction from the centerline of the tunnel was relatively large, with a maximum error of 0.13 m. However, the drilling and anchoring robot still successfully moved from a position 0.5 m away from the centerline of the tunnel to the centerline, with an error of 0.003 m, achieving steering correction, as shown in Figure 16c. During the steering correction process, the speed of the drilling and anchoring robot is not the main target; rather, it determines whether the drilling and anchoring robot can move to the centerline of the tunnel. Therefore, the drilling and anchoring robot can achieve steering correction control in a tunnel environment, meeting the requirements for safe and smooth operation of the drilling and anchoring robot.

To calculate the distances between the four corner points (A, B, C, D) of the drilling anchor robot and the sidewalls, ultrasonic sensors have been installed on the robot. The location of the right-side ultrasonic sensor is shown in Figure 17, and the left-side sensor is installed in a symmetric position on the robot's body. The conversion relationship between the distances from the four corner points to the sidewalls, and the ultrasonic sensor measurement values, are expressed by the following equation.

$$\begin{cases} S_A = US_{R1}\cos\theta - UL_{R1}\sin\theta \\ S_B = US_{R2}\cos\theta + UL_{R2}\sin\theta \\ S_C = US_{L1}\cos\theta + L_{L1}\sin\theta \\ S_D = US_{L2}\cos\theta - UL_{L2}\sin\theta \end{cases} \tag{26}$$

where L_{R1} represents the distance from the right front ultrasonic sensor to the front end, L_{R2} represents the distance from the right rear ultrasonic sensor to the rear end, L_{L1} represents the distance from the left front ultrasonic sensor to the front end, and L_{L2} represents the distance from the left rear ultrasonic sensor to the rear end. US_{R1}, US_{R2}, US_{L1}, and US_{L2} represent the measurement values of the right front, right rear, left front, and left rear ultrasonic sensors, respectively. S_A, S_B, S_C, and S_D represent the distances between points A, B, C, D and the side skirts, respectively, and θ represents the yaw angle.

Figure 15. Drilling and anchoring robot test platform.

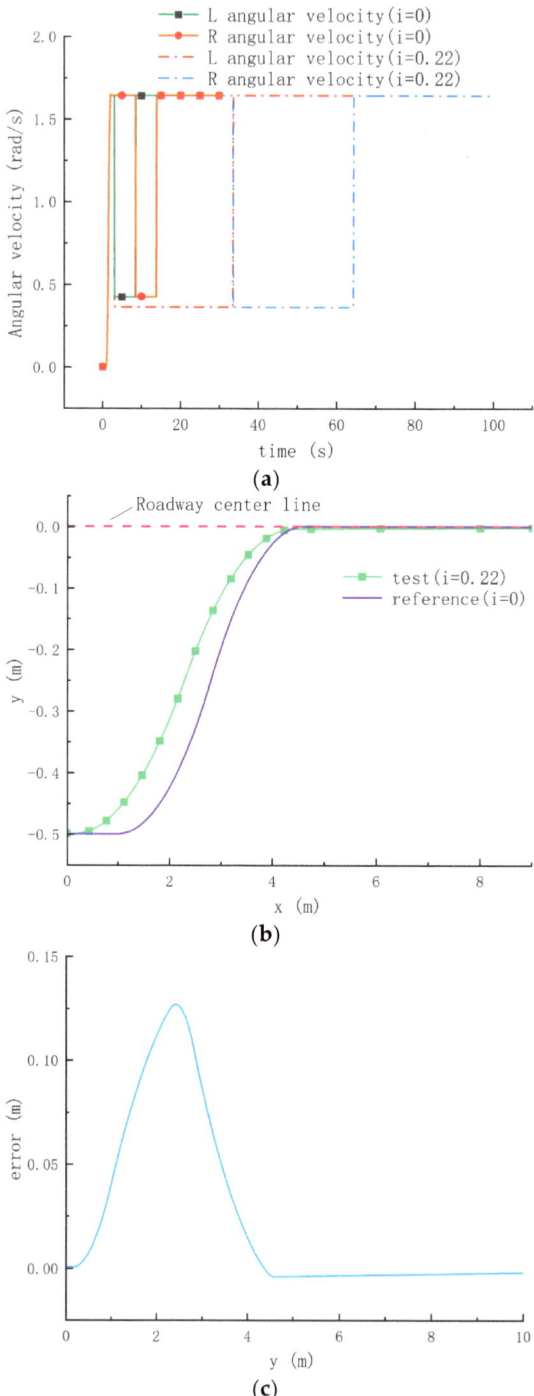

Figure 16. Curve plot of robot steering correction control: (**a**) drive wheel speed curve, (**b**) displacement curve, (**c**) real-time error y-direction.

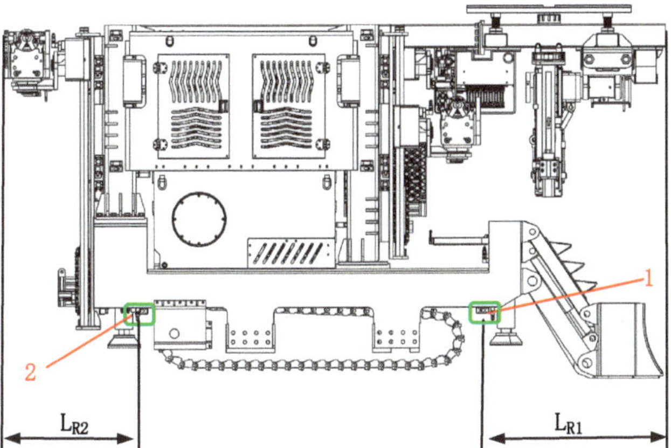

Figure 17. Schematic diagram of the installation position of the ultrasonic sensors (1 ultrasonic sensor US_{R1}, 2 ultrasonic sensor US_{R2}).

Based on the displacement monitoring of four corner points A, B, C, and D of the drilling and anchoring robot, as shown in Figure 8, the displacement curves of these four points are shown in Figure 18. From the figure, it can be observed that throughout the entire travel process, the drilling and anchoring robot maintains a minimum distance of 0.05 m from both sides of the roadway, with the nearest point being D1(23.7, 0.05) m. This distance exactly meets the requirement of the safety warning line, S_R = 0.05 m, thereby ensuring safe travel and smooth control of the drilling and anchoring robot.

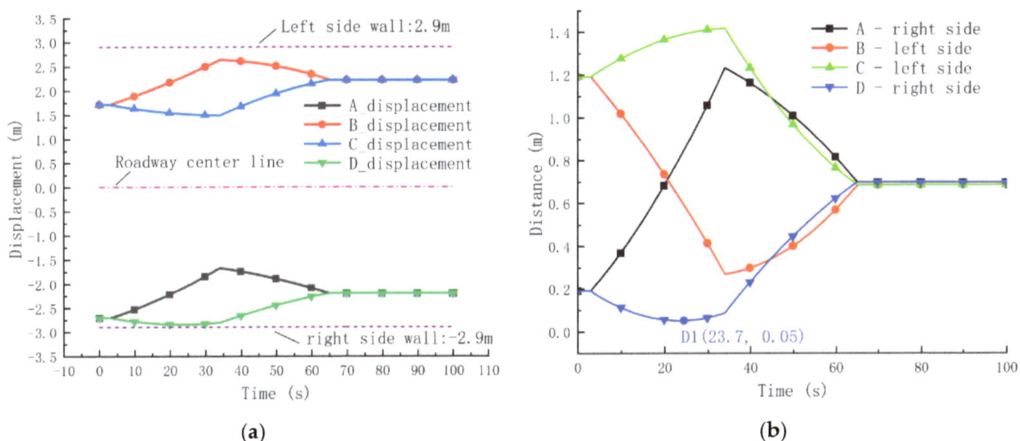

Figure 18. Curves of the displacement variation of the four corner points of the drilling and anchoring robot. (**a**) A, B, C, and D real-time displacement curves; (**b**) curves of the distance variation between points A, B, C, D and the sidewall.

7. Conclusions

In this paper, we investigated the slippage characteristics of a tracked drilling and anchoring robot during the steering process and proposed a steering control strategy that accounts for track slippage. A path tracking controller was designed, and its stability was verified through simulations and experiments. The general conclusions are as follows:

1. Steering Angle and Distance Relationship: The relationship between the steering angle of the drilling and anchoring robot and the distance to the sidewall in the roadway environment was determined. Based on this relationship, a corrective driving control strategy was formulated to enhance maneuverability and precision.

2. Path Planning and PID Controller: Path planning for the corrective driving of the drilling and anchoring robot in the roadway environment was completed. A PID kinematic controller was built, and path-tracking control simulation experiments demonstrated that the tracking error was minimal, indicating a well-designed control system.

3. Testing and Verification: A test platform for the corrective driving of the drilling and anchoring robot was established in a roadway environment. The performance of the corrective driving control was thoroughly tested, and the reliability of the proposed method was verified.

The results of this study provide a theoretical basis and technical support for the intelligent control of a tracked drilling and anchoring robot in tunnel environments, showcasing significant theoretical and engineering implications. Future research can focus on optimizing the control strategy to enhance performance under varying environmental conditions, as well as on real-world implementation and testing in diverse tunnel environments to validate the robustness and scalability of the proposed control strategy.

Author Contributions: Conceptualization, C.W. and H.M.; methodology, C.W., Q.M. and X.X.; software, C.W.; validation, C.W., J.S. and R.W. resources, J.S., R.W. and Q.L.; Manufacturing J.S. and R.W.; test, C.W., J.S., R.W. and Q.L.; writing—original draft preparation, C.W.; writing—review and editing, X.X. and Q.M. All authors have read and agreed to the published version of the manuscript.

Funding: This work was supported in part by the National Key Research Development Program of China under grants 2023YFC2907603 and 2022YFF0605300, the National Natural Science Foundation of China under grant 52374161, the Shaanxi Science and Technology Association under grant 2023-JC-YB-331, the Key Research and Development Projects of Shaanxi Province under grant 2023-LL-QY-03, and the Shaanxi Provincial Department of Education to Serve Local Special Program Projects under grant 22JC051.

Data Availability Statement: The data presented in this study are available upon request from the first author.

Conflicts of Interest: The authors declare that they have no known competing financial interests or personal relationships that could have influenced the work reported in this study.

References

1. Ge, S.R.; Hu, E.Y.; Li, Y.W. New progress and direction of robot technology in coal mine. *J. China Coal Soc.* **2023**, *48*, 54–73.
2. Ma, H.; Wang, S.; Mao, Q.; Shi, Z.W.; Zhang, X.H.; Yang, Z.; Cao, X.; Xue, X.; Xia, J.; Wang, C.; et al. Key common technology of intelligent heading in coal mine roadway. *J. China Coal Soc.* **2021**, *46*, 310–320.
3. Hao, X.D.; Jing, X.P.; Zhang, Z.P. Workspace analysis and trajectory planning of drill arm of roboticized bolting truck. *J. Cent. South Univ. (Sci. Technol.)* **2019**, *50*, 2128–2137.
4. Ma, H.W.; Wang, P.; Wang, S.B.; Mao, Q.H.; Shi, Z.W.; Xia, J.; Yang, Z.; Xue, X.S.; Wang, C.W. Intelligent parallel cooperative control method of coal mine excavation robot system. *J. China Coal Soc.* **2021**, *46*, 2057–2067.
5. Ma, H.W.; Sun, S.Y.; Wang, C.W. Key technology of drilling anchor robot with multi-manipulator and multi-rig cooperation in the coal mine roadway. *J. China Coal Soc.* **2023**, *48*, 497–509.
6. Zou, T.; Angeles, J.; Hassani, F. Dynamic modeling and trajectory tracking control of unmanned tracked vehicle. *Robot. Auton. Syst.* **2018**, *110*, 102–111. [CrossRef]
7. Jia, W.; Liu, X.; Zhang, C.; Qiu, M.; Zhang, Y.; Quan, Q.; Sun, H. Research on Steering Stability of High-Speed Tracked Vehicles. *Math. Probl. Eng.* **2022**, *2022*, 4850104. [CrossRef]
8. Xiong, H.; Chen, Y.; Li, Y.; Zhu, H.; Yu, C.; Zhang, J. Dynamic model-based back-stepping control design for-trajectory tracking of seabed tracked vehicles. *J. Mech. Sci. Technol.* **2022**, *36*, 4221–4232. [CrossRef]
9. Qin, H.; Shao, S.; Wang, T.; Yu, X.; Jiang, Y.; Cao, Z. Review of Autonomous Path Planning Algorithms for Mobile Robots. *Drones.* **2023**, *7*, 211. [CrossRef]
10. Sabiha, A.D.; Kamel, M.A.; Said, E.; Hussein, W.M. ROS-based trajectory tracking control for autonomous tracked vehicle using optimized backstepping and sliding mode control. *Robot. Auton. Syst.* **2022**, *152*, 104058. [CrossRef]

11. Saglia, J.A.; Tsagarakis, N.G.; Dai, J.S.; Caldwell, D.G. Inverse-kinematics-based control of a redundantly actuated platform for rehabilitation. *Proc. Inst. Mech. Eng. Part I—J. Syst. Control Eng.* **2009**, *223*, 53–70. [CrossRef]
12. Fang, Y.; Wang, S.; Cui, D.; Bi, Q.; Yan, C. Multi-body dynamics model of crawler wall-climbing robot. *Proc. Inst. Mech. Eng. Part K J. Multi-Body Dyn.* **2022**, *236*, 535–553. [CrossRef]
13. Ishikawa, T.; Hamamoto, K.; Kogiso, K. Trajectory tracking switching control system for autonomous crawler dump under varying ground condition. *Autom. Constr.* **2023**, *148*, 104740. [CrossRef]
14. Zhang, M.J.; Cheng, R.; Zhu, Y.; Zhang, M.; Zang, F.; Wu, M. Study on road header rectification running performance and control in the in-cline coalmine roadway. *J. China Coal Soc.* **2021**, *46* (Suppl. S1), 549–557.
15. Qu, Y.Y.; Song, L.K.; Wu, M.; Ji, X.D. Study on path planning and tracking of the underground mining road-header. *J. Min. Sci. Technol.* **2020**, *5*, 194–202.
16. Zhang, X.H.; Zhao, J.X.; Yang, W.J.; Zhang, C. Vision-based navigation and directional heading control technologies of boom-type road header. *J. China Coal Soc.* **2021**, *46*, 2186–2196.
17. Mao, Q.; Zhang, F.; Zhang, X.; Xue, X.; Wang, L. Deviation Correction Path Planning Method of Full-Width Horizontal Axis Road header Based on Improved Particle Swarm Optimization Algorithm. *Math. Probl. Eng.* **2023**, *2023*, 3373873. [CrossRef]
18. Jia, Z.; Fang, W.; Sun, C.; Li, L. Control of a Path Following Cable Trench Caterpillar Robot Based on a Self-Coupling PD Algorithm. *Electronics* **2024**, *13*, 913. [CrossRef]
19. Stefek, A.; Van Pham, T.; Krivanek, V.; Pham, K.L. Energy Comparison of Controllers Used for a Differential Drive Wheeled Mobile Robot. *IEEE Access* **2020**, *8*, 170915–170927. [CrossRef]
20. Xu, T.; Ji, X.; Liu, Y.; Liu, Y. Differential Drive Based Yaw Stabilization Using MPC for Distributed-Drive Articulated Heavy Vehicle. *IEEE Access* **2020**, *8*, 104052–104062. [CrossRef]
21. Wong, J.Y.; Chiang, C.F. A general theory for skid steering of tracked vehicles on firm ground. *Proc. Inst. Mech. Eng. Part D J. Automob. Eng.* **2001**, *215*, 343–355. [CrossRef]
22. Yang, J.; Ge, S.; Wang, F.; Luo, W.; Zhang, Y.; Hu, X.; Zhu, T.; Wu, M. Parallel tunneling: Intelligent control and key technologies for tunneling, supporting and anchoring based on ACP theory. *J. China Coal Soc.* **2021**, *46*, 2100–2111.
23. Macenski, S.; Singh, S.; Martín, F.; Ginés, J. Regulated pure pursuit for robot path tracking. *Auton. Robot.* **2023**, *47*, 669–685. [CrossRef]
24. Ahn, J.; Shin, S.; Kim, M.; Park, J. Accurate path tracking by adjusting look-ahead point in pure pursuit method. *Int. J. Automot. Technol.* **2021**, *22*, 119–129. [CrossRef]

actuators

Article

Design of a Tripod LARMbot Arm

Marco Ceccarelli [1,*], **Steven Beaumont** [2] **and Matteo Russo** [1]

1   Department of Industrial Engineering, University of Rome Tor Vergata, 00133 Rome, Italy; matteo.russo@uniroma2.it
2   SeaTech-School of Engineering, University of Toulon, 83130 La Garde, France; steven.beaumont.24@seatech.fr
*   Correspondence: marco.ceccarelli@uniroma2.it

Abstract: A new design for humanoid arms is presented based on a tripod mechanism that is actuated by linear servomotors. A specific prototype is built and tested, with the results of performance characterization verifying a possible implementation on the LARMbot humanoid. The design solves the main requirements in terms of a high payload ratio with respect to arm weight by using a tripod architecture with parallel manipulator behavior. The built prototype is assembled with commercial components to match the expectations for low-cost user-oriented features. The test results show satisfactory operation characteristics both in motion and force performance, which will ensure a future successful implementation in the LARMbot humanoid structure.

Keywords: humanoid robots; design; tripod design; LARMbot humanoid; prototype; performance analysis

Citation: Ceccarelli, M.; Beaumont, S.; Russo, M. Design of a Tripod LARMbot Arm. *Actuators* **2024**, *13*, 211. https://doi.org/10.3390/act13060211

Academic Editor: Ioan Doroftei

Received: 15 April 2024
Revised: 30 May 2024
Accepted: 3 June 2024
Published: 5 June 2024

1. Introduction

Humanoid robots, as used in research and applications, including market solutions, are designed with a mechanical structure that is strongly inspired by human anatomy. Since the first humanoid robot, WABOT-1, was designed by Ikiro Katao in the early 1970s at Waseda University in Tokyo, Japan [1], the limbs have been conceived with bulky anthropomorphic-shaped structures with actuators and sensors. Examples of these humanoid robots include famous solutions with very successful designs such as ASIMO [2], WABIAN [3], HRP [4], Johnnie and Lola [5], HUBO [6], ATLAS [7], and BHR [8]. These robots represent different experiences with similarities in activities worldwide.

Today, challenges for humanoid robots can be still recognized in terms of increasing humanoid capabilities, such as walking on various unstructured terrains, preventing falls and/or including suitable protections [9–12], and advanced cognitive-based interaction with the environment and humans [13–17]. In current humanoids, the arms are designed with anthropomorphic structures [18–20], which, although replicating the characteristics and capabilities of the human arm, show limitations, especially in load capacity, due to the serial kinematic structure with revolute joints [21,22].

Humanoid solutions have been attempted and investigated, spanning from very sophisticated and complex features to very simple low-cost operation-limited capabilities. Indeed, there is a growing interest in the community towards practical solutions, mainly in service fields, focusing on solutions with low-cost user-oriented designs, even if they have limited capabilities. Among these research interests, the LARMbot humanoid has been developed with different conceptual designs since the beginning of the 2000s, and a prototype was developed in 2016 [23–25]. The LARMbot humanoid is characterized by a low-cost efficient design that is based on parallel mechanism architectures using market components and 3D-printed parts. The parallel mechanism architectures used in all the main body parts are aimed at providing good performance in motion and force capacity, inspired by the muscle–skeleton human anatomy, despite the limited characteristics of the components and 3D-printed parts [24]. In particular, the limb structure has been designed using the simplest parallel tripod architecture by replicating the functioning of the muscle–skeleton human anatomy with synergistic and antagonistic features [25]. However, experiences with the built prototypes have highlighted possibilities for improvement by revising the original designs and attempting to have common structures for the limbs.

In this paper, considering previous designs and experiences reported, for example, in [23–25], a new tripod-based mechanism is presented for a new humanoid arm based on a 3-SPR joint mechanism. The tripod structure mechanism is actually well known and applied in robotics, but mainly in industrial manipulators, such as Tricept made by various robot manufacturers. However, in the case of humanoid robots, it is a novelty because the structures of the arms of humanoid robots are designed with an anthropomorphic serial kinematic chain structure, with parallel kinematic chain structures using the actuating part as a parallel kinematic chain being rare. A redesigned solution is developed with features that can be used for both legs and arms by looking at the similarities of human muscle–skeleton limb structures, with the functioning based on the synergistic and antagonistic operation of the muscles. A prototype solution for a humanoid arm is built with the LARMbot characteristics of low-cost user-oriented design and operation by using market components and 3D-printed parts. The feasibility of the prototype is tested successfully, including a characterization of the operation performance of the tripod-based arm, whose features are well suited for an improvement of the humanoid LARMbot.

2. Materials and Methods

The conceptual design of a tripod mechanism for a humanoid arm is developed based on the requirements and motivations, which also helps in defining a practical solution that is well-suited for future improvements of the LARMbot humanoid. The materials considered for this development include market components and 3D-printed parts, which are fundamental aspects to achieve the expected low-cost user-oriented solution. The methods for design definition are based on the kinematic design of a tripod parallel mechanism and its CAD modeling, leading to a prototype construction that can be useful for implementation in a new LARMbot humanoid. These methods are also useful for the validation and characterization of the obtained new humanoid arm design.

2.1. Design and Operation Requirements

The requirements for the design and construction of a prototype for a new humanoid arm were defined considering the functionality of the previous humanoid prototype, as shown in Figure 1a, with particular reference to the leg and arm structures subsequently tested, as shown in Figure 1b,c. Furthermore, the requirements and peculiarities of the structures and operations for humanoid robots, shown in Figure 2, were considered for the implementation of the tripod structure and also for the humanoid arm, as per improvements in functional performance as well as structural compactness.

In particular, in Figure 1a [23,24], we want to emphasize how the humanoid robot LARMbot is characterized by mechanisms with a parallel structure in the torso, based on a parallel cable actuation mechanism. This determines a wide range of movement for the torso, approximately 20 deg in all directions, as well as a high load capacity of approximately 10 kg along with its own weight of 800 g. At the time LARMbot was conceived, this torso solution suggested the implementation of a parallel manipulator for the legs, which was designed as shown in Figure 1a, with a mechanism using three actuators non-converging at the ankle point with operating complexity that was overcome with the solution shown in Figure 1b [25] by adopting the tripod architecture. The positive experience with the structure shown in Figure 1b for the legs, which ensured an improvement in the range of movement of approximately 30 degrees in all directions with a load capacity similar to that of the torso, suggested adopting the tripod architecture also for the arm structure, but with a hybrid solution represented in Figure 1c [23,24]. The first experience of this new humanoid arm with the hybrid parallel–serial kinematic structure is based on the tripod part of the arm that creates the shoulder joint, resulting in a larger load capacity and robustness of the full arm. The serial part of the forearm with the elbow joint still showed limited functionality both in the range of movement and in the load capacity, affecting the entire structure of the humanoid arm so much as to suggest the implementation of a tripod mechanism for the entire arm when adequately sized, with dimensions well-proportioned to the humanoid's torso.

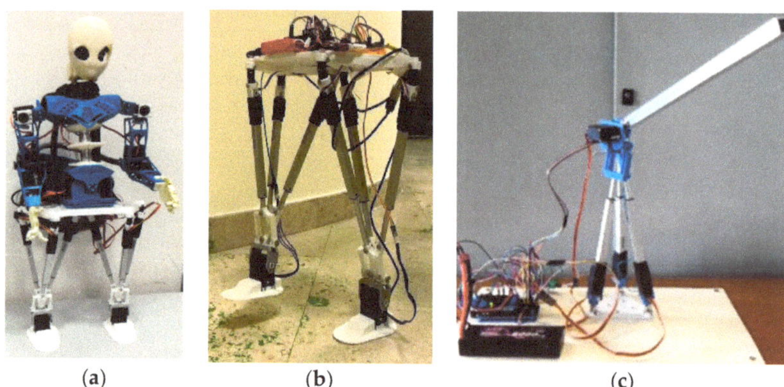

(a)   (b)   (c)

Figure 1. LARMbot humanoid: (**a**) 2018 full prototype [23,24]; (**b**) leg module prototype in 2020; (**c**) new arm prototype in 2021 [25].

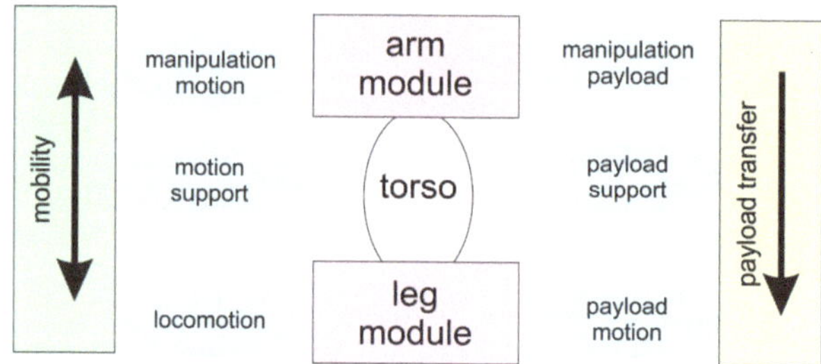

Figure 2. A scheme for design and operation requirements.

A revised consideration of the entire humanoid structure with its fundamental functional capabilities was necessary in order to define characteristic requirements for a new humanoid arm that will be integrated into the humanoid structure to ensure synergy and coordination typical for humanoid functioning. In Figure 2, the structural and functional characteristics are summarized by considering a modular structure of a humanoid with the subsystems of legs, torso, and arms that ensure the typical humanoid coordinated characteristics of manipulation and locomotion with adequate load capacity. In particular, it is indicated that the requirements and mobility characteristics of a humanoid can depend on the movement coordination of all three modules, requiring adequate proportionate design. Furthermore, it is emphasized that the load capacity of a humanoid depends on the three modules in a sequential manner; i.e., the load capacity of the arm system for manipulation depends on the support capacity of the torso and finally on the capability of the legs to support the entire structure loaded by the load. Therefore, the structure and functionality of a humanoid arm must be thought of and conditioned by the structural and functional characteristics according to the characteristics of the torso and the legs, as well as the function of the purposes designed for the entire structure.

In consideration of the summary presented in Figures 1 and 2, the requirements for designing a new humanoid arm with a tripod structure for the LARMbot humanoid can be summarized as follows: a length of approximately 50 cm, proportionate to the structure of the torso, with a mobility of approximately 30 degrees and a load capacity of approximately 5 kg, consistent with that ensured by the torso and legs.

2.2. Design Solution with Tripod Mechanism

The tripod structure of the new humanoid arm for LARMbot is presented in Figure 3. In particular, Figure 3a shows the conceptual design of the arm, including the wrist joint on which a two-fingered hand is installed. The scheme also shows the respective actuators m1 and m2, which are part of the load for the tripod mechanism that is operated by the three links with linear actuators L1, L2, and L3. The fixed platform creates an attachment plate on the torso structure as the shoulder frame, while the mobile platform is created with the plate on the convergence mechanism of the three links where the m1 wrist motor is installed. The shoulder frame is designed as the base frame of the tripod architecture with a specific symmetric configuration of an equilateral triangle to facilitate the assembly configuration as a function of the main expected performance direction. The functional kinematic design is represented by the model in Figure 3, which indicates the universal-type joints for the connections of the linear actuators to the plate of the shoulder platform, the linear joints for the actuated tripod links, and finally, the convergence joint of the three links with the original three-body mechanism connected by a system of five revolute pairs, as presented in [25].

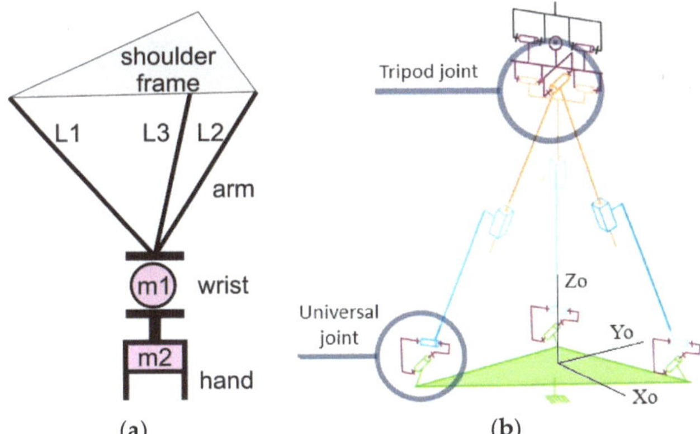

Figure 3. New design of humanoid arm based on tripod mechanism: (**a**) conceptual scheme; (**b**) kinematic design.

Figure 4 shows the CAD design developed for the realization of the kinematic design shown in Figure 3b, modeling commercial components for the universal joints and the linear actuators that create the tripod architecture, while the structural and support parts were modeled for manufacturing using lab 3D printing technology. These commercial components and 3D-printed parts have been selected and developed with the characteristics of low cost and easy use in both the assembly and operational functioning of the prototype. In particular, the geometries of the parts have been designed as a function of the dimensions that can ensure rigidity and solidity, as well as dimensional compactness, considering solutions that are easy to produce via 3D printer using commercial PLA.

2.3. Prototype Assembly

The CAD model shown in Figure 4 was also used to verify the structural and functional feasibility of the arm structure proposed in Figure 3 in order to produce the prototype shown in Figure 5a. Figure 4 shows the components used to build the prototype in Figure 5a using commercial products, and the parts were adequately produced using a 3D printer.

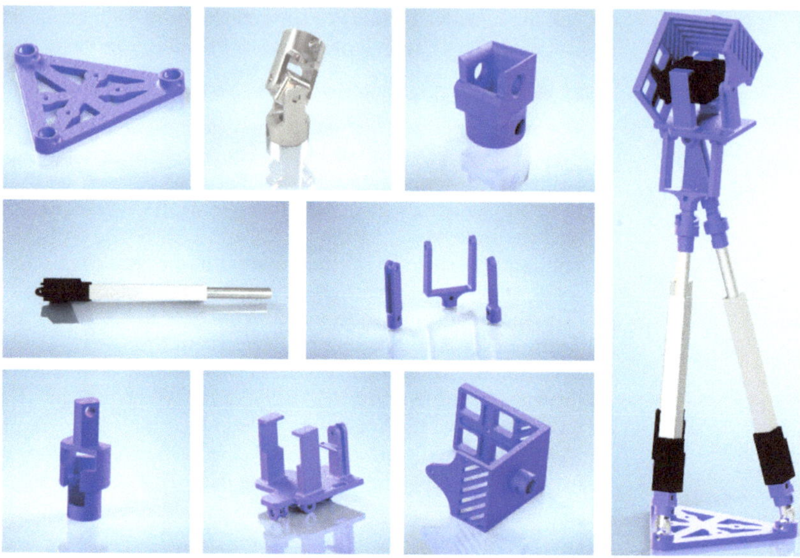

Figure 4. CAD design of the humanoid arm of the tripod arm shown in Figure 3 with 3D-printed parts and commercial linear actuators.

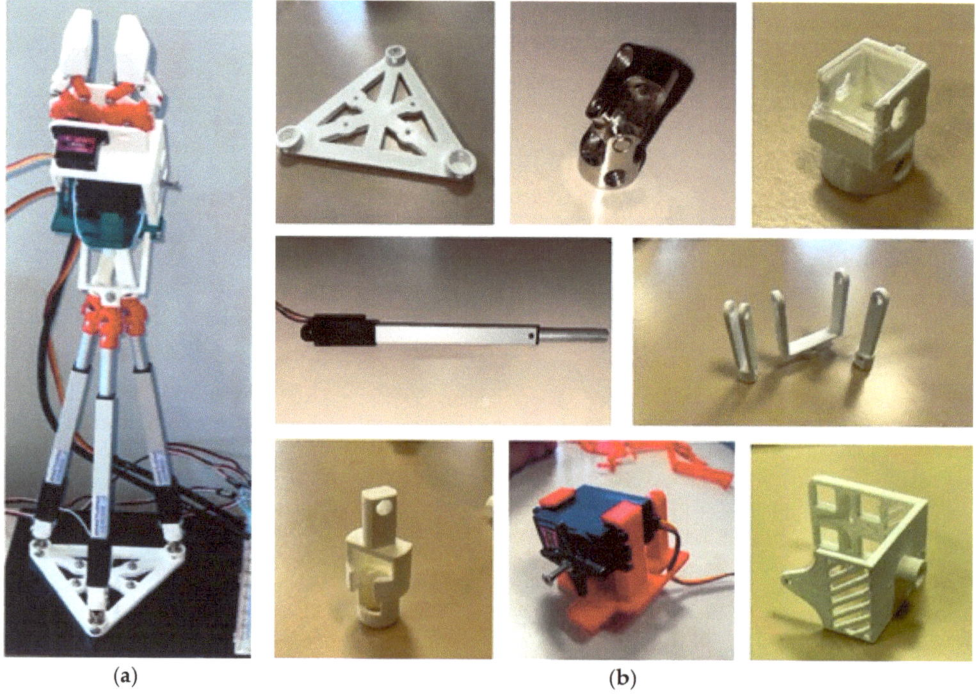

(a)　　　　　　　　　　　(b)

Figure 5. Prototype of the humanoid tripod arm of the arm shown in Figure 5: (**a**) full arm; (**b**) 3D-printed parts, and commercial linear actuators.

The prototype is characterized by a light structure with parts that are easily replaceable, and the assembly is based on simple connections fixed with small screws, facilitating both proper fixing and easy disassembly if necessary. The characteristic elements are the three linear actuators which create the tripod structure that is secured by the 3-axis converging mechanism. The mobile platform of the 3-axis converging houses the wrist actuator whose mobile frame is equipped with a two-finger gripper operated by a small servomotor.

The prototype thus constructed meets the requirements and expectations for a compact and efficient humanoid arm for LARMbot measuring approximately 40 cm in length, including the two-finger hand, with a shoulder attachment plate of $10 \times 10 \times 10 \text{ cm}^3$, for a total weight of approximately 300 g. The overall cost of the prototype can be estimated at less than EUR 400.

2.4. Modeling

The prototype can be moved by using a model-based algorithm that determines the position of the wrist joint from lengths $L1$, $L2$, and $L3$ of each linear actuator. To solve the forward kinematic problem with reference to the frame in Figure 3b, the position of the center of the tripod joint can be computed as the intersection of three spheres, each centered on one of the base universal joints (located at a distance of length a from the reference frame and equally spaced around the z-axis), with a radius equal to the length of the respective actuator:

$$\left(x + \frac{\sqrt{3}a}{2}\right)^2 + \left(y + \frac{a}{2}\right)^2 + z^2 = L_1{}^2 \tag{1}$$

$$\left(x - \frac{\sqrt{3}a}{2}\right)^2 + \left(y + \frac{a}{2}\right)^2 + z^2 = L_2{}^2 \tag{2}$$

$$x^2 + (y - a)^2 + z^2 = L_3{}^2 \tag{3}$$

These closed-form Equations (1)–(3) represent the kinematic problem of the system that has been solved to obtain the open-loop controlled motion in the desired paths for testing the prototypes' performance.

3. Results

Several laboratory tests were conducted to validate and characterize the structure and function of the arm prototype for the LARMbot humanoid. Tests were carried out to verify the ability to move and manipulate the hand and a load using the two-finger hand. The reported results refer to tests for the validation of the proposed design, including initial performance characterization. The tests were conducted to assess the maximum motion range with an open-loop controlled operation.

Figure 6 shows snapshots of the two experiments that highlight the satisfactory performance of the prototype in these modes. Figure 6a shows a manipulation test with no load on the hand, while Figure 6b shows a test with a load made of a piece of metal of about 200 g. The test results are shown in Figures 7–9, using the data acquired from the sensors appropriately installed to monitor the functionality of the prototype in terms of movement and power consumption. They are installed with a design that does not limit the arm motion and can also be used for future motion control.

An IMU sensor is installed on the wrist platform to monitor the motion in terms of angles and linear acceleration with respect to an internal reference system with the XY plane lying on the platform in the wrist. The angles are measured around the X- and Y-axes, while the linear accelerations are evaluated with respect to the reference axes. The IMU acquisitions are referred to as an XYZ reference system on the fixed platform, as shown in Figure 3b, with the home configuration labeled as XoYoZo.

A current acquisition sensor is used to evaluate power consumption to ensure the efficiency of future implementation on the LARMbot humanoid, considering the need to limit power usage according to the capacity of the humanoid's batteries to ensure its autonomy.

Figure 7 shows the movement results of the two tests in terms of the bending angles around the X- and Y-directions, which show a sufficiently gradual and regular movement with a maximum range of just over 60 deg. It should be noted that, in the case of manipulation with a load, as shown in Figure 7b, the characteristics of the angular motion are preserved even if a vibration is noted between 6.0 and 8.0 s, probably due to the movement of the load inside the hand. The range of motion of the built prototype reaches more than 60 deg, as indicated in the results shown in Figure 7, and is quite close to the theoretical possible motion range of the kinematic design.

The acceleration of the plots shown in Figure 8 refers to the acceleration sensed by the IMU that is installed on the wrist. Figure 8 shows the plots of the acceleration components acquired in the two tests, which confirm the fact that the load does not seem to significantly influence the characteristics of the motion since the time evolution and the numerical values are preserved, despite encountering a certain variation in the interval between 6.0 and 8.0 s. Considering the lack of excessive speed in the movement tested in cycles of approximately 5 s, the accelerations are modest in numerical value, showing an efficient smoothness of movement of the arm in both modes.

Finally, as shown in Figure 9, the acquisition of the actuation current for all the actuators of the tripod, including the wrist to keep it in a stationary position, shows a cyclical trend corresponding well to the tested movement of alternating inclination, especially along the Y direction as indicated in Figure 7, with a current variation limited to approximately a maximum of 350 mA. This trend of the power supply current, and therefore of power consumption, is not only acceptable for the actuators used but it is well-suited to the expected limited power consumption. The characteristics of the energy consumed by the prototype are investigated with the reported test results in Figure 9, in terms of servomotor feeding current at 12 V during open-loop controlled motion at the maximum range, swinging in all directions with and without a heavy payload. The good results can be represented by a maximum of 350 mA during motion with no payload against a maximum of 420 mA during motion with a payload.

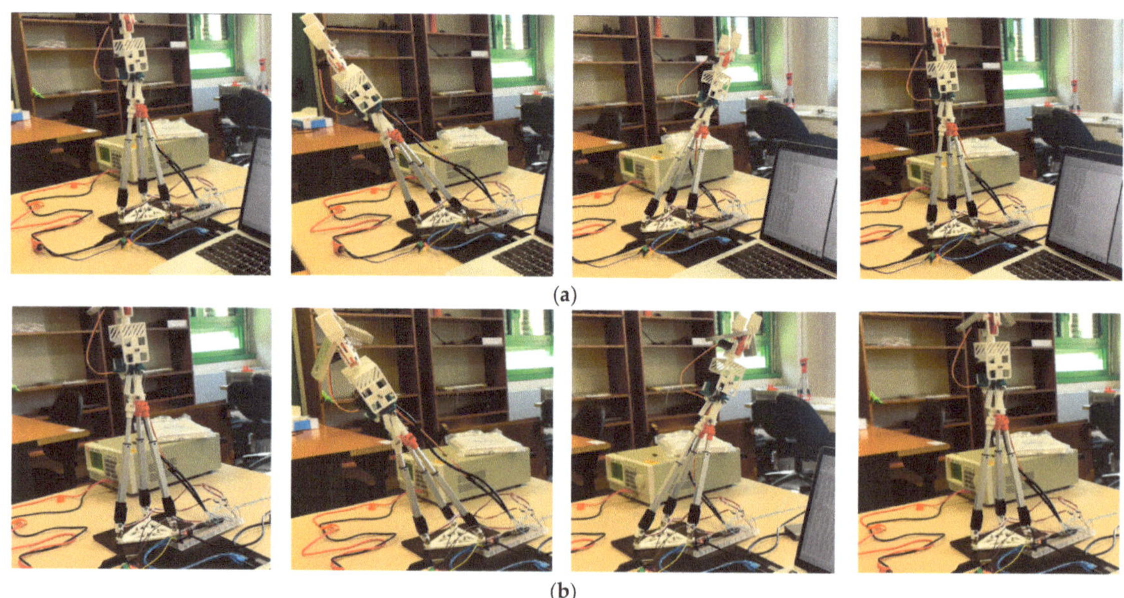

Figure 6. Snapshots of the test: (**a**) without payload; (**b**) with payload.

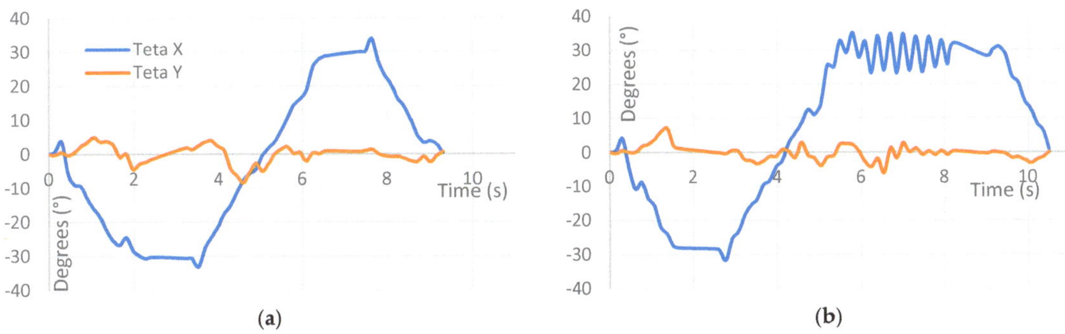

(a)   (b)

Figure 7. Acquired results during the test shown in Figure 6 in terms of angles: (**a**) without load; (**b**) with load.

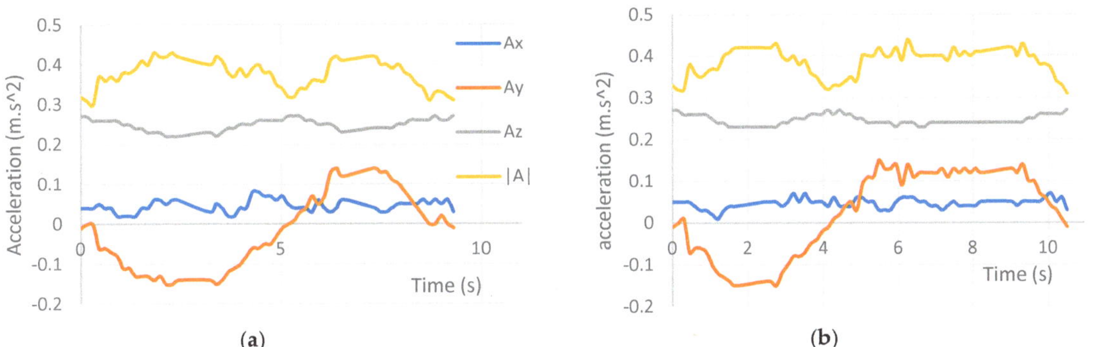

(a)   (b)

Figure 8. Acquired results during the test shown in Figure 6 in terms of acceleration of wrist point: (**a**) without load; (**b**) with load.

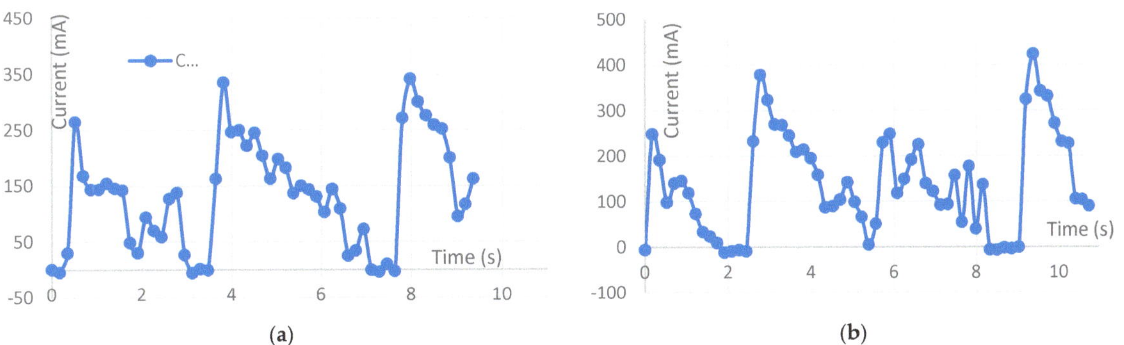

(a)   (b)

Figure 9. Acquired results during the test in Figure 6 in terms of current consumption: (**a**) without load; (**b**) with load.

In summary, the tested prototype, although not exhibiting high performance, works satisfactorily with the expected motion capability and load capacity (200 g against a full arm weight of about 300 g) that can be properly used when implementing the arm in the torso of the LARMbot humanoid. Future work can be planned to improve the motion range and payload capacity of the arm design with better linear actuators, both with increased stroke and force capacity.

4. Conclusions

The need to review the design of the arm for the LARMbot humanoid in order to have similar characteristics as other humanoid modules was addressed by revising the design requirements, aiming at a solution that satisfies the peculiar characteristics of the LARMbot in terms of compactness, motion capabilities, and payload by using mechanism structures with parallel architecture. The arm is designed with a tripod structure similar to the one already implemented for the legs of the LARMbot humanoid, with a design that is aimed at using low-cost components and non-commercial parts produced via 3D printing. The built prototype shows satisfactory characteristics, as reported by the results of the experimental tests conducted to validate and characterize the prototype in terms of performance in motion and load capacity. The motion range is detected with a range of more than 60 deg and the payload capacity is validated, with a ratio nearly equal to one with respect to the arm's own weight while using 420 mA for servomotor feeding.

Author Contributions: Conceptualization, M.C., M.R. and S.B.; methodology, M.C. and S.B.; software, S.B.; validation, M.R. and S.B.; formal analysis, M.C., M.R. and S.B.; investigation, M.C., M.R. and S.B.; resources, M.C.; data curation, M.R. and S.B.; writing—original draft preparation, M.C. and S.B.; writing—review and editing, M.C., M.R. and S.B.; visualization, M.C. and S.B.; supervision M.C.; project administration, M.C.; funding acquisition, M.C. All authors have read and agreed to the published version of the manuscript.

Funding: This research received no external funding.

Data Availability Statement: The original contributions presented in the study are included in the article, further inquiries can be directed to the corresponding author.

Acknowledgments: The second author wishes to thank SeaTech-School of Engineering, University of Toulon, France, for the support of an internship period at the laboratory of Robot Mechatronics of the University of Rome Tor Vergata during 2023.

Conflicts of Interest: The authors declare no conflicts of interest.

References

1. Lim, H.-O.; Takanishi, A. Biped Walking Robots Created at Waseda University: WL and WABIAN Family. *Philos. Trans. R. Soc. A Math. Phys. Eng. Sci.* **2007**, *365*, 49–64. [CrossRef] [PubMed]
2. Sakagami, Y.; Watanabe, R.; Aoyama, C.; Matsunaga, S.; Higaki, N.; Fujimura, K. The intelligent ASIMO: System overview and integration. In Proceedings of the IEEE/RSJ International Conference on Intelligent Robots and Systems, Lausanne, Switzerland, 30 September–4 October 2002; pp. 2478–2483. [CrossRef]
3. Ogura, Y.; Aikawa, H.; Shimomura, K.; Kondo, H.; Morishima, A.; Lim, H.O.; Takanishi, A. Development of a new humanoid robot WABIAN-2. In Proceedings of the 2006 IEEE International Conference on Robotics and Automation, ICRA 2006, Orlando, FL, USA, 15–19 May 2006; pp. 76–81. [CrossRef]
4. Kaneko, K.; Kanehiro, F.; Kajita, S. Humanoid Robot HRP-2. In Proceedings of the IEEE International Conference on Robotics and Automation, New Orleans, LA, USA, 26 April–1 May 2004; pp. 1083–1090. [CrossRef]
5. Buschmann, T.; Gienger, M. TUM Robots (e.g., Johnnie, Lola). In *Humanoid Robotics: A Reference*; Goswami, A., Vadakkepat, P., Eds.; Springer: Dordrecht, The Netherlands, 2017. [CrossRef]
6. Oh, J.-H.; Hanson, D.; Kim, W.-S.; Han, Y.; Kim, J.-Y.; Park, I.-W. Design of Android type Humanoid Robot Albert HUBO. In Proceedings of the 2006 IEEE/RSJ International Conference on Intelligent Robots and Systems, Beijing, China, 9–15 October 2006; pp. 1428–1433. [CrossRef]
7. Kuindersma, S.; Deits, R.; Fallon, M.; Valenzuela, A.; Dai, H.; Permenter, F.; Koolen, T.; Marion, P.; Tedrake, R. Optimization-based locomotion planning, estimation, and control design for the atlas humanoid robot. *Auton. Robot.* **2016**, *40*, 429–455. [CrossRef]
8. Yu, Z.; Huang, Q.; Ma, G.; Chen, X.; Zhang, W.; Li, J.; Gao, J. Design and development of the humanoid robot BHR-5. *Adv. Mech. Eng.* **2014**, *6*, 852937. [CrossRef]
9. Chignoli, M.; Kim, D.; Stanger-Jones, E.; Kim, S. The MIT humanoid robot: Design, motion planning, and control for acrobatic behaviors. In Proceedings of the 2020 IEEE-RAS 20th International Conference on Humanoid Robots (Humanoids), Munich, Germany, 19–21 July 2021; pp. 1–8.
10. Kashyap, A.K.; Parhi, D.R.; Muni, M.K.; Pandey, K.K. A hybrid technique for path planning of humanoid robot NAO in static and dynamic terrains. *Appl. Soft Comput.* **2020**, *96*, 106581. [CrossRef]

11. Ayala, A.; Cruz, F.; Campos, D.; Rubio, R.; Fernandes, B.; Dazeley, R. A comparison of humanoid robot simulators: A quantitative approach. In Proceedings of the 2020 Joint IEEE 10th International Conference on Development and Learning and Epigenetic Robotics (ICDL-EpiRob), Valparaiso, Chile, 26–30 October 2020; pp. 1–6.

12. Darvish, K.; Penco, L.; Ramos, J.; Cisneros, R.; Pratt, J.; Yoshida, E.; Ivaldi, S.; Pucci, D. Teleoperation of humanoid robots: A survey. *IEEE Trans. Robot.* **2023**, *39*, 1706–1727. [CrossRef]

13. Fox, J.; Gambino, A. Relationship development with humanoid social robots: Applying interpersonal theories to human–robot interaction. *Cyberpsychol. Behav. Soc. Netw.* **2021**, *24*, 294–299. [CrossRef] [PubMed]

14. Mara, M.; Stein, J.P.; Latoschik, M.E.; Lugrin, B.; Schreiner, C.; Hostettler, R.; Appel, M. User responses to a humanoid robot observed in real life, virtual reality, 3D and 2D. *Front. Psychol.* **2021**, *12*, 633178. [CrossRef] [PubMed]

15. Mukherjee, S.; Baral, M.M.; Pal, S.K.; Chittipaka, V.; Roy, R.; Alam, K. Humanoid robot in healthcare: A systematic review and future research directions. In Proceedings of the 2022 International Conference on Machine Learning, Big Data, Cloud and Parallel Computing (COM-IT-CON), Faridabad, India, 26–27 May 2022; Volume 1, pp. 822–826.

16. Yousif, J. Humanoid robot as assistant tutor for autistic children. *Int. J. Comput. Appl. Sci.* **2020**, *8*, 8–13.

17. Kiilavuori, H.; Sariola, V.; Peltola, M.J.; Hietanen, J.K. Making eye contact with a robot: Psychophysiological responses to eye contact with a human and with a humanoid robot. *Biol. Psychol.* **2021**, *158*, 107989. [CrossRef] [PubMed]

18. Yang, A.; Chen, Y.; Naeem, W.; Fei, M.; Chen, L. Humanoid motion planning of robotic arm based on human arm action feature and reinforcement learning. *Mechatronics* **2021**, *78*, 102630. [CrossRef]

19. Ramón, J.L.; Calvo, R.; Trujillo, A.; Pomares, J.; Felicetti, L. Trajectory optimization and control of a free-floating two-arm humanoid robot. *J. Guid. Control Dyn.* **2022**, *45*, 1661–1675. [CrossRef]

20. Wei, Y.; Zhao, J. Designing human-like behaviors for anthropomorphic arm in humanoid robot NAO. *Robotica* **2020**, *38*, 1205–1226. [CrossRef]

21. Huang, K.; Xian, Y.; Zhen, S.; Sun, H. Robust control design for a planar humanoid robot arm with high strength composite gear and experimental validation. *Mech. Syst. Signal Process.* **2021**, *155*, 107442. [CrossRef]

22. Chowdhury, D.; Park, Y.E.; Jung, I.; Lee, S. Characterization of Exterior Parts for 3D-Printed Humanoid Robot Arm with Various Patterns and Thicknesses. *Polymers* **2024**, *16*, 988. [CrossRef] [PubMed]

23. Ceccarelli, M. A History of LARMbot humanoid. In Proceedings of the of 2024 IFToMM Symposium on History of Machines and Mechanism, Ankara, Türkiye, 18–20 April 2024. Paper No. 23, *in print.*

24. Ceccarelli, M.; Russo, M.; Morales-Cruz, C. Parallel Architectures for Humanoid Robots. *Robotics* **2020**, *9*, 75. [CrossRef]

25. Fort, A.; Laribi, M.A.; Ceccarelli, M. Design and performance of a LARMbot PK arm prototype. *Int. J. Humanoid Robot.* **2022**, *19*, 2250009. [CrossRef]

Article

Design and Analysis of the Mechanical Structure of a Robot System for Cabin Docking

Ronghua Liu [1,2] and Feng Pan [1,*]

[1] School of Automation, Beijing Institute of Technology, Beijing 100081, China; ronghua5416@163.com
[2] Jiangsu Automation Research Institute, Lianyungang 222061, China
* Correspondence: panfeng@bit.edu.cn

Abstract: Aiming at the disadvantages of traditional manual docking, such as low assembly efficiency and large positioning error, a six-DOF dual-arm robot system for module docking is designed. Firstly, according to the operation tasks of the cabin docking robot, its functional requirements and key indicators are determined, the overall scheme of the robot is designed, and the composition and working principle of the robot joints are introduced in detail. Secondly, a strength analysis of the core components of the docking robot is carried out by finite element analysis software to ensure its load capacity. Based on the kinematics model of the robot, the working space of the robot mechanism is simulated and analyzed. Finally, the experimental platform of the docking robot is built, and the working space, repeated positioning accuracy, and motion control accuracy of the docking robot mechanism are verified through experiments, which meet the design requirements.

Keywords: cabin docking; robot design; strength analysis; work space

Citation: Liu, R.; Pan, F. Design and Analysis of the Mechanical Structure of a Robot System for Cabin Docking. *Actuators* **2024**, *13*, 206. https://doi.org/10.3390/act13060206

Academic Editor: Ioan Doroftei

Received: 19 April 2024
Revised: 23 May 2024
Accepted: 26 May 2024
Published: 30 May 2024

1. Introduction

Most aerospace products such as aircraft and rockets are composed of compartments with different structural characteristics. The assembly, testing, and testing process of aerospace products involve complex operations of module docking. At present, in the field of cabin assembly in China, manufacturers generally adopt the traditional assembly mode of "manual as the main, mechanical as the auxiliary". In this mode, the cabin components to be docked are manually hoisted to auxiliary tools, and the operator estimates the cabin attitude through eye observation. There are some problems such as low docking efficiency, poor docking quality, and unquantifiable docking accuracy. At the same time, when the diameter of the cabin increases to a certain level, the labor intensity increases sharply and the production efficiency decreases sharply. The traditional assembly mode can no longer meet the actual needs of manufacturing enterprises, so it is imperative to improve the automation level of cabin docking.

Regarding the cabin docking robot system, its level of development in foreign countries is relatively mature. Cabin docking mainly includes Three-coordinate POGO column assembly, Stewart platform assembly, Vertical assembly, Horizontal docking assembly, and other forms as follows:

(1) Three-coordinate POGO column assembly: The three-coordinate POGO column structure is used to adjust the position and pose of the workpiece [1–3]. The POGO column is an adjustable support system with three orthogonal translation degrees of freedom, which is composed of three mutually orthogonal translation units, and can adjust the supported parts in space with high precision. Aiming at the aircraft wing assembly problem, a six-DOF attitude docking system based on a three-DOF POGO column was proposed [4]. Each positioner is connected to the wing by a hemispherical end effector. The positioner and the wing mechanism form a three-PPPS (P is a moving joint, S is a spherical joint) redundant drive variable mechanism.

Assembly by the three-coordinate POGO column makes the production line have high flexibility. However, this assembly method has certain limitations; for example, the distributed POGO column is redundant in motion control, which will bring high difficulty and economic cost to the control.

(2) Stewart platform assembly: The Stewart platform is used to adjust the position and pose of the workpiece. A Stewart platform refers to a symmetrical parallel mechanism with six degrees of freedom [5,6]. Based on the six-SPS parallel mechanism, which is used to adjust the attitude of the module [7], a force sensor is integrated in each sliver chain of the parallel mechanism to achieve flexible docking control of the module through force feedback, so as to reduce the risk of damaged parts. Compared with distributed POGO columns, the Stewart platform is more compact and easier to control. However, this assembly method has certain limitations for the assembly of cabin parts with different diameters and lengths, and it is not suitable for continuous production in the engineering environment.

(3) Vertical assembly method: The components of the cabin are assembled vertically by means of lifting assembly. In the automatic docking of a satellite, a six-SPS parallel mechanism is used as the docking equipment to support the mobile module, and the mechanism can dock the mobile module with the suspended fixed module by adjusting its posture in the vertical direction [8]. However, the vertical assembly method requires enough space to be utilized in the vertical direction of the workshop, allowing multiple compartments to be placed vertically, which is not conducive to popularization and application in the general assembly workshop. At the same time, with an increase in the weight and size of the cabin, the vertical assembly method will become difficult to implement because of the difficulty of operation, high-risk factor, and increase in vertical assembly control.

(4) Horizontal docking method: By lifting or AGV transfer, the cabin components are transferred to the attitude adjustment mechanism of the assembly workstation. The attitude adjustment mechanism can move horizontally along the guide rail to complete the assembly and docking of the multi-stage cabins. Aiming at the docking problem of a certain type of rocket module, the automatic docking platform of the rocket module was built by the Shenyang Institute of Automation [9]. The six-DOF pose adjustment mechanism loaded with ring tooling is used as the pose adjustment platform of the docking module, and the transfer drive and the docking drive of the module are designed in an integrated way. Aiming to address the high precision and automation requirements of spacecraft module docking assembly, the Beijing Institute of Satellite Environmental Engineering proposed large module horizontal docking assembly technology based on automation means [10] and designed the module horizontal docking system structure.

Considering the advantages and disadvantages of the three-dimensional POGO column assembly method, Stewart platform assembly method, lifting vertical assembly method, and horizontal docking assembly method, a six-degree-of-freedom (DOF) cabin docking dual-arm robot system is designed, which has the characteristics of a large load, high flexibility, high precision, quick replacement, and so on. Based on the analysis of the functional requirements and key indexes of the robot system, the technical indexes of the development of the docking dual-arm robot system in the cabin are determined. The system composition and working principle of each joint in the module docking robot are studied in detail, and the strength of each joint core component is analyzed by software. Based on the kinematics model of the robot mechanism, the working space of the robot mechanism is simulated and analyzed. The working space, repeated positioning accuracy, and motion accuracy of the docking robot mechanism are verified by prototype testing.

This paper is organized as follows: Section 2 analyzes the functional requirements and key indicators of the module docking robot. Section 3 introduces the overall scheme of the docking robot system, the composition of each joint mechanism, and the strength analysis results. In Section 4, the kinematic model of the docking robot is established, and the kinematic simulation is carried out. Section 5 verifies the technical indicators of the docking robot in the module through the prototype. The results are summarized in Section 6.

2. Functional Requirements and Key Indicators of the Module Docking Robot

2.1. Functional Requirements of the Cabin Docking Robot

The main task of the module docking robot system is to realize the automatic attitude adjustment function of a large heavy-duty cabin by designing a six-degree-of-freedom robot mechanism and to complete the automatic docking function of multiple cabin workpieces with the aid of the visual measurement system. The specific functions include the following:

(1) Carrying a large load and large sized workpieces, and carrying weight not less than 500 kg;
(2) High precision, reliable and stable motion function;
(3) A certain product compatibility function that adapt to the compatibility function of different loads within a certain specification range.

2.2. Technical Indicators

In order to accomplish the above tasks, the docking robot has specific requirements for load capacity, working space, moving speed, and so on, as follows:

(1) Load capacity: including the motion accuracy of the robot platform, the degree of freedom of the robot, the load capacity of the robot, etc. These indicators determine the ability of the robot to perform the final task;
(2) Working space: the robot must have the working range required for the assembly of cabin workpieces;
(3) Motion speed: when the docking robot performs the assembly task, the docking robot is required to operate quickly and work efficiently.

The technical indicators of the designed cabin docking robot are shown in Table 1.

Table 1. Main technical indicators of the docking robot system.

Item	Specific Item	Value
Load capacity	DOF	6
	Rated load	500 kg
	Repeated positioning accuracy	± 0.05 mm
	Maximum load	1000 kg
Work space	Circumference roll range	$\pm 15°$
	Axial travel range	± 600 mm
	Vertical lift travel range	± 50 mm
	Radial travel range	± 50 mm
Speed	Circular rolling speed range	$0{\sim}6.8°/s$
	Axial movement speed range	$0{\sim}400$ mm/s
	Vertical lifting speed range	$0{\sim}4.5$ mm/s
	Radial movement speed range	$0{\sim}25$ mm/s
Acceleration	Acceleration range of circular roll	$0{\sim}240°/s^2$
	Acceleration range of axial movement	$0{\sim}100$ mm/s^2
	Acceleration range of vertical lift	$0{\sim}100$ mm/s^2
	Acceleration range of radial motion	$0{\sim}100$ mm/s^2

3. Mechanical Design and Strength Analysis of the Cabin Docking Robot System

3.1. Overall Scheme of the Cabin Docking Robot System

In order to meet the technical indicators in Table 1, a cylindrical cabin workpiece is taken as an example to design a cabin docking robot. The cabin docking robot is mainly composed of a left robotic arm, a right robotic arm, a base, and a control system. The three-dimensional model of the docking robot designed by NX2023 software is shown in Figure 1.

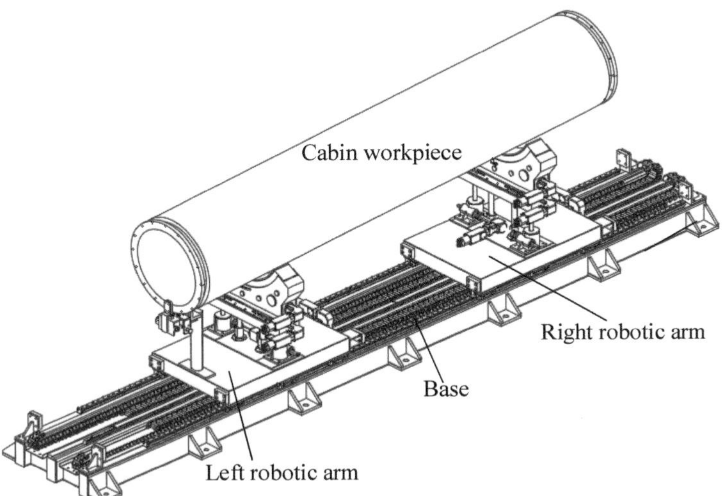

Figure 1. Composition and working diagram of the robot system for cabin docking.

In Figure 1, the docking robot is composed of a left robotic arm, a right robotic arm, and a base. The left robot arm and the right robot arm are connected to the base rail through a slider and move axially through a rack and pinion drive. The left robotic arm and the right robotic arm have the same structure, and each robotic arm includes four joints as follows: axial moving joint, vertical lifting joint, radial moving joint, and circular rolling joint. The composition of a single robotic arm is shown in Figure 2.

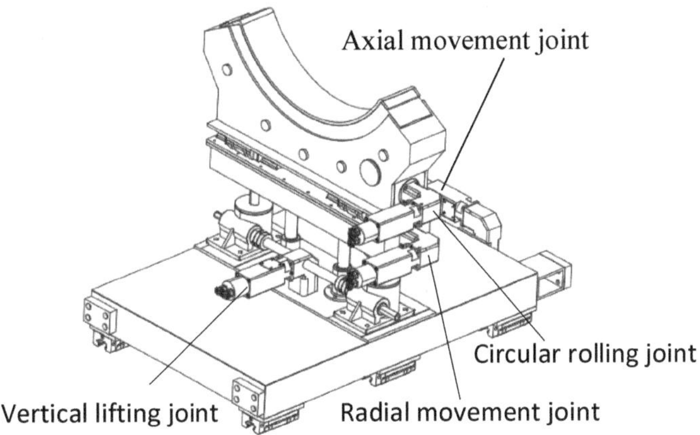

Figure 2. Diagram of a single robotic arm.

As shown in Figure 2, the width of a single mechanical arm is 820 mm, the length is 1200 mm, and the center of the circular rolling joint is 1100 mm above the ground. The docking robot can complete the adjustment of six degrees of freedom (pitch, yaw, roll, radial, lifting, and axial) of the workpiece in the whole cabin. The axial moving joint is driven by the servo motor to achieve axial horizontal action, the radial moving joint is driven by the servo motor to achieve horizontal action in the diameter direction of the screw structure, the vertical lifting joint is driven by the servo motor to achieve the height lifting as the spiral lift, and the circular rolling joint is driven by the servo motor to achieve the circular rotation action.

3.2. Joint Design of the Docking Robot

3.2.1. Axial Movement Joint

The axial movement joint mainly consists of a structural frame, drive gear, servo motor and reducer, slide block, bottom structural frame, and other parts. The servo motor and reducer drive the gear to rotate and then drive the mechanical arm to move the axis on the guide rail of the base, as shown in Figure 3.

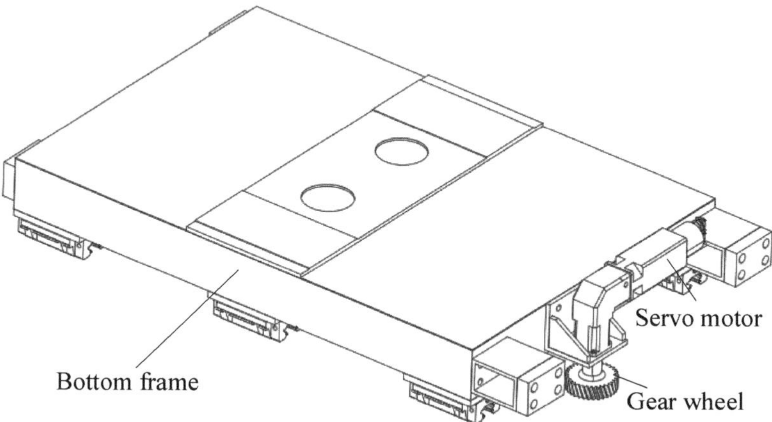

Figure 3. Structure diagram of the axial movement joint.

In Figure 3, the bottom of the axial movement joint is connected to the guide rail on the base by means of a slider. The bottom frame of the axial movement joint adopts a frame structure and Q345 steel plate welding molding. Four cushion pads are installed on the two ends of the structural frame to slow the impact of the collision between the mechanical arms. The structural mounting surface is used to directly mount the vertical lifting joint module. It is driven by an AC servo motor and equipped with a multi-turn absolute encoder and electromagnetic brake. The precision planetary reducer is selected as the reducer, and hard face gear can effectively improve the wear resistance and reliability of the reducer.

3.2.2. Vertical Lifting Joint

The vertical lifting joint drives two sets of spiral lifts through the servo motor and the reducer to realize the lifting action of the robot, including the servo motor, the reducer, the T-bevel gear steering box, the coupling, the spiral lift, the guiding mechanism, and other parts, as shown in Figure 4.

In Figure 4, the vertical lifting joint is driven by the structural form of a servo motor and a reducer driving two spiral elevators. The two ends of the spiral elevator are vertically installed on the top of the base structure module. The middle is connected through a T-shaped bevel gear steering box. The bottom of the vertical lifting joint is directly installed on the axial moving joint module through bolts. The spiral elevator achieve lifting through the servo motor drive reducer and coupling. The guide mechanism adopts the shaft sleeve structure, which is mainly composed of the guide seat, the guide shaft, the end cover, etc., which is used to provide reliable guidance for the screw lift in the lifting action and ensure the linearity and stability of the lifting action. The flange mounting surface and the end cover of the guide mechanism are directly connected with the vertical moving joint, which drives the radial moving joint and the circular rolling joint to lift simultaneously when the lift is lifted.

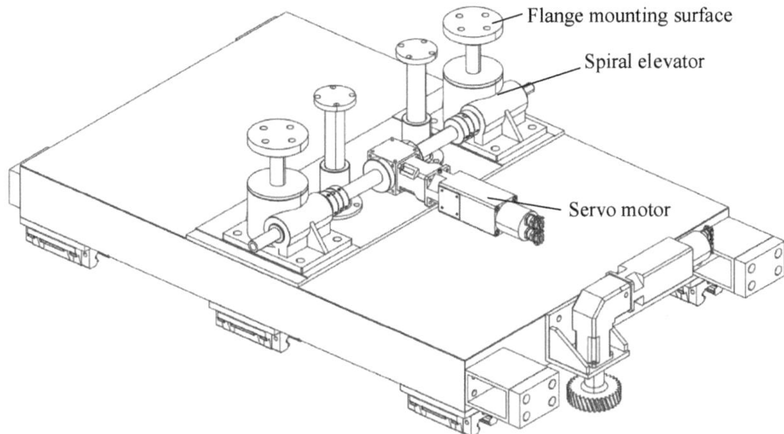

Figure 4. Structure diagram of the vertical lifting joint.

3.2.3. Radial Movement Joint

The radial movement joint drives the ball screw through the servo motor and the reducer to realize the radial movement function of the robot, including the servo motor, the reducer, the coupling, the ball screw, the connecting base, the top guide rail, the housing bracket, and other parts of the top view, as shown in Figure 5.

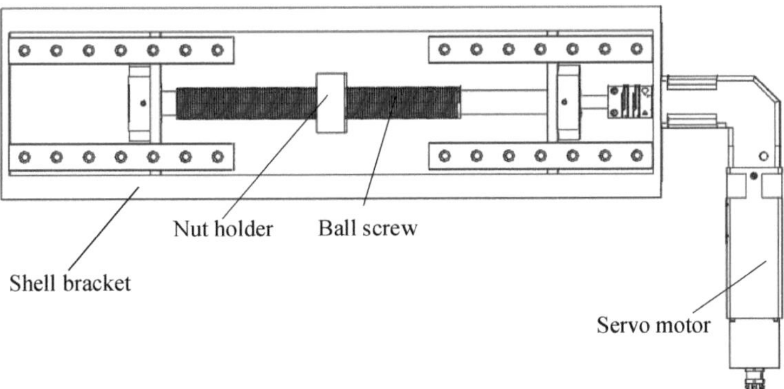

Figure 5. Top view of the radial prismatic joint structure.

In Figure 5, a lead screw is installed inside the shell bracket and a guide rail is installed on both sides of the top. The radial movement joint is connected to the bottom slider of the circular rolling joint through the top guide rail, and the middle is directly mechanically connected to the circular rolling joint through the nut holder. The servo motor drives the reducer to rotate, and the ball screw is driven by coupling. When the servo motor drives the ball screw to rotate, the circular rolling joint moves radially with the ball screw.

3.2.4. Circular Rolling Joint

The circular rolling joint realizes the circular rotation of the curved rolling body by driving the turbo worm mechanism and rack and pinion mechanism through the servo motor and the reducer, which mainly includes the servo motor and the reducer, turbo worm mechanism, curved rack, rolling bearing, curved rolling body, anti-static rubber, and rotation bracket, etc. The three-dimensional model of the circular rolling joint is shown in Figure 6.

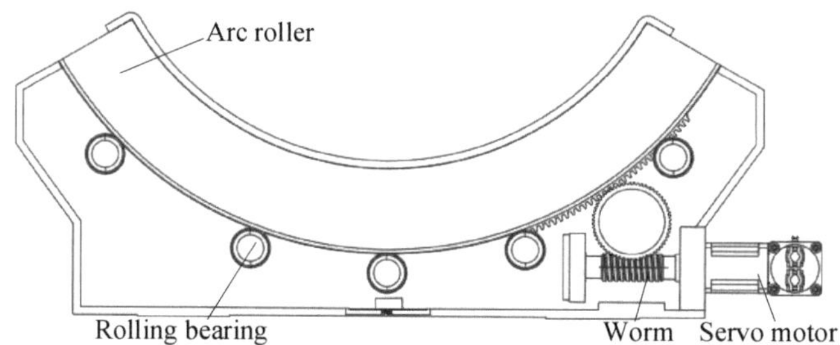

Figure 6. Frontal view of circular rolling joint.

In Figure 6, the arc roller is arranged on the top of the roll bracket as an arc structure matching the shape of the cabin workpiece, and the arc rack is arranged on the bottom side of the arc roller. The top circle diameter of the rack is larger than the top circle diameter of the turbine, and the rack engages with the turbine. When the servo motor and the reducer drive the worm gear to rotate, the rack action of the arc roller is driven by the gear of the turbine, and the circular rotation of the arc roller is realized. The rolling bearing is distributed in the inner cavity of the rolling bracket in an arc shape and rolls in contact with both sides of the bottom of the curved rolling body. The top of the arc roller is installed with curved anti-static material, which is in direct contact with the cabin workpiece.

3.3. Strength Analysis of Key Components

Strength analysis and strength checks of key components are indispensable links in the design of mechanical systems, aiming to theoretically ensure the stiffness and mechanical reliability of the overall mechanism. Strength analysis is required in the process of robot research and development and design [11–13]. Since the robot is self-heavy and the workpiece size and weight of the robot are large, it is necessary to analyze the strength of the main load-bearing components of the robot. The main load-bearing parts of the cabin docking robot include the base of the axial moving joint, the spiral elevator of the vertical lifting joint, the external shell of the radial moving joint, and the rolling bearing of the circular rolling joint.

The Simcenter Nastran 8031 software module of NX 2023 software is used to analyze the static stress of the main load-bearing components, assuming that the weight of the cabin workpiece is set to 1000 kg, and the weight of the actual key components is added as the working load. At the same time, the material of each key component is Q345 steel or 45 steel.

3.3.1. Strength Analysis of the Bottom Frame

The bottom frame of the axial moving joint is the load-bearing part of the whole robot, which is used to support the vertical lifting joint, the radial moving joint, the circular rolling joint, and the load workpiece. In order to simulate the stress of the axial moving joint frame, the force grid and nodes are divided, and the two mounting seats are set as the stress points, which is consistent with the actual situation. Six slider mounting positions are added at the bottom as fixed constraints, as shown in Figure 7. The bottom frame is 1200 mm long, 150 mm high, and 820 mm wide.

Taking the total gravity of 13 kN as the working load, the static stress analysis is carried out. The equivalent stress, elastic deformation, and total strain results of the bottom frame are shown in Figure 8.

As can be seen in Figure 8, the maximum equivalent stress on the bottom frame is 31.81 MPa, and the maximum total deformation is 0.125 mm. Because the bottom frame is made of Q345 steel, the yield limit is 345 MPa, and the safety factor of anti-instability is 5. The allowable stress of the bottom frame is much greater than its maximum equivalent stress, and the design is reasonable.

3.3.2. Strength Analysis of the Spiral Elevator

The spiral elevator of the vertical lifting joint supports the radial movement joint, the circular rolling joint, and the load weight through the mounting surface at the top, which is the main bearing part of the vertical lifting joint. In order to simulate the stress situation of the spiral elevator, the stress grid and nodes are divided, and the installation surface of the end flange of the trapezoid lead screw inside the elevator is taken as the stress point, which is consistent with the actual situation. Add the bottom of the lead screw as a fixed constraint, as shown in Figure 9. The stroke of the trapezoidal lead screw is 150 mm, and the diameter of the lead screw is 30 mm.

Taking the total gravity of 11 kN as the working load, the static stress analysis is carried out. The results of the effect force, elastic deformation, and total strain of the trapezoid lead screw inside the spiral elevator are shown in Figure 10.

As can be seen in Figure 10, the maximum equivalent stress of the trapezoidal screw inside the spiral lift is 28.86 MPa, and the maximum total deformation is 0.0177 mm. The trapezoidal screw is made of 45 steel, the yield limit is 355 MPa, and the safety factor of anti-instability is selected to be 5. The allowable stress of the spiral lift is much greater than its maximum equivalent stress, and the design is reasonable.

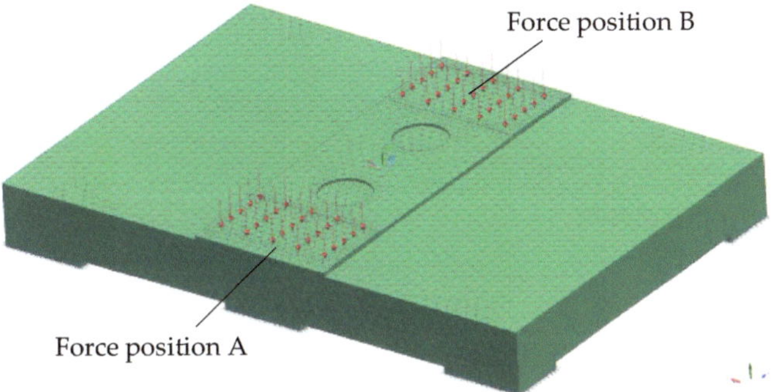

(**a**) Setting of force points on the bottom frame

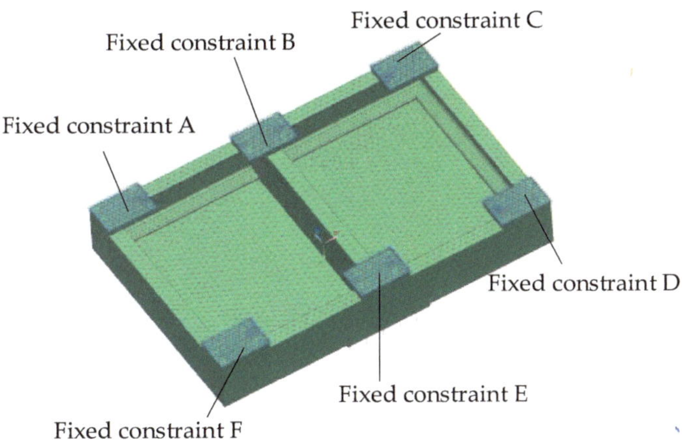

(**b**) Setting of fixed constraints on the bottom frame

Figure 7. Simulation setting of the bottom frame of the axial moving joint.

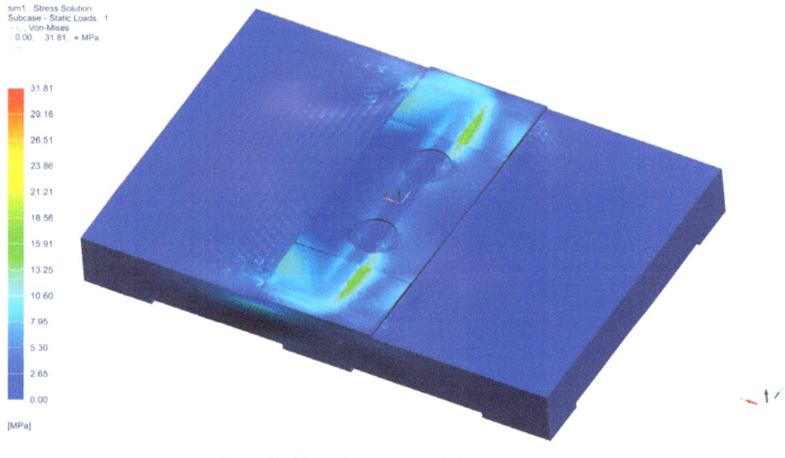

(**a**) Equal effect diagram of the bottom frame

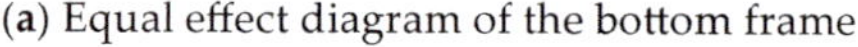

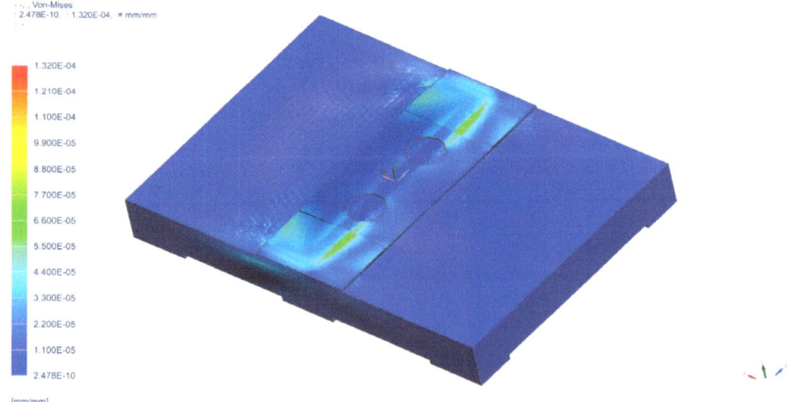

(**b**) Elastic deformation diagram of the bottom frame

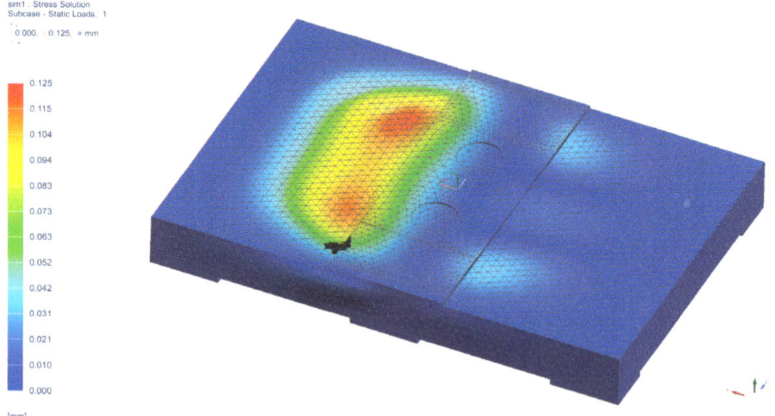

(**c**) Total strain diagram of the bottom frame

Figure 8. Static stress analysis results of the bottom frame.

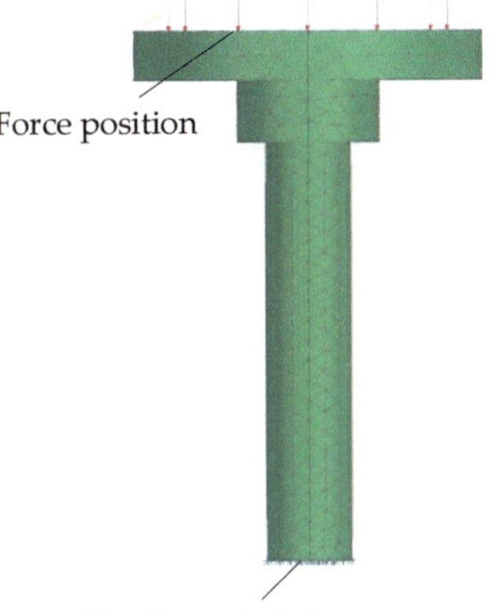

Figure 9. Setting of the force point and fixed constraint of the trapezoidal lead screw.

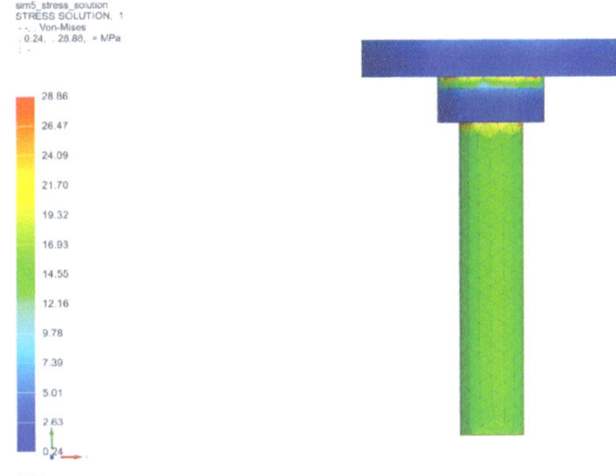

(**a**) Equal effect diagram of the trapezoidal lead screw in the spiral elevator

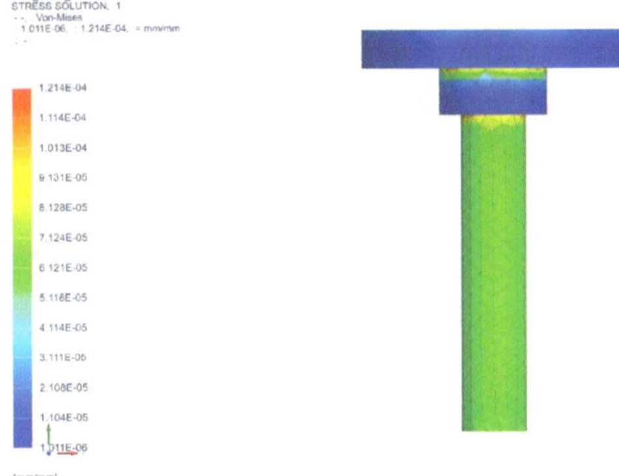

(**b**) Elastic deformation diagram of the trapezoid lead screw in the spiral elevator

Figure 10. *Cont.*

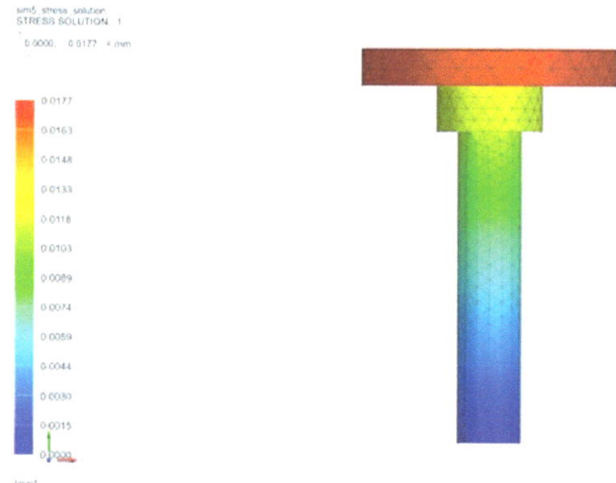

(**c**) Total strain diagram of the trapezoidal screw inside the spiral elevator

Figure 10. Static stress simulation results of the trapezoidal lead screw.

3.3.3. Strength Analysis of the Shell Bracket

The upper part of the shell bracket of the radial mobile joint is installed with a top guide rail, which is used to support the upper mounting mechanism and load, and is an important load-bearing part of the whole robot. In order to simulate the stress of the radial moving joint, the force grid and nodes are divided, and the installation positions of the four top guide rails are taken as the stress points, which are consistent with the actual situation. The top mounting position of the bottom two spiral lifts are added as a fixed constraint, as shown in Figure 11. The length of the shell bracket is 820 mm, the width is 240 mm, and the height is 115 mm.

Taking the total gravity of 10.2 kN as the working load, the static stress analysis is carried out. The equivalent stress, elastic deformation, and total strain results of the radial movement joint shell bracket are shown in Figure 12.

As can be seen in Figure 12, the maximum equivalent stress on the shell bracket is 23.13 MPa, and the maximum total deformation is 0.0182 mm. Because the shell bracket is made of Q345 steel, the yield limit is 345 MPa, and the safety factor of anti-instability is 5. The allowable stress of the shell bracket is much greater than its maximum equivalent stress, and the design is reasonable.

3.3.4. Strength Analysis of the Rolling Bearing

The rolling bearing of the circular rolling joint is connected to the curved rolling body to support the load and is an important bearing part of the robot. In order to simulate the stress situation of the rolling bearing with circular rolling joints, the stress grid and nodes are divided, and the contact position between the middle of the rolling bearing and the curved rolling body is taken as the stress point, which is consistent with the actual situation. The limit positions at both ends are added as fixed constraints, as shown in Figure 13. The diameter of the rolling bearing is 25 mm, and the length is 165 mm.

Taking the total gravity of 10 kN as the working load, the static stress analysis is carried out. The equivalent stress, elastic deformation, and total strain results of the circular rolling joint rolling bearing are shown in Figure 14.

As can be seen in Figure 14, the maximum equivalent stress on the rolling bearing is 9.153 MPa, and the maximum total deformation is 0.001195 mm. Because the rolling bearing is made of 45 steel, the yield limit is 355 MPa, and the anti-instability safety factor is 5. The allowable stress of the rolling bearing is much greater than its maximum equivalent stress, and the design is reasonable.

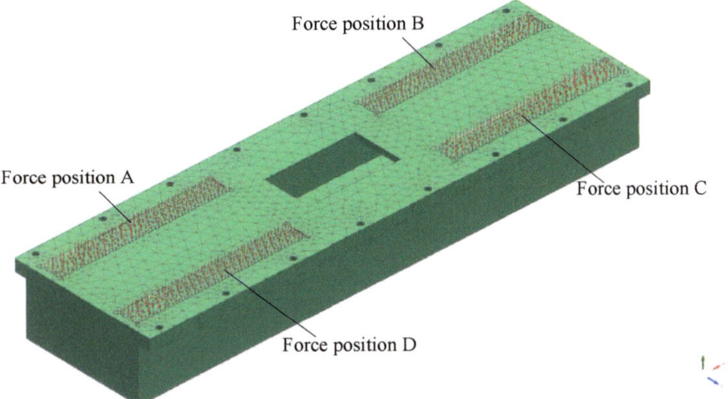

(**a**) Setting of the force points of the shell bracket

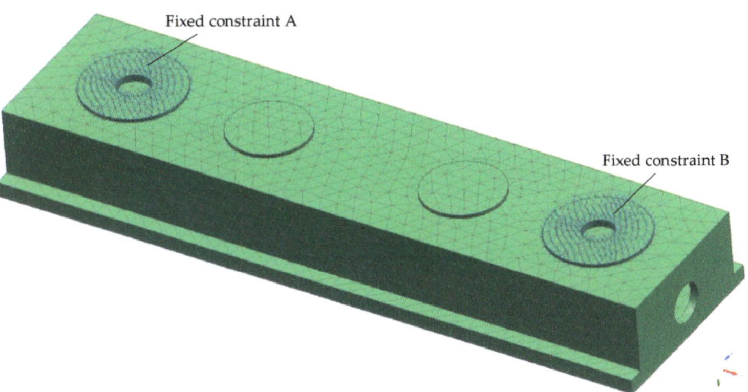

(**b**) Fixed constraint setting of the shell bracket

Figure 11. Loading point and fixed constraint setting of the shell bracket.

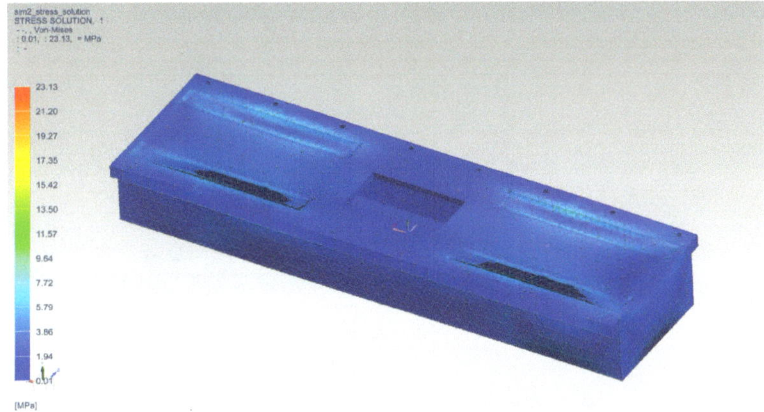

(**a**) Equal effect diagram of the shell bracket

Figure 12. *Cont.*

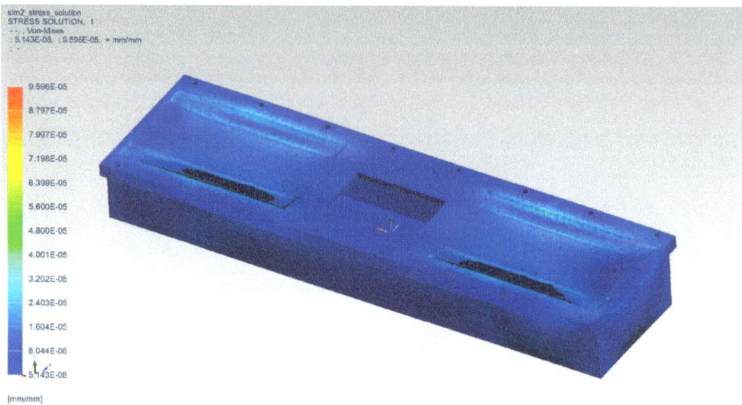

(**b**) Elastic deformation diagram of the shell bracket

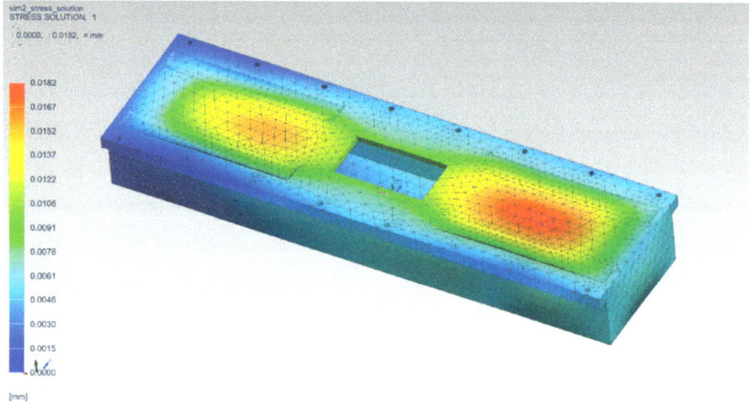

(**c**) Total strain diagram of the shell bracket

Figure 12. Static stress simulation of the shell bracket.

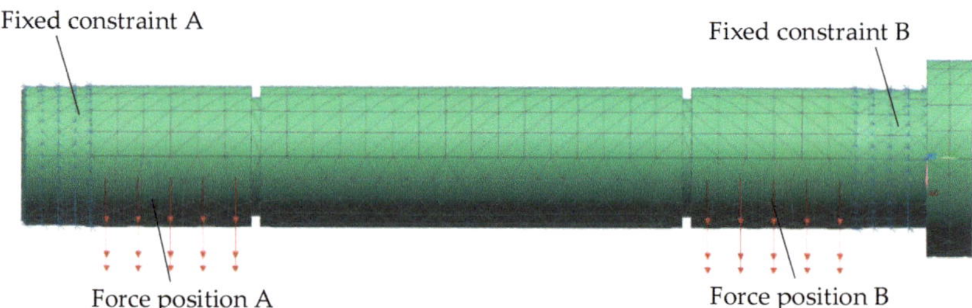

Figure 13. Setting of the force point and fixed constraint of the rolling bearing.

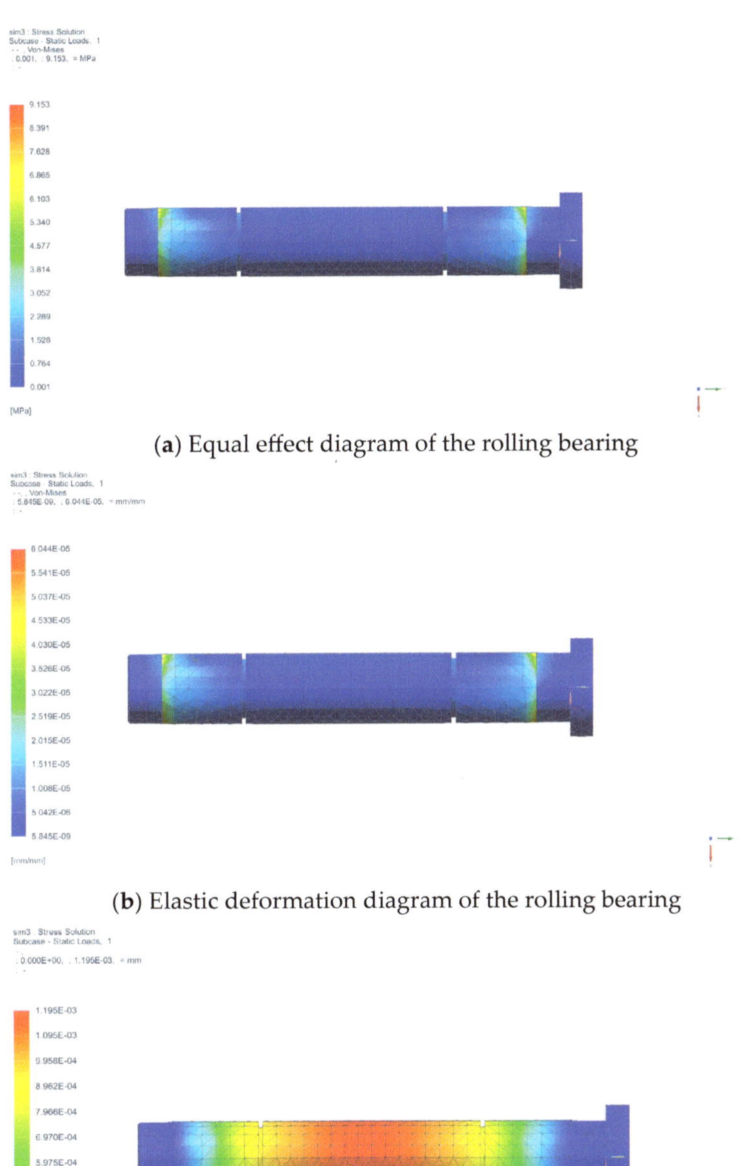

(**a**) Equal effect diagram of the rolling bearing

(**b**) Elastic deformation diagram of the rolling bearing

(**c**) Total strain diagram of the rolling bearing

Figure 14. Static stress simulation results of the rolling bearing.

4. Kinematics Modeling and Simulation Verification

4.1. Kinematics Modeling

4.1.1. Kinematics Modeling of the Cabin Docking Robot System

Kinematics analysis is the basis of robot design. In order to facilitate the kinematics analysis of the robot, a schematic diagram of the docking robot mechanism in the cabin is shown in Figure 15.

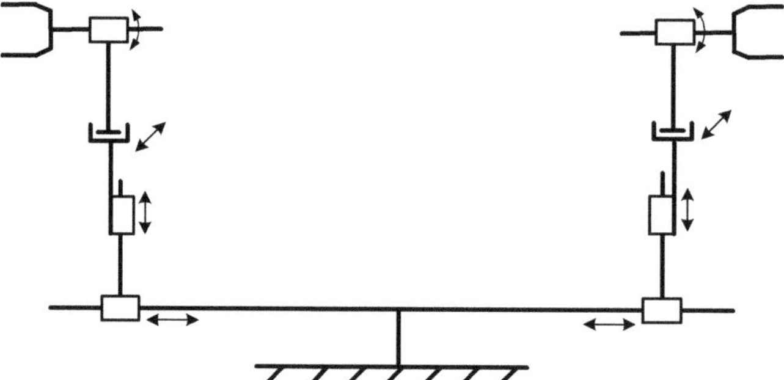

Figure 15. Schematic diagram of the docking robot mechanism in the cabin segment.

In Figure 15, the robot system is composed of left and right robotic arms, each of which is composed of four degrees of freedom, and from bottom to top are axial moving joints, vertical lifting joints, radial moving joints, and circular rolling joints. Two robotic arms are composed to coordinate the space attitude adjustment of the cabin workpiece.

The D-H (Denavit–Hartenberg) model [14,15] uses four parameters (rod length, zero position, bias, and torsion angle) to characterize the spatial coordinate positioning of a single joint, which is widely used in robot calibration, robot control, and trajectory planning [16–20]. According to the D-H modeling method, the joint coordinate system of the module docking robot is established, as shown in Figure 16.

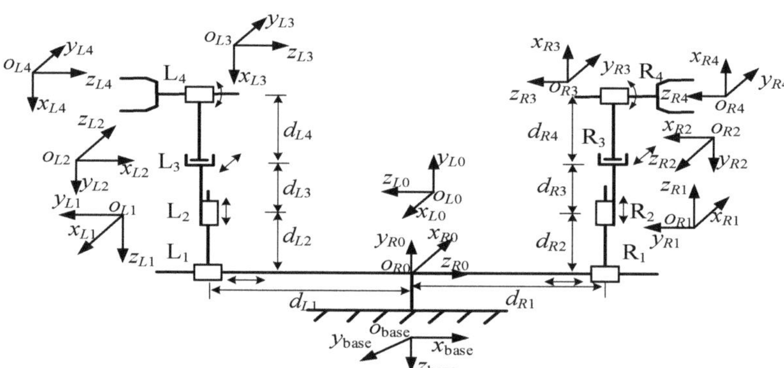

Figure 16. Schematic diagram of connecting rod coordinate system of the docking robot.

The base coordinate system is represented by $O_{base}(x_{base}, y_{base}, z_{base})$, and the coordinate system of the left robot arm $O_{L0}(x_{L0}, y_{L0}, z_{L0})$ and the coordinate system of the right robot arm $O_{R0}(x_{R0}, y_{R0}, z_{R0})$ are established in the middle position of the two robot arms on the base guide rail. Coordinate system $O_{L1}(x_{L1}, y_{L1}, z_{L1})$, coordinate system $O_{L2}(x_{L2}, y_{L2}, z_{L2})$, coordinate system $O_{L3}(x_{L3}, y_{L3}, z_{L3})$, and coordinate system $O_{L4}(x_{L4}, y_{L4}, z_{L4})$ are respectively established for the axial moving joint, vertical lifting joint, left and right moving joint, and circular rolling joint of the left mechanical arm.

Coordinate system $O_{R1}(x_{R1}, y_{R1}, z_{R1})$, coordinate system $O_{R2}(x_{R2}, y_{R2}, z_{R2})$, coordinate system $O_{R3}(x_{R3}, y_{R3}, z_{R3})$ and coordinate system $O_{R4}(x_{R4}, y_{R4}, z_{R4})$ are respectively established for the axial moving joint, vertical lifting joint, left and right moving joint and circular rolling joint of the left mechanical arm.

In Figure 16, each joint is described by the joint angle, connecting rod offset, connecting rod length, and connecting rod angle. The D-H parameters of the cabin docking robot are obtained, as shown in Table 2.

Table 2. D-H parameters of the docking robot.

Link	Joint Angle θ_i	Link Offset d_i	Link Length a_i	Link Angle α_i
Base-L_0	$\pi/2$	0	0	$-\pi/2$
L_1	0	$d_{L1} + 625$	0	$\pi/2$
L_2	$-\pi/2$	$d_{L2} + 140$	0	$\pi/2$
L_3	$\pi/2$	d_{L3}	600	$\pi/2$
L_4	θ_L	-470	0	0
Base-R_0	$\pi/2$	0	0	$\pi/2$
R_1	0	$d_{R1} + 625$	0	$-\pi/2$
R_2	$\pi/2$	$d_{R2} + 140$	0	$-\pi/2$
R_3	$-\pi/2$	d_{R3}	600	$-\pi/2$
R_4	θ_R	-470	0	0

The forward kinematics equations of the right robotic arm of the cabin docking robot are obtained according to the spatial pose matrix calculation, as shown in Equation (1). The forward kinematics equations of the left robotic arm of the cabin docking are shown in Equation (2).

$$
\begin{aligned}
p_{Rx} &= d_{R1} + 1095 \\
p_{Ry} &= -d_{R3} \\
p_{Rz} &= d_{R2} + 740 \\
y_R &= -\theta_R
\end{aligned}
\tag{1}
$$

where p_{Rx}, p_{Ry}, p_{Rz}, and y_R represent the coordinate variables of the X-axis direction, Y-axis direction, Z-axis direction, and circumferential direction of the operating space of the right robotic arm, respectively.

$$
\begin{aligned}
p_{Lx} &= -d_{L1} - 1095 \\
p_{Ly} &= -d_{L3} \\
p_{Lz} &= d_{L2} + 740 \\
y_L &= 180 + \theta_L
\end{aligned}
\tag{2}
$$

where p_{Lx}, p_{Ly}, p_{Lz}, and y_L represent the coordinate variables of the X-axis direction, Y-axis direction, Z-axis direction, and circumferential direction of the operating space of the left robotic arm, respectively.

According to the mapping relationship of Equations (1) and (2), the inverse kinematics transformation equations of the right robotic arm and the left robotic arm of the cabin docking can be established, respectively, as shown in Equations (3) and (4).

$$
\begin{aligned}
d_{R1} &= p_{Rx} - 1095 \\
d_{R3} &= -p_{Ry} \\
d_{R2} &= p_{Rz} - 740 \\
\theta_R &= -y_R
\end{aligned}
\tag{3}
$$

$$
\begin{aligned}
d_{L1} &= -p_{Lx} - 1095 \\
d_{L3} &= -p_{Ly} \\
d_{L2} &= p_{Lz} - 740 \\
\theta_L &= y_L - 180
\end{aligned}
\tag{4}
$$

4.1.2. Joint Kinematics Modeling of the Cabin Docking Robot and Cabin Workpiece

In order to analyze the attitude adjustment process of the cabin workpiece carried out by the cabin docking robot, the end coordinate system $O_L(x_L, y_L, z_L)$ of the left robotic arm for cabin docking, the end coordinate system $O_R(x_R, y_R, z_R)$ of the right robotic arm for cabin docking, the base coordinate system $O_{base}(x_{base}, y_{base}, z_{base})$, the center coordinate system $O_1(x_1, y_1, z_1)$ of the left end face of the cabin workpiece, and the center coordinate system $O_2(x_2, y_2, z_2)$ of the right end face of the cabin workpiece are established.

Suppose that the distance between coordinate system $O_1(x_1, y_1, z_1)$ and the origin of coordinate system $O_L(x_L, y_L, z_L)$ is D_1, the distance between coordinate system $O_L(x_L, y_L, z_L)$ and the origin of coordinate system $O_R(x_R, y_R, z_R)$ is D_2, and the distance between coordinate system $O_R(x_R, y_R, z_R)$ and the origin of coordinate system $O_2(x_2, y_2, z_2)$ is D_3, as shown in Figure 17.

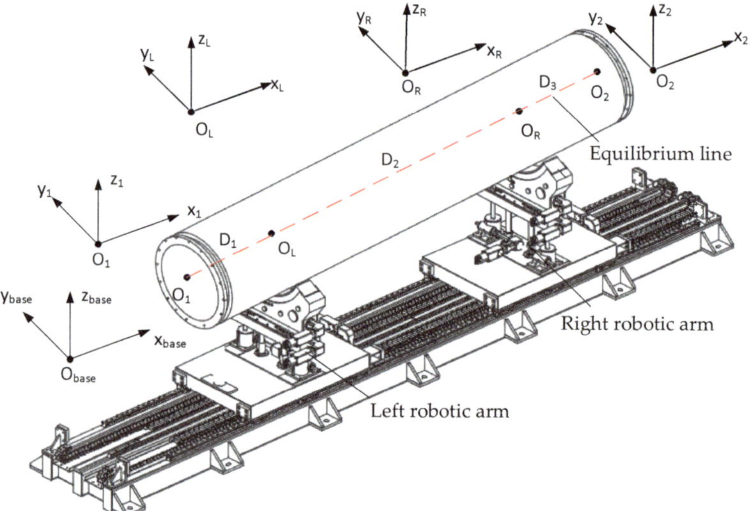

Figure 17. Distribution of each coordinate system of the docking two-arm robot in the cabin segment.

In Figure 17, the movement direction of the axial moving joints of the cabin docking robot is defined as the X-axis direction of the coordinate system, the movement direction of the left and right moving joints of the cabin docking robot is defined as the Y-axis direction of the coordinate system, and the movement direction of the vertical lifting joints of the cabin docking robot is defined as the Z-axis direction of the coordinate system. According to the coordinate system established in Figure 17, the definition of pitch angle, yaw angle, and rolling angle of the cabin workpiece is shown in Figure 18.

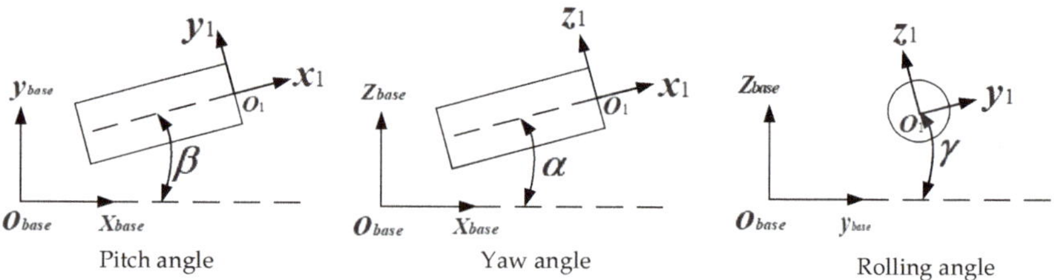

Figure 18. Definition of the attitude parameters of the cabin workpiece.

The relationship between the pose of the cabin workpiece and the coordinate system of the cabin docking robot can be analyzed in four situations as follows: the initial equilibrium position, the pitch angle is not at $0°$, the yaw angle is not at $0°$, and the roll angle is not at $0°$. The detailed description is as follows:

(1)   Initial equilibrium position

In the actual production process, it is generally required that the cabin docking robot is in a balanced spatial posture. The end coordinate system $O_L(x_L, y_L, z_L)$ of the left robotic arm for cabin docking, the end coordinate system $O_R(x_R, y_R, z_R)$ of the right robotic arm for cabin docking, the base coordinate system $O_{base}(x_{base}, y_{base}, z_{base})$, the center coordinate system $O_1(x_1, y_1, z_1)$ of the left end face of the cabin workpiece, and the center coordinate system $O_2(x_2, y_2, z_2)$ of the right end face of the cabin workpiece are established.

Their origin is in a straight line, and the pitch angle, yaw angle, and rolling ngle of the cabin workpiece are $0°$, as shown in Figure 19.

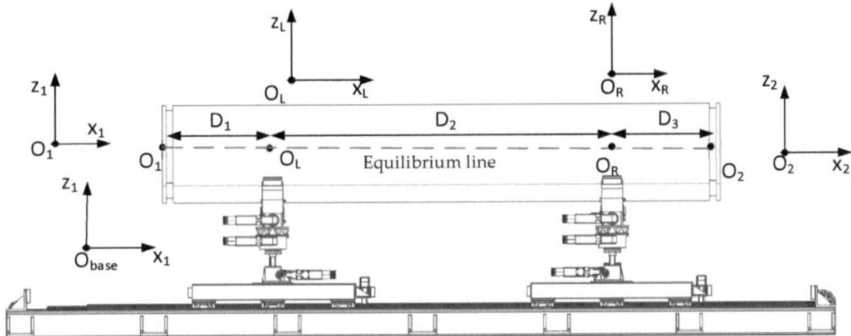

Figure 19. Coordinate system of the docking robot during the initial equilibrium position.

In Figure 19, when the cabin docking two-arm robot is in the initial equilibrium position, the values of coordinate systems $O_1(x_1, y_1, z_1)$, $O_L(x_L, y_L, z_L)$, $O_R(x_R, y_R, z_R)$, and $O_2(x_2, y_2, z_2)$ are equal in the X-direction and Y-direction. The position relationship of the coordinate system in the Z-direction is shown in Equation (5).

$$\begin{cases} (z_1 - z_L)^2 = D_1{}^2 \\ (z_L - z_R)^2 = D_2{}^2 \\ (z_R - z_2)^2 = D_3{}^2 \end{cases} \tag{5}$$

(2)   The pitch angle of the cabin workpiece is not $0°$

Because of the influence of friction between the docking robot and the cabin workpiece, and the limited stroke of the vertical lifting joint of the docking robot (± 50 mm), when the pitch angle α changes, the docking robot still makes contact with the cabin workpiece. Therefore, the origin of the coordinate system O_L at the end of the left robotic arm of the cabin docking, the coordinate system O_R at the end of the right robotic arm of the cabin docking, the center coordinate system O_1 at the left end face of the cabin workpiece, and the center coordinate system O_2 at the right end face of the cabin workpiece are approximately in a straight line, as shown in Figure 20.

In Figure 20, when the pitch angle α is not 0, the values of coordinate systems $O_1(x_1, y_1, z_1)$, $O_L(x_L, y_L, z_L)$, $O_R(x_R, y_R, z_R)$ and $O_2(x_2, y_2, z_2)$ in the Y-direction are equal, and the position relationship of each coordinate system is shown in Equation (6).

$$\begin{cases} (x_1 - x_L)^2 + (z_1 - z_L)^2 = D_1{}^2 \\ (x_1 - x_R)^2 + (z_L - z_R)^2 = D_2{}^2 \\ (x_R - x_2)^2 + (z_R - z_2)^2 = D_3{}^2 \end{cases} \tag{6}$$

The relationship between the pitch angle α and the coordinate system parameters is shown in Equation (7).

$$\sin(\alpha) = \frac{z_R - z_L}{D_2} \tag{7}$$

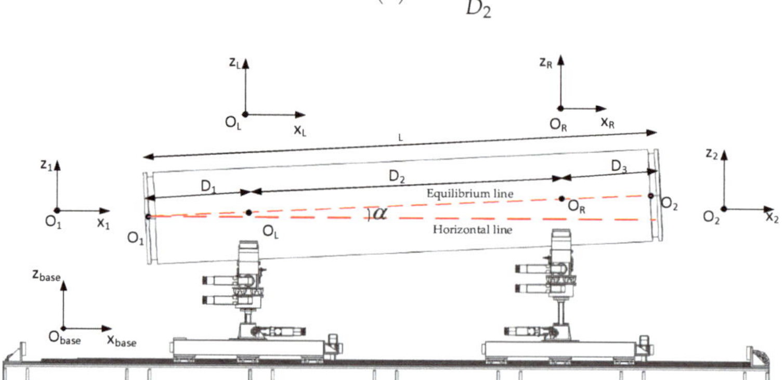

Figure 20. Case of the coordinate system when the pitch angle is not $0°$.

(3) The yaw angle of the cabin workpiece is not $0°$

Because of the influence of friction between the docking robot and the cabin workpiece, and the limited stroke of the radial movement joint of the docking robot (± 50 mm), when the pitch angle α changes, the docking robot still makes contact with the cabin workpiece. Therefore, the origin of the coordinate system O_L at the end of the left robotic arm of the cabin docking, the coordinate system O_R at the end of the right robotic arm of the cabin docking, the center coordinate system O_1 at the left end face of the cabin workpiece, and the center coordinate system O_2 at the right end face of the cabin workpiece are approximately in a straight line, as shown in Figure 21.

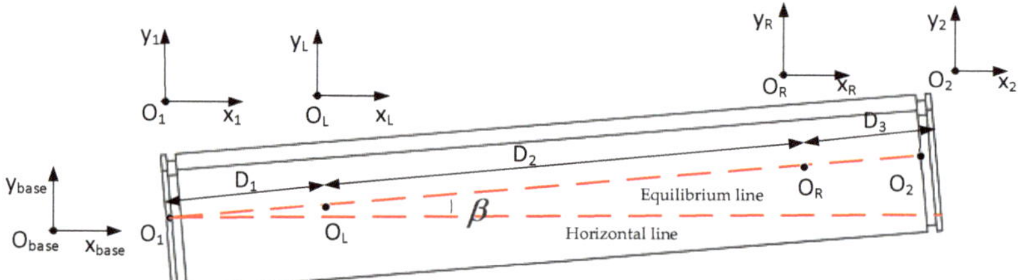

Figure 21. Case of the coordinate system when the yaw angle is not $0°$.

In Figure 21, when the yaw angle β is not 0, the values of coordinate systems $O_1(x_1, y_1, z_1)$, $O_L(x_L, y_L, z_L)$, $O_R(x_R, y_R, z_R)$, and $O_2(x_2, y_2, z_2)$ in the Z-direction are equal, and the position relationship of each coordinate system is shown in Equation (8).

$$\begin{cases} (x_1 - x_L)^2 + (y_1 - y_L)^2 = D_1^2 \\ (x_1 - x_R)^2 + (y_L - y_R)^2 = D_2^2 \\ (x_R - x_2)^2 + (y_R - y_2)^2 = D_3^2 \end{cases} \tag{8}$$

The relationship between the yaw angle and the coordinate system parameters is shown in Equation (9).

$$\sin(\beta) = \frac{y_R - y_L}{D_2} \tag{9}$$

(4)   The rolling angle of the cabin workpiece γ is not $0°$

Because the workpiece of the cabin belongs to rigid body, the position relation of each coordinate system is not affected when the rolling angle changes. Therefore, the origin of the coordinate system O_L at the end of the left robotic arm of the cabin docking, the coordinate system O_R at the end of the right robotic arm of the cabin docking, the center coordinate system O_1 at the left end face of the cabin workpiece, and the center coordinate system O_2 at the right end face of the cabin workpiece are approximately in a straight line, as shown in Figure 22.

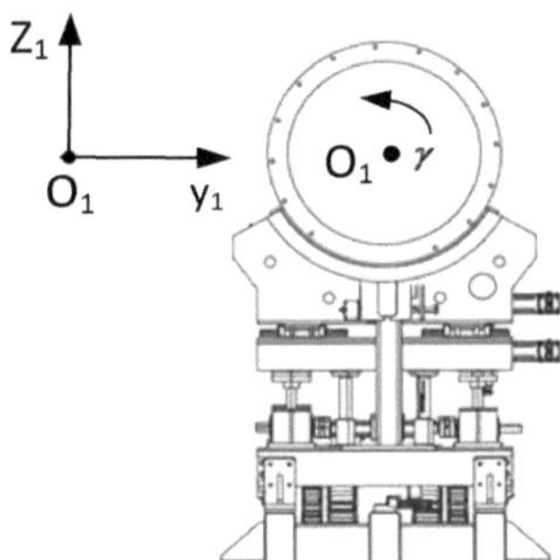

Figure 22. Case of the coordinate system when the rolling angle is not $0°$.

In Figure 22, when the rolling angle γ of the cabin workpiece is not 0, the position relationship of each coordinate system is the same as the initial equilibrium position.

Therefore, according to the above analysis results, the following conclusions can be drawn:

(1)   When the attitude of the module changes, the relative distance between the origin of the coordinate system at the end of the left robotic arm of the module docking and the origin of the coordinate system at the end of the right robotic arm of the module docking remains unchanged.

(2)   When the cabin roll angle changes, the position adjustment of the cabin docking robot is not affected, but the attitude of the cabin itself is affected.

Assuming that the motion trajectory of the end of the left robotic arm is represented by $K_L(x_L(t), y_L(t), z_L(t))$, and the motion trajectory of the end of the right robotic arm is represented by $K_R(x_R(t), y_R(t), z_R(t))$, the first equation of the joint kinematics model is obtained:

$$(x_L(t) - x_R(t))^2 + (y_L(t) - y_R(t))^2 + (z_L(t) - z_R(t))^2 = D_2{}^2 \tag{10}$$

In Equation (10), the adjustment of the space position of the cabin is based on the movement trajectory $K_L(x_L(t), y_L(t), z_L(t))$ of the end of the left robotic arm of the cabin docking. The movement trajectory $K_R(x_R(t), y_R(t), z_R(t))$ of the end of the right robotic arm of the cabin docking follows the movement of the left robotic arm, and the relative distance from the movement trajectory of the end of the left robotic arm is guaranteed to remain unchanged.

In addition to participating in the adjustment of the space position, the right robotic arm of cabin docking is also adjusted for the attitude of the cabin. In this way, the movement trajectory $K_R(x_R(t), y_R(t), z_R(t))$ of the end of the right robotic arm of cabin docking can be divided into two parts, respectively assumed to be $K_R{}^1(x_R(t), y_R(t), z_R(t))$ and $K_R{}^2(x_R(t), y_R(t), z_R(t))$. The relationship is as follows:

$$K_R(x_R(t), y_R(t), z_R(t)) = K_R{}^1\left(x_R(t)^1, y_R(t)^1, z_R(t)^1\right) + K_R{}^2\left(x_R(t)^2, y_R(t)^2, z_R(t)^2\right) \tag{11}$$

In Equation (11), A is used to follow the action of the left robotic arm, and B reflects the attitude change in the cabin segment. The following equation exists:

$$\begin{cases} x_R(t)^2 = x_L(t) + D_2 cos(\alpha(t))cos(\beta(t)) \\ y_R(t)^2 = y_L(t) + D_2 cos(\alpha(t))sin(\beta(t)) \\ \quad z_R(t)^2 = z_L(t) + D_2 cos(\alpha(t)) \end{cases} \tag{12}$$

where $\alpha(t)$ and $\beta(t)$ represent the yaw angle trajectory and the pitch angle trajectory of the cabin workpiece, respectively, and the motion trajectory of the yaw angle and pitch angle is reflected in the motion trajectory $K_R{}^2(x_R(t), y_R(t), z_R(t))$.

The coordinates of the end coordinate system $O_L(x_L, y_L, z_L)$ of the left robot arm are related to the coordinates $O_1(x_1, y_1, z_1)$ of the center of the end circle of the cabin workpiece. The second equation of the joint kinematics can be obtained as follows:

$$\begin{cases} x_L(t) = x_1 + D_1 cos(\alpha(t))cos(\beta(t)) \\ y_L(t) = y_1 + D_1 cos(\alpha(t))sin(\beta(t)) \\ \quad z_L(t) = z_1 + D_1 sin(\alpha(t)) \end{cases} \tag{13}$$

where $\alpha(t)$ and $\beta(t)$ are the pitch angle and yaw angle trajectories of the moving module, respectively.

4.2. Workspace Analysis

During the movement of the robotic arm, its travel is limited. The working space represents the reachable position of the reference point in the end mechanism of the robotic arm, and its size represents the spatial field of the movement of the robotic arm, which is one of the important indicators of the robotic arm [21]. According to the design parameters of the module docking robot in Table 1, and through the forward kinematics Equations (1) and (2) of the left and right robotic arms, the feasible domain in the working space is obtained, as shown in Table 3.

Table 3. Workspace analysis of the docking robot.

Serial Number	Item	Value
1	Left arm X-axis stroke	$-1695 \sim -495$ mm
2	Left arm Y-axis stroke	$-50 \sim 50$ mm
3	Left arm Z-axis stroke	$690 \sim 790$ mm
4	Left arm rolling angle range	$170 \sim 190°$
5	Right arm X-axis stroke	$495 \sim 1695$ mm
6	Right arm Y-axis stroke	$-50 \sim 50$ mm
7	Right arm Z-axis stroke	$690 \sim 790$ mm
8	Right arm rolling angle stroke	$-10 \sim 10°$

The Monte Carlo method, also known as statistical simulation test method, estimates and describes the unknown characteristic quantity by sampling random variables and then determines its distribution [22]. This method is widely used to solve robot workspaces [23–25].

After the strain sample data are obtained through calculation, the sample data distribution is processed, and then the sample data are randomly simulated by the Monte Carlo method. The Monte Carlo method consists of the following steps:

(1) Construct a probability model. In this paper, the probability distribution of the equivalent strain of the independent variable follows the normal distribution model.

(2) Obtain sub-samples by simulation. Random numbers are generated within the constraints of each joint in the workspace according to the normal distribution, and the position in the operation space is obtained through the coordinate transformation relationship.

In this paper, the Monte Carlo method is used to carry out relevant operations through MATLAB 2021 software, and the working space of the designed robot arm is obtained. The cloud image of the working space of the robot arm obtained by this method is shown in Figures 23 and 24.

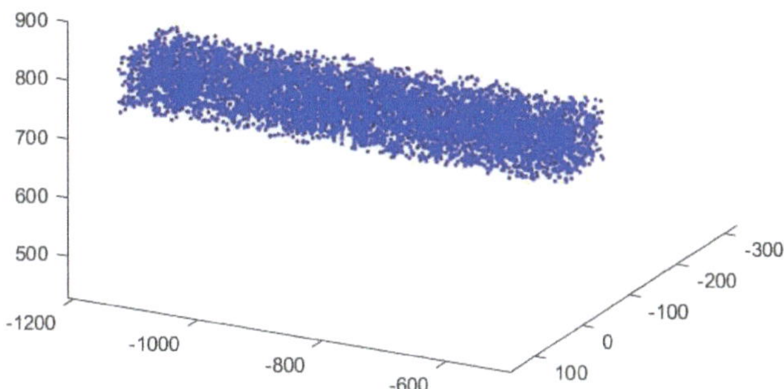

Figure 23. Working space cloud image of the left robotic arm.

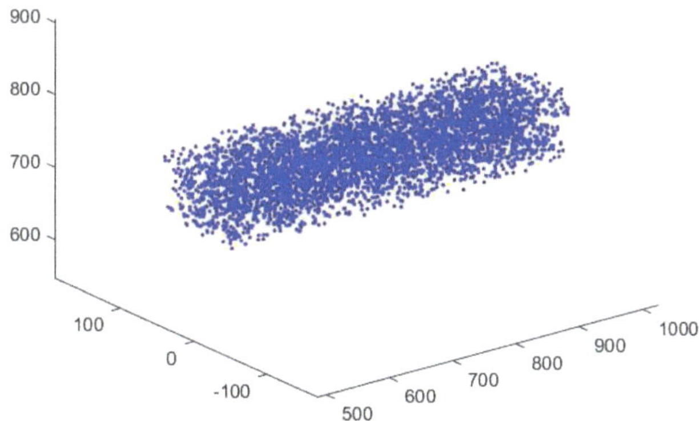

Figure 24. Working space cloud image of the right robotic arm.

4.3. Kinematics Simulation

4.3.1. Kinematics Simulation of Cabin Docking Robot System

Forward kinematics and inverse kinematics were verified by using Robotics Toolbox in MATLAB 2021 software.

(1)   Establishment of simulation model

The Robotics Toolbox in MATLAB 2021 software was used to model the robot. The modeling process of the toolbox requires the D-H parameters of the robot as the basis, the LINK function and SerialLink function are used to realize the robot modeling, and the teach function is used to simulate and drive the robot to realize the visualization of the robot movement. The robot control panel is shown in Figure 25.

In Figure 25, the pose parameters (x, y, z, R, P, Y) of the corresponding end operating space of the robot arm can be changed by adjusting the joint space variables of the left robot arm or the joint space variables of the right robot arm. Meanwhile, the joint space variables of the robot arm are displayed graphically, as shown in Figure 26. The joint space variables $q_2, q_3, q_4,$ and q_5 of the left robotic arm correspond to $d_{L1}, d_{L2}, d_{L3},$ and θ_L, respectively, in Equation (6). The joint space variables $q_2, q_3, q_4,$ and q_5 of the right robotic arm correspond to $d_{R1}, d_{R2}, d_{R3},$ and θ_R and, respectively, in Equation (4).

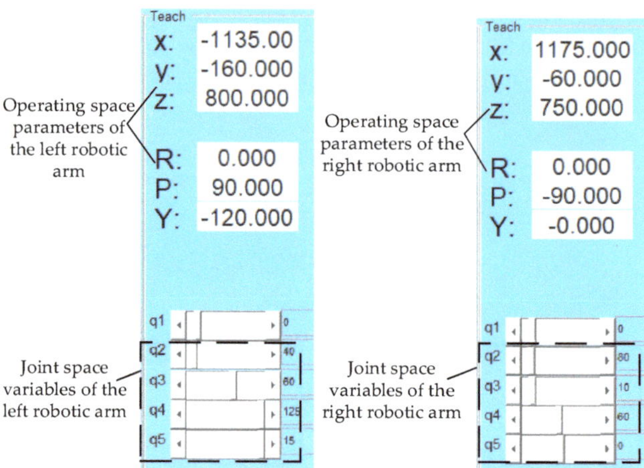

Figure 25. Control panel of the docking robot.

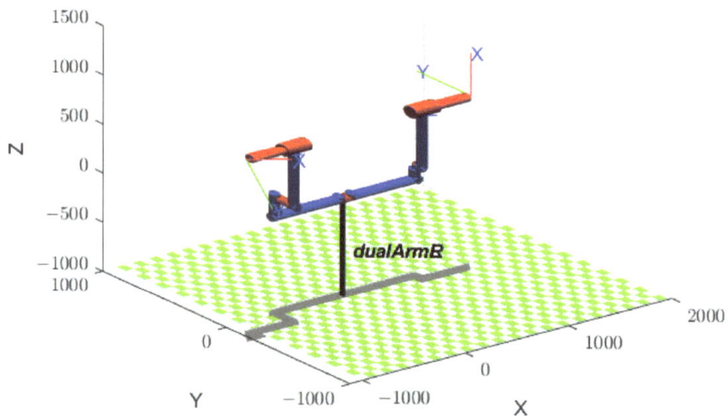

Figure 26. Three-dimensional space model of the docking robot.

(2)   Kinematics simulation

Kinematics simulation process: First, the joint variables of the manipulator are given specific values, and the space pose parameters of the end of the manipulator are calculated by the forward kinematics Equations (4) and (6). Then, the spatial pose and pose parameters of the end of the robot arm are calculated through the inverse kinematics Equations (7) and (8) to obtain a new joint variable of the robot arm. The new joint variable is compared to the original joint variable to determine if they are equal. If the new joint variable is equal to the original, the forward and inverse kinematics modeling is proved correct; otherwise, it is proved incorrect, as shown in Figure 27.

According to the simulation ideas in Figure 27, the simulation process is as follows:

(1)   The joint variable $(d_{L1}, d_{L2}, d_{L3}, \theta_L)$ of the left robotic arm is $(40, 60, 30, -130)$, and the joint variable $(d_{R1}, d_{R2}, d_{R3}, \theta_R)$ of the right robotic arm is $(80, 20, 10, 30)$;

(2)   Through forward kinematics calculation, the end space pose (x_L, y_L, z_L, R, P, Y) of the left robotic arm is $(-1135, -30, 800, 0, 90, 50)$, and the end space pose (x_R, y_R, z_R, R, P, Y) of the right robotic arm is $(1175, -10, 760, 0, -90, -30)$;

(3) The space pose of the end of the robot arm is brought into the inverse kinematics equation, and the calculated joint space variable $(d_{L1}, d_{L2}, d_{L3}, \theta_L)$ of the left robot arm is $(40, 60, 30, -130)$, and the joint variable $(d_{R1}, d_{R2}, d_{R3}, \theta_R)$ of the right robot arm is $(80, 20, 10, 30)$;

(4) The inverse kinematics solution results in the joint variable being equal to the initial assignment, which proves that the kinematics model is correct.

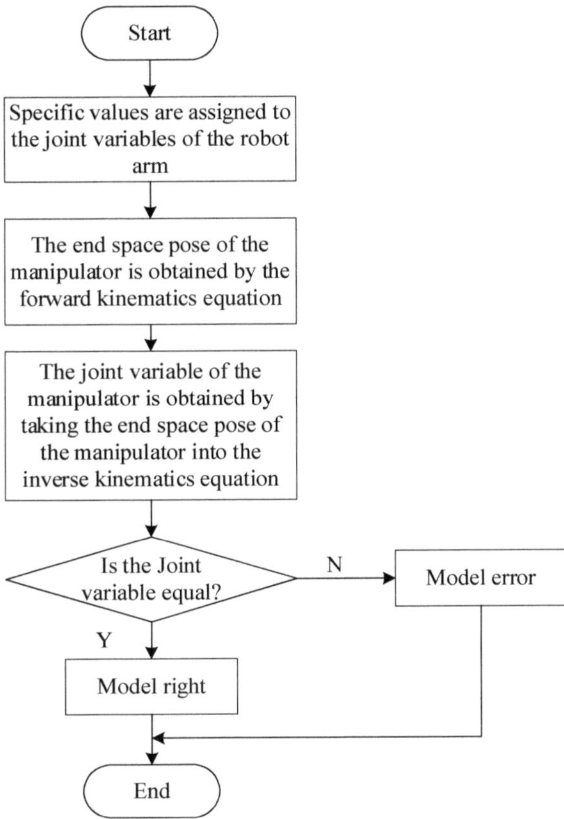

Figure 27. Kinematics simulation process.

4.3.2. Joint Kinematics Simulation of the Cabin Docking Robot and Cabin Workpiece

(1) Simulation parameter setting

The pose starting and ending points of the simulation process are

$$\begin{cases} p_o = [-1315, -35, 775, 0.2, -0.2, 190] \\ p_e = [-1515, -5, 745, 0, 0, 180] \end{cases} \tag{14}$$

The elements in the vector represent, from left to right, the X-axis position, Y-axis position, Z-axis position, pitch angle, yaw angle, and rolling angle of the cabin workpiece.

(2) Results of joint kinematics simulation

In order to verify the effectiveness of the joint kinematics modeling of the module docking robot and the module workpiece, the B-spline curve is used to plan the motion trajectory of the robot arm. The simulation results are shown in Figure 28.

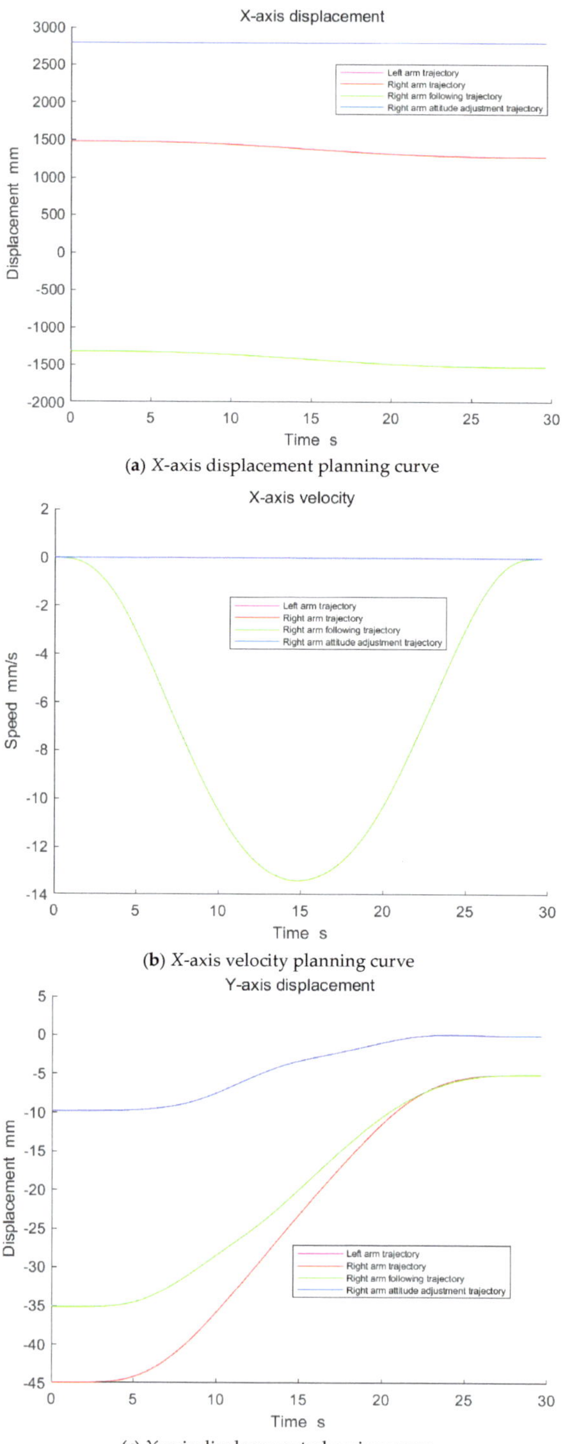

(**a**) *X*-axis displacement planning curve

(**b**) *X*-axis velocity planning curve

(**c**) *Y*-axis displacement planning curve

Figure 28. *Cont.*

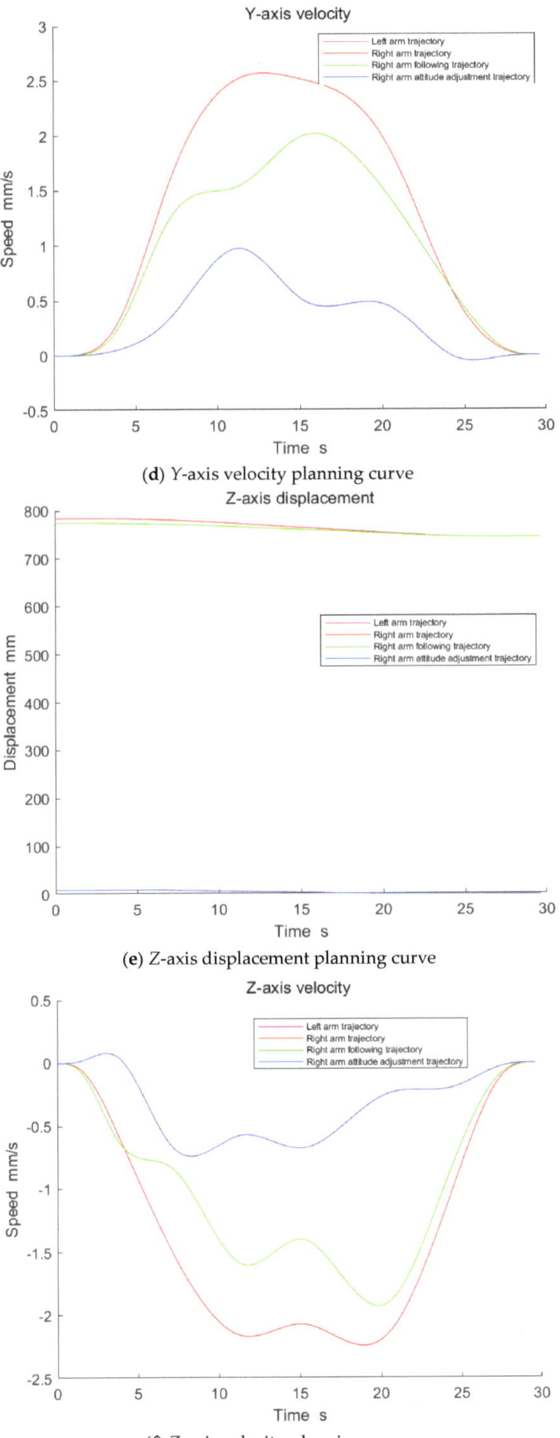

(**d**) Y-axis velocity planning curve

(**e**) Z-axis displacement planning curve

(**f**) Z-axis velocity planning curve

Figure 28. Results of the joint kinematic trajectory simulation.

It is assumed that the right robotic arm trajectory consists of the following two parts: the following trajectory of the right robotic arm and the attitude adjustment trajectory of the right robotic arm. As can be seen in Figure 28, when the left robotic arm runs from the starting position p_0 to the stopping position p_e, the planned displacement and planned speed of the left robotic arm are stable in the X-, Y-, and Z-axis directions. At the same time, the following trajectory of the right robotic arm and the trajectory of the left robotic arm coincide, and the running trajectory of the right robotic arm is equal to the synthesis of the following trajectory of the right robotic arm and the attitude adjustment trajectory of the right robotic arm. This verifies the effectiveness of the joint kinematic modeling of the docking arms in the cabin.

5. Experimental Verification

5.1. Development of a Prototype Robot System for Cabin Docking

The experimental platform of the cabin docking robot system is composed of a fixed cabin, a moving cabin, a fixed bracket, and the docking robot. The fixed cabin is placed on the fixed bracket, and the relative posture remains unchanged. The space pose adjustment of the moving cabin is realized by the docking robot. Finally, the docking between the mobile module and the fixed module is completed. The experimental platform of the cabin docking robot system is shown in Figure 29.

Figure 29. Experimental platform of the cabin docking robot system.

In Figure 29, the main specifications of the motor, encoder, driver and motion control system of the experimental platform are shown in Table 4.

Table 4. Main selection parameters of servo motor and reducer.

Name	Specification Parameter	Brand
Motor	EX310ESPB1511	Parker
Driver	C3S015V4F10I31T11M00	Parker
X-axis reducer	ABR060 L2-40-P2-S2	Newstart
Y-axis reducer	ABR060 L2-20-P1-S2	Newstart
Z-axis reducer	ABR060 L1-10-P2-S2	Newstart
Controller	CX5140-0135	Beckhoff
Digital output module	EL2809	Beckhoff
Digital input module	EL1808	Beckhoff
Analog input module	EXL3152	Beckhoff
Power module	EXL9560	Beckhoff

5.2. Workspace Test Experiment

5.2.1. Experimental Methods

The work space test experiments were carried out for the axial moving joint, radial moving joint, vertical lifting moving joint, and circular rotating joint of the cabin docking robot, and the experimental results were recorded for analysis.

The test method of working space of moving joint is as follows: set the starting position of the moving joint as D_0 (mm), move forward to the limit position, and record the value as D_{i+} (mm). Move to the limit position in the reverse direction, record the value as D_{i-} (mm), calculate the range of motion $D_i = (D_{i+} - D_{i-})$, and determine whether the range of movement covers the design index.

The test method of repeated positioning accuracy of rotational joint is as follows: set the starting position of the rotational joint as A_0 (°), move forward to the limit position, and record the value as A_{i+} (°). Move to the limit position in the reverse direction, record the value as A_{i-} (°), calculate the angle range $A_i = (A_{i+} - A_{i-})$, and determine whether the motion range of the rotational joint covers the design index.

5.2.2. Experimental Results

The working space test results of the axial movement joint, the vertical lifting joint, and the radial movement joint are shown in Table 5. The working space test result of the circular rolling joint is shown in Table 6.

Table 5. Range of motion of the moving joint.

Joint Name	D_{1+} (mm)	D_{1-} (mm)	Displacement (mm)
Axial movement joint	3150	−13	3163
Vertical lifting joint	55	−46	101
Radial movement joint	−1	−102	101

Table 6. Range of motion of the rotational joint.

Joint Name	A_{1+} (°)	A_{1-} (°)	Angle (°)
Circular rotating joint	10.1	−10.2	20.3

As can be seen in Tables 5 and 6, the motion displacement of the axial moving joint is 3163 mm, the motion displacement of the vertical lifting joint is 101 mm, the motion displacement of the radial moving joint is 101 mm, and that of the circular rolling joint is 20.3°. All of these values meet the design index requirements in Table 1.

5.3. Repeated Joint Positioning Accuracy Test

5.3.1. Experimental Methods

Repeated positioning accuracy tests were carried out for the axial moving joint, the radial moving joint, the vertical lifting joint, and the circular rotating joint, and the experimental results were recorded for analysis.

The test method of repeated positioning accuracy of moving joint is as follows: select an initial position on a movable joint and fix the dial indicator. Offset a distance L_0 (mm) and then return to the initial position and read the dial indicator value. The repeated positioning accuracy of each joint was obtained by statistical standard deviation of dial indicator values.

The test method of repeated positioning accuracy of rotational joint is as follows: install and fix the dial indicator at 400 mm of the center of circular rotating joint arc rolling body. Rotate a fixed angle A_0 (°) and then return to the initial position and read the dial indicator value. The repeated positioning accuracy of each joint was obtained by the statistical standard deviation of dial indicator values.

5.3.2. Experimental Results

The repeated positioning accuracy test results of the axial moving joint, the radial moving joint, the vertical lifting joint, and the circular rotating joint are shown in Table 7, Table 8, Table 9, and Table 10, respectively.

Table 7. Experimental results of repeated positioning accuracy of the axial movement joint.

Number	10 mm	30 mm	50 mm	70 mm	90 mm
1	2.879	2.85	2.81	2.82	2.83
2	2.86	2.83	2.81	2.83	2.83
3	2.87	2.85	2.83	2.82	2.84
Standard deviation	0.009504	0.011547	0.011547	0.005774	0.005774

Table 8. Experimental results of repeated positioning accuracy of the radial movement joint.

Number	5 mm	10 mm	15 mm	20 mm	25 mm
1	2.75	2.74	2.73	2.73	2.73
2	2.74	2.73	2.73	2.73	2.73
3	2.74	2.73	2.73	2.73	2.73
Standard deviation	0.005774	0.005774	0	0	0

Table 9. Experimental results of repeated positioning accuracy of the vertical lifting joint.

Number	5 mm	10 mm	15 mm	20 mm	25 mm
1	2.73	2.75	2.76	2.75	2.76
2	2.74	2.75	2.75	2.75	2.76
3	2.74	2.75	2.76	2.75	2.76
Standard deviation	0.005774	0	0.005774	0	0

Table 10. Experimental results of repeated positioning accuracy of the circular rotary joint.

Number	0.5°	1°	1.5°	2°	2.5°
1	2.46	2.54	2.61	2.63	2.63
2	2.46	2.54	2.62	2.63	2.63
3	2.46	2.54	2.62	2.63	2.62
Standard deviation	0	0	0.005773503	0	0.005773503

As listed in Tables 7–10, the maximum and minimum standard deviation of the measurement values of the axial mobile joint repeated positioning are 0.011547 mm and 0.005774 mm, respectively. The maximum and minimum standard deviations of the repeated positioning measurements of the radial shifting joint and the vertical lifting joint are 0.005774 and 0 mm, respectively. The maximum and minimum standard deviation of the repeated positioning measurement values are 0.005774 mm and 0 mm, respectively.

According to the above test results, the maximum standard deviation of the repeated positioning accuracy measurement value of each joint of the cabin docking robot is 0.011547 mm, which meets the design requirements.

5.4. Precision Experiment of Joint Motion Control

In order to test the precision of joint motion control and set the target moving displacement value of the joint, the actual moving displacement value of the joint was obtained with the joint positioning control experiment through the MC_MoveAbsolute function block of the controller. The displacement error of the joint motion was obtained through calculation, and whether the displacement error of the joint motion was within the allowable range was determined.

5.4.1. Control Accuracy of the Axial Movement Joint

The axial movement joint is controlled to move the fixed displacement of 40 mm. The displacement change curve in the moving process is collected, and the displacement control accuracy is calculated. The collected displacement curve is shown in Figure 30.

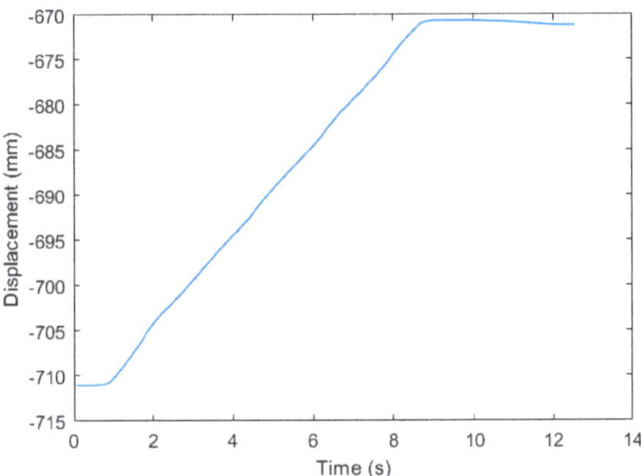

Figure 30. Displacement curve of 40 mm axial movement.

It can be seen in Figure 30 that when the axial moving joint moves 40 mm, the displacement motion is stable. The starting position of the joint movement is -711.027 mm, the target value of the stopping position of the joint movement is -671.027 mm, and the actual value of the stopping position of the joint movement is -671.085 mm. The motion error of the stopping position of the axial moving joint is 0.058 mm.

5.4.2. Control Accuracy of Vertical Lifting Joint

The vertical lifting joint was controlled to drop 5 mm. The displacement curve in the moving process was collected, and the displacement control accuracy was calculated. The collected displacement curve is shown in Figure 31.

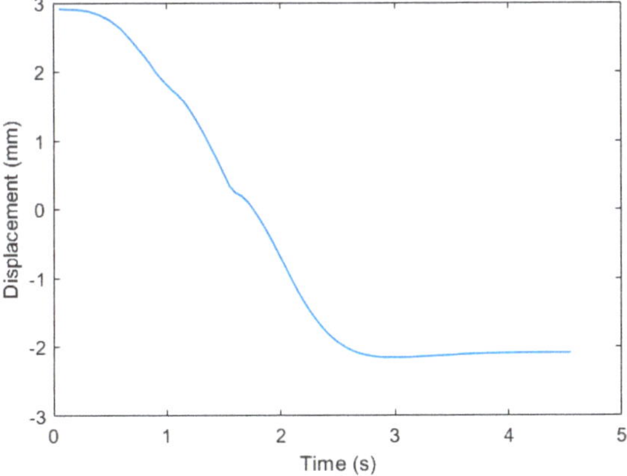

Figure 31. Displacement curve of vertical descent of 5 mm.

It can be seen in Figure 31 that when the vertical lifting joint moves 5 mm, the displacement movement process is stable. The starting position of joint motion is 2.914 mm, the target value of joint motion stop position is -2.086 mm, and the actual value of joint motion stop position is −2.085 mm. The motion error of vertical lifting joint stop position is 0.001 mm.

5.4.3. Control Accuracy of the Radial Movement Joint

The radial mobile joint was controlled to move the fixed displacement 10 mm. The displacement change curve in the moving process was collected, and the displacement control accuracy was calculated. The collected displacement curve is shown in Figure 32.

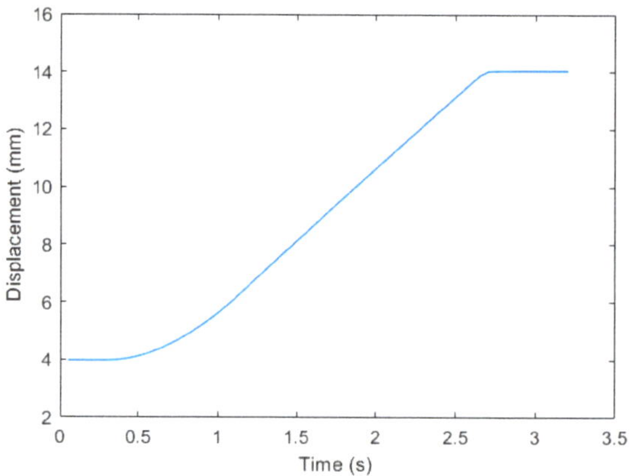

Figure 32. Displacement curve of radial movement of 10 mm.

It can be seen in Figure 32 that when the radial mobile joint moves 10 mm, the motion of displacement and velocity is stable. The initial position of the joint motion is 3.966 mm, the target value of the joint motion stop position is 13.966 mm, and the actual value of the joint motion stop position is 14.031 mm. The motion error of the radial moving joint stop position is 0.0065 mm.

5.4.4. Control Accuracy of the Circular Rolling Joint

The circular rolling joint was controlled to fix the angle of circular rolling 2°. The change curve of the angle in the moving process was collected, and the angle control accuracy was calculated. The collected angle curve is shown in Figure 33.

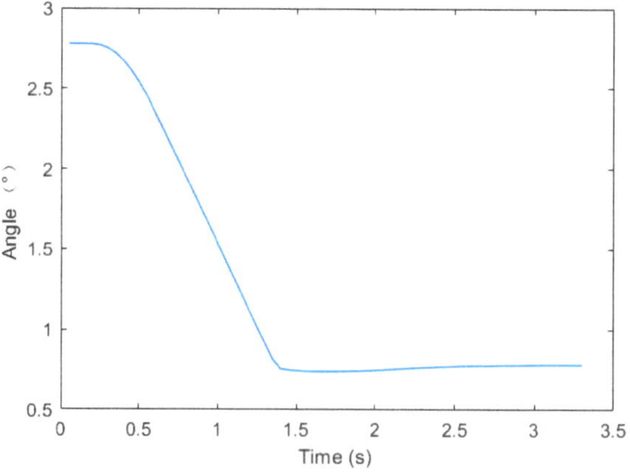

Figure 33. Angular curve of the 2° circular roll.

As can be seen in Figure 33, when the circular rolling joint is rotated by 2°, the angular motion process is stable. The initial angle of joint motion is 2.774°, the target value of joint motion stopping angle is 0.774°, and the actual value of joint motion stopping position is 0.778°. The motion error of radial moving joint stopping position is 0.004°.

5.5. Cabin Docking Experiment

According to the working principle of the cabin docking robot system, the control system block diagram of the cabin docking robot is designed, as shown in Figure 34.

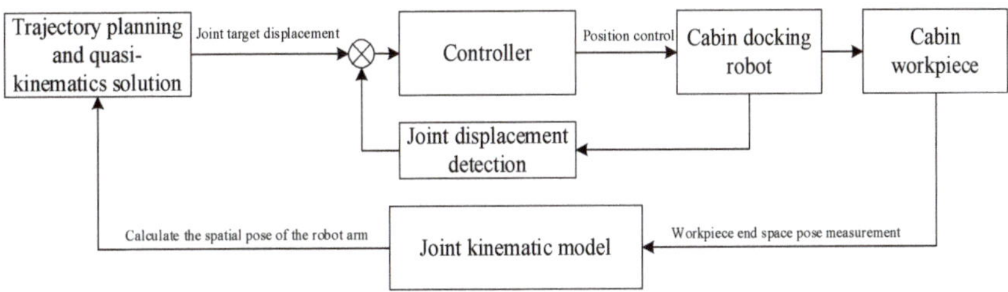

Figure 34. Block diagram of the docking control system of the docking robot.

In Figure 34, an external laser tracker is used to measure the six characteristic quantities of the workpiece end face center coordinate $O_1(x_1, y_1, z_1)$, pitch angle α, yaw angle β, and rolling angle γ and convert them into the end coordinate $O_L(x_L, y_L, z_L)$ of the docking robot arm by Equation (13). Then, the target displacement of each joint is calculated through trajectory planning and inverse kinematics. Finally, the MC_MoveAbsolute function block of the servo drive is invoked to control the position of the joint.

In the process of cabin docking, the starting position of the module docking robot in the Z-axis direction is 100 mm, and the stopping position is set to −1.99 mm. The planned target trajectory is compared with the actual trajectory collected, as shown in Figure 35.

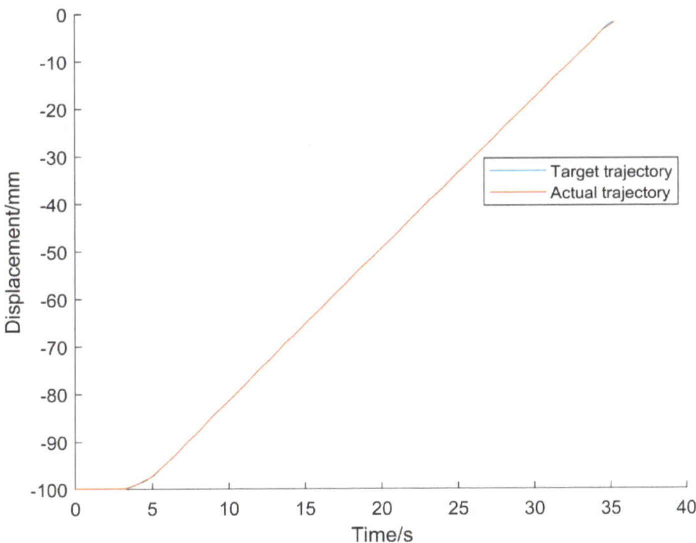

Figure 35. Trajectory control curve of the cabin docking robot in the Z-axis direction.

In Figure 35, the maximum displacement error between the target trajectory and the actual trajectory is 0.377 mm and the minimum error is 0 mm.

In the process of cabin docking, the start position of the cabin docking robot in the Y-axis direction is 30 mm, and the stop position is set to 0.63 mm. The planned target trajectory is compared with the actual trajectory collected, as shown in Figure 36.

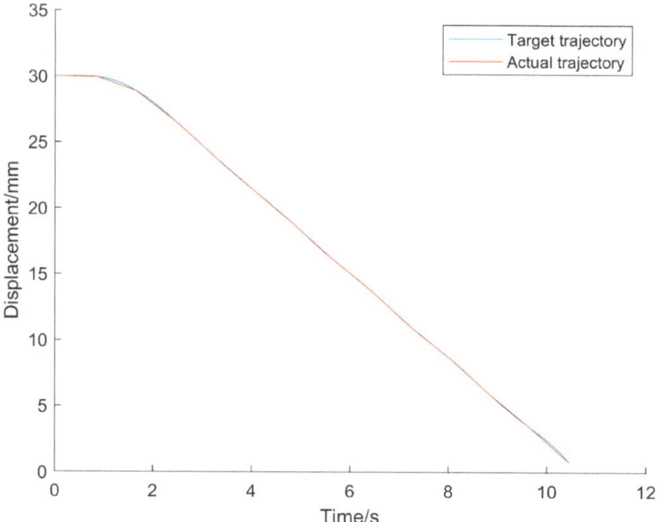

Figure 36. Trajectory control curve of the cabin docking robot in the Y-axis direction.

In Figure 36, the maximum displacement error between the target trajectory and the actual trajectory is 0.4655 mm and the minimum error is 3.333×10^{-5} mm.

5.6. Comparison of Robot Performance Indicators

In order to evaluate the performance advancement of the cabin docking robot system, the performance indicators of the module docking robot are compared with those of the standard heavy-load industrial robot (KR 1000 L950 TITAN PA). The comparison results are shown in Table 11.

Table 11. Comparison table of robot specification parameters.

Name	Cabin Docking Robot	Standard Industrial Robot	Comparison Result
Load	Max. 1000 kg	1000 kg	Same level
Degree of freedom	6	6	Same level
Repeated positioning accuracy	±0.05 mm	±0.2 mm	Higher level
Installation position	Ground	Ground	Same level
Work space	Circumference: ±15°; Axis: ±600 mm; Vertical: ±50 mm; Radial: ±50 mm;	Max. 3601 mm	Lower level

It can be seen from Table 11 that compared with standard industrial robots, the cabin docking robot is at the same level in terms of load, freedom, and installation position indicators. In terms of repeated positioning accuracy, the cabin docking robot is superior to the standard industrial robot. But in terms of scope of work, the cabin docking robot is lower than standard industrial robots.

6. Conclusions

In this paper, a cabin docking robot system is designed, the strength and kinematics of the key mechanical components of the robot are analyzed, and the key indexes of the robot system are verified by a prototype.

(1)   The cabin docking robot is designed by the horizontal docking mode, and the composition and working principle of each joint of the robot are analyzed. The degree of freedom of the cabin docking robot is six, and the maximum load capacity can reach 1000 kg.

(2)   The static stress analysis of the key components of the robot joints is carried out by finite element software. The maximum equivalent stress of the bottom frame is 31.81 Mpa and the maximum total deformation is 0.125 mm. The maximum equivalent stress of the trapezoidal lead screw is 28.86 Mpa, and the maximum total deformation is 0.0177 mm. The maximum equivalent stress of the shell bracket is 23.13 Mpa, and the maximum total deformation is 0.0182 mm. The maximum equivalent stress of the rolling bearing is 9.153 Mpa, and the maximum total deformation is 0.001195 mm. The simulation results are all within the reasonable range.

(3)   The kinematics model of the robot is established based on the D-H modeling method, the forward and inverse kinematics relationship of the robot are analyzed, and the correctness of the kinematics model of the robot is verified by simulation.

(4)   The cabin docking robot and the joint kinematics model of the module are established and the correctness of the joint kinematics model is verified through simulation.

(5)   The working space, repeated positioning accuracy, and motion control accuracy of each joint of the robot are verified through experiments, which proved the effectiveness of the robot design.

Based on the above analysis, the cabin docking robot system designed in this paper can meet the automatic attitude adjustment function of medium and large module workpieces. It has the advantages of a strong bearing capacity, a simple control algorithm, high stiffness, decoupling of degrees of freedom, and so on. However, the total structure of the module docking robot is large, and the area is also large, which brings certain restrictions to the field application.

In addition, it is necessary to strengthen the speed control algorithm in subsequent research.

Author Contributions: Conceptualization, R.L. and F.P.; methodology, R.L.; software, R.L.; validation, R.L.; formal analysis, R.L.; investigation, R.L.; resources, R.L.; data curation, R.L.; writing—original draft preparation, R.L.; writing—review and editing, F.P.; visualization, F.P.; supervision, F.P.; project administration, F.P.; funding acquisition, F.P. All authors have read and agreed to the published version of the manuscript.

Funding: We would like to express our gratitude to the National Natural Science Foundation of China (No. 61973036), the National Natural Science Foundation of China (No. 62261160575), and the National Natural Science Foundation of China (No. 61991414).

Data Availability Statement: The original contributions presented in the study are included in the article material, further inquiries can be directed to the corresponding author.

Conflicts of Interest: The authors declare no conflict of interest.

References

1.   Guo, Z.M.; Jiang, J.X.; Ke, Y.L. Posture Alignment for Large Aircraft Parts Based on Three POGO Sticks Distributed Support. *Acta Aeronaut. Astronaut. Sin.* **2009**, *30*, 1319–1324.

2.   Guo, Z.M.; Jiang, J.X.; Ke, Y.L. Design and accuracy for POGO stick with three-axis. *J. Zhejiang Univ. (Eng. Sci.)* **2009**, *43*, 1649–1654.

3.   Jiang, J.X.; Chen, Q.; Fang, Q.; Ke, Y. Analysis and experimental test on dynamic characteristic of 3-axis positioner system. *Comput. Integr. Manuf. Syst.* **2009**, *19*, 1004–1009.

4. Zhang, B.; Yao, B.; Ke, Y. A novel posture alignment system for aircraft wing assembly. *J. Zhejiang Univ. Sci. A* **2009**, *10*, 1624–1630. [CrossRef]
5. Yuan, X.; Tang, Y.; Wang, W.; Zhang, L. Parametric Vibration Analysis of a Six-Degree-of-Freedom Electro-Hydraulic Stewart Platform. *Shock. Vib.* **2021**, *2021*, 1–27. [CrossRef]
6. Jin, H.R.; Liu, D. Two-point Positioning and Pose Adjustment Method for Automatic Assembly of Barrel-type Cabin. *China Mech. Eng.* **2018**, *29*, 1467–1474.
7. Dai, W.B.; Hu, R.Q.; Yi, W.M. The flexible docking technology for large spacecraft cabins. *Spacecr. Environ. Eng.* **2014**, *31*, 584–588. [CrossRef]
8. Xiong, T. Automatic Docking Technology of Satellite. *Aeronaut. Manuf. Technol.* **2011**, *22*, 36–39. [CrossRef]
9. Wang, B.X.; Xu, Z.G.; Wang, J.Y.; Wang, Y.J. Design and Research of Missile General Assembly Automatic Docking Platform. *Mod. Def. Technol.* **2016**, *44*, 135–141+147.
10. Yi, W.M.; Duan, B.W.; Gao, F.; Han, X.G. Level docking technology in large cabin assembly. *Comput. Integr. Manuf. Syst.* **2015**, *21*, 2354–2360. [CrossRef]
11. Guo, J.; Zhang, X.; Li, H.; Sun, X. Mechanism Design and Strength Analysis of Key Components of Flight-climbing-slide Robot for High-voltage Transmission Line Inspection. In Proceedings of the 2017 IEEE 7th Annual International Conference on CYBER Technology in Automation, Control, and Intelligent Systems (CYBER), Honolulu, HI, USA, 1 July–4 August 2017.
12. Pao, J.C.; Banglos, C.A.; Salaan, C.J.; Guirnaldo, S.A. Simulation and Strength Analysis of Flippable Mobile Ground Robot for Volcano Exploration and Monitoring Application. In Proceedings of the 2022 IEEE 14th International Conference on Humanoid, Nanotechnology, Information Technology, Communication and Control, Environment, and Management (HNICEM), Boracay Island, Philippine, 1–4 December 2022.
13. Xu, M.; Zhang, Z.; Shao, L.; Zhang, Z.; Cao, C. Analysis and Calculation of Strength and Rigidity for Industrial Grinding Robot. In Proceedings of the 2022 12th International Conference on CYBER Technology in Automation, Control, and Intelligent Systems (CYBER), Baishan, China, 27–31 July 2022.
14. Denavit, J.; Hartenberg, R. A kinematic notation for lower-pair mechanisms based on matrices. *J. Appl. Mech.* **1955**, *22*, 215–221. [CrossRef]
15. Hayati, S.; Mirmirani, M. Improving the absolute positioning accuracy of robot manipulators. *J. Robot. Syst.* **1985**, *2*, 397–413. [CrossRef]
16. Ajayi, O.K.; Arojo, T.A.; Ogunnaike, A.S.; Adeyi, A.J.; Olanrele, O.O. Performance Enhancement with Denavit-Hartenberg (D-H) Algorithm and Faster-RCNN for a 5 DOF Sorting Robot. In *Mechatronics and Automation Technology, Proceedings of the 2nd International Conference (ICMAT 2023), Wuhan, China, 28–29 October 2023*; IOS Press: Amsterdam, The Netherlands, 2024; pp. 459–466.
17. Xiao, P.; Ju, H.; Li, Q.; Meng, J.; Chen, F. A new fixed axis-invariant based calibration approach to improve absolute positioning accuracy of manipulators. *IEEE Access* **2020**, *8*, 134224–134232. [CrossRef]
18. Zhu, Z.; Ma, G.; Liu, H.; Liu, B. D-H model and continuous trajectory planning for orbital welding robot of box-type steel structure. *Hanjie Xuebao/Trans. China Weld. Inst.* **2017**, *38*, 95–98.
19. Li, X.; Wang, H.; Lu, X.; Liu, Y.; Chen, Z.; Li, M. Neural network method for robot arm of service robot based on DH model. In Proceedings of the 2017 Chinese Automation Congress (CAC), Jinan, China, 20–22 October 2017.
20. Pan, Y.; Lu, Y.; Liu, C. Research on welding trajectory planning of MOTOMAN-MA1440 robot based on MATLAB robotics toolbox. In Proceedings of the Third International Conference on Control and Intelligent Robotics (ICCIR 2023), Changsha, China, 23–25 May 2023; Volume 12940.
21. Aliakbari, M.; Mahboubkhah, M.; Sadaghian, M.; Barari, A.; Akhbari, S. Computer integrated work-space quality improvement of the C4 parallel robot CMM based on kinematic error model for using in intelligent measuring. *Int. J. Comput. Integr. Manuf.* **2022**, *35*, 444–461. [CrossRef]
22. Li, X.P.; Yang, Y.; Tian, G.Q.; Li, S.Q.; Wang, J.Y. Life Calculation and Reliability Analysis of Turbine Wheel Based on Monte Carlo Method. *J. Chin. Soc. Power Eng.* **2023**, *43*, 1434–1439+1530.
23. Zhang, F.; Yu, G.; He, S.; He, W. An-improved α-Shapes algorithm for solving collaborative workspaces of dual SCARA robots based on Monte Carlo method. In Proceedings of the 2023 42nd Chinese Control Conference (CCC), Tianjin, China, 24–26 July 2023; Volume 5. [CrossRef]
24. Tomáš, S.; Jozef, S.; Štefan, O. Mapping Robot Singularities through the Monte Carlo Method. *Appl. Sci.* **2022**, *12*, 8330.
25. Gao, H.; Xu, Y. The solution of the volume of collaborative space of 6R two-arm robot based on the body-elemental algorithm. *Acad. J. Eng. Technol. Sci.* **2022**, *5*, 60–67.

Article

The Design and Analysis of a Tunnel Retro-Reflective Ring Climbing and Cleaning Robot

Yuhan Li [1], Shiqing Ye [1], Rongxu Cui [1] and Zhaoyu Shou [2,*]

1 School of Mechanical and Electrical Engineering, Guilin University of Electronic Technology, Guilin 541004, China; liyuhan038@163.com (Y.L.); yesq19@163.com (S.Y.); cui14749@163.com (R.C.)
2 School of Information and Communication, Guilin University of Electronic Technology, Guilin 541004, China
* Correspondence: guilinshou@guet.edu.cn; Tel.: +86-18707739284

Abstract: In response to the challenges posed by the difficult cleaning of tunnel retro-reflective rings and the unsuitability of existing climbing robots for ascending tunnel retro-reflective rings, a tunnel retro-reflective ring cleaning robot is proposed. Firstly, based on the analysis of the operational and environmental characteristics and functional requirements inside the tunnel, the design and planning of the robot's main framework, motion system, cleaning mechanism, and intelligent detection system are conducted to evaluate its walking ability under various working conditions, such as aluminum plate overlaps and rivet protrusions. Subsequently, stability analysis is performed on the robot. The static analysis explored conditions that can make the climbing robot stable, the dynamic analysis obtained the minimum driving torque and finally, verified the stability of the robot through experiments. After that, by changing the material and thickness of the main framework for deformation simulation analysis, the optimal parameters to optimize the design of the main framework are found. Finally, the three factors affecting the cleaning effect of the robot are discussed by the response surface method, and single factor analysis and response surface regression analysis are carried out, respectively. The mathematical regression model of the three factors is established and the best combination of the three factors is found. The cleaning effect is best when the cleaning disc pressure is 5.101 N, the walking wheel motor speed is 36.93 rad/min, and the cleaning disc motor speed is 38.252 rad/min. The development of this machine can provide equipment support for the cleaning of tunnel retro-reflective rings, reducing the requirement of manpower and material resources.

Keywords: tunnel retro-reflective ring; climbing and cleaning robot; stability condition; optimized design; response surface regression analysis

Citation: Li, Y.; Ye, S.; Cui, R.; Shou, Z. The Design and Analysis of a Tunnel Retro-Reflective Ring Climbing and Cleaning Robot. *Actuators* **2024**, *13*, 197. https://doi.org/10.3390/act13060197

Academic Editor: Ioan Doroftei

Received: 22 April 2024
Revised: 15 May 2024
Accepted: 20 May 2024
Published: 22 May 2024

1. Introduction

In recent years, tunnel retro-reflective rings [1], an effective measure to reduce the electricity consumption for tunnel lighting and induce optimal driving effects for motorists, have been widely employed in long and extra-long tunnels. Installed along the contours inside the tunnel, these retro-reflective rings utilize their reflective strips to reflect light received from vehicle headlights, creating a luminous arch-shaped halo within the tunnel, to guide the driver. However, prolonged usage leads to a decrease in the effectiveness of the reflective surface, or even complete loss of reflection, due to the accumulation of dust and fine particles on the retro-reflective ring, necessitating periodic cleaning. Currently, the cleaning process involves lane closures and manual cleaning using lift vehicles, resulting in poor cleaning effectiveness, low efficiency, high costs, and increased safety hazards within the tunnel.

The challenging issue of tunnel retro-reflective rings cleaning has received limited in-depth research and lacks effective solutions. Several methods have been disclosed in Chinese patents, such as the use of a small mobile vehicle equipped with an arc-shaped track matching the outer profile of the tunnel retro-reflective ring. This track features motors and cleaning devices, but requires manual pushing of the vehicle for cleaning, which is highly inconvenient [2]. Additionally, another patent has been disclosed which proposes that climbing and cleaning on the retro-reflective ring can be achieved through an adsorption-based approach [3]; however, the tunnel interior is often wet and slippery, and the intricate walking surface of the retro-reflective ring makes adhesion-based movement difficult.

Therefore, to address the challenge of cleaning the tunnel retro-reflective rings, it is imperative to design a robot capable of climbing on it and performing cleaning tasks. Currently, climbing robots are specifically designed to ascend and move on vertical or nearly vertical surfaces, enabling them to perform specialized tasks. Such robots are commonly employed in various fields, including wind turbine maintenance [4], inspection and repair of steel structures and towers [5], high-altitude operations, inspection [6], and cleaning. Climbing robots necessitate robust adhesion and stability to maintain steadiness during ascension. To fulfill this requirement, they are equipped with appropriate locomotion and adhesion mechanisms tailored to specific working environments [7]. Some climbing robots utilize the principles of foot adhesion, such as Waalbot II, which improves climbing performance by employing natural fiber adhesives [8]. Others mimic lizards, designing claws for adhesion [9,10], while some use suction cup structures for adhesion [11,12]. Some climbing robots employ multiple soft body and foot modules [13] to achieve the function of crossing large gaps, smooth surfaces, and rough terrain. Additional technologies include the use of electromagnetic units [14,15], grippers [16], micro-spines [17], tracks [18], assistance from external devices to aid climbing [19], and so forth. However, these climbing robots seem unsuitable for the cleaning of tunnel retro-reflective rings, as the methods they employ, such as suction cups, micro-spines, adhesives, and electromagnetic units, are not applicable for climbing on tunnel retro-reflective rings. Furthermore, the surface of the retro-reflective ring is relatively smooth, with protrusions or rivets on the back creating obstacles.

This study proposes an innovative solution to achieve efficient cleaning of tunnel retro-reflective rings, reducing the risks associated with manual cleaning and improving cleaning efficiency. As shown in Figure 1, firstly, by leveraging the retro-reflective rings and tunnel environment, a clamping-type wheeled motion mechanism is designed. This mechanism, combined with miniature photoelectric sensors, differential steering principles, and a PID controller, enables the robot to climb steadily. Secondly, a stability analysis of the robot's movement is conducted to establish stability conditions, facilitating the subsequent construction of a prototype for verification. Subsequently, structural optimization is performed using simulation analysis. Finally, response surface methodology is employed to analyze and determine the optimal combination of three parameters that affect cleaning performance. This robot helps reduce the burden on workers, while promoting the application of robotic control in tunnel retro-reflective ring cleaning.

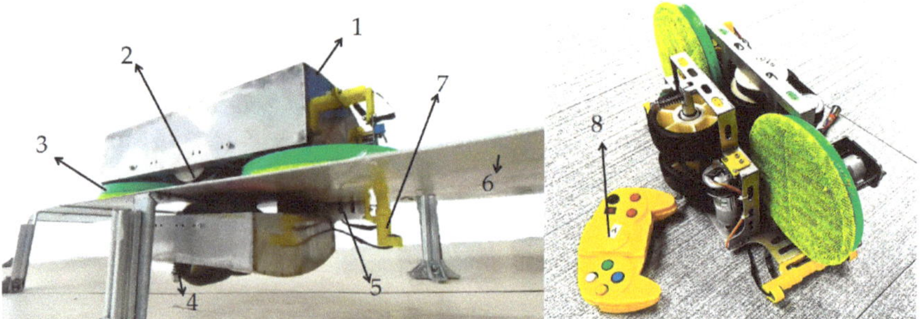

Figure 1. Physical prototype: 1—shell; 2—walking wheel; 3—cleaning device; 4—compression spring; 5—main framework; 6—tunnel retro-reflective ring; 7—miniature photoelectric sensor; 8—remote control.

2. Problem Description and System Design

2.1. The Tunnel Retro-Reflective Ring Exhibits Inherent Issues That Merit Attention

The reflective effect of the tunnel retro-reflective ring during normal operation is shown in Figure 2a. Figure 2b illustrates the retro-reflective panel after cleaning. After conducting on-site surveys in the tunnel, it was observed that the tunnel's arch apex is outfitted with cable trays, leading to a minimal clearance of approximately 5 cm between the side of the tunnel retro-reflective ring and the wire duct, as depicted in Figure 2c. This imposes heightened demands on the design of the climbing robot. To facilitate the climbing robot in traversing this specific location seamlessly, while undertaking the cleaning task for the tunnel retro-reflective ring, it is imperative to ensure the lateral dimensions of the main frame are suitably narrow.

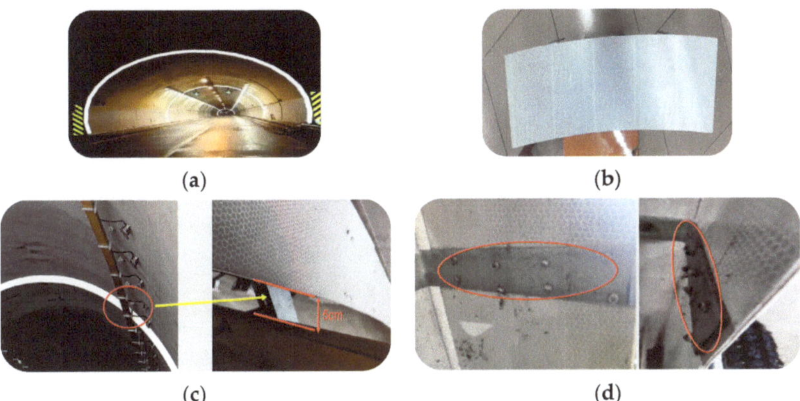

Figure 2. The internal situation of the retro-reflective ring in the tunnel. (**a**) Realistic installation effect of retro-reflective ring inside tunnels; (**b**) retro-reflective ring sample; (**c**) the distance between the wire duct and the retro-reflective ring; (**d**) installation status of the back of the retro-reflective ring.

Furthermore, the tunnel retro-reflective ring, assembled from multiple retro-reflective panels, encounters overlap during installation, due to construction nuances, thereby introducing variations in the overall thickness of the retro-reflective ring. To complicate matters, the retro-reflective panels are affixed to the tunnel wall using right-angled aluminum plates, and the juncture between the right-angled aluminum plates and the retro-reflective panels is secured through a multitude of rivets, as illustrated in Figure 2d. Consequently, the thickness of the right-angled aluminum plates and the protrusions from the rivets contribute to an augmented thickness of approximately 2.5–3.5 mm. Thus, it becomes imperative to devise a walking mechanism endowed with shock absorption and obstacle-crossing functionalities to accommodate fluctuations in the thickness of the retro-reflective ring.

2.2. Mechanical Structure Design

To address the aforementioned issues, a climbing robot was designed for cleaning tunnel retro-reflective rings. The robot integrates mechanical, electronic, and control components into a unified system, as depicted in Figure 3, which shows its internal rendering schematic.

The main body framework of the robot employs a Π-shaped structure, as illustrated in Figure 3, No. 8. The Π-shaped structure consists of two main boards, one sideboard, main board reinforcement ribs, and sideboard reinforcement ribs, all welded with 7075 aluminum alloy. This structure features compactness and high stiffness. The side profile of the Π-shaped frame occupies minimal space, and is narrow and high in strength, thereby smoothly passing through the position between the wire duct and the side of the retro-reflective ring. The 7075 aluminum alloy material has a lower density and higher strength, which is beneficial for making the entire machine lightweight. Additionally, as an aerospace aluminum alloy, it has a certain corrosion resistance, allowing it to be used for extended periods of time in relatively humid tunnel environments.

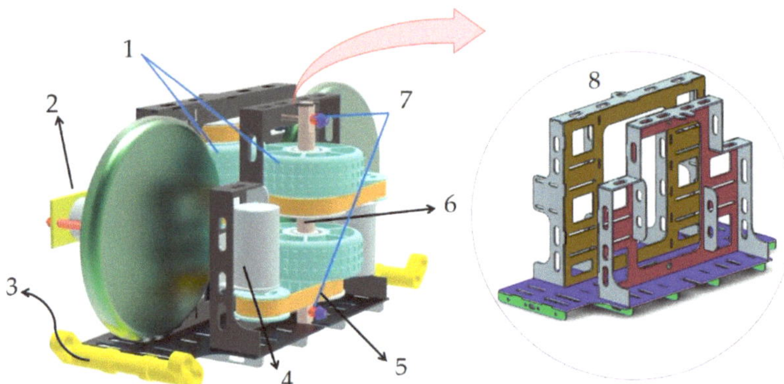

Figure 3. Machine model: 1—walking wheel; 2—cleaning device; 3—miniature photoelectric sensor; 4—encoder deceleration motor; 5—synchronous belt; 6—active spindle; 7—compression spring; 8—Π-shaped framework.

Due to the retro-reflective ring being securely fixed inside the tunnel by right-angled aluminum plates, and the retro-reflective ring itself being assembled from aluminum plates, it possesses a certain mechanical strength. To ensure stability during operation and prevent damage to the retro-reflective ring during climbing, the robot adopts a clamping-type wheeled movement mechanism. As shown in Figure 3, one main shaft is fixed at one end of the Π-shaped bracket, with a walking wheel installed. At the other end of the Π-shaped bracket, there is an adjustable pressure suspension mechanism, composed of an active main shaft, walking wheel, compression spring, a bolt, and nut, allowing the clamping pressure between the walking wheels to be adjusted by turning the nut. Block-patterned rubber wheels can increase friction with the tunnel retro-reflective ring and ensure smooth passage over the rivets embedded into the grooves on the rubber wheel's surface while rolling over the retro-reflective ring. This clamping-type wheeled movement structure ensures the stability of the robot's operation within the tunnel, enabling it to be unaffected by wind speed, and the use of rubber wheels prevents damage to the retro-reflective ring.

Similarly, during the installation process or prolonged use of the tunnel retro-reflective ring, there might be instances where one end of the connection between two retro-reflective panels becomes warped, or the panels do not completely adhere to each other, as shown in Figure 4. In such cases, the bell-mouth-shaped shell can be maneuvered to this location before the walking wheels, allowing the retro-reflective panels to fit together as closely as possible. This helps the transition over the warped sections of the retro-reflective ring, addressing this condition and enhancing the stability of the robot's climbing. Additionally, the shell is made of 1 mm stainless steel, with the steel plates welded together. This not only contributes to lightweight design, but also protects the internal structure of the robot from external factors within the tunnel.

The cleaning device consists of two brackets, a driving motor, a cleaning disk, compression springs, bolts, and nuts, each set up at both the front and rear. The cleaning disk of the cleaning device is made of soft material, and since the pressure exerted by the cleaning device on the tunnel retro-reflective ring is not particularly high, this ensures that it will not be damaged throughout the entire process.

As shown in Figure 5a, Bracket No. 8 remains consistently connected to the main framework in No. 7, while a portion of Bracket No. 4 is affixed to the motor, with its other end featuring a straight slot. Initially, the compression distance of the springs is adjusted by twisting the nuts, thereby regulating the pressure exerted by the cleaning disc on the tunnel retro-reflective ring. When the pressure is confirmed, the motor is firmly fixed on the main frame using the straight slot at the other side of the No. 4 bracket. This way, while ensuring that the pressure of the cleaning disc on the retro-reflective ring is unchanged, the motor will not be shaken. This approach facilitates subsequent adjustment of pressure parameters, enabling optimization to identify the most suitable pressure.

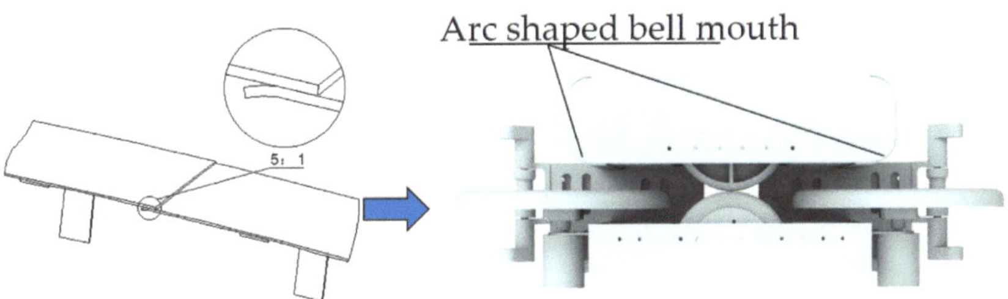

Figure 4. Machine appearance.

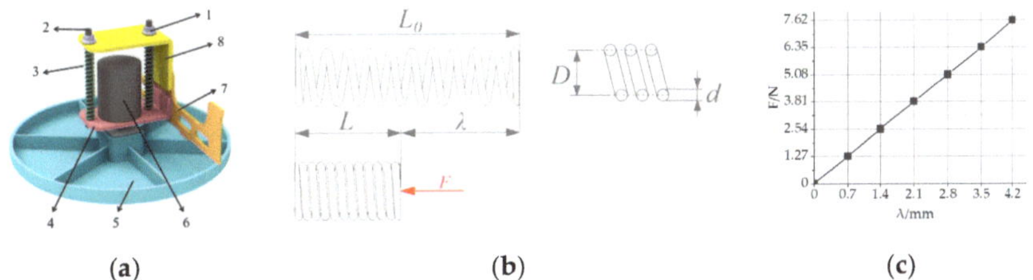

(a) (b) (c)

Figure 5. Cleaning institutions. (**a**) Clean model: 1—nut; 2—bolt; 3—compression spring; 4—support plate; 5—cleaning disc; 6—electrical machinery; 7—side panels of the main frame; 8—external extension plate; (**b**) Spring compression principle; (**c**) The relationship between spring elasticity and compression.

This type of cleaning mechanism can enhance cleaning effectiveness by setting appropriate pressure and speed, while also improving the stability of the equipment during climbing.

According to Figure 5b, under the action of axial load, the spring produces a deformation amount represented by "λ". There is a certain relationship between the deformation of the spring and the pressure generated [20].

$$\lambda = \frac{8FD^3n}{Gd^4} \tag{1}$$

F represents the axial pressure, λ denotes the compression amount, D stands for the mean diameter of the spring, n is the effective number of turns of the spring (here it is 16), d represents the wire diameter of the spring, and G represents the shear modulus of the spring material. The model of the compression spring is 6 mm × 8 mm × 50 mm (inner diameter × outer diameter × length), and it is made of carbon steel, with a G of 8×10^4 MPa. For further analysis, without affecting the results, a slight approximation is made to the second decimal place of the pressure value calculated from the original Hooke coefficient of the spring. The relationship between spring pressure and compression distance is shown in Figure 5c.

2.3. Hardware Design

The robot uses an STM32F103C8T6 microcontroller as its core controller, forming the microcontroller control circuit. The control system consists of the TB6612 driver module, HC-06 Bluetooth module, and a HM-GM37-3429 encoder reduction motor. Both the robot and the remote control are equipped with HC-06 Bluetooth modules, allowing the robot's movements to be controlled remotely. The TB6612 driver module drives the motor, while the Hall encoder on the motor provides feedback on the motor speed, which is then stabilized through a PID controller. The hardware control scheme, shown in Figure 6, is stable, operates well, and effectively controls the target.

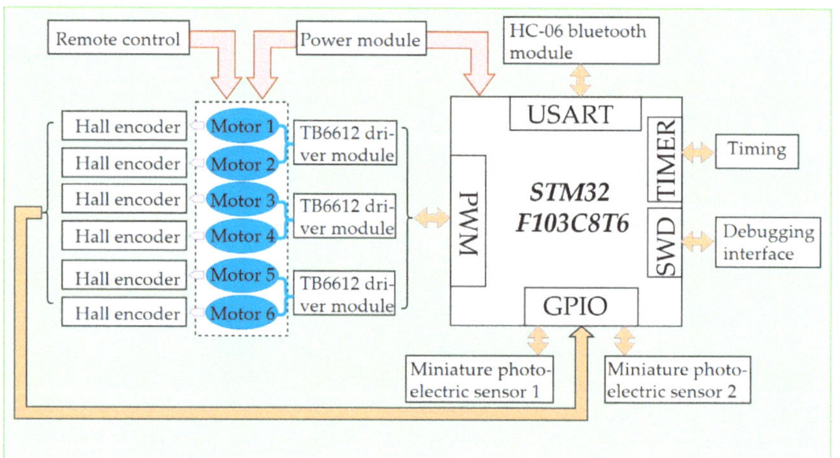

Figure 6. Hardware circuit scheme.

2.4. Microcontroller Programming

The program logic of the microcontroller includes the following processes. Firstly, the microcontroller performs system initialization, which includes clock initialization, GPIO pin definition initialization, serial port initialization, and timer initialization. After initialization, the system waits to receive control commands from the host computer. The Bluetooth module receives the remote control's commands and forwards them to the microcontroller. The microcontroller then determines the type of control command. If the control command is for movement, the initial desired speed of the movement is set, and the initial PWM value is output to drive the motor.

During timed intervals, the pulses output by the Hall encoder are recorded to calculate the robot's movement speed. The difference between this movement speed and the desired speed is then calculated, and the PWM output is updated via the PID controller. This process is repeated until the difference between the robot's movement speed and the desired speed approaches zero. While adjusting the speed using the PID controller, the microcontroller also receives measurement signals from the photoelectric sensor. Based on these signals, it determines whether the robot's movement has caused the body to deviate from the retro-reflective panel. If the body deviates from the retro-reflective panel, differential control of the robot's left and right walking wheels is used to correct the body.

The PID control algorithm is based on the error between the robot's current speed and the desired speed. It adjusts the controller's output through a combination of proportional, integral, and derivative terms, enabling the error to quickly and stably converge to zero, thereby achieving precise control. Proportional control adjusts the control amount based on the current error; integral control addresses the accumulated effect of the error, helping to eliminate steady-state error; derivative control focuses on the dynamic changes of the error, predicting future error magnitude by identifying the rate of error change, thus allowing for anticipatory adjustments of the control amount. The PID controller ensures the safety and speed of the climbing process. The principle of PID speed control is shown in Figure 7.

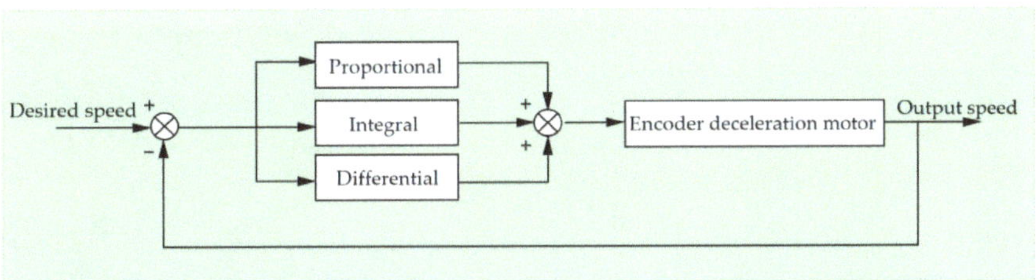

Figure 7. Principle of PID speed control.

2.5. Control of Miniature Photoelectric Sensor

As the tunnel retro-reflective ring is installed in an arc shape, to prevent the robot from falling off during operation, it must follow the arc of the retro-reflective ring. On the main framework of the robot, pairs of edge detection sensors are installed at the front and rear. Miniature through-beam NPN-type photoelectric switches are chosen for the edge detection sensors. When the infrared beam from one end is received by the other end, the sensor returns a high level; when the infrared beam is blocked by the retro-reflective ring, the sensor returns a low level. Sensors set up at the front and rear can detect whether the robot deviates. Upon deviation, the speed difference between the outer and inner wheels (I and II) is automatically adjusted to correct the wheel alignment, allowing the robot to move along the arc path of the entire retro-reflective ring. The principle of operation of the miniature through-beam NPN-type photoelectric switches is illustrated in Figure 8.

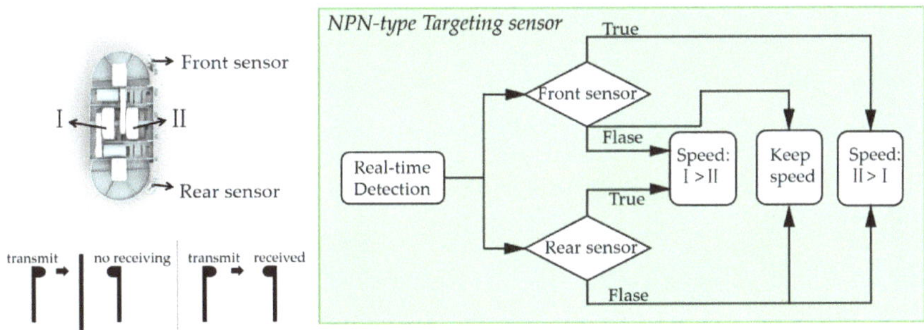

Figure 8. Principle of miniature photoelectric sensor.

3. Stability Analysis

A crucial metric for evaluating climbing robots is their stability, indicating their ability to climb or operate steadily on a wall. Specifically, the stability of climbing robots is influenced by contact forms and variations in the center of mass, leading to potential slipping due to gravity or adhesive forces [21]. Therefore, the following section will conduct static and dynamic analysis of the robot.

3.1. Static Force Analysis

The tunnel retro-reflective ring is assembled from retro-reflective panels, with a reflective film adhered to the front side, which is smooth. The back side is relatively rough, with aluminum plates overlay and rivet protrusions. Assuming the dynamic friction coefficient on the side with the reflective film is μ_1 and the backside is μ_2, with $\mu_1 < \mu_2$, a static force analysis of the entire system can be performed, as illustrated in Figure 9.

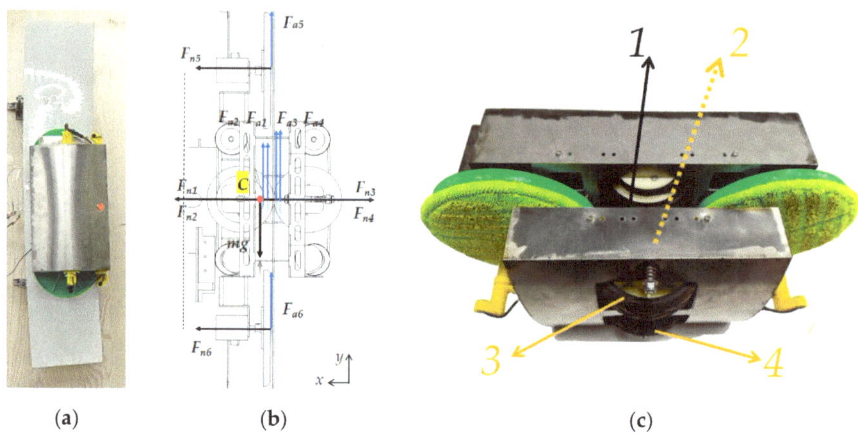

(a)   (b)   (c)

Figure 9. Static force situation. (**a**) The robot in a stationary clamping situation; (**b**) static force analysis; (**c**) distribution of identification numbers for the walking wheels: 1—the wheel closest to the tunnel inner wall on the working face; 2—the other wheel on the working face; 3—the wheel closest to the tunnel inner wall on the backside; 4—the other wheel on the backside.

As shown in Figure 9, F_{ni} ($i = 1\sim4$) represents the support force of the retro-reflective panel on each walking wheel, while F_{n5} and F_{n6} represent the support forces of the retro-reflective panel on the cleaning disks. F_{ai} ($i = 1\sim4$) indicates the static frictional force acting on each walking wheel, while F_{a5} and F_{a6} represent the static frictional forces acting on the cleaning disks. Since the support force on the cleaning disks is negligible, it can be disregarded. Therefore, the simplified mechanical equilibrium equation is as follows:

$$\begin{cases} \sum F_x = F_{n1} + F_{n2} - F_{n3} - F_{n4} = 0 \\ \sum F_y = F_{a1} + F_{a2} + F_{a3} + F_{a4} - mg = 0 \end{cases} \tag{2}$$

The sliding friction force is used for analysis instead of the maximum static friction force. For this robot to securely grip the tunnel retro-reflective ring, the overall maximum static friction force should be greater than the gravitational force.

$$\sum_{i=1}^{4} F_{ai} = \mu_1 F_{n1} + \mu_1 F_{n2} + \mu_2 F_{n3} + \mu_2 F_{n4} > mg \tag{3}$$

$$\sum_{i=1}^{4} F_{ni} > \frac{2mg}{\mu_1 + \mu_2} \tag{4}$$

When Equation (4) is satisfied, the robot will not slide downward.

3.2. Static Moment Analysis

Below is a static moment analysis of the entire machine. As the robot is front-to-back symmetric, but left-to-right asymmetric, and the side responsible for cleaning the tunnel retro-reflective ring is heavier, when it clamps onto the tunnel retro-reflective ring, its center of mass C will shift towards the working surface, a certain distance away from the center of the tunnel retro-reflective ring, as shown in Figure 10.

Where h_1 is the vertical distance from the center of mass C to the front of the retro-reflective ring and h_2 is the vertical distance from the center of mass C to the back of the retro-reflective ring, l_1 is the distance from the center of mass to the gravitational force on the working surface side and l_2 is the distance from the center of mass to the gravitational force on the opposite side, and $2b$ is the distance between wheels on the same side.

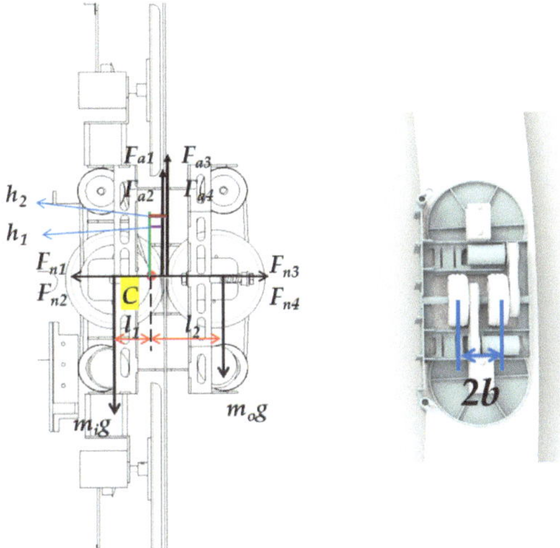

Figure 10. Static moment situation.

The moment caused by the pressure on the retro-reflective ring of the four sets of walking wheels will be offset, so the static moment equation for the center of mass C can be simplified as:

$$\sum M_C = m_i g l_1 - m_o g l_2 + (F_{a1} + F_{a2})h_1 + (F_{a3} + F_{a4})h_2 = 0 \tag{5}$$

In Equation (5), the first two terms represent the moment caused by the gravity on both sides separated by the retro-reflective plate, and the last two terms represent the moment caused by the static friction force of the four walking wheels. In fact, the gravity on the working side of the robot is greater than the gravity on the opposite side, and if the robot wants to maintain balance, it will cause the center of mass to shift.

Consequently, the pressure on the two walking wheels on the back side will be greater than on the working surface, introducing adverse effects due to the asymmetry of the robot. However, this will not affect the robot because it has a double-sided clamping mechanism that prevents tilting due to insufficient pressure. The front and rear configuration of the cleaning mechanism also provides auxiliary support and assistance for the robot's movement, eliminating the impact of the center of gravity deviation and ensuring the stability of the entire machine.

3.3. Dynamics Analysis

To investigate the robot's ability to move stably during its motion, without being immobilized due to insufficient driving torque, a simplified model is adopted. For dynamic analysis, it is assumed that the climbing robot travels straight along an inclined angle for a certain distance on the retro-reflective ring.

The climbing robot features four active wheels, each driven by an independent motor. The tunnel retro-reflective ring is vertically mounted within the tunnel. Here, v represents the robot's velocity and θ denotes the heading angle.

The following is an analysis of a single walking wheel, assuming F_{di} ($i = 1 \sim 4$) is the driving force, F_{ci} ($i = 1 \sim 4$) is the rolling friction force, F_{ei} ($i = 1 \sim 4$) is the sliding friction force due to the gravitational component, and F_{ni} ($i = 1 \sim 4$) is the pressure on the retro-reflective plate on each walking wheel, or the support force of the retro-reflective plate on the wheel. Based on the working condition of the robot on the reflector, the following assumptions are made [22]:

(1)   The external forces on the robot are evenly distributed;

(2) The robot does not exhibit wheel slippage, overturning, or lateral sliding;

(3) When the robot moves on the retro-reflective ring, its acceleration and rotational acceleration about the center of mass are zero, that is, $\dot{v} = 0$, $\dot{w} = 0$.

Figure 11 establishes the dynamic analysis equation for the robot:

$$\begin{cases} \sum\limits_{i=1}^{4} F_{di} - \sum\limits_{i=1}^{4} F_{ci} - mg\sin\theta = m\dot{v} + J\dot{w} \\ (F_{d2} + F_{d4})2b - (F_{c2} + F_{c4})2b - mg\sin\theta b = 0 \\ \sum\limits_{i=1}^{4} F_{ei} - mg\cos\theta = 0 \end{cases} \tag{6}$$

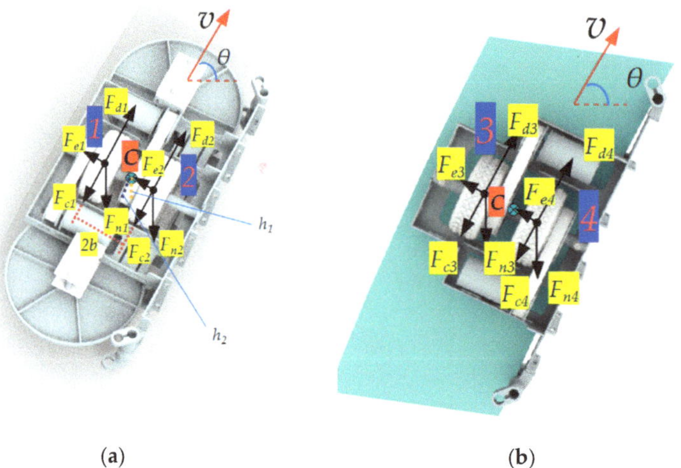

(**a**) (**b**)

Figure 11. Force situation of climbing robot movement. (**a**) The force situation of the two walking wheels on the working face, (**b**) the force situation of the two wheels on the back.

To solve the above equation, the driving torque of the robot drive wheel obtained is:

$$\begin{cases} T_1 = \left[\mu_f\left(F_{n1} - \frac{mg\cos\theta h_1}{4b}\right) + \frac{mg\sin\theta}{4}\right]r \\ T_2 = \left[\mu_f\left(F_{n2} + \frac{mg\cos\theta h_1}{4b}\right) + \frac{mg\sin\theta}{4}\right]r \\ T_3 = \left[\mu_f\left(F_{n3} + \frac{mg\cos\theta h_2}{4b}\right) + \frac{mg\sin\theta}{4}\right]r \\ T_4 = \left[\mu_f\left(F_{n4} - \frac{mg\cos\theta h_2}{4b}\right) + \frac{mg\sin\theta}{4}\right]r \end{cases} \tag{7}$$

μ_f is the rolling friction coefficient, r is the radius of the walking wheel, b is the center distance between walking wheels on the same side, h_1 is the vertical distance from the center of mass C to the front of the retro-reflective ring, and h_2 is the vertical distance from the center of mass C to the back of the retro-reflective ring. In theory, due to the lateral sliding friction caused by the component of gravity perpendicular to the direction of movement, as well as the special setting of the four sets of walking wheels clamping the retro-reflective ring, the pressure on the No. 2 and No. 3 walking wheels will increase to some extent, and the pressure on the No. 1 and No. 4 walking wheels will decrease to some extent, as shown in Equation (7).

In practice, although the gravity of the working face is slightly higher, the displacement of the center of mass is not large, and the ratio $\frac{\mu_f h_i}{b}$ $(i = 1, 2)$ is very small, so that the pressure effect caused by the gravity component on the term $\frac{mg\cos\theta h_i}{4b}$ $(i = 1, 2)$ in

Equation (7) can be ignored. Therefore, when θ takes $-90°$ and $90°$, the driving torque has a minimum or maximum value:

$$r\mu_f \sum_{i=1}^{4} F_{ni} - mgr < \sum_{i=1}^{4} T_i < r\mu_f \sum_{i=1}^{4} F_{ni} + mgr \tag{8}$$

The maximum torque demand on the motors is required when the robot is climbing upwards. When selecting motors and reducers, it is only necessary to refer to the required driving torque for climbing upwards, that is:

$$\sum_{i=1}^{4} T_i > r\mu_f \sum_{i=1}^{4} F_{ni} + mgr \tag{9}$$

When the total driving torque of the motor meets Equation (9), the entire machine can be cleaned back and forth on the retro-reflective ring of the tunnel, and there will be no situation where the driving torque is insufficient and cannot work during operation.

Given the mass of 4.7 kg, r = 0.047 m, μ_f is 0.1 and g is 9.8 m/s^2, and estimating the total pressure to be four times the weight to roughly find the minimum driving torque, we can use Equation (9) to calculate that the total minimum driving torque required for the machine is 3.03 N.m. Therefore, the HM-GM37-3429 encoder reduction motor is chosen, which has a rated torque of 2.4 N.m. Since the machine is equipped with four motors, the torque is sufficient.

4. Optimization Design

Section 2.1 analyzed the issues present in the tunnel retro-reflective ring. This climbing robot's ability to move stably on the tunnel retro-reflective ring depends on the strength and stiffness of the designed Π-shaped framework. Below, Solidworks Simulation 2022 is used to perform deformation simulation analysis on the Π-shaped bracket, to confirm the appropriate material and thickness.

Due to the structure being welded between the bottom plates, fixed constraints are set at the bottom. Compression spring constraints are applied at the ends of the shaft. Additionally, forces ranging from 0 N to 100 N, increasing in 10 N increments, are applied outward from the center of every wheel axle to simulate the increase in thickness of the clamped tunnel retro-reflective ring. The stress, strain, and displacement cloud map of 7075 aluminum alloy with a thickness of 2 mm is shown in Figure 12.

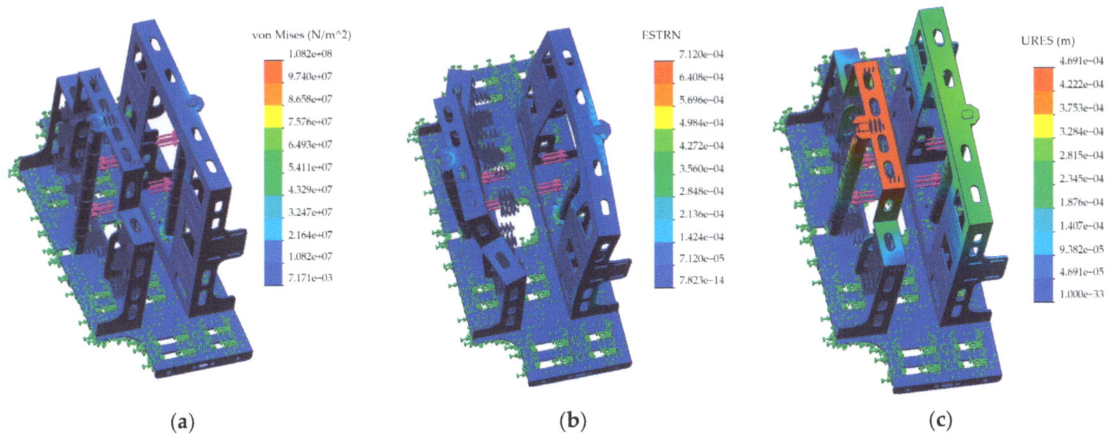

(a)   (b)   (c)

Figure 12. Cloud map of three indicators. (**a**) Stress cloud map, (**b**) strain cloud map, (**c**) displacement cloud map.

After confirming the shape of the main framework, the analysis of the main framework is as follows:

(1)   The influence of different materials on the deformation of the frame.

The thickness is uniformly set to 2 mm, and the materials selected are 201 stainless steel, QT600 cast iron, and 7075 aluminum alloy, with their respective parameters shown in Table 1.

Table 1. Material parameters.

	201 Stainless Steel	QT600 Cast Iron	7075 Aluminum Alloy
Density	$7859 \, \text{kg/m}^3$	$7200 \, \text{kg/m}^3$	$2830 \, \text{kg/m}^3$
Elastic modulus	$2.07 \times 10^{11} \, \text{Pa}$	$6.62 \times 10^{10} \, \text{Pa}$	$7.20 \times 10^{10} \, \text{Pa}$

Figure 13a–c depict the relationship between stress, strain, and displacement with increasing loads, after simulation analysis for each material. It can be observed that the deformation degree follows the order: 201 stainless steel < 7075 aluminum alloy < QT600 cast iron. Although 201 stainless steel exhibits the least deformation, its density is too high, leading to an increase in the overall weight of the robot. This necessitates greater pressure from the compression springs to ensure stability, and the increased weight also requires more energy, significantly reducing the robot's operational duration. The density and degree of deformation of QT600 are relatively high. Therefore, considering both the density and deformation degree, 7075 aluminum alloy is chosen as the material for the Π-shaped bracket.

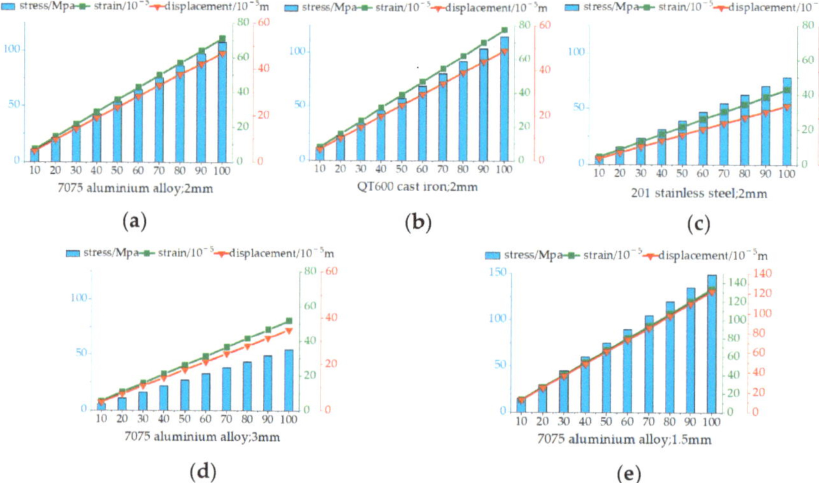

Figure 13. Changes in stress, strain, and displacement of different materials and thicknesses with loading load. (**a**) The relationship between stress, strain, and displacement of 2 mm 7075 aluminum alloy with increasing load; (**b**) the relationship between stress, strain, and displacement of 2 mm QT600 cast iron with increasing load; (**c**) the relationship between stress, strain, and displacement of 2 mm 201 stainless steel with increasing load; (**d**) the relationship between stress, strain, and displacement of 3 mm 7075 aluminum alloy with increasing load; (**e**) the relationship between stress, strain, and displacement of 1.5 mm 7075 aluminum alloy with increasing load.

(2)   The influence of sheet thickness on the deformation of the frame.

The material is uniformly set to 7075 aluminum alloy, with thicknesses of 2 mm, 3 mm, and 1.5 mm considered. The simulation results are depicted in Figure 13a,d,e. As the thickness increases, the total deformation displacement of the frame continuously decreases. However, with increasing thickness, the mass inevitably increases as well. A thickness of 3 mm does indeed provide high stability, but the mass is too large, and deformation at 1.5 mm is too large, making it unsuitable for use. Within an acceptable range of deformation, a thickness of 2 mm is chosen for the main framework.

Ultimately, 7075 aluminum alloy with a thickness of 2 mm is selected as the material for the Π-shaped bracket. From Figure 13a, it can be observed that at 100 N, the maximum displacement is only 469 μm, with a mass of only 0.9 kg.

5. Experiments and Analysis

5.1. Analysis of Stable Climbing Experiments

Inside the laboratory, the real reflective panels purchased are assembled and fixed using the same method as inside the tunnel. They are secured with right-angled aluminum plates, to simulate the real installation environment of the retro-reflective ring inside the tunnel. The climbing performance is illustrated in Figure 14. Figure 14a–c illustrate the stable working state and cleaning effect of the climbing robot on the tunnel retro-reflective ring, while Figure 14d provides an enlarged view of the cleaning effect. Figure 14e,f depict the clamping effect of the climbing robot on horizontal and vertical retro-reflective rings, respectively.

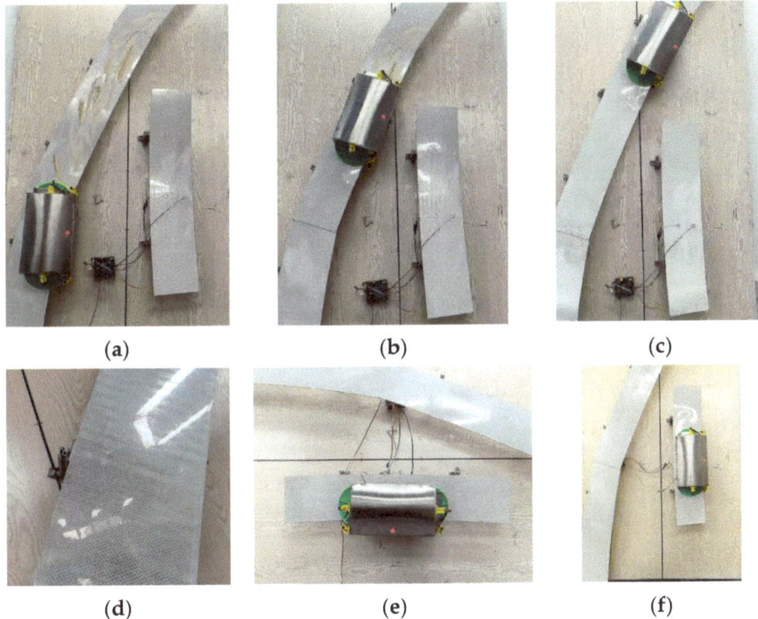

Figure 14. Climbing, cleaning, and hovering conditions. (**a**) Climbing starting position, (**b**) climbing the middle position, (**c**) climbing finishing position. (**d**) Effect after cleaning. (**e**) Verification of horizontal climbing stability, (**f**) verification of vertical climbing stability.

During the climbing process, four sets of thin film pressure sensors are placed at the locations where the four sets of walking wheels pass, to test the pressure. These tests can be conducted on different thicknesses, with a 5 mm thickness used to simulate a stepped retro-reflective ring and a 6 mm thickness used to simulate the stacking of aluminum plates. The pressure of each walking wheel is tested multiple times and the average value is taken, with the results shown in Figure 15.

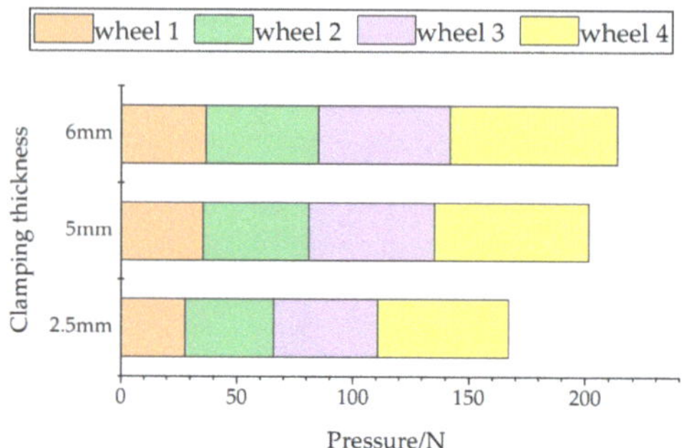

Figure 15. The pressure distribution and sum of four wheels with different clamping thicknesses.

From Figure 15, on the 2.5 mm thick retro-reflective ring, the pressure borne by the two wheels on the working side is relatively small. The pressure exerted by Wheel 1 and Wheel 2 on the reflective plate is 27.96 N and 37.91 N, respectively. The pressure borne by the rear wheels is higher, with Wheel 3 and Wheel 4 exerting pressures of 44.75 N and 56.45 N on the reflective plate, respectively. The reason for this is analyzed in the static moment analysis above; the lack of complete symmetry causes a shift in the center of mass, resulting in higher pressure on one side.

However, the spring-loaded structure allows the robot to grip retro-reflective rings of different thicknesses, whether it is a stepped retro-reflective ring or an aluminum plate overlay. Additionally, it has a double-sided clamping mechanism that prevents tilting due to insufficient pressure and the cleaning device also provides auxiliary support to keep the robot stable.

Next, the parameters will be plugged into the calculation. The known parameters are m = 4.7 kg, assuming μ_1 is 0.3, μ_2 is 0.4, μ_f is 0.1, and g is 9.8 m/s^2. Static analysis on a 2.5 mm reflective plate yields:

$$\sum_{i=1}^{4} F_{ni} = 167.07 \text{ N} > \frac{2 \cdot 4.7 \text{ kg} \cdot 9.8 \text{ m/s}^2}{0.3 + 0.4} = 131.6 \text{ N} \tag{10}$$

The reduction ratio of the reducer is 1.6 and the rated torque with the motor is 2.4 N·m. The minimum required driving torque and the driving torque that four motors can bring is:

$$\begin{cases} \sum_{i=1}^{4} T_i > 0.047 \text{ m} \cdot \left(0.1 \cdot 167.07 \text{ N} + 4.7 \text{ kg} \cdot 9.8 \text{ m/s}^2\right) = 2.95 \text{ N·m} \\ \sum_{i=1}^{4} T_i' = 4 \cdot 1.6 \cdot 2.4 \text{ N·m} = 15.36 \text{ N·m} \\ \sum_{i=1}^{4} T_i' > \sum_{i=1}^{4} T_i \end{cases} \tag{11}$$

Equations (10) and (11) indicate that the experimental test results meet the static and dynamic analysis results mentioned above, confirming that the sum of pressure and static friction is greater than gravity and that the selected motor model and reduction ratio provide sufficient driving torque. The results in Figure 14 also prove that it can grip, climb, and clean with stability.

5.2. Response Surface Methodology (RSM) Optimization Experiment

On the premise of the preliminary design of the robot and to ensure stability, the cleaning disc pressure, the walking wheel motor speed, and the cleaning disc motor speed are selected as experimental factors to analyze their effects on the cleaning performance. Firstly, single-factor experiments are conducted separately to explore the individual influences on the cleaning effect. Then, based on the principles of Box–Behnken [23] central composite design, with the number of cleaning times in a fixed area as the response variable, the optimization of cleaning is conducted through response surface analysis. The aim is to find the optimal combination of the three factors, to achieve the best cleaning performance.

5.2.1. Single-Factor Test

Fixing the other two factors simultaneously, only one factor is altered at a time. Especially since the walking wheel speed affects the driving torque, the selected speed of the walking wheel must be within a range that ensures the robot's stable driving capability. The experimental environment is created by using a roller brush covered with slightly wetted soil, to brush back and forth on the retro-reflective plate for a fixed number of times. Then, after waiting for it to air dry before experimenting, simulation of the actual situation of the tunnel retro-reflective ring, to the maximum extent possible, is performed. A section of the tunnel retro-reflective ring is selected as the testing area and the number of times the machine needs to clean this area is measured as the response. Figure 16 illustrates an example of cleaning three times. The results of the single-factor experiments are shown in Figure 17.

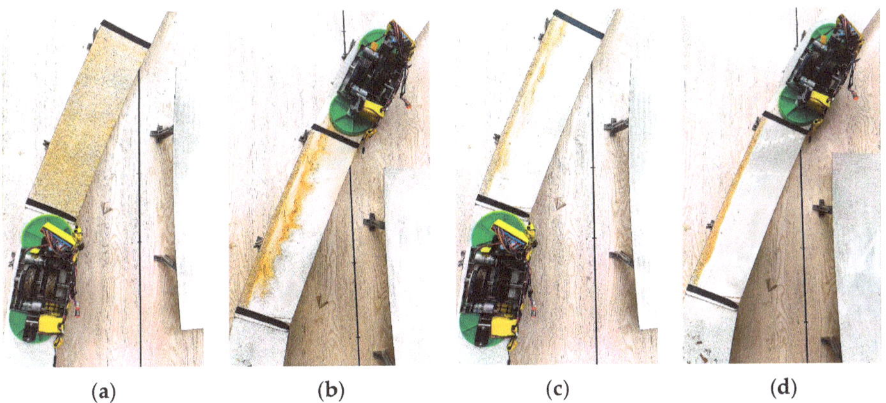

(a) (b) (c) (d)

Figure 16. Examples that require cleaning three times. (**a**) Cleaning starting point, (**b**) cleaning once, (**c**) cleaning twice, (**d**) cleaning three times.

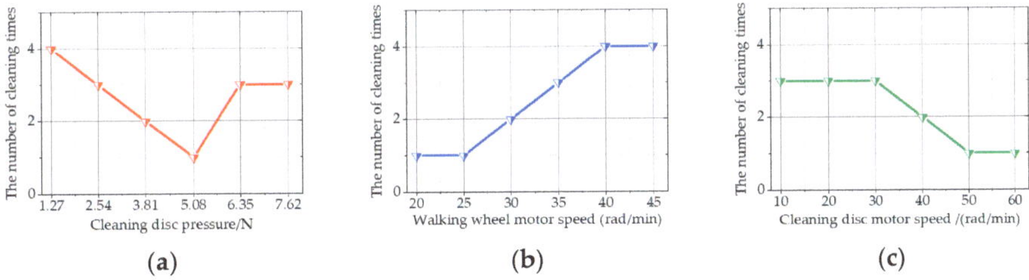

(a) (b) (c)

Figure 17. The results of the single-factor experiments. (**a**) Cleaning disc pressure single factor test, (**b**) walking wheel motor speed single factor test, (**c**) cleaning disc motor speed single factor test.

From Figure 17a, it can be observed that, with the walking wheel motor speed and the cleaning disc motor speed set at certain values, the number of cleanings initially decreases and then increases with increasing pressure. This indicates that when the pressure is too low, there is insufficient pressure between the cleaning disc and the retro-reflective ring, requiring multiple cleanings. On the other hand, when the pressure is too high, it may cause difficulty in the rotation of the cleaning disc and even damage the motor. Figure 17b reveals that a slower walking wheel motor speed results in better cleaning performance, without affecting climbing. However, excessively slow speed can prolong the working duration excessively, thereby affecting work efficiency. From Figure 17c, it is evident that a higher cleaning disc motor speed leads to better cleaning performance. Nevertheless, excessively high speeds can lead to increased energy consumption and thus may not be suitable for practical applications.

Moreover, both speed indicators show that once the speed rises or falls to a certain value, the number of cleaning times does not change significantly. This suggests that the optimal speed indicators are likely to occur at points where the slope of the curve changes.

Based on these factors, to achieve optimal cleaning performance with minimized cleaning times and energy consumption, while maintaining moderate machine speed, it is essential to identify the optimal combination of the three factors.

5.2.2. Response Surface Experimental Design

Based on the principle of Box–Behnken's central composite experimental design and the above single-factor experimental results, the experimental factors for the three-factor, three-level orthogonal experiment are shown in Table 2 as follows:

Table 2. Factors and factor levels of the experiment.

Factors	Level		
	−1	0	1
Cleaning disc pressure (A/N)	2.54	5.08	7.62
Walking wheel motor speed(B/(rad/min))	25	35	45
Cleaning disc motor speed(C/(rad/min))	20	40	60

The quadratic response surface regression analysis is performed on the experimental results using Design Expert 13 software, and the equation for the degree S is obtained as follows:

$$\begin{aligned} S = \ &1.2 + 0.875A + 0.625B - 0.25C \\ &+0.25AC - 0.25BC \\ &+1.9A^2 + 0.4B^2 + 0.15C^2 \end{aligned} \tag{12}$$

The experimental design and results are shown in Table 3:

Table 3. The experimental design and results.

NO.	A	B	C	S
1	5.08	25	60	1
2	5.08	35	40	1
3	5.08	35	40	1
4	2.54	35	20	3
5	5.08	35	40	2
6	7.62	45	40	5
7	5.08	45	60	2
8	5.08	35	40	1
9	2.54	45	40	3
10	7.62	35	20	4
11	5.08	35	40	1
12	5.08	25	20	1
13	7.62	35	60	4
14	7.62	25	40	4
15	2.54	25	40	2
16	5.08	45	20	3
17	2.54	35	60	2

5.2.3. Response Surface Experimental Analysis

The regression equation analysis [24] results are shown in Table 4. It is evident that the model has a significant effect on the number of cleaning times S, while the mismatch term is extremely insignificant. The coefficient of determination R^2 of the model is 0.9623, indicating that 96.23% of the change in S comes from the testing factor. The predicted R^2 of 0.8117 is in reasonable agreement with the adjusted R^2 of 0.9139; i.e., the difference is less than 0.2. The Adeq precision measurement signal-to-noise ratio is 13.0452, which is greater than 4, indicating that the model is good. Among them, A, B, and A^2 are extremely significant, and the order of influence of the experimental factors is A > B > C.

Table 4. Regression equation analysis of variance.

	Sum of Squares	Freedom of Degree	Mean Square	F-Value	*p*-Value	Significance
Model	26.83	9	2.98	19.88	0.0003	significant
A	6.13	1	6.13	40.83	0.0004	**
B	3.13	1	3.13	20.83	0.0026	**
C	0.5	1	0.5	3.33	0.1106	
AB	3.55×10^{-15}	1	3.55×10^{-15}	2.37×10^{-14}	1	
AC	0.25	1	0.25	1.67	0.2377	
BC	0.25	1	0.25	1.67	0.2377	
A^2	15.2	1	15.2	101.33	<0.0001	**
B^2	0.6737	1	0.6737	4.49	0.0718	
C^2	0.0947	1	0.0947	0.6316	0.4529	
Residual	1.05	7	0.15			
Lack of Fit	0.25	3	0.0833	0.4167	0.751	not significant
Pure Error	0.8	4	0.2			
Cor Total	27.88	16				

Notes: ** means the influence is highly significant, $p < 0.01$.

Figure 18 shows the impact of a single factor on the number of cleanings, while Figure 19 shows the impact of a combination of two factors on the number of cleanings.

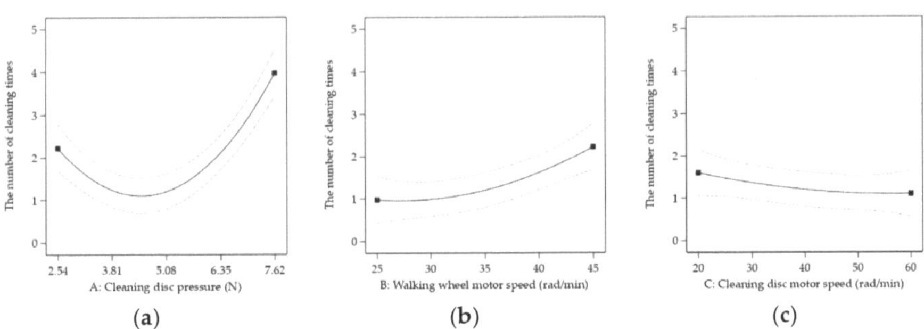

Figure 18. Response curve. (a) $S = f\,(A, 35, 40)$; (b) $S = f\,(5.08, B, 40)$; (c) $S = f\,(5.08, 35, C)$.

Specifically, as shown in Figures 18b,c and 19c, the increase in the speed of the cleaning disc motor has a smaller impact than the decrease in the speed of the walking wheel motor.

Finally, through the analysis using Design Expert 13 software, the optimal cleaning conditions are found to be a cleaning disc pressure of 5.101 N, a walking wheel motor speed of 36.93 rad/min, and a cleaning disc motor speed of 38.252 rad/min. In fact, to ensure that the tunnel retro-reflective ring is not damaged, the cleaning disc should not apply too much pressure. Additionally, considering that each cleaning cycle should not be too long, the speed of the walking wheels should not be too slow. The overall energy loss of the machine should not be too substantial; therefore, the speed of the cleaning disc motor should not be excessively high. By optimizing these factors, the best combination will meet the above requirements. Finally, physical validation experiments are conducted based on the optimal combination, with most cleaning cycles being performed only once, achieving satisfactory results.

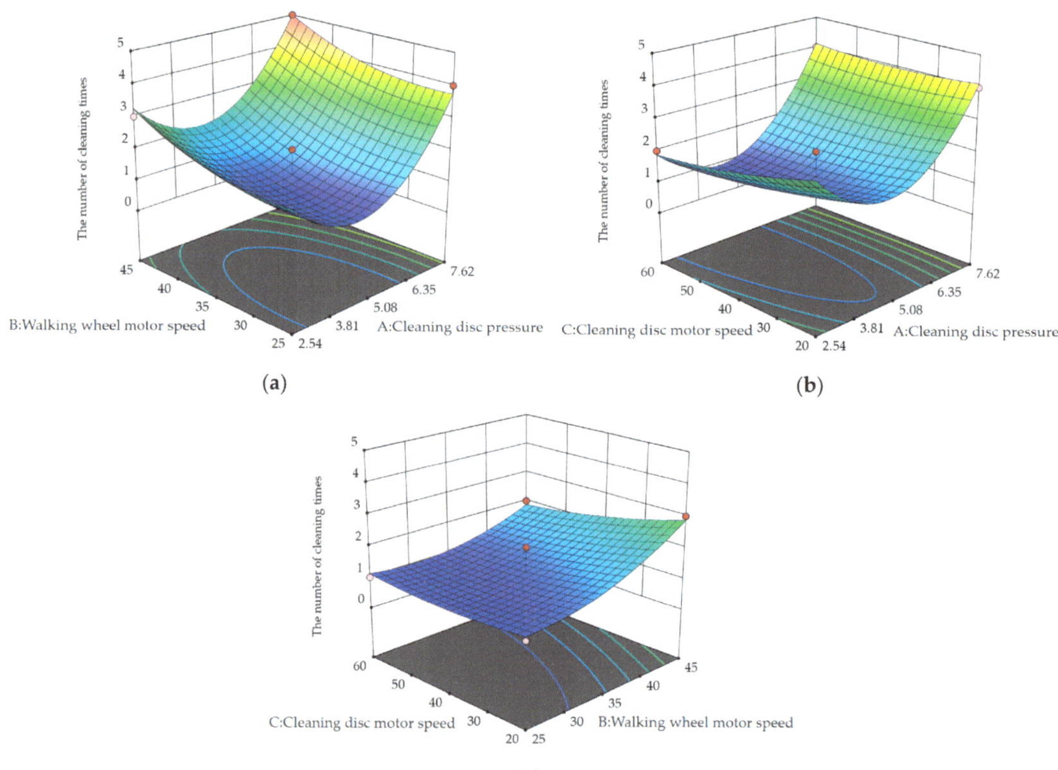

Figure 19. Response surface. (**a**) $S = f (A, B, 40)$; (**b**) $S = f (A, 35, C)$; (**c**) $S = f (5.08, B, C)$.

5.3. Discussion

Existing climbing robots are primarily designed to replace humans in completing tasks in complex environments [4–6]. They employ various climbing methods, such as magnetic adhesion [14,15], suction cups [11,12], adhesives [8], and so on. Previously, there has been no research on the application of climbing robots to tunnel retro-reflective rings. The main contribution of this paper is the proposal of a robot designed for climbing on tunnel retro-reflective rings, capable of overcoming challenging conditions such as stepped retro-reflective rings, stacked aluminum plates, and rivets. Additionally, using a response surface experiment, with three factors affecting the overall cleaning performance, the optimal parameter combination is found to achieve efficient cleaning of the tunnel retro-reflective rings. Furthermore, the robot is designed according to the most widely used size standards of retro-reflective panels inside Chinese tunnels, making it applicable for most tunnels.

However, the double-sided clamping and climbing method of this robot determines that it cannot climb arbitrarily, as with magnetic adhesion or suction cups. This robot is only suitable for clamping and climbing on thin and small surfaces, like tunnel retro-reflective rings, to accomplish cleaning tasks, indicating its incapability for cleaning other objects that require cleaning at heights. Moreover, it also fails to clean the blind spots of tunnel retro-reflective rings. These are the limitations of this robot, so the next step is to solve them.

6. Conclusions

This article proposes a climbing robot that can clean tunnel retro-reflective rings. Through the combination of mechanical and electronic systems, it can achieve automatic cleaning on tunnel retro-reflective rings, as well as manual remote controlled cleaning at designated points. It is especially capable of handling complex working conditions, such as stepped retro-reflective panels, aluminum plate overlays, and rivet protrusions.

Firstly, the mechanical structure, the hardware design, the microcontroller programming, and the principle of miniature photoelectric sensors are introduced, and its walking ability under different working conditions is analyzed. Then, the conditions that make the climbing robot stable are explored through static analysis, and the minimum driving torque is obtained through dynamic analysis. The stability of the robot is verified in the subsequent experimental section. After that, deformation simulation analysis is conducted on the Π-shaped main frame, using Solidworks Simulation 2022, to optimize the material and thickness of the frame. Finally, a mathematical regression model for cleaning disc pressure, walking wheel motor speed, and cleaning disc motor speed is established using response surface methodology. The optimal combination of the three factors is found. The cleaning disc pressure is 5.101 N, the walking wheel motor speed is 36.93 rad/min, and the cleaning disc motor speed is 38.252 rad/min, resulting in the best cleaning effect.

In the end, the robot can work stably on the retro-reflective rings of the tunnel and have good cleaning effects, without affecting the use of the tunnel.

Supplementary Materials: The following supporting information can be downloaded at https://www.mdpi.com/article/10.3390/act13060197/s1.

Author Contributions: Y.L. proposed the study. Y.L. and S.Y. designed the structure. S.Y. and R.C. performed and analyzed the simulations. Y.L. and S.Y. wrote the manuscript. Z.S. supervised and confirmed the research. All authors participated and contributed to the study. All authors have read and agreed to the published version of the manuscript.

Funding: This work was supported by the National Natural Science Foundation of China (No. 62177012) and the GUET Excellent Graduate Thesis Program (grant no. 19YJPYBS01).

Data Availability Statement: Data are contained within the article and Supplementary Materials.

Conflicts of Interest: The authors declare no conflicts of interest.

References

1. Zhao, X.; Zhang, C.; Ju, Y.; Li, J.; Bian, Y.; Ma, J. Evaluation of tunnel retro-reflective arch in an extra-long tunnel based on the matter-element extension method. *Accid. Anal. Prev.* **2021**, *150*, 105913. [CrossRef] [PubMed]
2. Deng, Z. A Tunnel Retro-Reflective Ring Cleaning Device. CN213086674U, 30 April 2021.
3. Yang, Y.; Cao, J. A Highway Tunnel Retro-Reflective Ring Self-Cleaning Device and Its Cleaning Method. CN110882957B, 14 July 2023.
4. Liu, J.H.; Padrigalan, K.E. The Kinematic Analysis of a Wind Turbine Climbing Robot Mechanism. *Appl. Sci.* **2022**, *12*, 1210. [CrossRef]
5. Lu, X.; Zhao, S.; Liu, X.; Wang, Y. Design and analysis of a climbing robot for pylon maintenance. *Ind. Rob.* **2018**, *45*, 206–219. [CrossRef]
6. Bogue, R. Climbing robots: Recent research and emerging applications. *Ind. Rob.* **2019**, *46*, 721–727. [CrossRef]
7. Chu, B.; Jung, K.; Han, C.S.; Hong, D. A survey of climbing robots: Locomotion and adhesion. *Int. J. Precis. Eng. Manuf.* **2010**, *11*, 633–647. [CrossRef]
8. Murphy, M.P.; Kute, C.; Mengüç, Y.; Sitti, M. Waalbot II: Adhesion recovery and improved performance of a climbing robot using fibrillar adhesives. *Int. J. Robot. Res.* **2011**, *30*, 118–133. [CrossRef]
9. Schultz, J.T.; Beck, H.K.; Haagensen, T.; Proost, T.; Clemente, C.J. Using a biologically mimicking climbing robot to explore the performance landscape of climbing in lizards. *Proc. R. Soc. B* **2021**, *288*, 20202576. [CrossRef] [PubMed]
10. Li, R.; Feng, S.; Yan, S.; Liu, X.; Yang, P.A.; Yang, X.; Shou, M.; Yu, Z. Wind Resistance Mechanism of an Anole Lizard-Inspired Climbing Robot. *Sensors* **2022**, *22*, 7826. [CrossRef]

11. Zhu, H.; Guan, Y.; Wu, W.; Zhang, L.; Zhou, X.; Zhang, H. Autonomous pose detection and alignment of suction modules of a biped wall-climbing robot. *IEEE/ASME Trans. Mechatron.* **2014**, *20*, 653–662. [CrossRef]

12. Guan, Y.; Zhu, H.; Wu, W.; Zhou, X.; Jiang, L.; Cai, C.; Zhang, L.; Zhang, H. A modular biped wall-climbing robot with high mobility and manipulating function. *IEEE/ASME Trans. Mechatron.* **2012**, *18*, 1787–1798. [CrossRef]

13. Hu, Q.; Dong, E.; Sun, D. Soft modular climbing robots. *IEEE Trans. Robot.* **2022**, *39*, 399–416. [CrossRef]

14. Enjikalayil Abdulkader, R.; Veerajagadheswar, P.; Htet Lin, N.; Kumaran, S.; Vishaal, S.R.; Mohan, R.E. Sparrow: A magnetic climbing robot for autonomous thickness measurement in ship hull maintenance. *J. Mar. Sci. Eng.* **2020**, *8*, 469. [CrossRef]

15. Zhao, Z.; Tao, Y.; Wang, J.; Hu, J. The multi-objective optimization design for the magnetic adsorption unit of wall-climbing robot. *J. Mech. Sci. Technol.* **2022**, *36*, 305–316. [CrossRef]

16. Spenko, M.J.; Haynes, G.C.; Saunders, J.A.; Cutkosky, M.R.; Rizzi, A.A.; Full, R.J.; Koditschek, D.E. Biologically inspired climbing with a hexapedal robot. *J. Field Robot.* **2008**, *25*, 223–242. [CrossRef]

17. Liu, Y.; Sun, S.; Wu, X.; Mei, T. A wheeled wall-climbing robot with bio-inspired spine mechanisms. *J. Bionic Eng.* **2015**, *12*, 17–28. [CrossRef]

18. Kim, H.; Kim, D.; Yang, H.; Lee, K.; Seo, K.; Chang, D.; Kim, J. Development of a wall-climbing robot using a tracked wheel mechanism. *J. Mech. Sci. Technol.* **2008**, *22*, 1490–1498. [CrossRef]

19. Seo, K.; Cho, S.; Kim, T.; Kim, H.S.; Kim, J. Design and stability analysis of a novel wall-climbing robotic platform (ROPE RIDE). *Mech. Mach. Theory* **2013**, *70*, 189–208. [CrossRef]

20. Zhu, Y.; Wang, Y.; Huang, Y. Failure analysis of a helical compression spring for a heavy vehicle's suspension system. *Case Stud. Eng. Fail. Anal.* **2014**, *2*, 169–173. [CrossRef]

21. Akinfiev, T.; Armada, M.; Prieto, M.; Uquillas, M. Concerning a Technique for Increasing Stability of Climbing Robots. *J. Intell. Robot. Syst.* **2000**, *27*, 195–209. [CrossRef]

22. Zhang, X.; Wu, Y.; Liu, H.; Zhong, D. Design and Analysis of wheel-foot Magnetic Adsorption Barrier Climbing Robot. *J. Mech. Eng.* **2024**, *60*, 248–261.

23. Wang, D.; Xu, S.; Li, Z.; Cao, W. Analysis of the influence of parameters of a spraying system designed for UAV application on the spraying quality based on Box–Behnken response surface method. *Agriculture* **2022**, *12*, 131. [CrossRef]

24. Benyounis, K.Y.; Olabi, A.G.; Hashmi, M.S.J. Multi-response optimization of CO_2 laser-welding process of austenitic stainless steel. *Opt. Laser Technol.* **2008**, *40*, 76–87. [CrossRef]

Article

Sensor-Based Identification of Singularities in Parallel Manipulators

Jose L. Pulloquinga [1,*,†], Marco Ceccarelli [2,†], Vicente Mata [3] and Angel Valera [1]

1   Department of Systems Engineering and Automation, Universitat Politècnica de València, 46022 Valencia, Spain; giuprog@isa.upv.es
2   Department of Industrial Engineering, University of Rome Tor Vergata, 00133 Rome, Italy; marco.ceccarelli@uniroma2.it
3   Department of Mechanical Engineering and Materials, Universitat Politècnica de València, 46022 Valencia, Spain; vmata@mcm.upv.es
*   Correspondence: jopulza@doctor.upv.es
†   These authors contributed equally to this work.

Abstract: Singularities are configurations where the number of degrees of freedom of a robot changes instantaneously. In parallel manipulators, a singularity could reduce the mobility of the end-effector or produce uncontrolled motions of the mobile platform. Thus, a singularity is a critical problem for mechanical design and model-based control. This paper presents a general sensor-based method to identify singularities in the workspace of parallel manipulators with low computational cost. The proposed experimental method identifies a singularity by measuring sudden changes in the end-effector movements and huge increments in the forces applied by the actuators. This paper uses an inertial measurement unit and a 3D tracking system for measuring the end-effector movements, and current sensors for the forces exerted by the actuators. The proposed sensor-based identification of singularities is adjusted and implemented in three different robots to validate its effectiveness and feasibility for identifying singularities. The case studies are two prototypes for educational purposes—a five-bar mechanism and an L-CaPaMan parallel robot—and a four-degree-of-freedom robot for rehabilitation purposes. The tests showcase its potential as a practical solution for singularity identification in educational and industrial robots.

Keywords: robotics; parallel manipulators; kinematics; singularities; experimental analysis

Citation: Pulloquinga, J.L.; Ceccarelli, M.; Mata, V.; Valera, A. Sensor-Based Identification of Singularities in Parallel Manipulators. *Actuators* **2024**, *13*, 168. https://doi.org/10.3390/act13050168

Academic Editor: Ioan Doroftei

Received: 20 March 2024
Revised: 26 April 2024
Accepted: 28 April 2024
Published: 1 May 2024

1. Introduction

In robotics, a singularity is a configuration where the number of degrees of freedom (DOF) of the robot changes instantaneously [1,2]. In serial manipulators, a singularity constrains the mobility of the end-effector in at least one direction and generally appears at the boundary of the workspace [3]. However, parallel manipulators, or Parallel Kinematic Mechanisms (PKMs), have various types of singularities due to their closed kinematic chain architecture [4–6].

Gosselin and Angeles [7] classified the singularities of PKMs into three types by analysing the input-output kinematic velocity relationship. Type I singularities are analogous to the singular configurations in serial manipulators, i.e., the PKMs lose at least one DOF of the end-effector or mobile platform. In Type II singularities, despite the actuator of the PKM being locked, the mobile platform gains at least one uncontrollable motion [1]. Type III singularities combine the two previous types simultaneously and only appear under certain geometrical parameters of the links [7,8]. The singularities of a PKM have been classified in different manners [9–13], always stressing the loss of mobility or uncontrolled movements in the end-effector.

The consequences of singularities pose significant challenges to PKM design and control. Type I singularities amplify the forces required by the actuators to move the mobile platform, increasing electric current demands [2]. Conversely, Type II singularities generate uncontrolled motion that risks the integrity of the user and the PKM itself [14,15], and they degrade the model-based control laws, resulting in infinite control actions [16,17].

In non-redundant PKMs, singularities are identified using graphical procedures, analytical methods, or a combination of both. Graphical methods are usually based on PKM geometry [18,19]. Analytical methods can use Jacobian matrices from the input-output kinematic velocity relationship [20–22], Screw theory [23–25], or motion/force transmissibility [26]. These theoretical methods primarily rely on precise kinematic modelling of the PKM under study with offline implementations because of their substantial mathematical calculations. In [27], experimental measurements of the position and orientation (pose) reached by the end-effector and the force applied by the actuators were used to improve singularity identification based on kinematic models. In real PKMs, the kinematic model-based singularity identification described above faces the following challenges:

- Real-time implementation in resource-limited control units is hindered by the substantial mathematical calculations required. Combining singularity identification with control laws, such as computed torque control [17] or model predictive control [28], requires considerable computational resources.
- Modelling complexity varies across mechanical architectures, complicating the establishment of a general approach applicable to different PKMs. A comprehensive examination of PKM modelling and its associated challenges is provided [29].
- Non-model effects, such as joint clearances, add complexity to determining the appropriate singularity proximity thresholds. Some approaches for measuring the closeness to a singularity are provided in [30,31].

In order to address the challenges faced by existing methods, this paper proposes a novel approach to identifying Type I and Type II singularities in non-redundant PKMs using only sensor measurements. The sensor-based singularity identification method is designed to detect unexpected movements of the end-effector and sharp oscillations in the forces applied by the actuators. By utilising inertial or vision sensors to measure the motion of the end-effector and current sensors to measure the actuator forces, the proposed method bypasses the need for accurate kinematic modelling. The avoidance of kinematic modelling allows for real-time implementation across resource-limited control units. The singularity identification threshold is set by the average pose error and actuator current during normal operations of the real PKM, making the proposed method robust to non-model effects. Thus, the proposed singularity identification provides high adaptability to different PKMs with low complexity in implementation. The effectiveness of this method is validated through experiments conducted on two educational PKMs and a PKM designed for knee rehabilitation.

The remainder of this manuscript is organised as follows. Section 2 presents the singularities in a PKM and the problems for users and robots. Section 3 describes the proposed sensor-based procedure for identifying Type I and Type II singularities. Section 4 uses the proposed sensor-based procedure to identify Type I singularities in a 3-DOF PKM known as L-CaPaMan. Section 5 describes the sensor-based identification of Type II singularities in a five-bar planar mechanism. In both cases, the motion of the end-effector is detected by an inertial measurement unit, and current sensors measure the force applied by the actuators. Section 6 describes the identification of Type II singularities using a 4-DOF PKM for knee rehabilitation as a case study. In this case, the motion of the end-effector is measured by a 3D tracking system. Finally, the main conclusions are presented in Section 7.

2. Singularities in Parallel Manipulators

A parallel robot, or Parallel Kinematic Mechanism (PKM), is a mechanism that controls the location and orientation of the end-effector or mobile platform using at least two open kinematics chains [1]. The position and orientation of the mobile platform, often referred to as the pose of the end-effector, are represented by a vector $x = \begin{bmatrix} x_p & y_p & z_p & \varphi & \theta & \psi \end{bmatrix}$. The translational DOFs of the PKM are represented by x_p, y_p, and z_p, and the rotational

DOFs are represented by φ, θ, and ψ. The pose x is constrained by the kinematic arrangement of the open kinematic chains, which dictate the permissible motions of the mobile platform [2].

Each open kinematic chain, also known as a limb, consists of interconnected links joined by joints. The relative motion between the links within a limb is described using generalised coordinates, denoted as q_{ij}, which may represent translational or rotational motion at any given instance [2]. The pose x is controlled by a subset of active or actuated joints q_{ind}.

The kinematic structure of a PKM typically includes key elements such as the mobile reference frame $\{P - X_p Y_p Z_p\}$ attached to the end-effector and the fixed reference frame $\{O - X_0 Y_0 Z_0\}$ affixed to the base or fixed platform. Figure 1 illustrates a schematic representation of a PKM, highlighting these main kinematic elements.

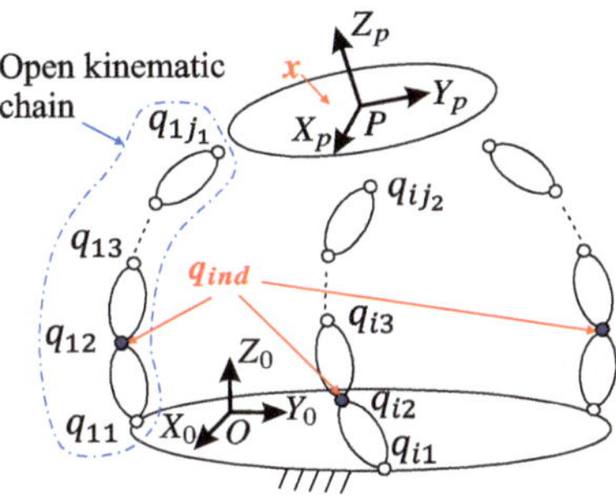

Figure 1. A general scheme of a PKM and its main kinematic elements. The pose of the end-effector and the active joints are shown in red.

The inverse and forward kinematics problems of the PKM are defined by the input-output constraint equations Φ

$$\Phi(x, q_{ind}) = 0 \tag{1}$$

where x and q_{ind} are vectors that represent the input and output variables of the PKM, respectively.

Taking the time derivatives of Equation (1), the relationship between the output and input velocities is

$$J_I \dot{q}_{ind} + J_D \dot{x} = 0 \tag{2}$$

where $\dot{q}_{ind}$ stands for the velocities in the actuators, $\dot{x}$ represents the velocities of the end-effector, and J_I and J_D represent the inverse and forward Jacobian matrices, respectively. For non-redundant PKMs, J_I and J_D are square $F \times F$ matrices, with F as the DOFs of the mobile platform.

Based on the rank deficiency of the Jacobian matrices, Gosselin and Angeles [7] defined three types of singularities:

Type I: The J_I matrix becomes rank-deficient, i.e., the determinant of the J_I matrix is zero ($|J_I| = 0$). The mobile platform of the PKM loses mobility in at least one direction despite having a set of non-zero velocities in the actuators, $\dot{q}_{ind} \neq 0$. Figure 2a shows a five-bar planar mechanism, or 5R PKM (where R stands for revolute joint), in this singular configuration.

Type II: Here, the Jacobian matrix J_D becomes rank-deficient ($|J_D| = 0$). In this case, despite all actuators being locked ($\dot{q}_{ind} = 0$), the mobile platform of the PKM experiences at least one uncontrollable motion. Figure 2b depicts a Type II singularity configuration. Under this condition, if an external action is applied to the mobile platform, the PKM moves despite the actuators being locked.

Type III: Both J_I and J_D become rank-deficient simultaneously. This configuration occurs only for specific geometric parameters of the links. Figure 2c shows an example using a four-bar mechanism, or 4R PKM.

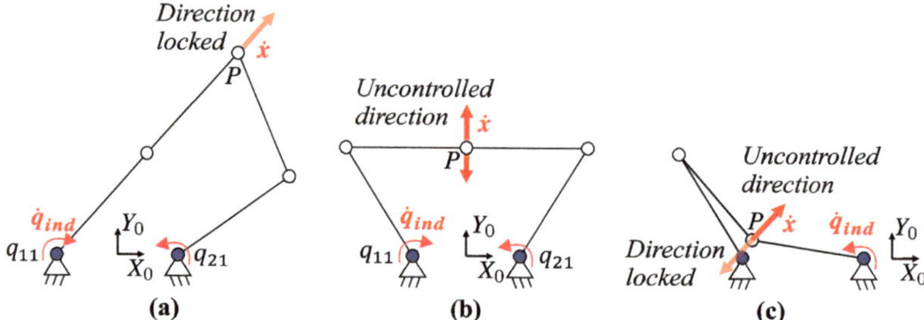

Figure 2. Examples of singularities in a 5R PKM: (**a**) Type I; (**b**) Type II; (**c**) Type III in a 4R PKM.

3. Sensor-Based Identification of Singularities

Type I and Type II singularities cause instantaneous changes in the end-effector motion [1,2]. Hence, a motion sensor can identify singularities by identifying disruptions in the end-effector motion. Such disruptions necessitate an increase in the force applied to the actuators. Since force correlates with the electric current consumed by the actuators, incorporating current sensors could measure this increment in force. Therefore, this work proposes an experimental identification of the singularities of a PKM based on sudden changes in the end-effector motion and the corresponding electric current consumed by the actuators.

Figure 3 shows a block diagram of the proposed sensor-based procedure. The procedure requires the reference pose of the mobile platform x_r, the measured pose x_m, and the current consumed by the actuators i_m. Singularities are identified by calculating the pose error e_x and the time derivative of the electric current consumption $\dot{i}_m$.

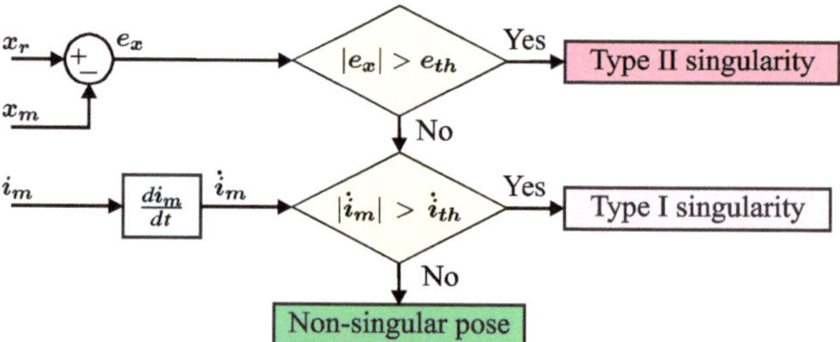

Figure 3. Scheme of the proposed sensor-based procedure for identifying PKM singularities.

The closeness to a singularity is identified when e_x exceeds the threshold of the permitted motion disturbance e_{th}. If the end-effector exceeds the permitted motion disturbance $|e_x| > e_{th}$, the PKM reaches a Type II singularity. This singularity is considered potentially dangerous because the end-effector presents uncontrolled motions. If e_x is within the range of motion ($|e_x| < e_{th}$), and i_m increases suddenly over a certain limit, $|\dot{i}_m| > \dot{i}_{th}$, it is a singularity Type I. This paper considers a Type I singularity non-dangerous because a PKM maintains its stiffness despite the reduction in the mobility of the end-effector. In contrast, Type II singularities are potentially dangerous because the end-effector becomes unstable, reducing its controllability.

The thresholds e_{th} and $\dot{i}_m$ are defined based on the operating specifications of the PKM. The threshold e_{th} is defined as the average pose error admissible in the F DOFs for the fundamental movements of the PKM. In contrast, $\dot{i}_{th}$ is defined as the instantaneous change in the average current consumed by the actuators i_a during the fundamental movements of the PKM. Considering that the control unit is discrete, i_{th} could be calculated for a constant sample time as

$$\dot{i}_{th} = \frac{i_a}{t_s} \tag{3}$$

where t_s is the sample time from the control unit.

The proposed identification method is based on the following assumptions:

- The working conditions of the PKM remain constant during its principal or fundamental movements. The fundamental movements represent the most commonly performed actions by the PKM within a given task. Variations in payload and wear and tear over the PKM's lifetime are not taken into account.
- The sensor signals are assumed to be minimally affected by noise, either through hardware or software filtering mechanisms.
- The control unit works at a constant sample time.

The proposed procedure identifies singularities based solely on motion and electric current measurements. Therefore, this work introduces a rapid sensor-based method to identify singularities of PKMs, which is suitable for offline and online applications.

If the thresholds e_{th} and i_{th} are defined for the principal movements of the PKM, the proposed method provides robust detection of singularities that involve sudden movement changes or drastic current increments. Moreover, the proposed method can detect constraint singularities, which occur when the constraint wrenches from the limbs disappear, resulting in effects similar to Type II singularities [11].

The exclusive reliance on sensor measurements ensures that the proposed identification procedure is computationally efficient, rendering it suitable for control units with limited resources, such as robotics prototypes in engineering education. Moreover, by circumventing the need for kinematic models, the proposed method offers broad applicability, reducing the time investment required for implementation. In cases with enough computation resources, the proposed sensor-based identification procedure could be used for fast initial identification of singularities prior to verifying them using an analytical method.

The current research focuses on analysing the features of the proposed method under constant working conditions across different PKMs. The robustness of the proposed technique under varying working conditions, payload changes, and noise sensor signals is not analysed.

The following sections describe the setup and implementation of the proposed sensor-based identification of singularities in two PKMs developed for educational purposes and a prototype of a PKM for knee rehabilitation.

4. Case Study: L-CaPaMan

L-CaPaMan is a prototype of a PKM designed for earthquake simulation, featuring three DOFs [32]. This PKM operates by controlling translation along the Z_p axis and rotations φ and ψ using three angular actuators, denoted as α_1, α_2, and α_3, as illustrated in Figure 4a.

The positioning of the end-effector along X_p and Y_p, along with the lengths of the three strokes s_i, is intricately tied to the three DOFs Z_p, φ, and ψ, rendering them passive variables. Geometric lengths represented by a_i, b_i, d_i, and h_i determine the specific connection points between the end-effector and the fixed platform. The coordinates frame $\{O_i - X_i Z_i\}$ stands for the local system at each limb $i = 1 \ldots 3$.

For experimental validation, a cost-effective prototype of L-CaPaMan, featuring a 3D-printed structure [32], was employed, as depicted in Figure 4b. This figure illustrates the physical realisation of the prototype, providing insights into its practical implementation.

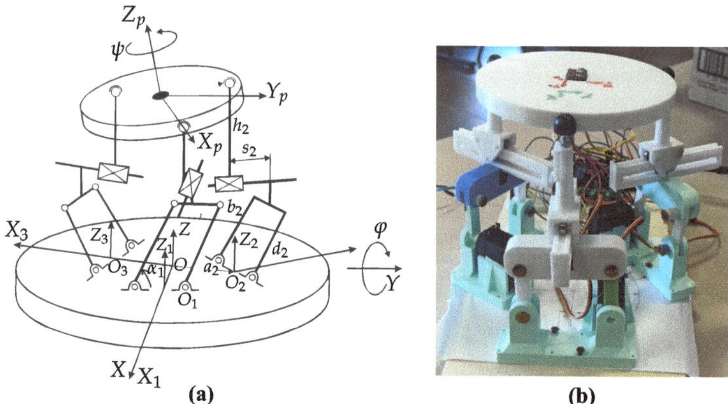

Figure 4. L-CaPaMan PKM [32]: (**a**) kinematic design; (**b**) current prototype.

The proposed sensor-based identification shown in Figure 3 was applied to detect Type I singularities in the current L-CaPaMan prototype shown in Figure 4b. The sensor-based identification of singularities was carried out online using Matlab code during the execution of two singular trajectories, which are defined as follows:

- TM1: Crossing two Type I singularities in the middle of two desired configurations.
- TM2: Starting from a Type I singularity to reach a non-singular pose.

The reference trajectories (TM1 or TM2) are sent from Matlab R2023a code to the control unit of the L-CaPaMan PKM via serial communications at a baud rate of 115,200. The control unit is an Arduino Mega, manufactured in Italy by Arduino.cc, that modifies the pose of the end-effector using three servomotors in the PKM. All the measurements are sent from the Arduino Mega to Matlab for numerical identification of the singularities. Figure 5 shows a schematic diagram of the sensor-based identification of Type I singularities for the L-CaPaMan PKM, and Algorithm 1 details the code executed in Matlab.

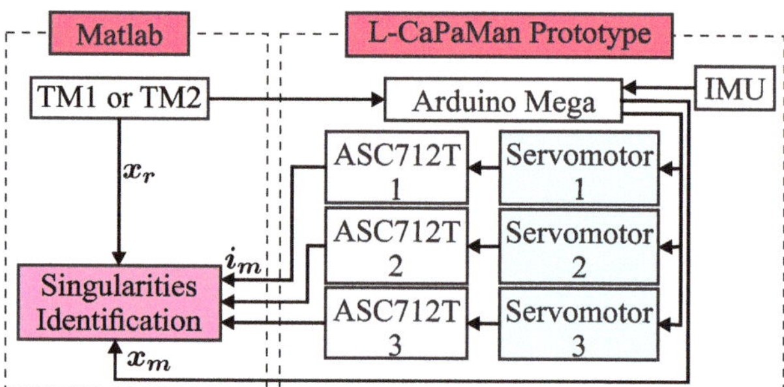

Figure 5. Scheme for sensor-based identification of singularities in the L-CaPaMan PKM.

The current consumed by the servomotors is measured by three current sensors (ASC712T), which provide an analogue voltage proportional to the current at 120 kHz. These sensors have a range of ± 5 A, an output sensitivity of 185 mV/A, and a supply voltage of 4.5 V to 5.5 V. The linear acceleration and the angular velocity of the mobile platform are measured by an inertial measurement unit (IMU) attached to point H. The IMU sensor is an MPU6050 powered at 5 V 800 Hz, with an acceleration range of ± 2 g and a gyroscope range of ± 125 degrees/s.

The fundamental motion for the L-CaPaMan PKM consists of sinusoidal translations in Z_p with simultaneous sinusoidal rotations in φ and ψ. This fundamental motion is executed ten times in the L-CaPaMan PKM to set the thresholds e_{th} and i_{th}. The current prototype of the L-CaPaMan PKM has an average pose error of 2 degrees for φ and ψ, and 3 mm along Z_p. The actuators require an average current of 150 mA, and the control unit runs at 50 Hz (20 ms). Thus, for the L-CaPaMan PKM, the parameters for identifying singularities are set as $e_{th} = \begin{bmatrix} 3\,\text{mm} & 2\,\text{deg.} & 2\,\text{deg.} \end{bmatrix}$ and $\dot{i}_{th} = \begin{bmatrix} 7.5 & 7.5 & 7.5 \end{bmatrix}$ A/s.

Algorithm 1: Matlab pseudo-code for sensor-based identification of singularities in the L-CaPaMan PKM

Data: Reference Trajectories from TM1 or TM2
Result: Numerical Identification of Singularities
initialise threshold e_{th} and $\dot{i}_{th}$
initialise serial communication with Arduino Mega
i = 0
while *running* **do**

    send x_r at instant i from TM1 or TM2 to Arduino Mega
    if *i > trajectory end* **then**
        break
    end
    while *i $\leq$ trajectory end* **do**
        receive current measurement i_m and pose x_m from Arduino
        calculate pose error $e_x = x_r - x_m$
        if $|e_x| > e_{th}$ **then**
            Identify Type II singularity
        end
        else if $|e_x| < e_{th}$ *and* $|\dot{i}_m| > \dot{i}_{th}$ **then**
            Identify Type I singularity
        end
        else
            No singularity detected
        end
        i++
    end
end

Results

The initial pose, expected singularity, and final pose of the L-CaPaMan PKM for testing trajectories TM1 and TM2 are listed in Table 1.

Table 1. Data of the trajectories executed with L-CaPaMan.

	TM1					TM2			
Location	Z_p (m)	φ (deg.)	ψ (deg.)	Time (s)		Z_p (m)	φ (deg.)	ψ (deg.)	Time (s)
Start	0.13	0	−22	0		0.13	−6	−7	0
Singularity	0.14	0	22	0.18	1.22	0.13	−6	−7	0
End	0.13	0	−22	3.1		1.8	0	0	2

Since TM1 mainly executes a vertical movement along Z_p, the motion error is shown only for this DOF. Figure 6a shows that the absolute position error in Z_p ($e_x[1]$) remained under the corresponding threshold ($e_{th}[1]$) for detecting Type I singularities during TM1. This means that the mobile platform experienced no sudden movement changes.

Although the mobile platform executed a controlled motion during TM1, the current consumed by the actuators presented two sudden changes. Figure 6b shows the rate of change of the current consumed by actuator 1. It verifies that $|\dot{i}_m|[1]$ detected the two peaks of the current corresponding to a Type I singularity. Figure 7 shows that the two peaks of the current detected by $|\dot{i}_m|[1]$ occurred when $|J_I|$ crossed zero.

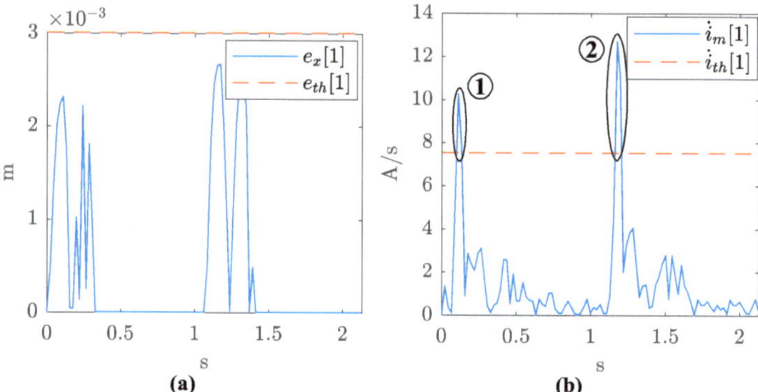

Figure 6. Results of testing TM1 with the L-CaPaMan PKM: (**a**) absolute position error along Z_p; (**b**) time derivative of the current consumed by actuator 1. The detected Type I singularities are enclosed in two black circles.

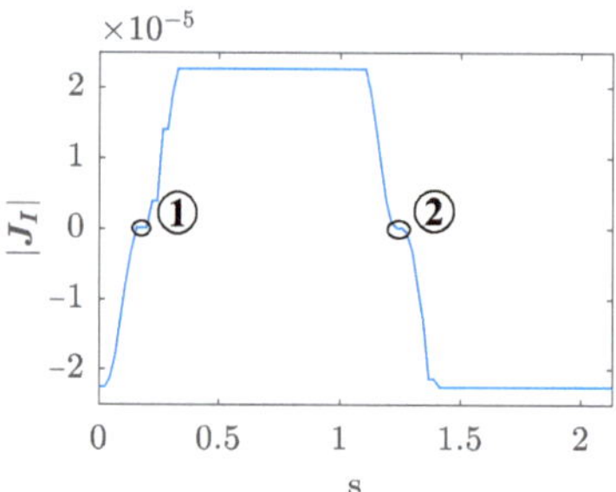

Figure 7. Computed $|J_I|$ during TM1 with the L-CaPaMan PKM. The detected Type I singularities are enclosed in two black circles.

The testing of TM2 demonstrated a controlled motion ($|e_x| < e_{th}$) with a current peak at the beginning ($|\dot{i}_m| > \dot{i}_{th}$). Figure 8 shows the absolute error in Z_p and the current consumed by actuator 1 during TM2. These results verify that the proposed experimental procedure is suitable for detecting Type I singularities.

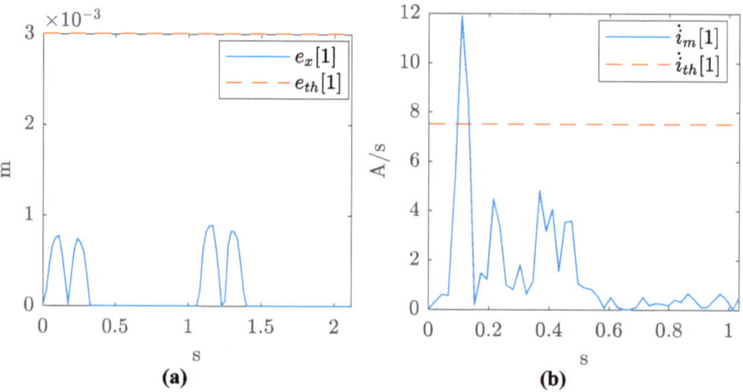

Figure 8. Results of testing TM2 with the L-CaPaMan PKM: (**a**) absolute position error along Z_p; (**b**) time derivative of the current consumed by actuator 1.

5. Case Study: Five-Bar Mechanism

The five-bar mechanism, also known as the 5R mechanism, is a planar PKM featuring two DOFs designed to position a point P within a defined plane $O - X_p Y_p Z_p$. The name 5R originates from the five revolute joints denoted by the letter "R" that connect the links. Of these joints, two are actuated, typically connected to the fixed platform. The positions x_p and y_p of the point P are driven by two angular actuators represented by the generalised coordinates q_{11} and q_{21} (Figure 9a). The locations of the passive revolute joints, represented by q_{12} and q_{22}, are calculated based on the end-effector pose.

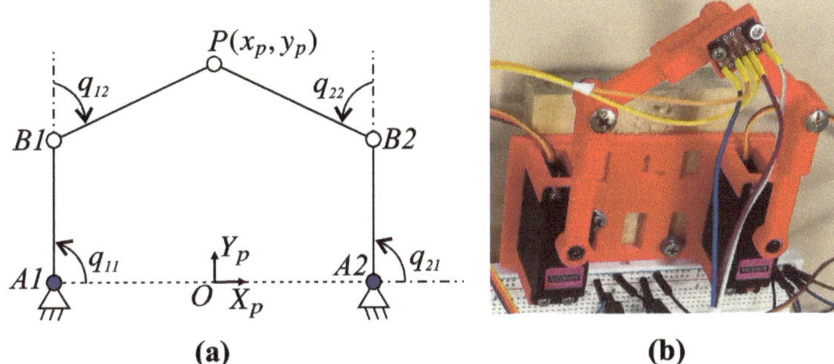

Figure 9. 5R mechanism: (**a**) kinematic design; (**b**) current prototype.

The kinematic analysis of the 5R mechanism requires defining a working mode. In this work, the 5R mechanism has a symmetrical architecture with the working mode $-+$. This working mode specifies that the actuator at $A1$ corresponds to the negative solution for inverse kinematics, while the actuator at $A2$ corresponds to the positive solution. For a detailed kinematic model of the 5R PKM, readers are referred to [33]. Figure 9b depicts the 5R PKM utilised in our research, developed specifically to demonstrate the impact of singularities in pick-and-place tasks to engineering students. Table 2 provides a list of the geometrical parameters for the current 5R PKM prototype.

Table 2. Geometrical parameters for the 5R PKM prototype.

$OA1, OA2$ (m)	$A1B1, A2B2$ (m)	$B1P, B2P$ (m)
0.04	0.06	0.05

In this case, the experimental identification of singularities was carried out online in Matlab during the execution of two singular trajectories. Trajectory TM3 crosses a Type II singularity, and trajectory TM4 starts from a non-singular position and stops in a Type II singularity.

Analogous to Section 4, reference trajectories TM3 or TM4 are sent from Matlab code to the Arduino Mega for controlling the location of the 5R PKM via serial communications with a baud rate of 115,200. The linear acceleration and angular velocity of the B1P link are measured by an IMU MPU6050 in point P, and two ASC712T sensors measure the current consumed by the actuators. The ASC712T sensors provide a voltage proportional to the current at 120 kHz, with a range of ± 5 A and a sensitivity at 185 mV/A. The MPU6050 measures linear acceleration at ± 2 g and angular velocity at ± 125 degrees/s with a frequency of 800 Hz.

A schematic diagram of the sensor-based identification of Type II singularities for the 5R PKM is shown in Figure 10, and its code executed in Matlab is depicted in Algorithm 2.

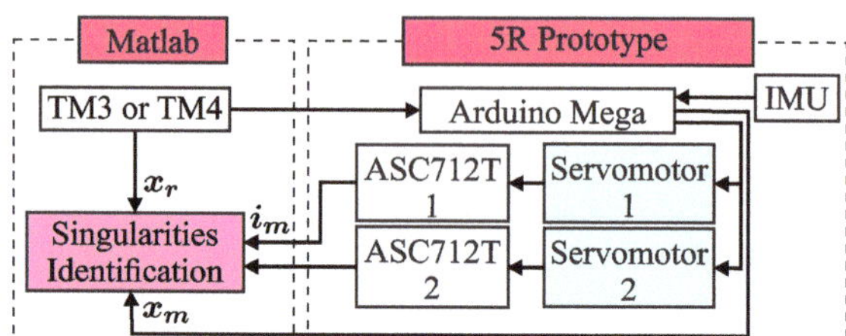

Figure 10. Scheme for sensor-based identification of singularities in the 5R PKM prototype.

Algorithm 2: MATLAB pseudo-code for sensor-based identification of singularities in the 5R PKM

Data: Reference Trajectories from TM3 or TM4
Result: Numerical Identification of Singularities
initialise threshold e_{th} and $\dot{i}_{th}$
initialise serial communication with Arduino Mega
i = 0
while *running* **do**
    send x_r at instant i from TM3 or TM4 to Arduino Mega
    if *i > trajectory end* **then**
        | break
    end
    while *i ≤ trajectory end* **do**
        receive current measurement i_m and pose x_m from Arduino
        calculate pose error $e_x = x_r - x_m$
        if $|e_x| > e_{th}$ **then**
            | Identify Type II singularity
        end
        else if $|e_x| < e_{th}$ *and* $|\dot{i}_m| > \dot{i}_{th}$ **then**
            | Identify Type I singularity
        end
        else
            | No singularity detected
        end
        i++
    end
end

The 5R PKM under study develops a translation from $(-0.02, 0.06)$ to $(0.02, 0.06)$ on the $X_p Y_p$ plane as its fundamental movement. After executing the fundamental movement ten times, the 5R PKM exhibits an average position error of 1.5 mm for x_p and y_p, and the actuators require an average current of 150 mA. Considering that the control unit runs at 20 ms, the parameters for identifying singularities are set as $e_{th} = \begin{bmatrix} 1.5 & 1.5 \end{bmatrix}$ mm and $\dot{i}_{th} = \begin{bmatrix} 7.5 & 7.5 \end{bmatrix}$ A/s.

Results

The initial pose, expected singularity, and final pose for testing trajectories TM3 and TM4 are listed in Table 3.

Table 3. Data of the trajectory executed with the 5R PKM prototype.

	TM3			TM4		
Location	x_p (m)	y_p (m)	Time (s)	x_p (m)	y_p (m)	Time (s)
Start	0	0.09	0	0	0.09	0
Singularity	−0.03	0.05	2.8	−0.03	0.05	3
End	−0.04	0.03	3.1	−0.03	0.05	3

During the execution of TM3, the 5R mechanism lost control of the position of point P in the closeness to the expected Type II singularity. Figure 11a shows that the absolute position error for the vertical DOF y_p ($e_x[2]$) exceeded the corresponding threshold ($e_{th}[2]$) starting from the time instant 2.4 s. This figure verifies that the PKM was unable to track the reference frame after 2.4 s, i.e., a Type II singularity occurred. During the execution of TM3, there were no sudden changes in the current consumed by the actuators. Figure 11b shows the absolute derivative of the current consumed by actuator 1 ($|\dot{i}_m|[1]$).

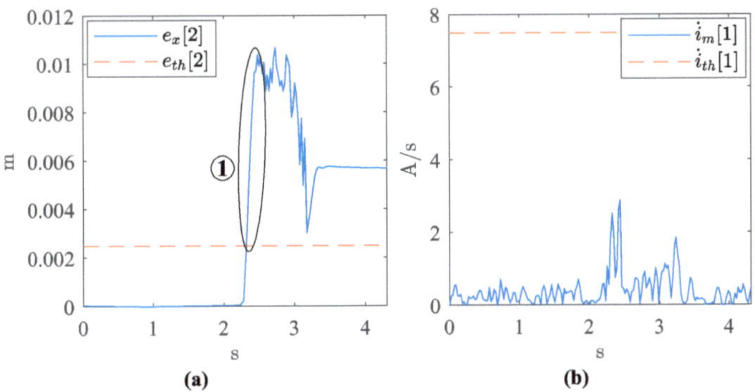

Figure 11. Results of testing TM3 with 5R PKM: (**a**) absolute position error for y_p; (**b**) time derivative of consumed current by actuator 1. The detected Type II singularity is enclosed in a black circle.

According to Table 3, the Type II singularity was expected to appear at 2.8 s. However, the proposed sensor-based procedure identified the Type II singularity at 2.4 s. $|J_D|$ verified that the proposed sensor-based procedure anticipated the Type II singularity (see Figure 12). This is because non-modelled effects, such as clearances in joints, generate regions with singularities in actual PKMs [34].

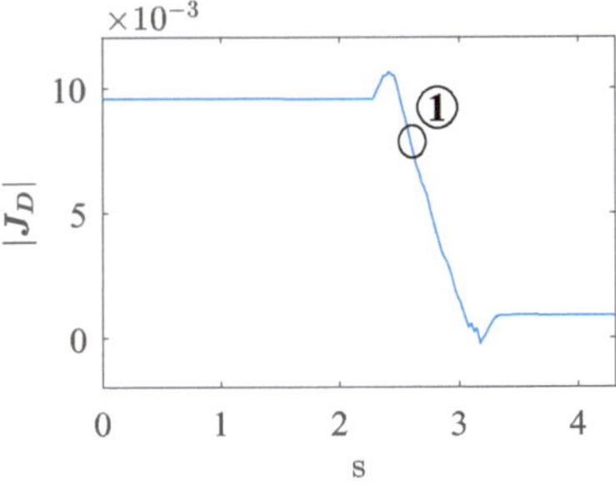

Figure 12. Computed $|J_D|$ during TM3 with the 5R PKM. The detected Type II singularity is enclosed in a black circle.

The degeneracy of the end-effector motion with non-sudden changes in actuator currents was observed again towards the end of TM4 (Figure 13). Thus, the proposed procedure was able to identify Type II singularities in the 5R mechanism.

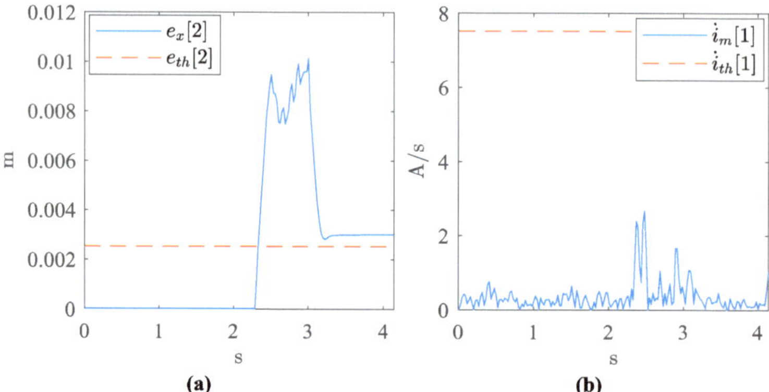

(a)
(b)

Figure 13. Results of testing TM4 with the 5R PKM: (**a**) absolute position error for y_p; (**b**) time derivative of the current consumed by actuator 1.

6. Case Study: 4-DOF Parallel Manipulator

The 4-DOF parallel manipulator under analysis is a PKM built for knee rehabilitation and diagnosis purposes at Universitat Politècnica de València [35]. This PKM is named 3UPS+RPU due to the three external limbs having a UPS configuration and the central one having an RPU configuration. Figure 14 shows a kinematic representation of the 4-DOF PKM and its current prototype. The letters R, U, S, and P stand for revolute, universal, spherical, and prismatic joints, respectively, and "_" identifies the actuated joint.

The four DOFs of the PKM consist of two translational movements (x_m, z_m) in the tibiofemoral plane, along with rotations (ψ, θ) around the coronal and tibiofemoral planes, respectively [35]. These four DOFs are controlled by four linear actuators, denoted as q_{13}, q_{23}, q_{33} and q_{42}, as shown in Figure 14a.

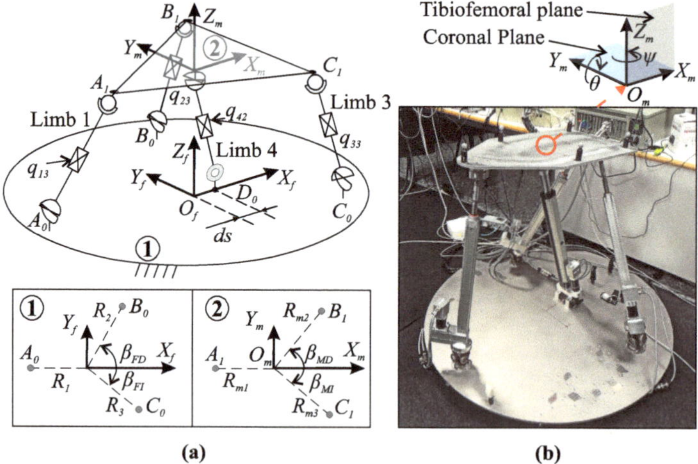

(a)
(b)

Figure 14. 3UPS+RPU PKM: (**a**) kinematic design; (**b**) current prototype. The geometrical parameters of the fixed and mobile platform are shown in Table 1 and Table 2, respectively.

The connections between the limbs and the fixed platform $(A_0, \ldots, D_0)$ are defined by geometrical variables such as R_1, R_2, R_3, β_{FD}, β_{FI}, and ds. Conversely, the connections between the limbs and the mobile platform (A_1, B_1, C_1, O_m) are determined by variables like R_{m1}, R_{m2}, R_{m3}, β_{MD}, and β_{MI}. These geometrical parameters are referenced to the fixed frame $\{O_f - X_f Y_f Z_f\}$ for the fixed platform and the moving frame $\{O_m - X_m Y_m Z_m\}$

for the mobile platform. Detailed geometrical parameters for the 3U$\underline{P}$S+R$\underline{P}$U PKM are provided in Table 4.

Table 4. Geometrical parameters for the 3U$\underline{P}$S+R$\underline{P}$U PKM.

R_1, R_2, R_3 (m)	β_{FD} (deg.)	β_{FI} (deg.)	ds (m)	R_{m1}, R_{m2}, R_{m3} (m)	β_{MD} (deg.)	β_{MI} (deg.)
0.4	90	45	0.15	0.3	50	90

The experimental identification presented in Section 3 was applied to identify Type II singularities in the current 3U$\underline{P}$S+R$\underline{P}$U PKM prototype during the execution of two trajectories that crossed a Type II singularity, TM5 and TM6.

The sensor-based singularity identification method was developed on an industrial computer using ROS2, as reported in Figure 15. Reference trajectories TM5 or TM6 are sent to a PID controller that drives the four prismatic actuators of the PKM (q_{13}, q_{23}, q_{33}, and q_{42}) using an ESCON 50/5 current amplifier powered at 24 V, manufactured in Sachseln, Switzerland by Maxon motor ag. The current amplifier provides a nominal power of 250 W and accurate feedback of the current consumed by the actuators using an analogue output channel with a 12-bit resolution at 53.6 kHz. The actual pose of the mobile platform of the 3U$\underline{P}$S+R$\underline{P}$U PKM is measured by a 3D tracking system (3DTS) manufactured by Optitrack, with an accuracy of 0.1 mm. The 3DTS system consists of 10 infrared Flex 13 cameras with a resolution of 1.3 Megapixels at 120 Hz.

The code implemented in ROS2 is depicted in Algorithm 3.

Algorithm 3: ROS2 pseudo-code for sensor-based identification of singularities in the 3U$\underline{P}$S+R$\underline{P}$U PKM

Data: Reference Trajectories from TM5 or TM6
Result: Numerical Identification of Singularities
initialise threshold e_{th} and $\dot{i}_{th}$
initialise communication with Optitrack 3DTS
i = 0
while *running* **do**
    compute control actions from the PID controller for x_r at instant i using
      actuators' encoder feedback
    send control actions to ESCON 50/5
    if *i > trajectory end* **then**
      | break
    end
    while *i $\leq$ trajectory end* **do**
      read current measurement i_m
      receive pose x_m from Optitrack 3DTS
      calculate pose error $e_x = x_r - x_m$
      if $|e_x| > e_{th}$ **then**
      | Identify Type II singularity
      end
      else if $|e_x| < e_{th}$ *and* $|\dot{i}_m| > \dot{i}_{th}$ **then**
      | Identify Type I singularity
      end
      else
      | No singularity detected
      end
      i++
    end
end

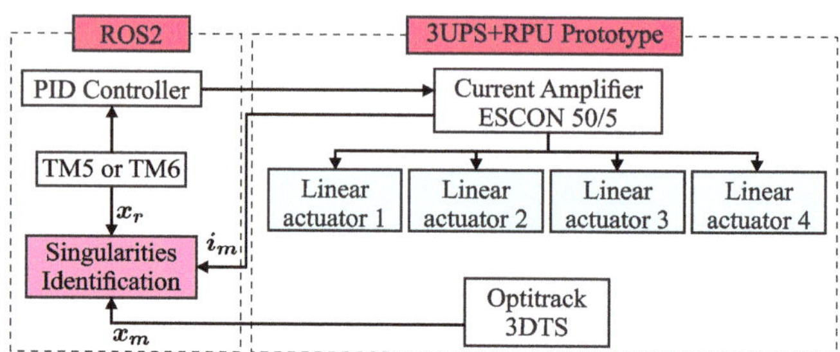

Figure 15. Scheme for sensor-based identification of singularities in the 3U$\underline{P}$S+R$\underline{P}$U PKM.

The 3U$\underline{P}$S+R$\underline{P}$U PKM performs three fundamental movements: (i) flexion of the hip, (ii) flexion-extension of the knee, and (iii) internal-external rotation of the knee [35]. After executing the fundamental movements ten times, the actual 3U$\underline{P}$S+R$\underline{P}$U PKM prototype exhibits an average position error of 9 mm for x_m and z_m, and 3 degrees for θ and ψ, with an average of 5A for the current consumed by the actuators. Considering that the control unit runs at 10 ms, the parameters for the sensor-based identification procedure are set as $e_{th} = \begin{bmatrix} 9\,\text{mm} & 9\,\text{mm} & 3\,\text{deg.} & 3\,\text{deg.} \end{bmatrix}$ and $\dot{i}_{th} = \begin{bmatrix} 50 & 50 & 50 & 50 \end{bmatrix}$ A/s.

Results

The initial pose, expected Type II singularity, and final pose of the 3U$\underline{P}$S+R$\underline{P}$U PKM for executing trajectories TM5 and TM6 are listed in Table 5.

Table 5. Data of the trajectory executed with the 3U$\underline{P}$S+R$\underline{P}$U PKM prototype.

			TM5					TM6		
Pose	x_m (m)	z_m (m)	θ (deg.)	ψ (deg.)	t (s)	x_m (m)	z_m (m)	θ (deg.)	ψ (deg.)	t (s)
Start	−0.05	0.63	5	0	0	0	0.62	8	−6	0
Singularity	−0.05	0.73	5	34	14	0.08	0.72	−3	15	15
End	−0.05	0.73	5	44	17	0.16	0.76	−16	41	24

During the execution of TM5 and TM6, the 3U$\underline{P}$S+R$\underline{P}$U PKM lost control over the mobile platform in the closeness to the expected Type II singularities. Figure 16a shows that the PKM under study increased the error tracking on x_m ($e_x[1]$) at time instant 13.9. When the 3U$\underline{P}$S+R$\underline{P}$U PKM lost control over the end-effector, the current consumed by the actuators exhibited no sudden increments. Figure 16a shows the absolute derivative of the current consumed by actuator 1 ($|\dot{i}_m|[1]$) during the execution of TM5.

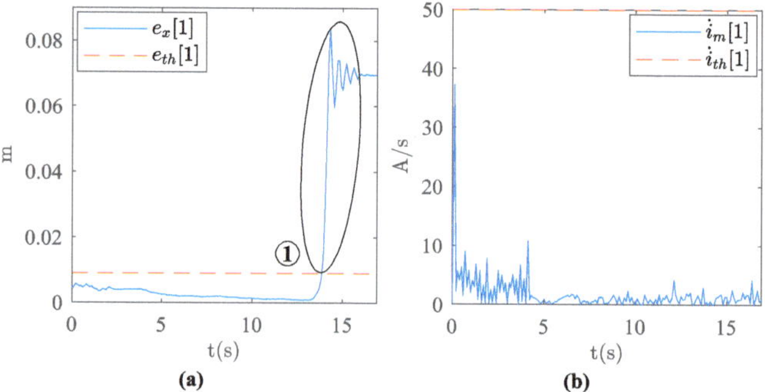

Figure 16. Results of testing TM5 with the 3U$\underline{P}$S+R$\underline{P}$U PKM: (**a**) absolute position error for x_m; (**b**) time derivative of the current consumed by actuator 1. The detected Type II singularity is enclosed in a black circle.

Figure 17 shows that the instant of uncontrolled motion over the mobile platform appeared in the closeness to zero crossing of $|J_D|$, i.e., a Type II singularity was detected.

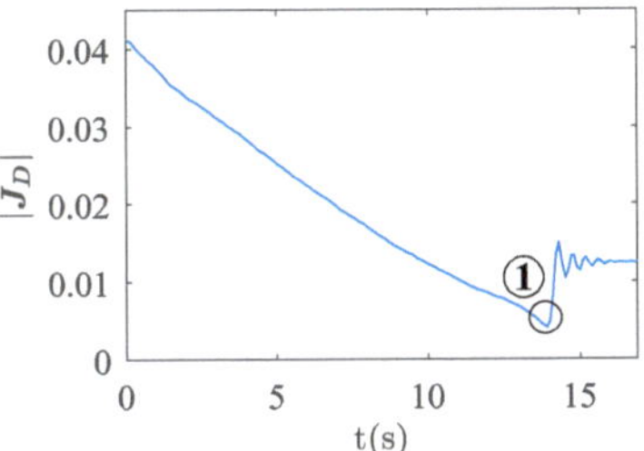

Figure 17. Computed $|J_D|$ during TM5 with the 3U$\underline{P}$S+R$\underline{P}$U PKM. The detected Type II singularity is enclosed in a black circle.

In TM6, the Type II singularity was detected by an increment in the error tracking at time instant 13.0 s (Figure 18). Thus, the proposed sensor-based procedure demonstrated its capability to identify Type II singularities in spatial PKMs. In this work, the interaction between the 3U$\underline{P}$S+R$\underline{P}$U PKM and a human was avoided in order to reduce the external forces that could modify the expected singular configuration.

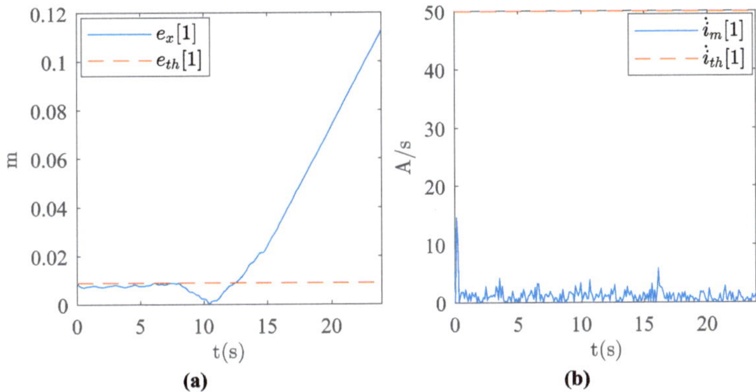

Figure 18. Results of testing TM6 with the 3U$\underline{P}$S+R$\underline{P}$U PKM: (**a**) absolute position error for x_m; (**b**) time derivative of the current consumed by actuator 1.

7. Conclusions

The proposed sensor-based procedure has been successfully applied to identify Type I and II singularities in three non-redundant PKMs with different applications. The proposed sensor-based identification of singularities only requires measuring the actual position and orientation (pose) of the mobile platform and the current consumed by the actuators. In the closeness to a Type I singularity, the current consumed by the actuators increases suddenly with no changes in the pose tracking error. In contrast, a Type II singularity is identified by a sudden increment in the pose tracking error without abrupt changes in the electric current consumed by the actuators. The novel sensor-based identification of singularities could be extended to constraint singularities because the identification depends only on sensor measurements.

The sensor-based procedure for detecting singularities requires moderate time for adjustment. The sensor-based procedure is adjusted through the average pose tracking error and the average current consumed by the actuators during the fundamental movements of the PKM under study. Indirectly, the adjustment considers non-modelled effects, such as friction and clearances in joints, because all measurements are taken through experimentation on the actual PKM. Therefore, the proposed sensor-based procedure provides practical singularity identification for different PKMs.

The effectiveness of singularity identification depends on the accuracy of the sensor used to measure the motion of the end-effector and the current consumed by the actuators. An IMU requires numerical integration for measuring the pose of the end-effector. In contrast, a 3DTS directly measures the pose of the end-effector with high accuracy. Thus, a vision system is preferred over an IMU sensor for pose tracking.

The simplicity of the proposed sensor-based identification method of singularities reduces the computational cost. Thus, the proposed sensor-based singularity identification method is suitable for low-resource control units commonly used in educational robots and prototypes, especially in the first stage. Moreover, in industrial PKMs with high-performance control units, the proposed sensor-based procedure could be used for preliminary identification of singularities, which could then trigger an analytical method to verify the singularities. The use of the proposed singularity identification procedure for preliminary identification could allow the high-performance control units to focus on other tasks.

In future work, the authors will analyse the robustness of the proposed technique under varying working conditions, payload changes, and noise sensor signals. Potential solutions may involve refining the threshold definitions for detecting singularities, implementing advanced filtering techniques to mitigate noise effects, incorporating additional sensor data for improved reliability, and integrating artificial intelligence (AI) into the sensor-based singularity identification procedure. Moreover, the proposed sensor-based identification method will be leveraged to develop a singularity avoidance algorithm tailored for low-computation resource controllers, using the 5R and 3UPS+RPU PKMs as case studies. Additionally, the sensor-based identification method will be extended to detect constraint singularities in 3-DOF PKMs.

Author Contributions: Conceptualisation, M.C. and V.M.; methodology, V.M. and A.V.; software, A.V.; validation, V.M., M.C. and A.V.; formal analysis, J.L.P.; investigation, J.L.P.; resources, A.V.; data curation, J.L.P.; writing—original draft preparation, J.L.P.; writing—review and editing, J.L.P.; visualisation, J.L.P.; supervision, M.C.; project administration, V.M.; funding acquisition, V.M. and A.V. All authors have read and agreed to the published version of the manuscript.

Funding: This research was partially funded by Fondo Europeo de Desarrollo Regional (PID2021-125694OB-I00), cofounded by Erasmus+ project: Mechatronics for Improving and Standardizing Competences in Engineering (2022-1-ES01-KA220-HED-00089155).

Data Availability Statement: Data are contained within the article.

Conflicts of Interest: The authors declare no conflicts of interest.

Abbreviations

The following abbreviations are used in this manuscript:

DOF	Degrees of freedom
PKM	Parallel Kinematic Mechanisms
ISA	Instantaneous Screw Axis
IMU	Inertial measurement unit

References

1. Briot, S.; Khalil, W. *Dynamics of Parallel Robots—From Rigid Links to Flexible Elements*, 1st ed.; Mechanisms and Machine Science; Springer: Cham, Switzerland, 2015; pp. 19–33. [CrossRef]
2. Ceccarelli, M. *Fundamentals of Mechanics of Robotic Manipulation*; Springer: Cham, Switzerland, 2022; Volume 112, pp. 1–381. [CrossRef]
3. Karger, A. Singularity Analysis of Serial Robot-Manipulators. *J. Mech. Des.* **1996**, *118*, 520–525. [CrossRef]

4. Theingi; Chen, I.M.; Angeles, J.; Li, C. Management of parallel-manipulator singularities using joint-coupling. *Adv. Robot.* **2007**, *21*, 583–600. [CrossRef]
5. Conconi, M.; Carricato, M. A New Assessment of Singularities of Parallel Kinematic Chains. *IEEE Trans. Robot.* **2009**, *25*, 757–770. [CrossRef]
6. Di Gregorio, R. Instantaneous Kinematics and Singularity Analysis of Spatial Multi-DOF Mechanisms Based on the Locations of the Instantaneous Screw Axes. *Mech. Mach. Theory* **2024**, *196*, 105586. [CrossRef]
7. Gosselin, C.; Angeles, J. Singularity analysis of closed-loop kinematic chains. *IEEE Trans. Robot. Autom.* **1990**, *6*, 281–290. [CrossRef]
8. Taghirad, H.D. *Parallel Robots*, 1st ed.; CRC Press: Boca Raton, FL, USA, 2013. [CrossRef]
9. Park, F.C.; Kim, J.W. Singularity Analysis of Closed Kinematic Chains. *J. Mech. Des.* **1999**, *121*, 32–38. [CrossRef]
10. McAree, P.R.; Daniel, R.W. An Explanation of Never-Special Assembly Changing Motions for 3–3 Parallel Manipulators. *Int. J. Robot. Res.* **1999**, *18*, 556–574. [CrossRef]
11. Zlatanov, D.; Bonev, I.; Gosselin, C. Constraint singularities of parallel mechanisms. In Proceedings of the 2002 IEEE International Conference on Robotics and Automation (Cat. No. 02CH37292), Washington, DC, USA, 11–15 May 2002; Volume 1, pp. 496–502. [CrossRef]
12. Altuzarra, O.; Pinto, C.; Aviles, R.; Hernandez, A. A Practical Procedure to Analyze Singular Configurations in Closed Kinematic Chains. *IEEE Trans. Robot.* **2004**, *20*, 929–940. [CrossRef]
13. Wu, C.; Liu, X.J.; Xie, F.; Wang, J. New measure for 'Closeness' to singularities of parallel robots. In Proceedings of the 2011 IEEE International Conference on Robotics and Automation, Shanghai, China, 9–13 May 2011; pp. 5135–5140. [CrossRef]
14. Voglewede, P.; Ebert-Uphoff, I. Measuring "closeness" to singularities for parallel manipulators. In Proceedings of the IEEE International Conference on Robotics and Automation, New Orleans, LA, USA, 26 April–1 May 2004; Proceedings, ICRA '04; IEEE: Piscataway, NJ, USA, 2004; Volume 5, pp. 4539–4544. [CrossRef]
15. Liang, D.; Mao, Y.; Song, Y.; Sun, T. Kinematics, dynamics and multi-objective optimization based on singularity-free task workspace for a novel SCARA parallel manipulator. *J. Mech. Sci. Technol.* **2024**, *38*, 423–438. [CrossRef]
16. Choudhury, P.; Ghosal, A. Singularity and controllability analysis of parallel manipulators and closed-loop mechanisms. *Mech. Mach. Theory* **2000**, *35*, 1455–1479. [CrossRef]
17. Six, D.; Briot, S.; Chriette, A.; Martinet, P. A controller for avoiding dynamic model degeneracy of parallel robots during type 2 singularity crossing. In *New Trends in Mechanism and Machine Science*; Springer: Cham, Switzerland, 2017; Volume 43, pp. 589–597. [CrossRef]
18. Monsarrat, B.; Gosselin, C.M. Singularity Analysis of a Three-Leg Six-Degree-of-Freedom Parallel Platform Mechanism Based on Grassmann Line Geometry. *Int. J. Robot. Res.* **2001**, *20*, 312–328. [CrossRef]
19. Merlet, J.P. Singular Configurations of Parallel Manipulators and Grassmann Geometry. *Int. J. Robot. Res.* **1989**, *8*, 45–56. [CrossRef]
20. Corinaldi, D.; Callegari, M.; Angeles, J. Singularity-free path-planning of dexterous pointing tasks for a class of spherical parallel mechanisms. *Mech. Mach. Theory* **2018**, *128*, 47–57. [CrossRef]
21. Pond, G.; Carretero, J.A. Formulating Jacobian matrices for the dexterity analysis of parallel manipulators. *Mech. Mach. Theory* **2006**, *41*, 1505–1519. [CrossRef]
22. Merlet, J.P. Jacobian, Manipulability, Condition Number, and Accuracy of Parallel Robots. *J. Mech. Des.* **2006**, *128*, 199–206. [CrossRef]
23. Joshi, S.A.; Tsai, L.W. Jacobian Analysis of Limited-DOF Parallel Manipulators. *J. Mech. Des.* **2002**, *124*, 254–258. [CrossRef]
24. Ben-Horin, P.; Shoham, M. Singularity condition of six-degree-of-freedom three-legged parallel robots based on Grassmann–Cayley Algebra. *IEEE Trans. Robot.* **2006**, *22*, 577–590. [CrossRef]
25. Slavutin, M.; Shai, O.; Sheffer, A.; Reich, Y. A novel criterion for singularity analysis of parallel mechanisms. *Mech. Mach. Theory* **2019**, *137*, 459–475. [CrossRef]
26. Liu, X.J.; Wu, C.; Wang, J. A New Approach for Singularity Analysis and Closeness Measurement to Singularities of Parallel Manipulators. *J. Mech. Robot.* **2012**, *4*, 041001. [CrossRef]
27. Hesselbach, J.; Maaß, J.; Bier, C. Singularity prediction for parallel robots for improvement of sensor-integrated assembly. *CIRP Ann. Manuf. Technol.* **2005**, *54*, 349–352. [CrossRef]
28. Wang, Y.; Liu, Y.; Leibold, M.; Buss, M.; Lee, J. Hierarchical Incremental MPC for Redundant Robots: A Robust and Singularity-Free Approach. *IEEE Trans. Robot.* **2024**, *40*, 2128–2148. [CrossRef]
29. Russo, M.; Zhang, D.; Liu, X.J.; Xie, Z. A review of parallel kinematic machine tools: Design, modeling, and applications. *Int. J. Mach. Tools Manuf.* **2024**, *196*, 104118. [CrossRef]
30. Hubert, J.; Merlet, J.P. Static of Parallel Manipulators and Closeness to Singularity. *J. Mech. Robot.* **2009**, *1*, 011011. [CrossRef]
31. Kapilavai, A.; Nawratil, G. Singularity distance computations for 3-RPR manipulators using extrinsic metrics. *Mech. Mach. Theory* **2024**, *195*, 105595. [CrossRef]
32. Titov, A.; Russo, M.; Ceccarelli, M. Design and Performance of L-CaPaMan2. *Appl. Sci.* **2022**, *12*, 1380. [CrossRef]
33. Liu, X.J.; Wang, J.; Pritschow, G. Kinematics, singularity and workspace of planar 5R symmetrical parallel mechanisms. *Mech. Mach. Theory* **2006**, *41*, 145–169. [CrossRef]

34. Dai, Y.; Fu, Y.; Li, B.; Wang, X.; Yu, T.; Wang, W. Clearance effected accuracy and error sensitivity analysis: A new nonlinear equivalent method for spatial parallel robot. *J. Mech. Sci. Technol.* **2017**, *31*, 5493–5504. [CrossRef]
35. Vallés, M.; Araujo-Gómez, P.; Mata, V.; Valera, A.; Díaz-Rodríguez, M.; Page, Á.; Farhat, N.M. Mechatronic design, experimental setup, and control architecture design of a novel 4 DoF parallel manipulator. *Mech. Based Des. Struct. Mach.* **2018**, *46*, 425–439. [CrossRef]

actuators

Article

On the Relative Kinematics and Control of Dual-Arm Cutting Robots for a Coal Mine

Peng Liu [1,2,*], Haochen Zhou [1,2], Xinzhou Qiao [1] and Yan Zhu [1]

1   School of Mechanical Engineering, Xi'an University of Science and Technology, Xi'an 710000, China; sentaurry@outlook.com (H.Z.); qiaoxinzhou@xust.edu.cn (X.Q.); zhuyan1803@163.com (Y.Z.)
2   Shaanxi Key Laboratory of Mine Electromechanical Equipment Intelligent Monitoring, Xi'an University of Science and Technology, Xi'an 710054, China
*   Correspondence: liupeng@xust.edu.cn; Tel.: +86-180-9286-5343

Abstract: There is an unbalanced problem in the traditional laneway excavation process for coal mining because the laneway excavation and support are at the same position in space but they are separated in time, consequently leading to problems of low efficiency in laneway excavation. To overcome these problems, an advanced dual-arm tunneling robotic system for a coal mine is developed that can achieve the synchronous operation of excavation and the permanent support of laneways to efficiently complete excavation tasks for large-sized cross-section laneways. A dual-arm cutting robot (DACR) has an important influence on the forming quality and excavation efficiency of large-sized cross-section laneways. As a result, the relative kinematics, workspace, and control of dual-arm cutting robots are investigated in this research. First, a relative kinematic model of the DACR is established, and a closed-loop control strategy for the robot is proposed based on the relative kinematics. Second, an associated workspace (AW) for the DACR is presented and generated, which can provide a reference for the cutting trajectory planning of a DACR. Finally, the relative kinematics, closed-loop kinematic controller, and associated workspace generation algorithm are verified through simulation results.

Keywords: dual-arm coal cutting robot; kinematics; workspace; feedback control

Citation: Liu, P.; Zhou, H.; Qiao, X.; Zhu, Y. On the Relative Kinematics and Control of Dual-Arm Cutting Robots for a Coal Mine. *Actuators* **2024**, *13*, 157. https://doi.org/ 10.3390/act13050157

Academic Editors: Ioan Doroftei and Zhuming Bi

Received: 5 March 2024
Revised: 14 April 2024
Accepted: 22 April 2024
Published: 24 April 2024

1. Introduction

Coal is the main form of non-renewable energy in the world, and coal mining and laneway excavation are the two main aspects of coal production. However, there is a big gap between the intelligence levels for coal mining and laneway development. Laneway excavation and support are crucial factors affecting the development efficiency of coal mine laneways. The present situation of traditional laneway excavation, to the best of the authors' knowledge, is that the single-arm tunneling machine is the main equipment [1–3] for laneway excavation. The consequence of using the machine above is that in space, laneway excavation and support are in the same position, but they are separated in time, consequently leading to problems of low efficiency in laneway excavation. It is well known that coal mining efficiency far exceeds laneway excavation efficiency, further leading to the unbalanced problem of coal mining and laneway excavation. To overcome these problems, an advanced dual-arm tunneling robotic system for a coal mine is developed in this research. This system can achieve the synchronous operation of excavation and support of laneways to efficiently complete excavation tasks for large-section coal/rock laneways. The major advantage of the proposed dual-arm tunneling robotic system for coal mines is that the laneway excavation and support are separated in space but synchronized in time, consequently shortening laneway excavation time and improving laneway excavation efficiency. In more detail, the proposed dual-arm tunneling robotic system for a coal mine is composed of a dual-arm cutting robot (DACR), a temporary support robot, a drilling and anchoring robot, an electro-hydraulic control platform, a transportation system, and a ventilation and dust removal system.

The proposed DACR uses two cutting arms to efficiently complete excavation tasks for large-section coal/rock laneways. Many challenges arise from advanced dual-arm tunneling robotic systems for coal mines, among which the accurate motion control of the DACR is a crucial issue. Some researchers have discussed kinematics models [4–7] and control strategies [8–12] for dual-arm robots. Lee et al. [13] transformed the inverse kinematics problem of the robots into a mathematical optimization problem to replace traditional numerical methods, such as the Jacobian matrix. Yang et al. [14] established a special optimization objective function based on the Generalized Relative Jacobian Matrix (GRJM) of dual-arm space robots, which is employed to plan the coordinated motion of two arms. Lei et al. [15] proposed a control algorithm based on the kinematic model of dual-arm robots to prevent arm collisions and introduced a sensitivity index to measure the accuracy of the algorithm. Du and Hu [16] established the kinematic model of a lower limb exoskeleton robot and calculated the angle relationship between the hip and knee joints in the dual-arm structure. In the study, an adaptive controller was also designed to control the motion of the robot, and the effectiveness of the controller was validated through simulation. Ahmed et al. [17] proposed a new analytical quaternion based on the axis vector with the tangent of the rotation angle, which was used to establish the kinematic model of the robot arm. Jiang et al. [18] proposed a combination of geometric and algebraic methods for the calculation of the inverse kinematics of a mine cart robot arm. Wan et al. [19] used Adams to generate the kinematic model of a six-degree-of-freedom intelligent collaborative robot, solved the forward kinematics problem for the robot, and generated the motion trajectory and the limits of motion space for the end-effector based on the proposed kinematic model. Wei [20] established the kinematic model of a four-degree-of-freedom industrial robot based on the Denavit–Hartenberg (D-H) parameter method and solved the pose representation matrix of the robot using the RPY angles. The aforementioned research on robots was conducted based on a common premise, which was that the base of the dual-arm robot was in a fixed state. Compared with the existing literature on dual-arm robots, there is a co-shared movement for the two cutting arms, which will inevitably have an important influence on the movement of the two cutting arms. However, due to the complexity of constraints, the motion of the two cutting arms of a DACR is influenced by the base of the DACR. Existing research on the kinematics, dynamics, and control of dual-arm robots has been presented, but the relative kinematics and control of a proposed DACR with a co-shared movement for the two cutting arms, to the best of the authors' knowledge, have not yet been investigated. As a result, the motivation behind this research is to investigate the relative kinematics, closed-loop kinematic control strategy, and workspace for a DACR.

It is common knowledge that the workspace of a robot is a 2D or 3D space that the end-effector of the robot can approach with different poses [21–25]. This is an important parameter for designing, controlling, and applying a robot. Wang and Ding [26] proposed a surface enveloping overlay (SEO) algorithm for identifying the workspace of multi-joint serial robots. The algorithm was suitable for analyzing the characteristics of holes in a robot's workspace. Zeng et al. [27] employed a multi-island genetic algorithm to optimize the workspace of a parallel robot, and they analyzed the factors influencing the volume of the robot's workspace. Li et al. [28] employed an improved Monte Carlo method to generate clearer boundaries for the dynamic workspace of a multi-robot collaborative towing system. Boanta and Brișan [29] combined robot kinematics with feedforward neural networks to estimate the workspace volume of a robot and solve for the Cartesian coordinates of the robot's end-effector in a Cartesian space. Xu et al. [30] divided the Monte Carlo algorithm into two stages, namely, subspace generation and subspace expansion, to generate the workspace of a robot. This algorithm addressed the issue of insufficient accuracy in the workspace generated by traditional methods. Determining the workspace of the proposed DACR is more complicated than for traditional single-arm cutting robots because of the complexity of the constraints as well as the co-shared movement for the two cutting arms. The two cutting arms are installed on the same mobile platform, so they cannot be regarded as two single arms that are controlled separately. This makes the workspace of the DACR more complex than that of typical single-arm and dual-arm robots with a fixed base; therefore, a new workspace, the associated workspace (AW) for the DACR, is presented and generated in this research, and it can provide a reference for the cutting trajectory planning of the two cutting arms. To summarize, it can be seen that controlling a DACR has always presented a challenge for dual-arm tunneling robotic systems. Reviewing various scientific literature references, a DACR, to the best of the authors' knowledge, has not been investigated so far. For this reason, the novelty of this work comes from the following aspects:

1. An advanced dual-arm tunneling robotic system for coal mines is developed in this research. This system can achieve the synchronous operation of excavation and the permanent support of laneways to efficiently complete excavation tasks for large-section coal/rock laneways.
2. The proposed relative kinematic model of the DACR counteracts the influence of shared motion by integrating the kinematics of both arms into a unified framework. It can simultaneously describe the motion states of the two cutting arms. Additionally, a closed-loop kinematic controller for the robot is developed based on this relative kinematics, enabling control of both cutting arms through a single parameter.
3. The AW for the DACR is presented, and this can provide a reference for cutting trajectory planning of the two cutting arms.

The paper is organized as follows. Section 2 presents the advanced dual-arm tunneling robotic system for a coal mine. In Section 3, the relative kinematic model of the DACR is established, and the AW generation algorithm is presented. In Section 4, a closed-loop kinematic controller for the DACR is proposed based on the relative kinematics. Section 5 delineates the simulation examples and results. The paper concludes with discussion and comments in Section 6.

2. Proposed Dual-Arm Tunneling Robotic System

The proposed dual-arm tunneling robotic system for a coal mine (Figure 1a) is composed of a DACR (Figure 1b), a temporary support robot (I and II), a drilling and anchoring robot, an electro-hydraulic control platform, a transportation system, and a ventilation and dust removal system. As can be seen in Figure 1a, the DACR is installed on the lower shield of the temporary support robot, and it can move forward along the direction of laneway development. When cutting large-sized cross-sections, the movements of the two cutting arms of the DACR are coordinated, and the mobile platform beneath the DACR provides them with a co-shared movement, granting the DACR redundant degrees of freedom, enabling it to move forwards and backwards along the rail direction. The shovel plate beneath the two cutting arms is used to collect the crushed rock. The DACR is employed to complete the task of large-sized cross-section laneway development without moving the body of the robot in the direction perpendicular to the laneway development. This type of robot has a potentially large reachable workspace relative to a traditional single-arm cutting robot. A major challenge in DACR study is the intrinsic mutual influence between the two cutting arms because of the co-shared movement of the robot, which must be addressed simultaneously. The temporary support robot, which includes two identical temporary support shields, is used to support the laneway roof and achieve the progress of the dual-arm tunneling robotic system via the push-and-pull effect between the front and back temporary support shields. Specifically, the back temporary support shield pushes the front shield and the front shield pulls the back shield. The drilling and anchoring robot is used to complete the permanent support of the laneways with a truss bolting technique. The electro-hydraulic control platform supplies power for the dual-arm tunneling robotic system. The transportation system is responsible for transporting the raw coal produced by laneway development to the ground. The ventilation and dust removal system is used to provide fresh air and eliminate dust for the driving face in the coal tunnel.

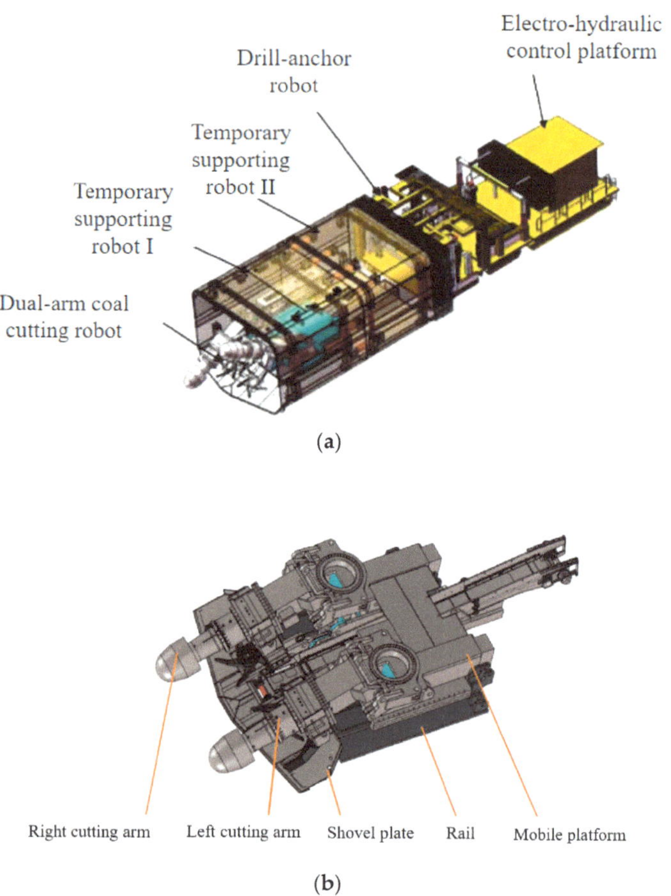

Figure 1. The dual-arm tunneling robotic system and dual-arm coal cutting robot. (**a**) The dual-arm tunneling robotic system; (**b**) The dual-arm coal cutting robot.

The proposed dual-arm tunneling robotic system for a coal mine can realize the simultaneous operation of the excavation and permanent support of large-sized cross-section coal/rock laneways, consequently shortening the laneway excavation time and improving the laneway excavation efficiency. However, there are two critical disadvantages in designing and employing the dual-arm tunneling robotic system; one disadvantage is space and the other is complexity. From the perspective of time, the proposed dual-arm tunneling robotic system can significantly shorten the laneway excavation time because of the simultaneous operation of the excavation and permanent support. From the perspective of space, additional space is needed for this system due to the space separation of the excavation and permanent support. With regard to the proposed DACR, because of the co-shared movement for the two cutting arms, the kinematics analysis and control, as has been mentioned above, are much more complex than those for conventional dual-arm robots. For this reason, this issue is considered to be one of the most important aspects in the field of DACRs. An additional challenge for DACRs is the possibility of the two cutting arms colliding with each other, which leads to a constraint in the optimal design and motion control of the robot. The relative kinematics, associated workspace, and control of DACRs are also investigated, as discussed in the next sections.

3. Relative Kinematics of the DACR

As discussed above, there are two critical disadvantages to the control of dual-arm cutting robots. One of the main concerns in modeling and controlling a DACR is dealing with the intrinsic mutual influence between the two cutting arms due to the co-shared movement of the dual-arm cutting robot's body, which is pivotal to the operation of the entire DACR. The other issue is that the control of the DACR cannot treat the two cutting arms as separate entities, and the two cutting arms require a clear relative positional relationship.

The presented DACR consists of two cutting arms connected to the same mobile platform (Figure 2). Each arm is composed of yaw joint, pitch joint, and prismatic joint. Additionally, the end-effector of each cutting arm is a rotating mechanism used for coal cutting. However, the size of this rotational joint does not affect the relative positions of the end-effectors of the two cutting arms. Therefore, in this research, the rotating mechanism is treated as a point mass. Consequently, each arm of the DACR can be regarded as an arm with three degrees of freedom. The D-H method is commonly used to analyze the kinematic characteristics of the robot and establish the relationship between joint angles and end-effector positions of the two arms in order to determine the transformation relationship between the joint link coordinate systems. As a result, the coordinate system $O_i X_i Y_i Z_i$ is established at the ith joint.

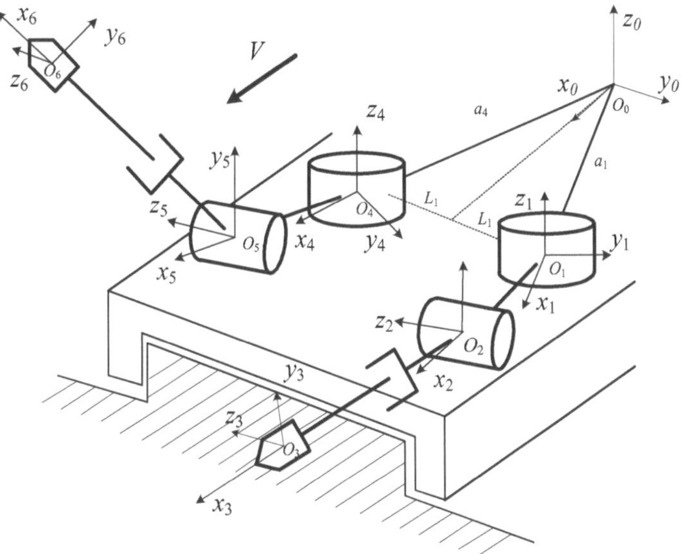

Figure 2. Dual-arm cutting robot structure.

3.1. Relative Kinematics Model of the DACR

D-H parameters are used to describe the kinematic chain of a robotic arm. With this method, every variable represents a specific geometric or angular relationship within the robotic arm. In particular, a denotes the distance from the z_{i-1} axis to the intersection of the x_i and z_{i-1} axes along the x_i axis, and α represents the angle from the z_{i-1} axis to the z_i axis measured about the x_i axis. In addition, d represents the distance from the x_{i-1} axis to the intersection of the x_i and z_i axes along the z_i axis, and q denotes the angle from the x_{i-1} axis to the x_i axis measured about the z_i axis. These parameters collectively define the transformation between consecutive links in a serial robot, thus allowing for the computation of the end-effector's position and orientation based on the joint variables. D-H parameters of the DACR are shown in Table 1. The mobile platform of the DACR provides additional degrees of freedom to both cutting arms simultaneously, with the platform being considered as virtually connected to two yaw joints and the ground coordinate system. Therefore, in Table 1, the parameters q_1 and q_4 are of equal magnitude but opposite in sign, while the parameters a_1 and a_4 are equal. The distance L_1 between the yaw joint and the axis of the DACR is 1 m, the lengths a_5 and a_2 between the pitch joint and the yaw joint are both 1 m, and the length variation range of the prismatic joint is the interval [4.08, 4.88].

Table 1. D-H parameters of the two arms.

Link	q_i/rads	d_i/m	a_i/m	α_i/rads
$i = 1$	q_1	0	a_1	0
$i = 2$	q_2	0	a_2	$\pi/2$
$i = 3$	q_3	0	a_3	0
$i = 4$	q_4	0	a_4	0
$i = 5$	q_5	0	a_5	$\pi/2$
$i = 6$	q_6	0	a_6	0

D-H, Denavit–Hartenberg.

The homogeneous transformation matrices $^0T_3$ and $^0T_6$ for each arm of the DACR are expressed as follows:

$$^0T_3 = {}^0T_1\,{}^1T_2\,{}^2T_3$$
$$^0T_6 = {}^0T_4\,{}^4T_5\,{}^5T_6 \tag{1}$$

The homogeneous transformation matrix $^{i-1}T_i$, using the D-H method, can be expressed as follows:

$$^{i-1}T_i = \begin{bmatrix} \cos q_i & -\sin q_i \cos \alpha_i & \sin q_i \sin \alpha_i & a_i \cos q_i \\ \sin q_i & \cos q_i \sin \alpha_i & -\cos q_i \sin \alpha_i & a_i \sin q_i \\ 0 & \sin \alpha_i & \cos \alpha_i & d_i \\ 0 & 0 & 0 & 1 \end{bmatrix}. \tag{2}$$

Substituting the D-H parameter into Equation (1), the homogeneous transformation matrices of two arms are calculated as follows:

$$^0T_3 = \begin{bmatrix} n_{x_3} & o_{x_3} & a_{x_3} & p_{x_3} \\ n_{y_3} & o_{y_3} & a_{y_3} & p_{y_3} \\ n_{z_3} & o_{z_3} & a_{z_3} & p_{z_3} \\ 0 & 0 & 0 & 1 \end{bmatrix}, \quad {}^0T_6 = \begin{bmatrix} n_{x_6} & o_{x_6} & a_{x_6} & p_{x_6} \\ n_{y_6} & o_{y_6} & a_{y_6} & p_{y_6} \\ n_{z_6} & o_{z_6} & a_{z_6} & p_{z_6} \\ 0 & 0 & 0 & 1 \end{bmatrix}, \tag{3}$$

in which

$$\begin{cases} n_{x_3} = \cos(q_1 + q_2)\cos q_3 \\ n_{y_3} = \sin(q_1 + q_2)\cos q_3 \\ n_{z_3} = \sin q_3 \\ o_{x_3} = -\cos(q_1 + q_2)\sin q_3 \\ o_{y_3} = -\sin(q_1 + q_2)\sin q_3 \\ o_{z_3} = \cos q_3 \\ a_{x_3} = \sin(q_1 + q_2) \\ a_{y_3} = -\cos(q_1 + q_2) \\ a_{z_3} = 0 \\ p_{x_3} = a_2\cos(q_1 + q_2) + L_1 \cos q_1 + a_3\cos(q_1 + q_2)\cos q_3 \\ p_{y_3} = a_2\sin(q_1 + q_2) + L_1 \sin q_1 + a_3\sin(q_1 + q_2)\cos q_3 \\ p_{z_3} = a_3 \sin q_3 \end{cases}, \quad \begin{cases} n_{x_6} = \cos(q_4 + q_5)\cos q_6 \\ n_{y_6} = \sin(q_4 + q_5)\cos q_6 \\ n_{z_6} = \sin q_6 \\ o_{x_6} = -\cos(q_4 + q_5)\sin q_6 \\ o_{y_6} = -\sin(q_4 + q_5)\sin q_6 \\ o_{z_6} = \cos q_6 \\ a_{x_6} = \sin(q_4 + q_5) \\ a_{y_6} = -\cos(q_4 + q_5) \\ a_{z_6} = 0 \\ p_{x_6} = a_5\cos(q_4 + q_5) + L_1 \cos q_4 + a_6\cos(q_4 + q_5)\cos q_6 \\ p_{y_6} = -a_5\sin(q_4 + q_5) - L_1 \sin q_4 - a_6\sin(q_4 + q_5)\cos q_6 \\ p_{z_6} = a_6 \sin q_6 \end{cases}.$$

Further, utilizing matrices $^0\boldsymbol{p}_3 = \begin{bmatrix} p_{x_3} & p_{y_3} & p_{z_3} \end{bmatrix}^{\mathrm{T}}$ and $^0\boldsymbol{p}_6 = \begin{bmatrix} p_{x_6} & p_{y_6} & p_{z_6} \end{bmatrix}^{\mathrm{T}}$ in Equation (3), the Jacobian matrix for each cutting arm is established as follows:

$$\frac{d\boldsymbol{p}}{dt} = J(\boldsymbol{q})\frac{d\boldsymbol{q}}{dt}. \tag{4}$$

For the DACR, the mobile platform provides a common motion for its dual arms. When the DACR is in operation, the position of the mobile platform must be able to satisfy the positions of both end-effectors simultaneously. This means that the motion of the mobile platform needs to be coordinated simultaneously with the movement of the dual arms, and this results in the kinematics of the two cutting arms being interrelated. Therefore, the dual arms cannot be regarded as two separate robots, but rather require a new kinematic model to integrate the kinematics of both arms together. The DACR may deviate from the expected trajectory because of manufacturing errors and rock hardness, and this can result in collisions between the two cutting arms. It is crucial to have clear and controllable relative position and velocity relationships between the end-effectors of the two cutting arms for the DACR in order to prevent accidents. The relative Jacobian matrix of the DACR is derived to establish the relationship between the relative velocities of the two end-effectors and their joint velocities.

The relationship between the rotation–translation composite transformation from the base to the end-effector of the DACR is established with the following equation:

$$\begin{aligned} {}^0R_3{}^3R_6 &= {}^0R_4{}^4R_6 \\ {}^0p_3 + {}^0R_3{}^3p_6 &= {}^0p_4 + {}^0R_4{}^4p_6 \end{aligned} \tag{5}$$

in which ${}^jR_i$ represents the rotation of coordinate system i with respect to coordinate system j, and ${}^jp_i$ represents the position of coordinate system i relative to coordinate system j.

The relative Jacobian matrix for the DACR is obtained by taking the derivative of Equation (5), and this can be expressed as:

$$J_R(q) = \begin{bmatrix} \mathbf{\Psi}_R{}^3\mathbf{\Omega}_0 J_L & {}^3\mathbf{\Omega}_0 J_{R1} \end{bmatrix}, \tag{6}$$

in which $\mathbf{\Psi}_R = \begin{bmatrix} I & -S(p_R) \\ 0 & I \end{bmatrix}$, ${}^j\mathbf{\Omega}_i = \begin{bmatrix} {}^jR_i & 0 \\ 0 & {}^jR_i \end{bmatrix}$, I is an identity matrix, J_L and J_{R1} are the Jacobian matrices of the left arm and right arm of the DACR, respectively. $S(p_R)$ is a skew–symmetric matrix composed of elements from the relative position vectors ${}^3p_6$, used to replace the cross-product operation of vectors. $S(p_R)$ is expressed in the following form:

$$S(p_R) = \begin{bmatrix} 0 & -z_R & y_R \\ z_R & 0 & -x_R \\ -y_R & x_R & 0 \end{bmatrix}.$$

The relative Jacobian matrix integrates the kinematic models of the two arms together, establishing the mapping relationship between the relative velocities of the two end-effectors and the joint velocities, which can be expressed as:

$$\dot{X}_R = J_R(q)\dot{q} \tag{7}$$

in which $\dot{X}_R$ represents the relative velocity of the two end-effectors, and $\dot{q}$ represents the joint velocities of the DACR.

Based on Equation (3), the mapping relationship between the position of the single-arm end-effector of the DACR and the joint angles/lengths can be obtained through forward kinematics, laying the foundation for generating the workspace of the DACR. The size and contour of the DACR's workspace determine the dimensions of the cross-sections it can cut, ultimately determining the cutting trajectory of the DACR. The relative Jacobian matrix indicates the mapping relationship between the relative velocities of the two end-effectors of the DACR and the joint velocities. Based on the relative Jacobian matrix, a trajectory tracking control algorithm applicable to the cutting trajectories of the DACR can be designed.

3.2. AW Generation Algorithm

The DACR uses the two cutting arms to complete the task of large-sized cross-section laneway development without moving the body of the robot in the direction perpendicular to laneway development. However, for the DACR, the workspaces of the two cutting arms partially overlap, leading to uncertainty about the size of cross-sections that the AW can accommodate. This will affect the motion trajectories of the two cutting arms within the AW, ultimately impacting the control effectiveness. Therefore, it is necessary to investigate the projection of the AW in the forward direction of the DACR, as well as the motion of the two cutting arms in the AW. The Monte Carlo algorithm [31] is a typical method to generate the workspace of a robot, and it is characterized by using a large number of random points to plot the possible positions that the robot's end-effector can reach in Cartesian space, thereby obtaining the contour of the robot's workspace. The Monte Carlo algorithm has high computational efficiency, and the number of the robot's degrees of freedom does not affect the computational error of the Monte Carlo algorithm. As a result, the Monte Carlo algorithm is employed to generate the AW for the DACR. The steps for performing the computation follow:

1. According to the actual size of the DACR, the variable range of the kth joint of the DACR is specified as $(q_{\min k}, q_{\max k})$.
2. N random values in the interval $(0, 1)$ can be generated by using the function rand(N, 1). The random step size generated by each joint is denoted as $q_k = q_{\min k} + \mathrm{rand}(N, 1)(q_{\max k} - q_{\min k})$.
3. The end-effectors of the DACR are plotted in Cartesian space at random positions by substituting q_k into ${}^0p_3$ and ${}^0p_6$ in Equation (3). The contour of the AW is generated when the number of random samples N is sufficiently large. It should be noted that the larger the value of N is, the more accurate the depiction of the contour of the AW for the DACR.

The flowchart of the AW generation algorithm for the DACR is shown in Algorithm 1.

Algorithm 1. AW generation algorithm of the DACR.

Input: Dimensional data of the DACR, random number N, and the angle/length range of the kth joint of the DACR ($q_{\min k}, q_{\max k}$).
Output: The AW of the DACR.
R = rand(N,1);
For m = 1: N

    $q_k = q_{\min k} + \mathrm{R}(\mathrm{m}) * (q_{\max k} - q_{\min k})$; % Generate random postures for each joint of the DACR.

    ${}^0p_3 = \begin{bmatrix} p_{x_3} & p_{y_3} & p_{z_3} \end{bmatrix}^{\mathrm{T}}$; ${}^0p_6 = \begin{bmatrix} p_{x_6} & p_{y_6} & p_{z_6} \end{bmatrix}^{\mathrm{T}}$; % Generate the positions of the end-effectors.

    plot ${}^0p_3$; plot ${}^0p_6$; % Plot the positions of the end-effectors in Cartesian space.

hold on
End

---

4. Proposed Kinematics Controller

The primary objective of controlling most dual-arm robots is to enable them to grasp, manipulate, or lift objects, forming a kinematic closed chain where the relative positions of the end-effectors remain constant. However, for the DACR, during the process of cutting through large-sized cross-sections, the relative positions of the end-effectors change over time. This implies that the scenario where the dual-arm robot forms a kinematic closed chain is merely a special case during the motion of the DACR's arms. Therefore, for cases where the relative positions of the end-effectors are time-varying, a more general controller needs to be proposed. A closed-loop relative kinematic controller is proposed to control the relative position of the end-effectors for the DACR. The expected relative position and the actual relative position are assumed to be X_{Rd} and X_R, respectively. Therefore, the relative position error of the two end-effectors can be expressed as follows:

$$e = X_{Rd} - X_R. \tag{8}$$

By substituting Equation (7) into the derivative of Equation (8) with respect to time, the expression of the derivative of error e can be expressed as:

$$\dot{e} = \dot{X}_{Rd} - J_R(q)\dot{q}. \tag{9}$$

The DACR exhibits kinematic redundancy due to the co-shared motion provided by the mobile platform of the DACR to two end-effectors. This implies that the relative Jacobian matrix of the DACR is not a square matrix, and, therefore, the control laws based on relative kinematics for the DACR are designed in the following form:

$$\dot{q} = J_R^{+}(q)\left(\dot{X}_{Rd} + K_1 e\right) + (I - J_R^{+}J_R)\dot{q}_N, \tag{10}$$

where K_1 is a symmetric positive definite matrix that denotes the gain matrix of the relative position error, and the superscript "+" represents the pseudo-inverse, while $\dot{q}_N$ represents the kinematic redundancy of the DACR, which means that when the DACR tracks the trajectory, the controller may generate multiple possible movements of the DACR, all of which can satisfy the expected relative velocity $\dot{X}_{Rd}$ of the end-effectors at this time.

The position-level motion controller based on the relative kinematics model of the DACR is shown in Figure 3, where the feedback loop takes the desired relative position and actual relative position of the two end-effectors as input. By employing Equation (10), the joint velocity vector of the two cutting arms $\dot{q}$ can be obtained for any desired relative position of the two cutting arms and the tracking error along the given trajectory converges to zero with a suitable gain K_1. Due to the fact that the relative Jacobian matrix is not square, when solving the joint velocity based on Equation (10), multiple sets of solutions may arise, corresponding to the red dashed lines in Figure 3. While these solutions can simultaneously satisfy the relative positions of the two end-effectors and the co-shared motion provided by the mobile platform to the dual arms, they may potentially result in joint angles/lengths exceeding their limits, ultimately leading to singular arm postures. Therefore, it is necessary to simulate the proposed relative kinematic controller.

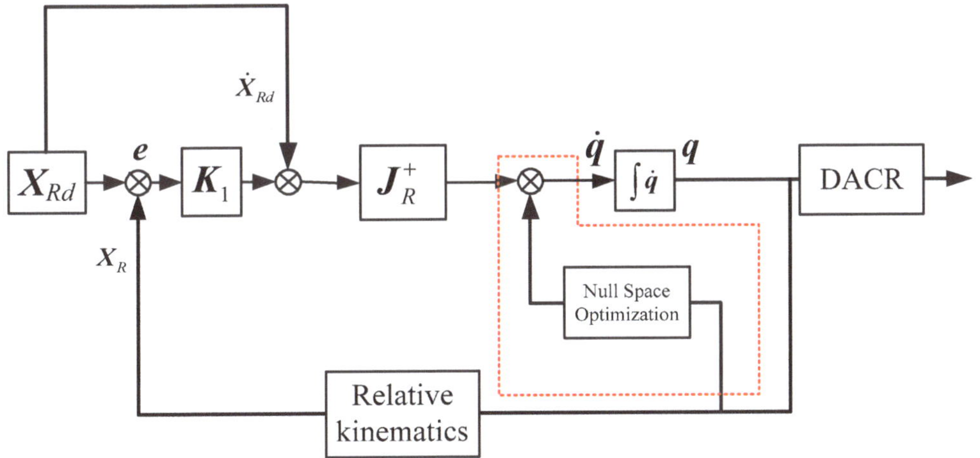

Figure 3. Controller based on relative position error. DACR, dual-arm cutting robot.

Substitute the control input $\dot{q}$ into Equation (9), then:

$$K_1 e + \dot{e} = 0. \tag{11}$$

The choice of K_1 can guarantee that the error uniformly converges to zero since K_1 is a symmetric positive definite matrix. According to the Lyapunov theorem, the proposed controller, as a result, can stabilize the system, and the relative position error of the two end-effectors for the DACR is ultimately uniformly limited.

5. Simulation Analysis and Discussion

Two working modes are available for the proposed DACR from the viewpoint of practical application. In the first working mode, the co-shared movement of the two cutting arms is usually locked up to reduce the control difficulty and improve the motion accuracy of the two end-effectors for the DACR, which is defined as mode 1 in this paper. In the second working mode, the cutting trajectories of the two cutting arms are generated by the coordinated movement of the body, which is defined as mode 2. With regard to mode 1, the base of the DACR remains stationary, and thus, there is no co-shared movement that affects the movement of the two cutting arms. To reduce the empty distance from the front end of the laneway to the temporary support robot, it is necessary to cooperate with the co-shared movement of the two cutting arms to generate the cutting trajectories of the two cutting arms. With regard to mode 2, the co-shared movement will have a key influence on the movement of the two cutting arms for the DACR. Subsequently, for the above two typical working conditions, the relative kinematic control and AW generation for the DACR are investigated.

5.1. Motion Continuity of the DACR

The continuity of robot motion is a measure of whether the robot can function properly and achieve smooth continuous movements [32]. This section describes how a random rectangle and random circle in Cartesian space are generated to verify whether the two cutting arms of DACR can achieve continuous motion. The trajectory of the circle is as follows:

$$\begin{cases} x = 5.222 \\ y = -2 + \cos(t) \\ z = 1.739 + \sin(t) \end{cases},$$

The expressions for the four sides of the rectangle are as follows:

$$\begin{cases} x = 5.222 \\ y = 1 + 2t, t \in [0,1], \\ z = 2.739 \end{cases} \begin{cases} x = 5.222 \\ y = 3 \\ z = 2.739 - 2t \end{cases}, t \in [0,1], \begin{cases} x = 5.222 \\ y = 3 - 2t, t \in [0,1], \\ z = 0.739 \end{cases} \begin{cases} x = 5.222 \\ y = 1 \\ z = 0.739 + 2t \end{cases}, t \in [0,1].$$

To ensure that the co-shared motion of the two cutting arms is unique, the mobile platform is fixed, and in this case, this implies that the variables a_1 and a_4 are fixed values. The expressions for the positions ${}^0p_3$ and ${}^0p_3$ of the two end-effectors relative to the base coordinate system in the single-arm kinematic model are known. Based on the analytical method, the changes in joint angles/lengths of the single arm can be solved through the end-effector positions. Therefore, the inverse kinematic model of the single arm can be used to solve the changes in joint angles/lengths corresponding to the end-effector when tracking trajectories. The tracking times for the rectangle trajectory and the circle trajectory are 20 s and 10 s, respectively, with one second divided into 10 steps. The angles/lengths of the joints of the two cutting arms of the DACR, obtained through inverse kinematics, are shown in Figure 4. Figure 4a illustrates the tracking process of the left arm of the DACR for a rectangular-shaped trajectory, while Figure 4b depicts the tracking process of the right arm for a circular-shaped trajectory. The joint parameters of the DACR change continuously and smoothly over time when tracking the continuous trajectories, and there are no sudden changes in the angles or extensions of the DACR's joints. The results of the inverse kinematics solution indicate that the two cutting arms of the DACR exhibit good motion continuity and do not produce singular poses.

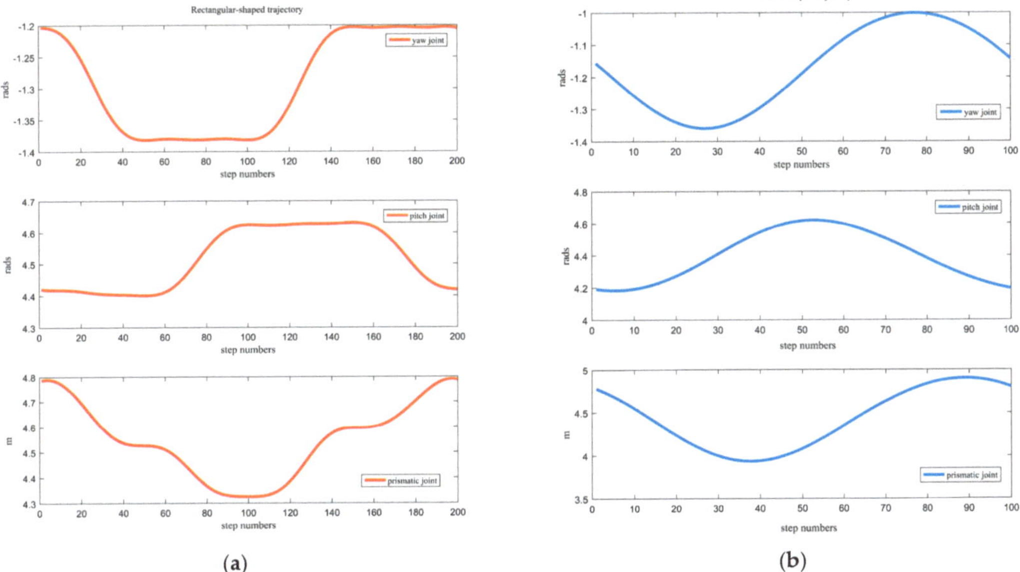

Figure 4. Changes in the angles/lengths of the joints when the DACR is tracking continuous trajectories.

To demonstrate the correctness of the above conclusion, and also to validate the correctness of the single-arm kinematic model, it is necessary to verify the forward kinematic model of the single arm. The model of the DACR is generated in the MATLAB Robotic Toolbox. The two cutting arms of the DACR are considered to be connected at the same rotary joint in the MATLAB Robotic Toolbox due to the base of the DACR being locked. The joint angles/lengths corresponding to each step in Figure 4 are input into the MATLAB Robotic Toolbox for forward kinematic calculation, obtaining the motion animations of the end-effectors of the two cutting arms in Cartesian space, as shown in Figure 5a–d. For the rectangular-shaped trajectory, when the cutting arm is in the states shown in Figure 5a,c, the motion of the end-effector is primarily determined by the yaw joint of the left arm. Therefore, the motion states shown in Figure 5a,c correspond to the angular changes that occur in the yaw joint in Figure 4a. The motion states shown in Figure 5b,d are primarily determined by the pitch joint of the left arm, corresponding to the angular changes of the pitch joint in Figure 4a. For the circular-shaped trajectory, the joint angles/lengths of the cutting arm change uniformly. The simulation shows that the two cutting arms of the DACR do not generate singular postures when tracking the trajectories. This indicates that the DACR has good motion continuity.

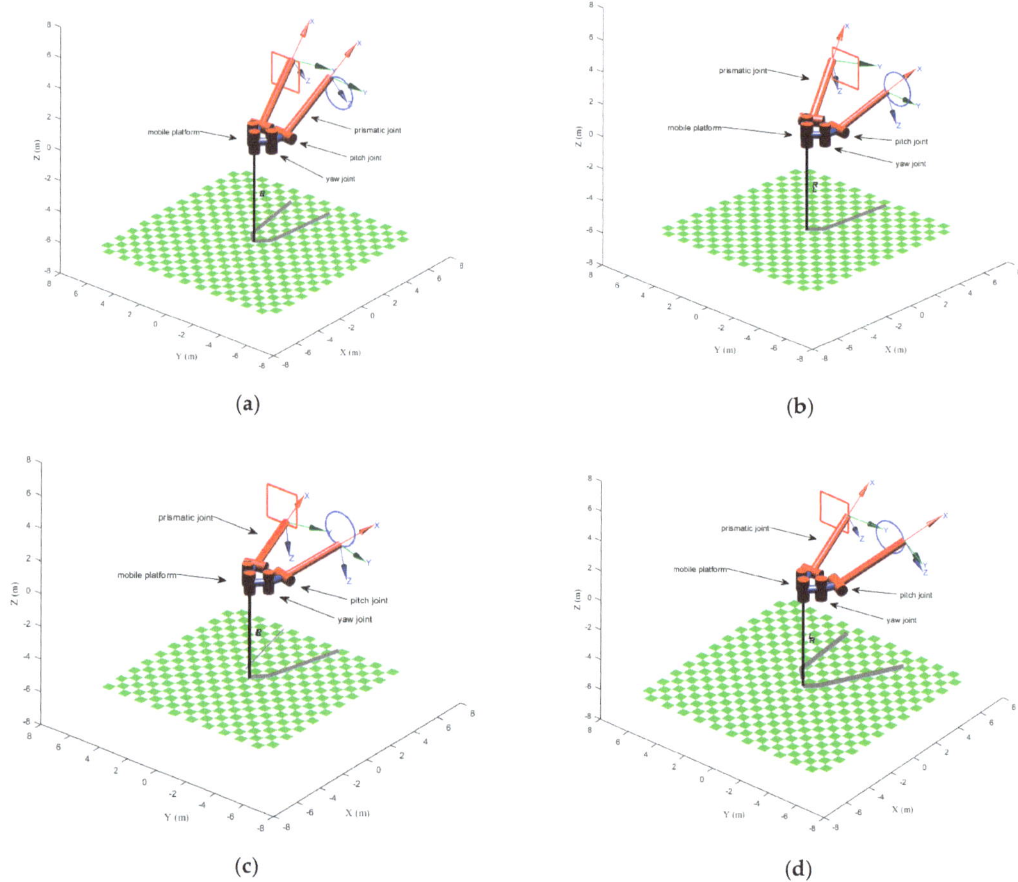

(a)

(b)

(c)

(d)

Figure 5. The motion of the two end-effectors of the DACR in Cartesian space.

5.2. AW of the DACR

For the DACR, there is an overlapping portion in the workspace of the two cutting arms. Therefore, calculating the projection of the AW in the robot's forward direction can determine whether the DACR's working range can accommodate large-sized cross-sections, upon which the cutting trajectories of the two cutting arms can be planned. This is a prerequisite for the DACR to simultaneously complete the cutting of large-sized cross-sections and forms the foundation for controlling the DACR. The DACR has two working modes that provide the DACR with two different AWs. With respect to mode 1, the position of the DACR's base is locked, and the two cutting arms perform the cutting. After completing the cutting work inside the AW, the base moves forward, and the double cutting arms start a new round of cutting. With respect to mode 2, the base maintains a uniform forward progression while the two cutting arms cyclically cut along the trajectory.

In this simulation, the dimensions and operating conditions of the DACR are as follows. The distance from joint 1 to the centerline of the DACR, L_1, is one meter. The link lengths a_2 and a_5 are both one meter. For the DACRs in modes 1 and 2, the yaw angles q_2 and q_5 are in the intervals $[-\pi/2, 0]$ and $[0, \pi/2]$, respectively, while the pitch angles q_3 and q_6 are both in the interval $[0, \pi/3]$. The lengths of prismatic joints 3 and 6, denoted as a_3 and a_6, are both in the interval $[4.08, 4.88]$, and the random number N is set to 10,000. Further, for working mode 2, the base maintains a uniform forward velocity V set to 0.02 m/s. This results in a_1 and a_4 being represented as follows:

$$a_1(t) = a_4(t) = \sqrt{(Vt)^2 + L_1^2} \approx 0.0001411t^2 + 0.001482t + 0.993$$

As shown in Figure 6a,b, the workspace of a single cutting arm of the DACR is generated for working modes 1 and 2, with the workspace in both modes appearing as a spherical shell. It should be noted that the volume of the workspace of the single arm of the DACR in mode 2 is 174.2% larger than in mode 1. The comparison of the workspaces corresponding to the two working modes is shown in Figure 6c. The simulation indicates that the workspace in mode 2 can accommodate laneways with a larger cross-section.

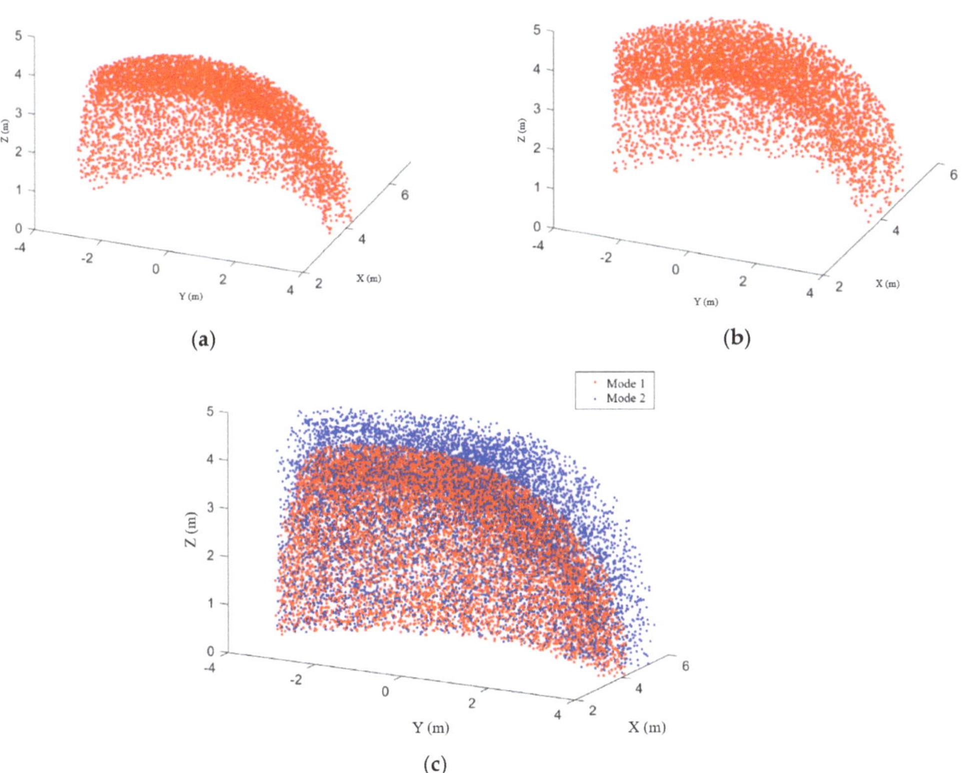

Figure 6. Workspace of a single arm of the DACR in working modes 1 and 2 (relative to the base coordinate system). (**a**) Workspace of the single arm in mode 1; (**b**) Workspace of the single arm in mode 2; (**c**) The workspace comparison of the two modes.

As shown in Figure 7, when the DACR is in working mode 2, which is the continuous mode, the volume of the generated AW is significantly larger than the AW corresponding to working mode 1. This means that in working mode 2, the DACR's end-effectors are able to reach farther positions in Cartesian space compared to working mode 1. In addition, this implies that in working mode 2, the AW accommodates a larger volume of the rock wall, which allows the DACR to excavate greater depths compared to mode 1.

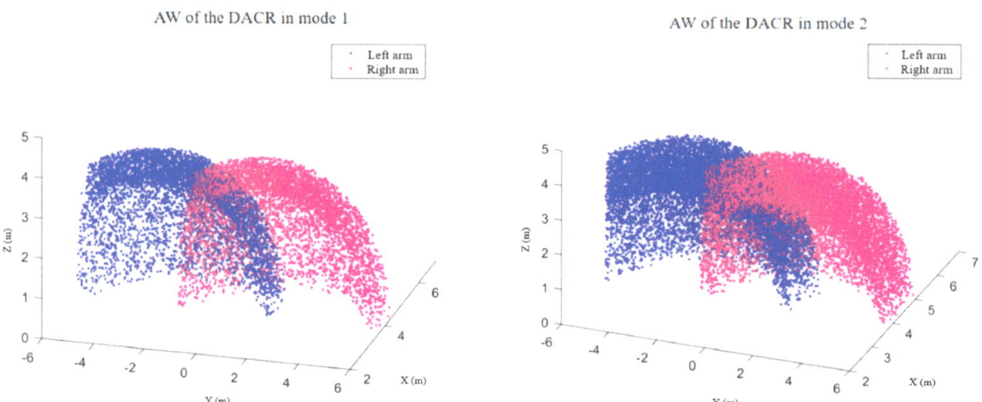

Figure 7. AW of the DACR in working modes 1 and 2. AW, associated workspace (relative to the base coordinate system).

The projection of the workspaces in the direction perpendicular to the advancement of the DACR is shown in Figure 8, in which the red dots represent the projection of the workspace of the DACR's single arm, while the blue dots and pink dots denote the projection of the AW formed by the left and right arms. The black contour lines represent the outline of the sections that the DACR can accommodate. For the single cutting arm of the DACR, its workspace can only envelop small-sized cross-section laneways (with a net height not less than 2.5 m and a net width not less than 2.6 m). However, the large-sized cross-section laneways, generally speaking, have a net height greater than 4 m and a net width greater than 5.2 m. This means that a single cutting arm cannot complete the cutting task of large-sized cross-section laneways within one work cycle. However, the DACR with two cutting arms can resolve this issue, enabling the DACR's AW to fully cover large-sized cross-section laneways. When both cutting arms of the DACR work simultaneously, large-sized cross-section laneways can be cut in one cycle of the DACR. Compared with other dual-arm robots, the two arms of the DACR are kinematically interrelated, and the mobile platform provides a common motion for the two cutting arms. Therefore, the mobile platform simultaneously affects the shape and size of the working space of both cutting arms, making the DACR's AW more complex compared to the workspace of dual-arm robots with a fixed base, ultimately affecting the trajectory planning of the two end-effectors in the overlapping portion of the AW.

The working procedure of the single cutting arm tunnel boring machine for cutting large-sized cross-section laneways perpendicular to the direction of advance is shown in Figure 9. The red trajectory on the left depicts the first cut of the large section by the single cutting arm tunnel boring machine, and the blue trajectory on the right represents the second cut. The single cutting arm needs to excavate two tunnels when cutting the large-sized cross-section laneway. First, the single-arm tunnel boring machine operates in the left tunnel, cutting along the red trajectory. The space at the rear of the machine is left vacant for support and anchoring. After completing the cutting in the left tunnel, the single-arm tunnel boring machine is paused and moved to the right tunnel, and the second cut begins along the blue trajectory. During the cutting of the large-sized cross-section laneway, the single cutting arm tunnel boring machine alternates between cutting and moving. Although this process is continuous in terms of space, due to the limited workspace of the single cutting arm, it is necessary to alternate work in the two tunnels to complete the cutting of the large-sized cross-section laneway. Therefore, this procedure is discontinuous in terms of time and the workflow is quite cumbersome. Frequent movement of the tunnel boring machine not only increases the workload but also leads to other issues, such as the cross-sectional shapes of the tunnel being inconsistent between the front and rear ends.

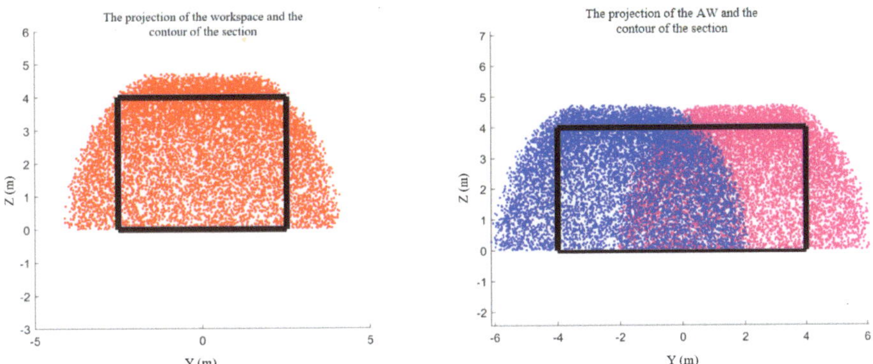

Figure 8. The workspace projection of the DACR's single cutting arm and dual arms in the forward direction.

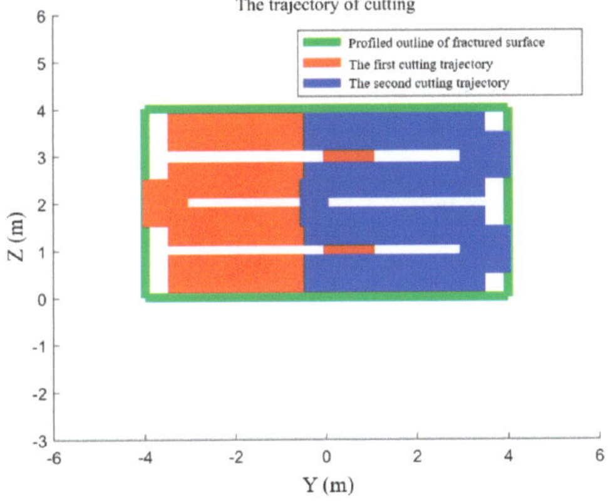

Figure 9. Process of the single arm tunnel boring machine cutting the large-sized cross-section laneway.

The mechanical limits of the two cutting arms are removed in order to ensure that the cutting trajectories of the two end-effectors of the DACR can fully cover the cross-section of the tunnel. The cutting trajectory of the DACR is designed to be a centrally symmetric "S" shape, as shown in Figure 10.

There is a partial overlap in the trajectories of the left and right arms to cut the overlapping rock walls in the AW. The two cutting arms are started asynchronously to ensure that the overlapping portion in the AW is only passed through by one of the two cutting arms during operation. The large rotating cutting mechanism at the end of the two cutting arms fills the gap portion of the cutting trajectory. Compared to mode 1, mode 2 provides a larger AW for the DACR. Therefore, mode 2 is adopted as the main motion mode of the DACR. In mode 2, the trajectories of the two arms of the DACR are divided into seven segments. The cutting trajectory along the X axis of the inertial coordinate system is divided into 100 steps, while the cutting trajectory along the Z axis of the inertial coordinate system is divided into 20 steps, and marks are placed at the starting and ending points of each segment. The DACR joint angles/lengths corresponding to the generated eight endpoints are listed in Table 2. Due to the inflection points of the "S"-shaped trajectory, the direction of joint acceleration at these points undergoes abrupt changes. For DACRs with joints of significant mass, these discontinuities in joint acceleration can affect the dynamics of the joints, causing motor vibrations and impacts, ultimately affecting the operational stability of the DACR. To constrain the accelerations of the joints of the DACR, fifth-order polynomial interpolation is used for trajectory planning for each joint of the DACR, with the constraint that the velocities and accelerations at the inflection points of the joints are zero.

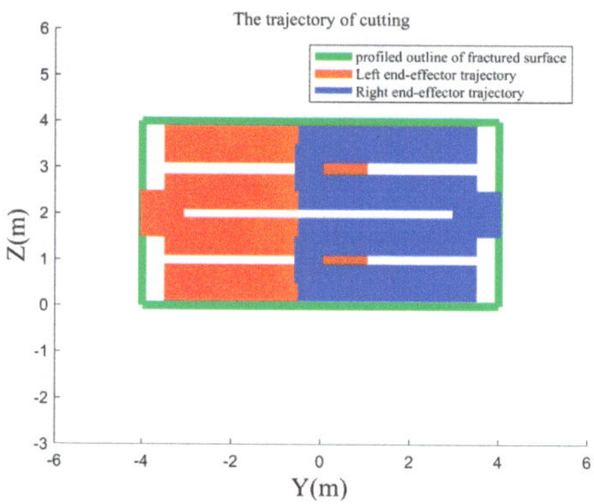

Figure 10. Process of the DACR cutting the large-sized cross-section laneway.

Table 2. The joint variables of the DACR corresponding to each point on the "S"-shaped trajectory in mode 2.

Point	Base/m	q_2/rads	q_3/rads	a_3/m	q_5/rads	q_6/rads	a_6/m
1	0	0	$\pi/3$	4.88	0	$\pi/3$	4.88
2	0.2174	$-\pi/2$	$\pi/3$	4.08	$\pi/2$	$\pi/3$	4.08
3	0.2609	$-\pi/2$	$2\pi/9$	4.08	$\pi/2$	$2\pi/9$	4.08
4	0.4783	0	$2\pi/9$	4.88	0	$2\pi/9$	4.88
5	0.5217	0	$\pi/9$	4.88	0	$\pi/9$	4.88
6	0.7391	$-\pi/2$	$\pi/9$	4.08	$\pi/2$	$\pi/9$	4.08
7	0.7826	$-\pi/2$	0	4.08	$\pi/2$	0	4.08
8	1	0	0	4.88	0	0	4.88

Based on Table 2 and the fifth-order polynomial motion planning of the joints, the trajectories of the two end-effectors of the DACR are shown in Figure 11. The end-effector moves in the direction indicated by the black arrows during cutting, and numbers 1–8 denote the turning points of the trajectory. By reasonably planning the movements of the two cutting arms, the DACR is capable of cutting large-sized cross-sections without collisions, and the DACR does not require moving the mobile platform along a direction perpendicular to the rail. Further, mode 2 of the DACR maintains the movement of the base, enabling the DACR to advance while cyclically cutting. A temporary support robot installed at the rear of the DACR synchronously supports the tunnel during cutting. In comparison to the single cutting arm tunnel boring machine, the DACR, as a dual-arm tunneling robotic system, maintains cyclic and continuous cutting at the front while temporary support robots located at the rear of the DACR immediately support the top of the tunnel after cutting. Subsequently, the drill–anchor robot drives anchors into the tunnel walls to secure the tunnel's shape. The work of each robot in the dual-arm tunneling robotic system is separated in space but synchronized in time, ultimately expediting the excavation, support, and shaping of large-sized cross-section laneways. This simplifies the "cut–move–cut" workflow of the single cutting arm tunnel boring machine. In addition, the single cutting arm tunnel boring machine may need to excavate two tunnels to complete the cutting of large-sized cross-section laneways, potentially causing non-parallel axes between the two tunnels and leading to dimensional errors in the resulting cross-sections. The base of the DACR provides co-shared movement for its two cutting arms, ensuring that the orientation of the DACR's AW remains constant in any direction. The dimensions of the resulting cross-sections remain consistent from front to back through reasonable planning of the movements of the two cutting arms.

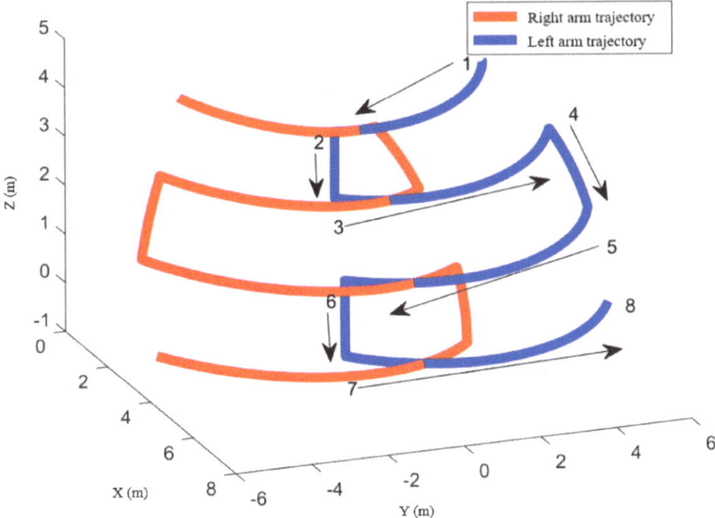

Figure 11. Trajectories of the two end-effectors when cutting a large-sized cross-section laneway (relative to the base coordinate system).

5.3. Trajectory Tracking Control System

To conform to the proposed operational mode 2 of the DACR, the actual motion of the DACR's mobile platform is assumed to be consistent with the expected motion, maintaining a constant forward motion along the rail, therefore the redundant terms in Equation (10) (i.e., the red dashed line in Figure 3) are removed and only the minimum norm joint velocity solution to the inverse velocity problem is generated. When the DACR is initiated in mode 2, the mobile platform of the DACR is uniformly propelled along the rail, and the two cutting arms of the DACR are both at arbitrary positions in the AW, resulting in the relative positions of the two end-effectors of the DACR being random. Therefore, the positions of the two end-effectors of the DACR in Cartesian space are assumed to be $[0.878\ -2.0265\ 4.2225]^{\mathrm{T}}$ and $[1.834\ 4.82\ 3.66]^{\mathrm{T}}$, and the two cutting arms of the DACR are set to start simultaneously. These values are substituted into matrix ${}^{3}p_{6}$ in Equation (5) to represent the initial relative positions of the two end-effectors of the DACR, yielding the vector ${}^{3}p_{6}$ as $[4\ 3\ -2.9]^{\mathrm{T}}$, which is used to indicate the initial relative positions of the DACR's two end-effectors. At the start of the DACR, the two end-effectors are treated as two particles with a zero relative rotational angle. Hence, the value of p_{R_0} is as follows:

$$p_{R_0} = \begin{bmatrix} {}^{3}p_6 \\ {}^{3}\varphi_6 \end{bmatrix} = \begin{bmatrix} 4 & 3 & -2.9 & 0 & 0 & 0 \end{bmatrix}^{\mathrm{T}}.$$

The gain matrix K_1 is diag $[0.08\ 0.068\ 0.083\ 0.07\ 0.065\ 0.08]$, and the tracking time for X_{Rd} is 47 s, with 1 s divided into 10 steps. The relative kinematic controller produces a unique solution because the movement of the DACR's mobile platform in mode 2 is deterministic, and the actual motion of the mobile platform is assumed to be consistent with the expected motion, with the movement speed of the DACR's mobile platform locked at 0.02 m per second. Due to the mapping relationship between the forward pushing movement of the mobile platform and q_1 in Table 1, the motion of the mobile platform can be represented by $\dot{q}_1$, as shown in Figure 12a. It can be determined from Figure 12b–g that the controlled joint velocities, with the action of the controller based on the relative Jacobian matrix, eventually exhibited regular changes within 100 steps. Figure 12b–g indicate that within 100 steps, there were abrupt changes in the pitch and prismatic joint velocities of the DACR. This is because by this time, both yaw joints have already achieved trajectory tracking, while there were still errors in the angles and lengths of the pitch and prismatic joints. At this stage, the controller only acted on the pitch and prismatic joints, resulting in abrupt changes in their velocities.

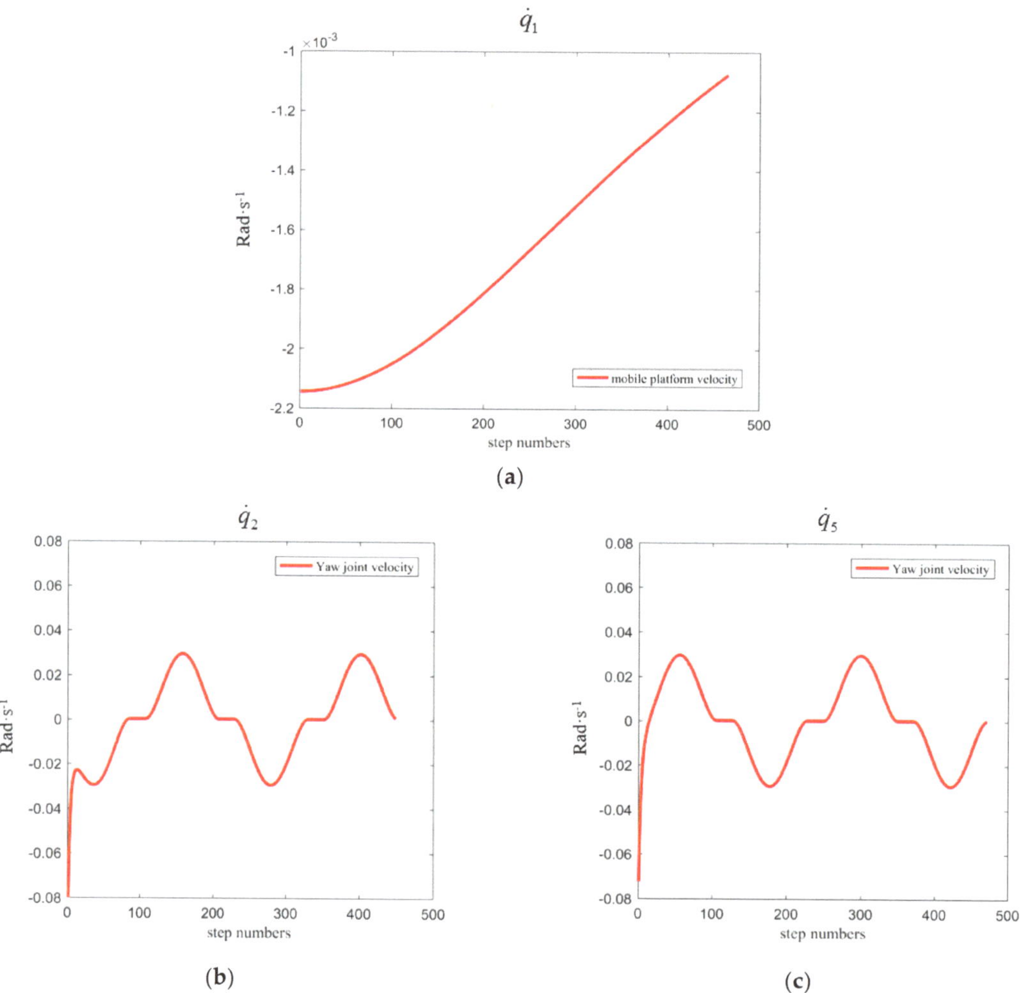

(a)

(b)

(c)

Figure 12. *Cont.*

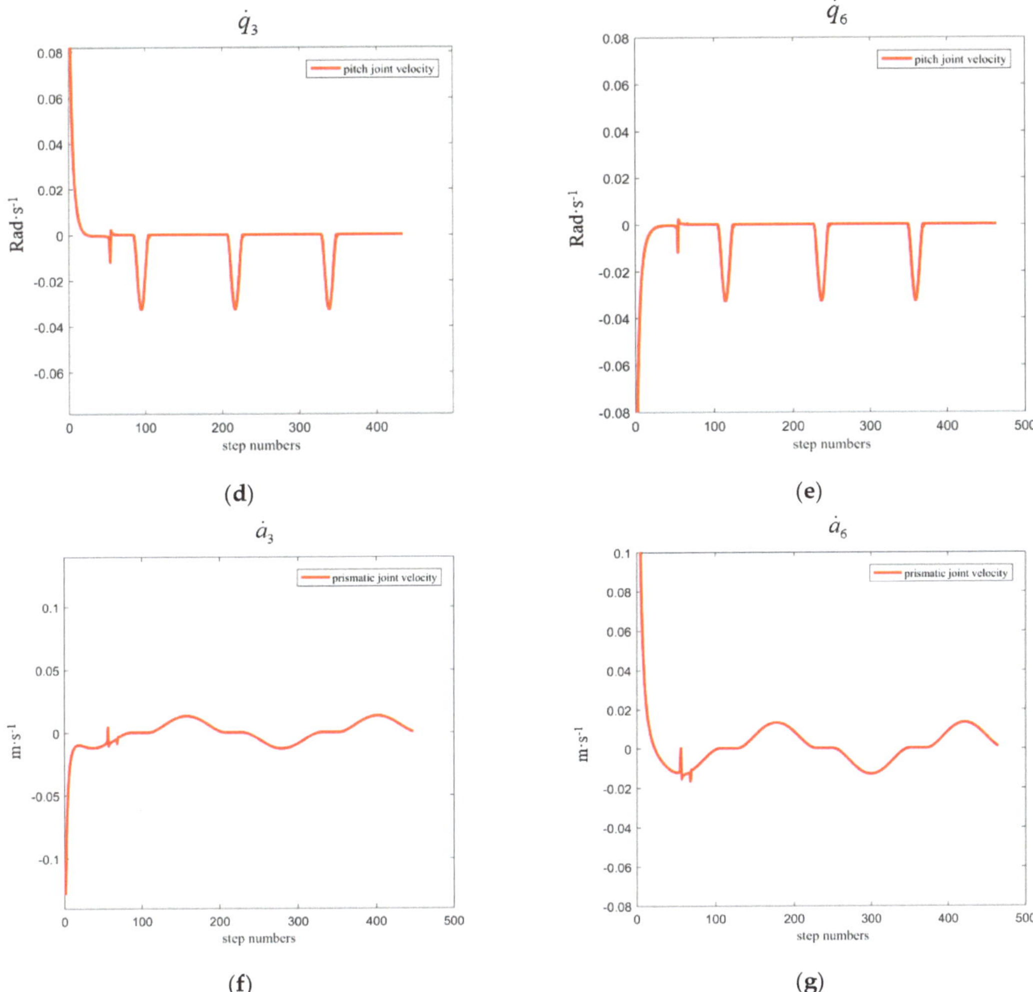

Figure 12. The controlled variable of the DACR. (**a**) The velocity of the mobile platform; (**b**) The velocity of the yaw angle of the DACR's left arm; (**c**) The velocity of the yaw angle of the DACR's right arm; (**d**) The velocity of the pitch angle of the DACR's left arm; (**e**) The velocity of the pitch angle of the DACR's right arm; (**f**) The velocity of the prismatic joint of the DACR's left arm; (**g**) The velocity of the prismatic joint of the DACR's right arm.

The two cutting arms of the DACR are required to start from a stationary state and cut the rock wall along the red and blue trajectories in Figure 11. The actual motion of the DACR's mobile platform matches the expected motion, both locked to move uniformly forward, and the mapping relationship between q_1 in Table 1 and the forward motion of the mobile platform can be represented by q_1 as shown in Figure 13a, which causes the relative kinematic controller to generate only the minimum norm solution for joint velocities, thus resulting in matrix $\left(I - J_R{}^+ J_R\right)\dot{q}_N$ being a zero matrix in Equation (10) at this time. From Figure 13b–g, it can be seen that when the two cutting arms of the DACR track the S-shaped trajectory, the variations in joint angles/lengths are continuous and smooth, further demonstrating the continuity of the DACR's motion. The angle/length variations of the joints in the DACR's two arms are shown in Figure 13b–g, where the blue lines (i.e., the expected values) come from the joint angle/length variations of the DACR as shown in Table 2. Figure 13b,c represent the yaw joint angles of the DACR's two arms, Figure 13d,e show the pitch joint angles, and Figure 13f,g display the length variations of the prismatic joints in the DACR's two arms. It can be observed from Figure 13b–g that the angle/length variations of the DACR's two arms converge to the desired values within 100 steps, thus indicating the effectiveness of the relative kinematic controller. It should be noted that the desired motions of the DACR's two arms are asynchronous, which causes the expected value of the yaw joint angle of the DACR's right arm to remain constant for some time after the simulation begins, while the two arms of the DACR in the simulation are started synchronously. This indicates that the two arms of the DACR are effectively controlled to approach the motion state when working asynchronously.

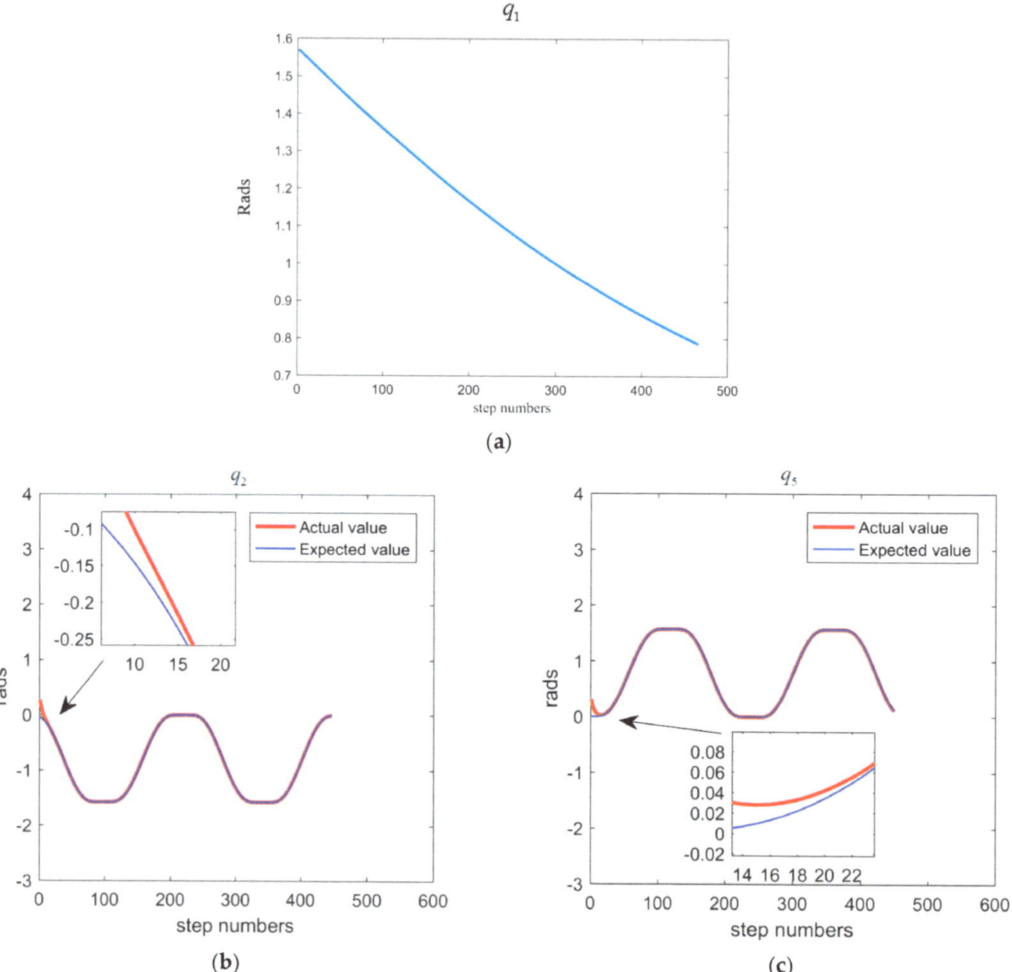

Figure 13. *Cont.*

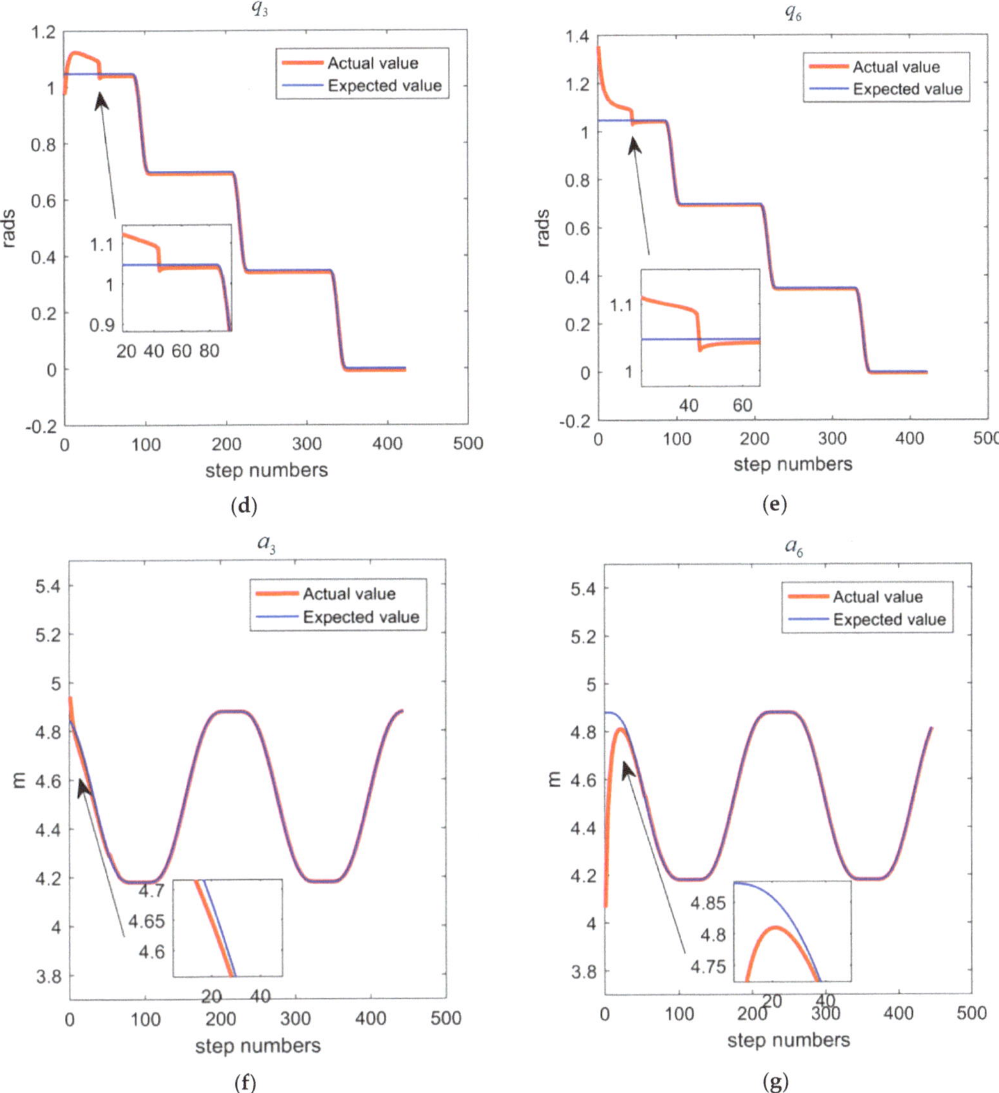

Figure 13. The angle/length changes of each joint of the DACR. (**a**) The movement of the mobile platform of the DACR in mode 2; (**b**) The yaw angle of the left arm of the DACR; (**c**) The yaw angle of the right arm of the DACR; (**d**) The pitch angle of the left arm of the DACR; (**e**) The pitch angle of the right arm of the DACR; (**f**) The length of the prismatic joint of the DACR's left arm; (**g**) The length of the prismatic joint of the DACR's right arm.

With the action of the relative kinematic controller, after the synchronous start of the two cutting arms, the X_R generated by the two end-effectors gradually approaches the expected values, as shown in Figure 14, where the blue lines (i.e., the expected relative positions) are calculated by substituting the parameters shown in Table 2 into the forward relative kinematics model (i.e., $^3p_6$ in Equation (5)). The positions of the two end-effectors are controlled to the expected values within 100 steps.

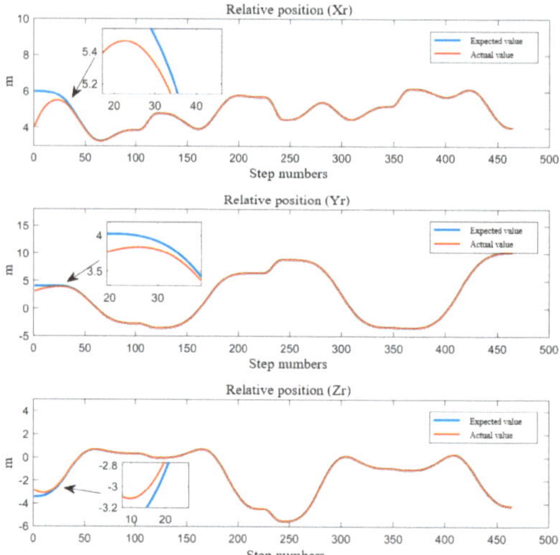

Figure 14. The convergence process of relative position.

As shown in Figure 15, the errors along the X, Y, and Z axes converge to zero within 100 steps. It can be observed from the figure that the relative kinematic controller can adjust the two cutting arms of the DACR with a synchronous start of the two cutting arms to follow the motion pattern of asynchronous operation. The controller shows good performance in reducing errors with a fast feedback velocity.

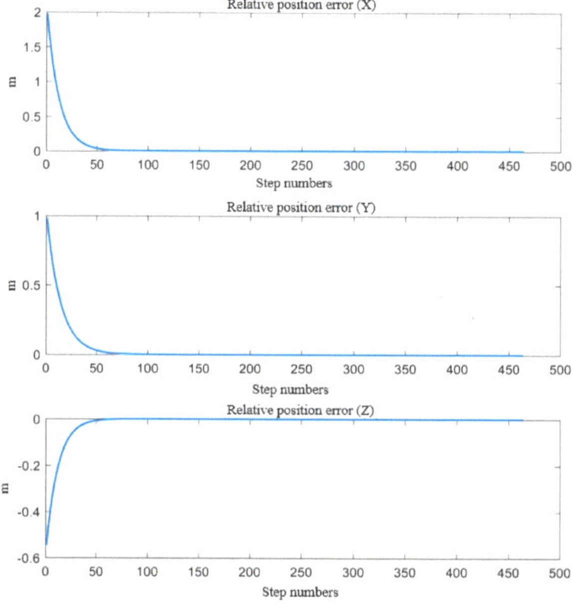

Figure 15. The convergence process of the relative position error.

The control effect of the relative kinematic controller is shown in Figure 16, where the blue and red trajectories represent the actual positions and expected positions of the two end-effectors of the DACR in Cartesian space, respectively. With correction from the controller, the end-effectors of the left and right cutting arms gradually approach the desired red trajectory and eventually remain close to the desired S-shaped trajectory. Because the dual arms of the DACR work asynchronously, after stably tracking the desired trajectory, the two end-effectors of the DACR pass through the overlapping section of the AW one after another.

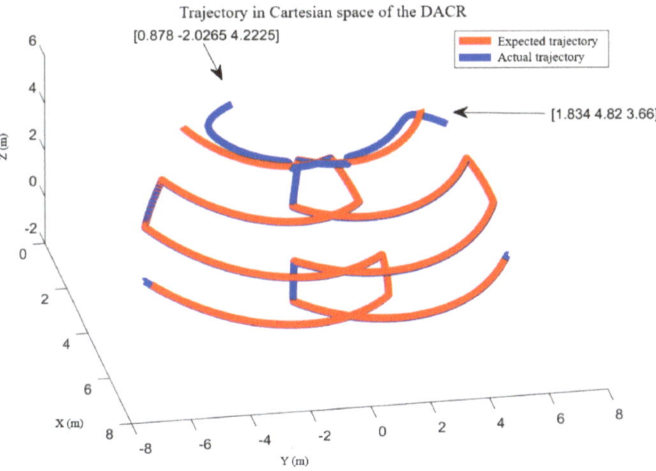

Figure 16. The trajectories of the two end-effectors of the DACR (relative to the base coordinate system).

The absolute error is defined as the two-norm of the difference between the relative position vectors of the two end-effectors of the DACR after tracking the desired trajectory stabilizes and the expected relative position vector. The absolute error of the relative position of the DACR's end-effectors can be obtained as follows:

$$e = \|X_{R\max} - X_{Rd}\| = 0.01837 \text{ m}.$$

The root mean square error (*RMSE*) is a statistical measure of the difference between observed values and expected values. In this simulation, the *RMSE* is defined as follows. After the expected trajectory tracking stabilizes, the angle/length error of each joint at each step by the DACR is calculated. This results in each joint generating N errors in angle/length. Compute the sum of squares of errors for each joint, take the average, and, finally, calculate the square root to obtain the root mean square error for each joint. The expression for the *RMSE* is as follows:

$$RMSE = \sqrt{\frac{1}{N}\sum_{i=1}^{N}(q_d - q)^2}$$

The *RMSE* for each joint of the DACR is shown in Table 3.

Table 3. The *RMSE* of each joint.

	q_2/**Rads**	q_3/**Rads**	a_3/**m**	q_5/**Rads**	q_6/**Rads**	a_6/**m**
RMSE	0.0032	0.0149	0.0136	0.0053	0.0127	0.0125

The simulation results and error calculations for the controller indicate that tracking of the expected relative positions is effective, and the tracking performance of the angles/lengths of each joint under control is good. Solely from the perspective of simulation results, the controller is capable of effectively controlling the motion of the DACR's dual arms using only the relative position as the unique variable. However, this research only focuses on the kinematic control of the DACR and does not consider other factors that may affect its control effectiveness, such as vibrations generated during coal cutting, dynamic characteristics of the system, and the stiffness of each joint. Therefore, establishing a more precise model and conducting experiments on the DACR are the future research directions for enhancing its control effectiveness.

6. Conclusions

DACRs are one of the most significant concerns for proposed dual-arm tunneling robotic systems because DACRs have an important influence on the forming quality and excavation efficiency of large-sized cross-section laneways. However, the critical issues related to DACRs have not been reported, so the goal of this work is to address the main issues related to the presented DACR. The following conclusions can be drawn:

1. An advanced dual-arm tunneling robotic system for a coal mine is developed, and the main components and characteristics of the robotic system are presented in this paper. The major benefit of this type of robotic system is the achievement of the synchronous operation of excavation and permanent support of laneways.
2. The relative kinematic model of the DACR is established. This model integrates the independent kinematics of both arms into a unified framework, allowing the simultaneous description of the motion states of both arms using only one variable. Additionally, a control strategy is proposed based on relative kinematics, enabling simultaneous control of both cutting arms by using a single variable. Furthermore, the AW of the DACR is generated and proposed by a Monte Carlo algorithm.
3. The simulation of the motion continuity of the DACR validates the correctness of its relative kinematics model. The status of the DACR is studied for two typical working modes, and the simulation of the AW verifies that the DACR has a larger workspace and reduces the excavation process of large-sized cross-section laneways, thereby improving mining efficiency. Finally, the simulation of relative kinematic control selects the more suitable mode 2 and demonstrates the effectiveness of the proposed control strategy. The DACR converges to the desired trajectory within 50 steps, with an absolute error in the relative positions between the two arms and the desired relative positions of less than 0.01837 m, and the root mean square error of the angle/length of each joint is maintained at a small value. This indicates that the DACR can achieve precise feedback control with good error correction effect and response speed, with minimal fluctuations after trajectory tracking stabilization. This provides a theoretical foundation for the design, analysis, and future implementation of dynamic control for the proposed DACR.
4. Compared to other dual-arm robots, the advantage of the DACR's control strategy lies in the proposed relative position controller's ability to track changes in the relative positions of the end-effectors. In contrast, conventional dual-arm robots primarily focus on operations such as transporting, gripping, and lifting objects, where the relative positions of the two end-effectors remain constant. Therefore, the proposed controller exhibits greater versatility and can serve as a reference for future dual-arm robot control systems.

There are some intrinsic limitations of the advanced dual-arm tunneling robotic system for a coal mine and the DACR in this work. Firstly, the proposed controller can simultaneously control the motion of both arms by using a single variable (i.e., the relative position of the dual-arm end-effectors). However, the controller may yield multiple sets of feasible solutions. These solutions can satisfy both the expected motion of both arms and the co-shared motion induced by the mobile platform, but they cannot guarantee that the angles/lengths of other joints remain within specified ranges, ultimately leading to singular arm poses. This research only considers solutions under the condition of deterministic motion of the mobile platform. Therefore, further research is needed to address the issues of the controller yielding multiple solutions and avoiding singular arm poses. Secondly, the effects of the relative dynamics of the DACR on the trajectory tracking accuracy of the two cutting arms, which are of great importance and challenging, still need to be investigated in detail. The consideration of the relative dynamics of the DACR is a topic for further research. Therefore, in the future, an experimental platform will be established to validate the proposed relative kinematic model and controller of the DACR. Additionally, there will be an in-depth investigation into the dynamic model of the DACR, particularly focusing on the relationship between the output force of the DACR and the joint force, which is crucial for the DACR's ability to cut rock walls. Finally, other control algorithms based on the dynamic model of the DACR will be proposed, and the effectiveness of the DACR's dynamic controller will be verified through experiments.

Author Contributions: Conceptualization, P.L. and H.Z.; methodology, P.L.; software, H.Z. and Y.Z.; validation, P.L. and H.Z.; formal analysis, H.Z.; investigation, H.Z.; resources, Y.Z.; data curation, H.Z.; writing—original draft preparation, P.L.; writing—review and editing, P.L. and X.Q. All authors have read and agreed to the published version of the manuscript.

Funding: The authors gratefully acknowledge the financial support of Shaanxi Province Natural Science Basic Research Program Project under Grant No. 2024JC-YBMS-324, Education Department of Shaanxi Provincial Government Service Local Special Project under Grant No. 23JC053, and Bilin District Applied Technology Research and Development Projects in 2022 under Grant No. GX2228.

Data Availability Statement: The datasets used and analyzed during the current study are not available for public disclosure and are available from the corresponding authors upon reasonable request.

Acknowledgments: The authors would like to thank the support from Open Fund of Key Laboratory of Electronic Equipment Structure Design (Ministry of Education) at Xidian University.

Conflicts of Interest: The authors declare no conflicts of interest.

References

1. Tang, Y.; Tang, J.; Wang, S.; Wang, S. Analysis of rock cuttability based on excavation parameters of TBM. *J. Geomech. Geophys. Geo-Energy Geo-Resour.* **2023**, *9*, 93. [CrossRef]
2. Zheng, Y.; He, L. TBM tunneling in extremely hard and abrasive rocks: Problems, solutions and assisting methods. *J. Cent. South Univ.* **2021**, *28*, 454–480. [CrossRef]
3. Zhang, X.; Zheng, X.; Yang, W.; Li, Y.; Ma, B.; Dong, Z.; Chen, X. Research on path planning method of underground mining robot. *J. Coal Geol. Explor.* **2024**, *52*, 1–13. [CrossRef]
4. Zhang, Y.; Huang, Q.; Liao, Q.; Yang, X.; Wei, S. Research on the novel method for inverse kinematics analysis of spatial 6R serial manipulators. *J. Mech. Eng.* **2022**, *58*, 1–11. [CrossRef]
5. Vavro, J.; Vavro, J.; Marček, L.; Taraba, M.; Klimek, L. Kinematic and dynamic analysis and distribution of stress for six-item mechanism by means of the SolidWorks program. *IOP Conf. Ser. Mater. Sci. Eng.* **2021**, *1199*, 012047. [CrossRef]
6. Kheylo, S.V.; Garin, O.A.; Terekhova, A.N.; Prokhorovich, V.E.; Dukhov, A.V. Solving problems of the dynamics of a manipulator with six degrees of freedom. *J. Mach. Manuf. Reliab.* **2022**, *51*, 30–36. [CrossRef]
7. Sun, X.; Wang, H.; Guo, X.; He, X. Kinematics study on redundant manipulator with serial-parallel hybrid mechanism of bolter-drilling rig. *Mech. Sci. Technol. Aerosp. Eng.* **2023**, *42*, 493–503. [CrossRef]
8. Wang, X.; Shi, L.; Katupitiya, J. Robust control of a dual-arm space robot for in-orbit screw-driving operation. *Acta Astronaut.* **2022**, *200*, 139–148. [CrossRef]
9. Ouyang, Y.; Sun, C.; Dong, L. Actor–critic learning based coordinated control for a dual-arm robot with prescribed performance and unknown backlash-like hysteresis. *ISA Trans.* **2022**, *126*, 1–13. [CrossRef]
10. Shen, Y.; Fan, G.; Tang, X.; Dong, Y.; Deng, L. Gaussian process based differential kinematics control of the soft robot. *Control Eng. China* **2021**, *28*, 1360–1365. [CrossRef]
11. Guo, J.; Liu, J. Workspace analysis and structure optimization of parallel robot. *Manuf. Technol. Mach. Tool* **2022**, *11*, 20–26. [CrossRef]
12. Wang, J.; Xu, J. Kinematic modeling and simulation of dual-arm robot. *J. Robot. Netw. Artif. Life* **2021**, *8*, 56–59. [CrossRef]
13. Lee, S.H.; Lee, Y.J.; Kim, D.H. Extension of inverse kinematic solution for a robot to cope with joint angle constraints. *Int. J. Control Autom. Syst.* **2023**, *21*, 1899–1909. [CrossRef]
14. Yang, S.; Wen, H.; Hu, Y.; Jin, D. Coordinated motion control of a dual-arm space robot for assembling modular parts. *Acta Astronaut.* **2020**, *177*, 627–638. [CrossRef]
15. Lei, M.; Wang, T.; Yao, C.; Liu, H.; Wang, Z.; Deng, Y. Real-time kinematics-based self-collision avoidance algorithm for dual-arm robots. *Appl. Sci.* **2020**, *10*, 5893. [CrossRef]

16. Du, H.; Hu, N. Kinematics and dynamics analysis of lower limb robot based on human motion data. *J. Mech. Transm.* **2022**, *46*, 128–134. [CrossRef]

17. Ahmed, A.; Ju, H.; Yang, Y.; Xu, H. An improved unit quaternion for attitude alignment and inverse kinematic solution of the robot arm wrist. *Machines* **2023**, *11*, 669. [CrossRef]

18. Jiang, Z.; Yu, Z.; Zhang, X. Research on the inverse kinematic algorithm of the robot arm of the minecart. *J. Phys. Conf. Ser.* **2023**, *2483*, 012012. [CrossRef]

19. Wan, Y.; Li, H.; Chen, X. Kinematics analysis and visualization simulation of 6-DOF intelligent cooperative robot. *J. Phys. Conf. Ser.* **2022**, *2283*, 012013. [CrossRef]

20. Wei, Y. Kinematics analysis and modelling based on 4 DOF industrial robot. *J. Phys. Conf. Ser.* **2023**, *2562*, 012076. [CrossRef]

21. Meng, Z.; Cao, W.; Ding, H.; Chen, Z. A new six degree-of-freedom parallel robot with three limbs for high-speed operations. *Mech. Mach. Theory* **2022**, *173*, 104875. [CrossRef]

22. Huang, Y.; Li, Z.; Xing, K.; Gong, H. A manipulator size optimization method based on dexterous workspace volume. *Cobot* **2022**, *1*, 3. [CrossRef]

23. Bader, A.M.; Maciejewski, A.A. The kinematic design of redundant robots for maximizing failure-tolerant workspace size. *Mech. Mach. Theory* **2022**, *173*, 104850. [CrossRef]

24. Meng, Q.; Fei, C.; Jiao, Z.; Xie, Q.; Dai, Y.; Fan, Y.; Shen, Z.; Yu, H. Design and kinematical performance analysis of the 7-DOF upper-limb exoskeleton toward improving human-robot interface in active and passive movement training. *Technol. Health Care* **2022**, *30*, 1167–1182. [CrossRef]

25. Liu, C.; Gao, B.; Yu, C.; Tapus, A. Self-protective motion planning for mobile manipulators in a dynamic door-closing workspace. *Ind. Rob.* **2021**, *48*, 803–811. [CrossRef]

26. Wang, X.-Y.; Ding, Y.-M. Method for workspace calculation of 6R serial manipulator based on surface enveloping and overlaying. *J. Shanghai Jiaotong Univ. Sci.* **2010**, *15*, 556–562. [CrossRef]

27. Zeng, Q.; Zhou, G.; Shi, J.; Li, Y.; Liu, Y.; Chen, T.; Wang, L. Robot workspace optimization based on Monte Carlo method and multi island genetic algorithm. *Mechanics* **2022**, *28*, 308–316. [CrossRef]

28. Li, J.; Zhao, Z.; Zhang, S.; Su, C. Dynamics and workspace analysis of a multi-robot collaborative towing system with floating base. *J. Mech. Sci. Technol.* **2021**, *35*, 4727–4735. [CrossRef]

29. Boanta, C.; Brișan, C. Estimation of the kinematics and workspace of a robot using artificial neural networks. *Sensors* **2022**, *22*, 8356. [CrossRef]

30. Xu, Z.; Zhao, Z.; He, S.; He, J.; Wu, Q. Improvement of Monte Carlo method for robot workspace solution and volume calculation. *Opt. Precis. Eng.* **2018**, *26*, 2703–2713. [CrossRef]

31. Su, C.; Li, J.; Ding, W.; Zhao, Z. Static workspace analysis of a multi-robot collaborative towing system with floating base. *Int. J. Model. Simul. Sci. Comput.* **2021**, *12*, 2150009. [CrossRef]

32. Li, T.; Jia, Q.; Chen, G.; Sun, H. Motion reliability assessment of robot for trajectory tracking task. *Syst. Eng. Electron.* **2014**, *36*, 2556–2561. [CrossRef]

Article

On Combining Shape Memory Alloy Actuators and Pneumatic Actuators for Performance Enhancement in Soft Robotics

Florian-Alexandru Brașoveanu and Adrian Burlacu *

Department of Automatic Control and Applied Informatics, Faculty of Automatic Control and Computer Engineering, "Gheorghe Asachi" Technical University of Iasi, Str. Dimitrie Mangeron, 700050 Iasi, Romania; florian-alexandru.brasoveanu@academic.tuiasi.ro
* Correspondence: adrian.burlacu@academic.tuiasi.ro

Abstract: Through soft robotics, flexible structures confer an elevated degree of protection and safety in usage, as well as precision and reliability. Using theoretical models while combining different types of soft components opens a wide variety of possibilities for the development of new and better alternatives to rigid robots. Modeling and controlling soft robotic structures is still a challenge and is presented in different ways by the scientific community. The present scientific work aims to combine two of the most popular types of soft actuators, specifically shape memory alloy and pneumatic actuators. The purpose is to observe the interaction between individual entities and the resulting combined dynamics, highlighting the distinctive effects and influences observed in the combined system. An evaluation is conducted from a numerical simulation perspective in the MATLAB environment using representative mathematical models. The tests prove that a structure combining these particular actuators benefits from the advantages of both components and even compensates for individual downsides.

Keywords: soft robotics; mathematical modeling; hybrid structures; hybrid actuation; robotics

Citation: Brasoveanu, F.-A.; Burlacu, A. On Combining Shape Memory Alloy Actuators and Pneumatic Actuators for Performance Enhancement in Soft Robotics. *Actuators* **2024**, *13*, 127. https://doi.org/10.3390/act13040127

Academic Editor: Giulia Scalet

Received: 13 February 2024
Revised: 14 March 2024
Accepted: 17 March 2024
Published: 3 April 2024

1. Introduction

In the context of permanently evolving robotics, the area of soft and flexible robots distinctly appears to be increasingly popular. The fundamental driving reason for every technological innovation is the necessity to offer people an easier, more comfortable life. In the current era of engineering, a significant portion of rapidly developing technology is, naturally, oriented toward people's interactive needs, as researchers and engineers always keep in focus the social and economic impact of their developments. The healthcare sector is one of the few fields that combines technological advances with the diverse range of needs, from basic to complex, of patients and medical personnel to enable an environment where impressive medical procedures can be performed. Retrospectively, the medical field has received important help from the technology field through the adoption of tools and equipment capable of increasing the efficiency of medical staff, the speed of rehabilitation, and even professional at-home healthcare.

As described in [1], the integration of robots in medicine practice started with the neuro-surgical robot in the 1980s, leading to the voice-controlled endoscope in 2007 and the more recent da Vinci SP incision flexible arm and Monarch robot for bronchoscopy. In this manner, the development of medical robots uses a different approach, separating different types of robots and actuation for specific procedures or domains. Rigid robots are mainly used in heavy-wielding and mobility tasks, whereas medical procedures use flexible robots and structures. Increasing the safety of the operation of these devices led to the idea of different types of actuation, such as ones with reduced contact damage risk.

The need for small and powerful devices implies the need for specific types of actuation, including volume-reduced and high-precision robots and actuators. Some categories of soft actuators are electric and piezoelectric actuators [2], chemically and magnetically driven actuators [3], and pneumatic/ fluidic actuators [4], but considering the development speed of the field, new actuators continue to emerge.

As presented in [5], continuum robots have various medical applications classified by body area and actuation type. These robotic components (soft actuators) can enhance safety through their usage, but the control strategies necessitate the development of complex models and control structures. While mechanical and task-oriented approaches lead to fast development and implementation, even with many intermediary versions, the modeling area struggles to maintain mathematical representations' simulation capabilities and accuracy.

Soft actuators include artificial muscles [6], sensing skin-imitating devices [7], and other flexible robots. The authors of [8] provided a detailed perspective on using flexible structures for actuation and sensing. An important subdivision of flexible robots is the soft robotics area and soft actuation technologies, which have been applied in developing rehabilitation devices, as presented in [9].

Performing actuations in soft robotics requires movement predictability, precision, and consistent performance. More precisely, the control structure driving the robotic soft actuator needs to be specifically designed based on a mathematical model for which classical and modern control strategies can provide predictable performance.

Most actuators in the literature are specialized, using basic actuation methods and simplified models (based mainly on physical design) to determine expected displacements. But while rigid robotics can combine electrical rotational joints, which provide angular precision, with pneumatic cylinders, which offer power and force stability, soft actuation can be used similarly, combining the benefits of both components.

Rigid–soft hybrids are one possibility of combining different actuations, using the rigid component to sustain the soft actuation or, as presented in [10], using a rigid structure to encapsulate the control of the soft robot. The resulting actuators and methods could provide more complex dynamics that enhance performance due to the use of both elements. Combining different types of soft actuation can also provide capabilities applicable in various domains or yield better outcomes. The components in the soft robotics class are fundamentally different from each other but are nevertheless compatible and interconnectable.

When considering the two types of soft actuation, one can define two types of combined soft structures: hybrid and mixed systems. Mixed structures use two or more different actuation methods and actuators in parallel, and the result is a sum of effects applied to a theoretical target point. The hybrid structure, on the other hand, represents a chain of interactions between the soft structures, with the output being the motion of the last segment.

The reduced number of mixed or hybrid soft actuators in the literature and the physical implementations or mathematical approaches used leave a widely unexplored research direction. This is a first step in simulating hybrid soft actuators and can be considered a starting point for future endeavors.

The idea of hybrid soft structures is a recent research concept, and exploring it could provide novel insights. The already realized studies regarding this subject are limited in number and only cover certain aspects. A first example can be found in [11], which utilizes parallel actuation of shape memory alloy (SMA) wires in a spring form and a pneumatic bladder. This forms an actuation system aimed at the rehabilitation of upper limb movements. The actuations are realized separately and are used for specific individual purposes. It represents a hybrid system but not a hybrid actuation.

In [12], the authors presented an approach similar to a mixed structure. A soft cylinder with intrinsic passive elastic proprieties is covered with a separately driven SMA short wires network. When a particular arrangement of wires is driven, the cylinder develops a specific bending motion. The elastic proprieties are used as a reshaping method, and the cylinder body also serves as an actuator interaction body. This approach confirms the current research direction as a potential actuation field with unexplored possibilities.

Performing actuations in soft robotics requires movement predictability, precision, and known performances. More precisely, the control structure driving the robotic soft actuator needs to be specifically designed based on the mathematical model. Most actuators illustrated in the literature are specialized, using the basic actuation method and the known model (based on physical design) to determine the outcome displacements. Mixed actuators and methods could provide more complex dynamics that combine the benefits of the individual elements.

A first step guiding toward the mentioned area can be a simple but concrete mathematical model compatibility, starting from the already created models and developing a methodology to combine and test the numerical interaction between the outcomes. In this manner, one soft actuator model gives the driving input to another soft instance, creating a deterministic outcome based on both serial actuator models.

To the authors' knowledge, the ongoing research differs from the state of the art and does not replicate or test any previous approaches. The limited state of the art argues that the research direction is yet to be explored with a broad potential spectrum.

The novelty brought to the research field by the present work is an efficient and easy-to-handle manner of combining different soft actuation models proposed, which can describe complex soft dynamics, which is essential in designing hybrid actuators. The mentions regarding compatibility strategies and testing methods are of the same importance. One necessary and significant contribution resulting from the overall research is the usage of fundamentally different simulation strategies, proving that the mathematically represented soft structures are data-compatible, regardless of the testing method.

The chosen design uses SMA wires and a PneuNet structure to underline the potential of hybrid soft robots. A system with an SMA actuator is complicated to control precisely due to its complex electrical–thermal–mechanical characteristics. Depending on the temperature and the external load, the behavior varies, and vibrations may occur due to the wires' elasticity. Therefore, SMA actuators must be controlled robustly in response to environmental changes, modeling errors, and vibration suppression. Also, considering hysteresis as a nonlinear element in the control system, the controllers should be able to handle both position and force control. The development of a simulation model can start with approaches based on experimental data [13].

Both SMA and PneuNet analytical models are considered in a hybrid architecture, which is implemented via simulation in a MATLAB framework. The experimental results show how the performance of the hybrid architecture surpasses the performance of individual components.

The following sections of the paper cover all the developments in hybrid architecture. The second section describes the development and testing strategies and the general connection of the elements, from physical inspiration to the simulated product. The third section covers the mathematical principles at the foundations of the separated actuators along with each integration method and results, covering the development from concept to data. The following section presents the interconnection of the actuators and the constraints and limitations of the resulting hybrid structure. The testing results are presented in Section 5, along with discussions for each case, followed by a final section containing conclusions and perspectives for the future.

2. Methodology

In the context described above, we propose a solution inspired by two essential actuation elements, air pressure and electrical power, for soft materials represented by the SMA wire and the pneumatic network (PneuNet) actuator. The proposed hybrid soft structure uses the dynamics of the SMA wire to drive a classical pneumatic network, combining the electro-thermal fast response of the first with the elastic proprieties of the second.

The PneuNet presents a shape-recovering elasticity propriety that allows the actuator to move in the initial position once the actuation is stopped. he volume is maintained constant by variation of length to diameter ratio from low temperature (blue shape) to high temperature (red shape). As represented in Figure 1, where Δx is the SMA actuation displacement, the SMA liner displacement can be used to pull the extended silicon robot, a process similar to the string actuation used for the homogeneous silicon actuators.

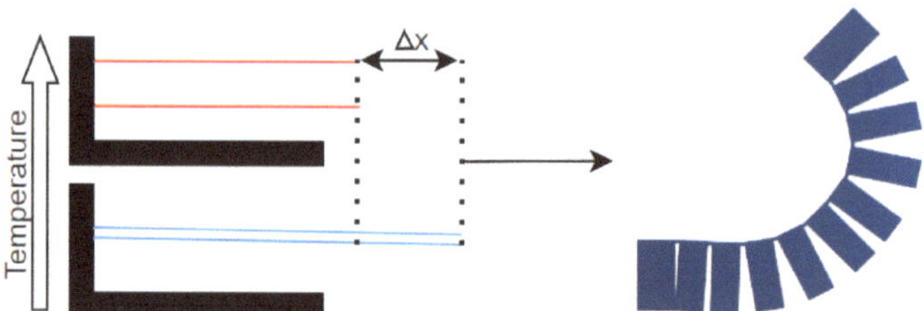

Figure 1. From SMA displacement to PneuNet bending connection.

The downside of the SMA thermal-driven behavior is its environmental dependency. Once the actuator is heated to the desired temperature, the cooling process depends on the environmental temperature, not on a controllable measure. On the other hand, the PneuNet has an internal shape recovery process due to its elastic properties and the resistive component included in the actuator body. In this manner, the pneumatic actuator performs a fast shape-recovery speed when the pressure actuation stops or, in the current case, when the pulling force starts to decrease.

Combining the fast and stable actuation of the SMA and the recovery proprieties of the PneuNet can output a fast actuation in both directions, resulting in a reliable actuator with speed, stability, and force performance.

A numerical simulation of the two separated actuations resulting in a hybrid structure is realized in the MATLAB environment to prove the hybrid actuator and the mentioned hypothesis. The resulting interconnection of the expected mathematical outcomes in a graphical representation can be found in Figure 2. A mathematical model is implemented for the SMA, and the force-resulting force motion is used as a negative pressure input. P is noted at different pressure inputs in the figure. The PneuNet is simulated in a soft robotics specialized toolbox, SoRoSim, using the force mentioned as input.

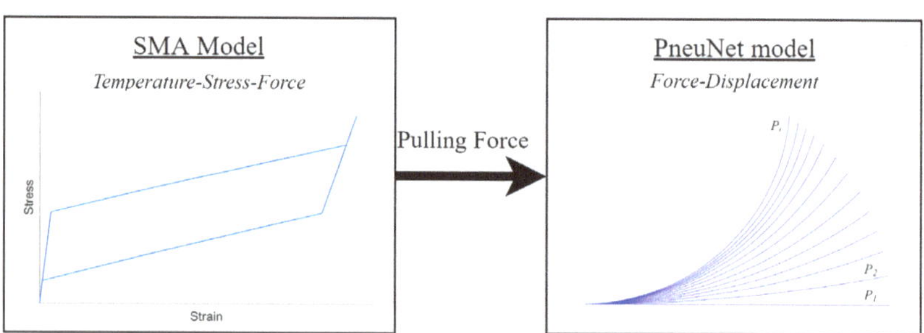

Figure 2. From SMA numerical model towards the PneuNet simulation.

3. Modeling and Integration

The authors of [14] strongly argue about the complexity of the soft robotics model and the computational power required to accurately describe a soft behavior numerically, especially in online control structures. A simple solution is to design the control systems offline using an approximate representation of the desired actuator, which is also valid for mixed or hybrid soft structures.

The current section contains a comprehensive description of the physical principles based on the soft components and the practical integration process of the simulated models.

The subsections also include the numerical representations of the physical proprieties and the numerical data used for the experimental simulations. The description provided is realized according to the mathematical model used in the numerical description.

3.1. SMA Preliminaries

The shape memory alloys are conductive metal composites, mostly found in nickel–titanium mixtures, which offer a smart-material propriety called the shape memory effect. This thermo-dynamical characteristic allows the alloy to shift from one molecular crystal structure to another when the temperature changes, reacting to the influence of the environment [15]. The alloy composition determines the range of shifting temperature and the state shapes resulting from preemptive 'teaching'. The shape memory alloys are used in various actuators and can be driven by a wide range of actuation. As shapes, one of the most popular types of actuators is the linear wire, with the dynamics produced being a linear longitudinal displacement. For varying the actuating temperature, the conductive metal can react to chemical components, electromagnetic field changes, or even changes in the surrounding air temperature [16]. To change the internal temperature directly and obtain a faster response, an electrical current can power the wire, with the residual thermal energy activating the shape memory effect.

Considering the state of the art, different data-driven techniques were proposed to estimate the model parameters. According to the published results, the SMA model is a relation between the parameters and experimentally obtained stress, strain, and resistance data at various constant temperatures. For modeling the SMA wire, the shape memory effect and elastic and conductive properties are also considered. On the other hand, the temperature-shifting values are to be chosen according to the pre-training process and alloy composition (interdependent characteristics) [16].

3.2. SMA Model Implementation

Considering the complex behavior described above, it is clear that intertwined internal characteristics and structure variation as a response to external factors return a complex model description. For the current case exploration, the modeling considered will be the one proposed in [17]. The paper describes a coupled model of a generic shape memory alloy bar close to the SMA wire but with a detailed internal structure and material data.

The authors offer two approaches to determine the SMA internal manners, one having the stress-induced command returning the internal strain and one opposite, having the strain determining the stress. For the modeling necessary in the current development, the strain-driven model, the observable outcome is the stress of the metallic wire and the force present, both having similar forms.

The data found in the technical documents [18] was insufficient to have a complete model of the real SMA, so the available data was combined with various results from research papers, including [16,19,20].

Firstly, the material and environmental data must be specified with the critical consideration that the transitions between austenite and martensite and back determine the material propriety changes. Another aspect is related to the shape of the SMA and its size [21]. For the current implementation, Table 1 shows the material data and the phase separate proprieties.

The model is based on Gibbs energy, clearly representing the thermo-mechanical process inside the alloy and the material reactions. Computing the model requires the discretization of the SMA wire. The effect of the input is propagated along the wire, successively affecting the entire length of time.

The model uses the discretization of the SMA bat that keeps a consecutive step state change on the length. The surface area to volume ratio represents the physical dimensions and is included in the algorithm as the thermo-mechanical coupling.

Table 1. Shape memory alloy numerical data.

Heat capacity H_c	3.2 MPa/K
Transformation strain Tr_{St}	4.7 %
Maximum strain	5.5%
Environmental temperature T_0	297 K
Time-scale ratio λ	320
Expansion coefficient α	$11 \times 10^{-6}\,\mathrm{K}^{-1}$

Austenite		**Martensite**	
Data measurement	Value	Data measurement	Value
Young modulus Y_A	80 GPa	Young modulus Y_M	28 GPa
Austenite start temperature A_s	343 K	Martensite start temperature M_s	323 K
Austenite finish temperature A_f	363 K	Martensite finish temperature M_f	338 K

The initial data and command signals need to be specified. Inspired by Equations (17), (24a,b), and (30) from [17], we can assume the initial values for force, the mean value of state-shifting temperatures $du0$, the differential mean temperature for coupled state temperatures w, and different prescribed strain shapes, used as a starting point for the current implementation. Also, the initial temperature is the room temperature specified in Table 1.

Each discreet segment of the wire is transitioning from one state to another depending on the strain value given as input. By transitioning, the internal variables of the wire are changing and need to be iteratively actualized, depending on the instantaneous phase.

Consider the following notations: ξ is the state shifting length proportion, St is the strain input, T_0 is the room temperature, Y_M and Y_A are the Young modulus at the shifting states, Tr_{St} is the transfer strain, and Th_{St} is the thermal stress.

The internal temperature T shifting in time, containing all the material transformations, is given by (1), where k is the time moment for strain command. The stress S_s is given in (2), and force F is given in (3).

$$T_{k+1} = \frac{\frac{-\alpha}{H_c}\frac{T_k}{\frac{\xi_{k+1}}{Y_M} \frac{1-\xi_{k+1}}{Y_A}}(St_{k+1} - St_k) - \lambda(T_k - T_0)dt}{1 - \frac{\alpha^2}{H_c}\frac{T_k}{\frac{\xi_{k+1}}{Y_M} + \frac{1-\xi_{k+1}}{Y_A}}} + T_k, \tag{1}$$

$$Ss_{k+1} = \frac{(St_{k+1} - \xi_{k+1}T_{St} - \alpha(T_{k+1} - T_0))}{\frac{\xi_{k+1}}{Y_M} + \frac{1-\xi_{k+1}}{Y_A}}, \tag{2}$$

$$F_{k+1} = Ss_{k+1}Tr_{St} - T_{k+1}Th_{St}, \tag{3}$$

Now, the phase shifting is to be analyzed by comparing the updated value of the force with the previous one. If the segment analyzed is transiting one state, the force is between the stress–temperature boundaries of the state. At this point, depending on the state, the stress is computed using (4) for martensite or (5) for austenite:

$$Ss_{k+1} = \frac{\xi_{k+1}(M_s - M_f)Th_{Ss} - (M_s - T_{k+1})Th_{Ss}}{Tr_{St}}, \tag{4}$$

$$Ss_{k+1} = \frac{\xi_{k+1}(A_f - A_s)Th_{Ss} - (A_f - T_{k+1})Th_{Ss}}{Tr_{St}}. \tag{5}$$

In both cases, the temperature is required to be recalculated using (6):

$$T_{k+1} = \frac{-\alpha(Ss_{k+1} - Ss_k)}{H_c}T_k + \frac{Ss_{k+1}Tr_{St} + du0 - w(2\xi_{i+1} - 1)}{H_c}(\xi_{k+1} - \xi_k) - \lambda(T_k - T_0)dt + T_k. \tag{6}$$

Depending on the phase-shifting state of the SMA, the next step is to compute the new shifting proportion with (7):

$$\tilde{\zeta}_{k+1} = \frac{St_{k+1} - Ss_{k+1}\frac{\tilde{\zeta}_{k+1}}{Y_M} + \frac{1-\tilde{\zeta}_{i+1}}{Y_A} - \alpha(T_{k+1} - T_0)}{Tr_{St}}. \tag{7}$$

After the iterative phase-shifting process is realized, the force will be updated using (3). This process is to be applied for each discrete part of the considered wire, in this manner, the strain used as input is gradually applied to each one, and the transition is treated fluently and dynamically.

The modeled process is coupled, indicating that the stress–strain diagram suggests the type of alloy described, with the thermo-mechanical interaction having a critical impact. For the current approach, the stress–strain diagram is portrayed in Figure 3, in which the hysteresis gap describes the characteristic fount on a commercial SMA produced by DYNALLOY Inc., Irvine, CA, USA.

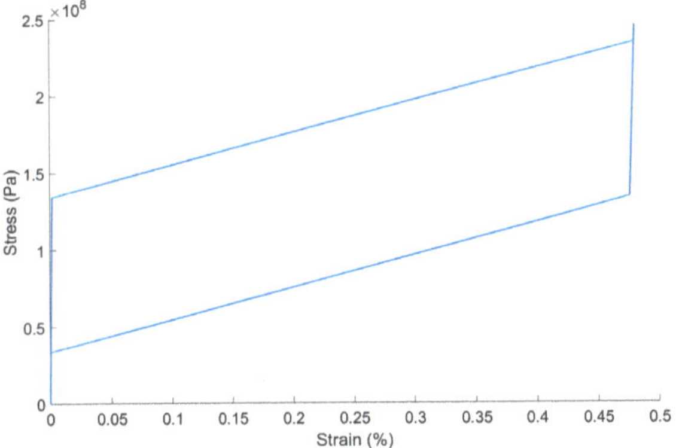

Figure 3. Shape memory alloy stress–strain diagram.

3.3. PneuNet Functional Principle

The secondary actuator is a fast PneuNet pneumatic actuator, a succession of interconnected elastic silicon chambers. This type of actuator is prevalent in the literature and can be considered a standard choice for the current application. The proprieties of this actuator are explored in terms of design and material in [22].

From the actuation point of view, the PneuNet is driven by air pressure inflating the individual chambers and creating consecutive separating forces between chambers. The constraining layer at the base of the actuator has the decency to determine the shape of the motion.

The speed of the actuation and overall displacement are denoted by the shape, the elastic material used in the fabrication, and the wall thickness. Therefore, creating a universal interpretation in the form of a mathematical model or simulation environment is a challenge. In the literature, the modeling of the PneuNet is developed overwhelmingly in mechanical-oriented environments for the study of forces and elastic proprieties.

3.4. PneuNet Implementation

For the current approach, the actuator in the case is an 11-chamber PneuNet, as displayed in Figure 4. The model of the simulated body accurately represents the physical dimensions of the actuator, the overall elastic properties, and the motion restrictions on the base layer.

Figure 4. Physical PneuNet used for modeling.

For development, the MATLAB R2023b toolbox SoRoSim is used due to its facilities for interconnecting soft, rigid, and combined structures. The only downside for the problem at stake is the non-homogeneity of the PneuNet; as the toolbox is designed for homogeneous robotic structures, the problem proved to be solvable in [23]. The material data presented and used for the integration are already modified to describe a non-homogeneous actuator, based on the same principle that the actuators used are similar.

The toolbox uses the interaction pop-up windows to create the composing elements of the PneuNet and the complete body, interconnections, and actuation. The elements have individual dimensions and material properties to reproduce the actual segment and, when connected, accurately reproduce the physical actuator.

Geometric Variable Strain (GVS) formalism is used for mathematical modeling to create similar support for rigid, soft, and mixed actuators. The modeling procedure is described in [24] and is resumed in the following paragraphs.

The technique used to map a shape reconfiguration of the robot uses the Cosserat roads description theory. This is known as micropolar elasticity, which incorporates the elastic behavior (twisting, bending, etc.) and defines the dynamics of a continuous segment. From this theory, the authors in [24] propose the interpretation of a continuous soft link as being from the initial configuration, or the Cosserat road base, $(X_i = 0)$ to the final configuration at the end of the road $(X_i = L_i)$.

Starting from a chain of consecutive segments composing a Cosserat rod, where g_i is an element in the Special Euclidian group (SE(3)) or a curve for flexible structures and is represented by a homogeneous transformation containing a rotational matrix R_i and a position vector r_i in (8), where the body-frame at X_i also includes joint constraints. A specific g_i can be used to express the relation between a soft or rigid body segment and the previous chain of segments, as for a particular X_i, from the Cosserat rod $[0, L_i]$, a transformation has the form from (8):

$$g_i(\cdot) : X_i \in [0, \ L_i] \in SE(3),$$

$$g_i(X_i) = \begin{pmatrix} R_i & r_i \\ 0 & 1 \end{pmatrix}. \tag{8}$$

To construct a chain of bodies describing consecutive spatial transformations, a product of exponentials (PoE) is used, based on the partial derivatives of space $(\acute{\cdot})$ and time $(\dot{\cdot})$ as in (9):

$$g_i'(X_i) = g_i \widehat{\tilde{\xi}}_i,$$
$$\dot{g}_i(X_i) = g_i \widehat{\eta}_i^{\Upsilon}. \tag{9}$$

The strain twist body-frame can be further described as a component of the Euclidean space or in a vectorial representation in (10), where $\widetilde{k}_i$ and k_i are representations of angular strains, and p_i represents the linear strain, with $\mathfrak{se}(3)$ being the operations added to the Euclidean group:

$$\widehat{\xi}_i(X_i) = \begin{pmatrix} \widetilde{k}_i & p_i \\ 0 & 0 \end{pmatrix} \in \mathfrak{se}(3) \quad or \quad \xi_i(X_i) = (k_i^T, p_i^T)^T \in \mathbb{R}^6. \tag{10}$$

Similarly, the velocity twist body-frame is described in the Euclidean space and in the vectorial representation in (11), where $\widetilde{w}_i^r$ and w_i^r are representations of angular velocity, and v_i represents the linear velocity:

$$\widehat{\eta}_i^r(X_i) = \begin{pmatrix} \widetilde{w}_i^r & v_i^r \\ 0 & 0 \end{pmatrix} \in \mathfrak{se}(3) \quad or \quad \eta_i^r(X_i) = (w_i^{rT}, v_i^{rT})^T \in \mathbb{R}^6. \tag{11}$$

Using the equations in (9), the construction of a PoE can be done regardless of the rigidity of the analyzed segment. For the rigid body, the X_i is similar to the body frame attached to i. In this manner, the product of exponentials describing the configuration of link i for the spatial frame can be developed using the Magnus expansion of $\widehat{\xi}_i$ at $X_i = L_i$:

$$g_i(1) = exp(\widehat{\xi}_i), \tag{12}$$

$$g_i(L_i) = exp(\widehat{\Omega}_i(L_i)). \tag{13}$$

Therefore, the PoE results in the form of (14)

$$g_{si}(X_i) = g_{s0}exp(\widehat{\xi}_0)g_{01}exp(\widehat{\xi}_1 \ or \ \widehat{\Omega}_1(L_1)) \ldots exp(\widehat{\Omega}_i(X_i)). \tag{14}$$

Starting from the product of exponentials, the kinematic chain of rigid, soft, or mixed links has the development described in [24] from Equation (8) to (19). The dynamic model covers the classical Lagrangian representation in (15), where q is the vector of coordinates based on the considered strain. $\mathbf{M}$ represents the mass matrix, $\mathbf{C}$ is the Coriolis matrix, $\mathbf{K}$ is the stiffness matrix, $\mathbf{D}$ is the damping matrix, all of $n x n$ order, $\mathbf{B}$ is the actuation matrix dependent on the number of actuators, and $\mathbf{F}$ is the vector of generalized external forces. All matrices have specific representations as found in [24].

$$\mathbf{M}\ddot{\mathbf{q}} + \mathbf{C}\dot{\mathbf{q}} + \mathbf{K}\mathbf{q} + \mathbf{D}\dot{\mathbf{q}} = \mathbf{B}\mathbf{u} + \mathbf{F}. \tag{15}$$

Concerning the computational implementation, some restrictions occur regarding the finite representations of components. Therefore, the product of exponentials implemented has the form of (16), considering the hth interval of a soft link i for which the value of the configuration g_{si} is computed. The Magnus expansion also requires an approximation in the form detailed in the [24], along with the rest of the computational integration of the model.

$$g_{si}(X_i + h_k) = g_{si}(X_i)exp(\widehat{\Omega}_i^k) \tag{16}$$

These mathematically described components describe the kinematic and dynamic behavior of the complete elastic chain that can be created at the toolbox level. Working with the SoRoSim required the abstract definition of the individual links and their particular proprieties to define a complete linkage.

When the link is created, the model defines a chain of interaction between the consecutive geometric variables, describing the possible deformation in the form of linear and twist motion. As the discretization defines the motion model precision, the created links must reflect the deformation generated or applied during actuation. In the physical actuation, as mentioned in the previous subsection, the chambers' interaction creates the pushing force, therefore the segments have separate effects.

For the current PneuNet, in the toolbox, three different links representing the principal types of segments are created:

- Ending link: representing the first and last chamber, with larger chambers and least powerful deformations;

- Chamber link: representing the intermediary chambers, which cause the deformation and the shape restoration;
- Channel link: representing the small segments interconnecting the other links, encapsulating the majority of the deformation.

The three link types are combined in a structure of 21 links, each keeping the material proprieties. The linkage structure also determines the links' interaction and actuation possibilities. The toolbox's default actuation type is string actuation. Although the PneuNet is generally actuated pneumatically, with pressurized air, a capability of the toolbox already proven possible in [25], the default string actuation accurately represents the SMA actuation.

The PneuNet propriety on focus is the shape recovery characteristic of elastic material. For the effect of the propriety to reach maximum potency, the pulling force outputted by the SMA model must be uniformly distributed along the actuated body. This is achieved using a distributed actuator (one for each link) with equal pulling force amplitude applied in the same manner as the air pressure [25].

The physical data used in the SoRoSim toolbox to describe PneuNet can be found in Table 2. The data describes each segment by the exterior dimensions, and the material data is already modified to represent the non-homogeneous elastic behavior as described in [25].

Table 2. Soft robot physical dimensions and material characteristics.

Data Measurement	Value
PneuNet length	103 mm
Rigid layer length	110 mm
Interior chamber (L|l|h)	3|14|13 mm
Chamber exterior (L|l|h)	8|20|15 mm
Common channel (L|l|h)	65|2|2 mm
Inter-chamber distance	6 mm
Wall thickness (top|lateral)	2|3 mm
Rigid layer thickness	5 mm
Young modulus	10^6 Pa
Sheer modulus	3.33×10^5 Pa
Poison ratio	0.5
Material density	500–1000 kg/(m^3)
Material damping	11,200 Pa·s

Another critical aspect of the PneuNet linkage model is the presence of gravitational force. As the expected dynamic displacement evolves on the XOZ plane, or around the Y-axis, in the form of a rising on the Z-axis, the gravitational force is oriented in the negative Y direction. The motion of the PneuNet aims to be antagonistic to the gravitational pull, amplifying the shape recovery speed. Figure 5 presents the spatial representation of the PneuNet linkage created. The gravitational force is marked by the black arrow pointing in the negative Z-axis direction.

The actuation functions used as input are provided to the simulation part of the toolbox as mathematical time-dependent expressions. The actuator's initial condition must also be specified. The default posture is used in all of the following simulations, with no previous displacement on any robot link. The integration resulting linkage has the dimensional representation shown in Figure 6, with the physical constraint layers on the upper side.

The validity of the integrated model is given by the existing models found in the literature in the form of the displacement shown as the final posture and Z and X-axis time-displacement graphics, as the PneuNet was realized with the data provided in the research mentioned. The functions used for testing are a constant step function and a sigmoidal evolution, both with the same maximum value, as proven in [25].

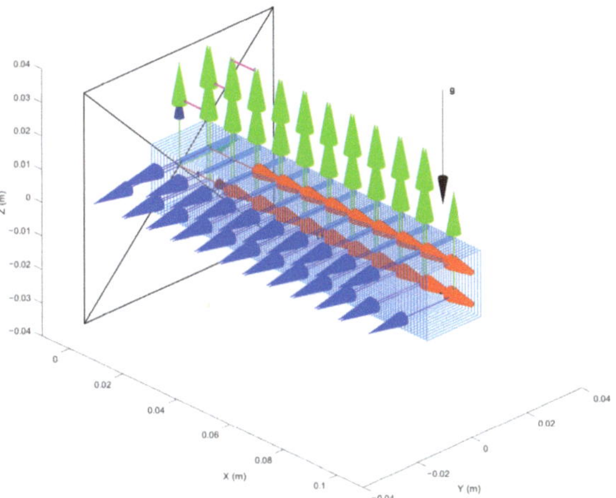

Figure 5. Schematic PneuNet resulting from modeling.

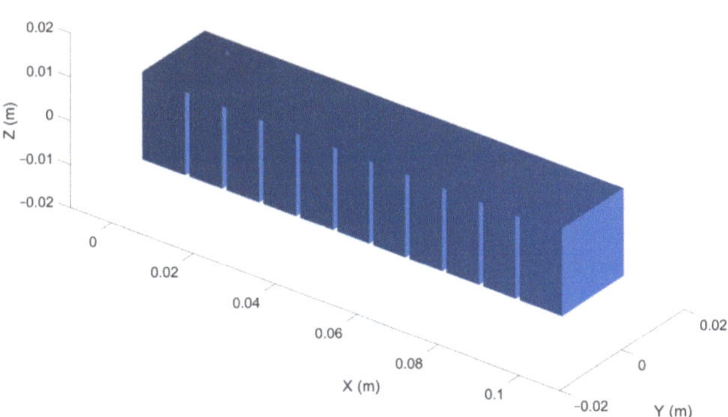

Figure 6. PneuNet integrated body in default configuration.

4. Models Interconnection and Testing

The mathematical models of both actuators require and return specific data based on the modeling approach but are not necessarily simulation-compatible. The present section covers the procedures needed to convert the SMA model output, represented by force–time variation, into an input compatible with the PnueNet integration, represented by air pressure, in the toolbox. As mentioned above, the driving data used in the dynamic simulation of the PneuNet is a pulling force applied equivalently on each segment of the actuator.

The data obtained from the SMA model are a numerical representation of the force variation in time, being a deterministic result of a given temperature variation. The resulting force evolution in time does not respect a simple characteristic; it can be a succession of complex shapes united by discontinuity points. An example is depicted in Figure 7, showing a force output used for the first experiment.

On the other hand, the dynamic simulation in the SoRoSim toolbox requires a time-dependent solvable shape to interpret and drive the PneuNet. Considering this, several procedures are required to create numerical compatibility. Firstly, the SMA resulting shape must be separated into continuum intervals using the discontinuity points for discrimination. To increase the precision, the discontinuity points are determined using the second-order gradient for the evolution. In this manner, points are numerically determined based on the evolution, not approximated.

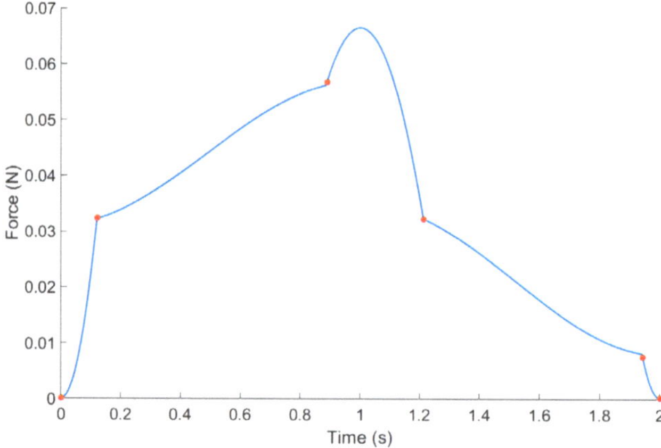

Figure 7. Example of SMA model output.

Each interval, containing numerically continuous partial evolution from the complete force, needs to be approximated using polynomial characteristics with a certain precision. The resulting time-dependent functions approximating the entire evolution of the SMA model force are compatible with the SoRoSim dynamic simulation input.

A critical aspect of the procedure is that a different simulation is realized for each approximated force interval. The dynamic simulation requires an initial condition for the geometric strain for each interval. The condition is the PneuNet state at the end of the previous simulation. In the case of the first interval, the initial condition is the default geometric strain state.

5. Results

This section covers the dynamics returned by the composed topology described above and the modeling principles presented, and it analyzes the final shape and the movement evolution. The expected displacement is shown distinctively in planar evolution (on the XOZ plane) and on the principle axis. It is important to note that the simulated experiments are realized in an open loop, focusing on the usefulness of the combined structure's effect rather than on a controlled outcome.

Firstly, we need to analyze the effect of the SMA output shown in Figure 7 of the modeled PneuNet. As observed, the evolution presents five continuum intervals, offering a rapid variation over a 2 s period, which, according to [25], can be considered a shock actuation, therefore resulting in oscillatory potential behavior. The resulting outcomes for the described simulated experiment are shown in Figure 8.

The oscillatory behavior of the actuated PneuNes is a simple and natural response of the elastic body to a rapidly increasing actuation force. It is a consequence that further proves the accuracy of the model. The physical reaction to a sudden pressure input in the pneumatic chambers is determining the fast contact interaction.

The compatibility forces and dynamics need to be ensured in the measurable environment provided by the PneuNet integration. The initial approaches correlating the inputs and outcomes for the model in a compatible manner were flawed but even more necessary in the development road, returning useful information about the compatibility status.

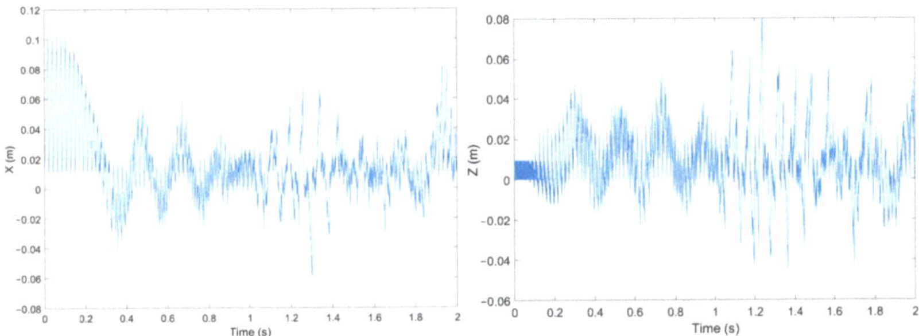

Figure 8. PneuNet displacement on a rapid driving force (X—left; Z—right).

One possibility to reduce the shock effect is to expand the actuation over a more significant period, giving a more extended actuation period for each continuum interval. The following experiment uses the exact strain actuation of the SMA model but extends over a longer period, from 2 s to 20 s as in Figure 9. The figure below depicts the strain input on the alloy model and the force outputted to be used as PneuNet's driving force.

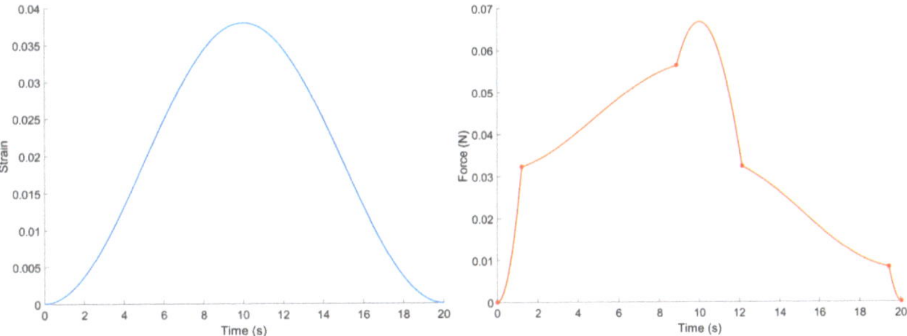

Figure 9. SMA signals: Slower strain command (**left**) and outputted force (**right**).

As proven in [25], the expected behavior is an evolution with minimum oscillations, avoiding elastic shock–vibrational proprieties. A consequence of the extended intervals is a slower axial evolution, returning a more complex motion. In addition, the actuation returns a stable dynamic around the 10 s moment as a response to a negative oscillatory evolution by increasing the force. It is clear that the SMA response is directly reactive to the SMA outputted force and contains the specific PneuNet proprieties. The resulting displacements on the axis are displayed in Figure 10.

Another aspect is represented by the observable fast response, presented by the experiments in Figure 10, where the Z-axis displacement has a constant initial increase, similar to an S-shaped function. The PneuNet outcome is an elastically damped response to the first interval (Figure 9 Right), resulting in a combined effect of the soft hybrid actuator.

Another experiment is realized with a strain input at half amplitude, returning a continuum pulling force. As displayed in Figure 11, the force evolution presents only two discontinuity points necessary for three interval approximations. Still, more importantly, it describes a steady force increase but at half the amplitude.

By comparing with the experiments described in [25], the evolution on the first interval is expected to be also oscillatory; the time given for the force to reach a specific value, after which a slower variation occurs, is almost instantaneous. Although this can cause an unwanted oscillation, the effect should be damped by the slow and steady increase over time (over 10 s) before the start of the decreasing ramp.

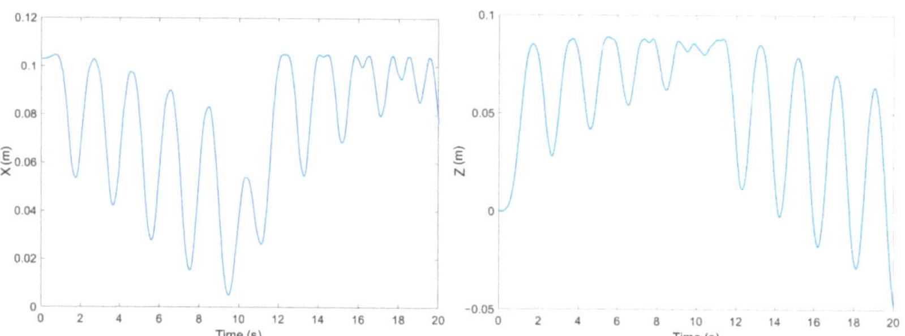

Figure 10. PneuNet displacement on a slower driving force (X—left; Z—right).

On the second interval, the force is quickly decreasing, and considering also the gravitational pull on the PneuNet, it is expected to perform an amplified fall-like behavior. As the fall occurs, the final force's level will rise after a decrease below the force's resulting value. For the last interval, the displacement is expected to be stabilize at a particular position, oscillating around it. The results are observable in Figure 12.

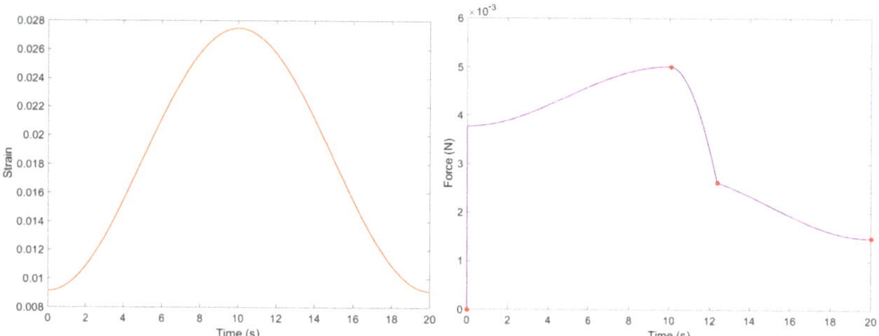

Figure 11. SMA signals: Wider strain command (**left**) and outputted force (**right**).

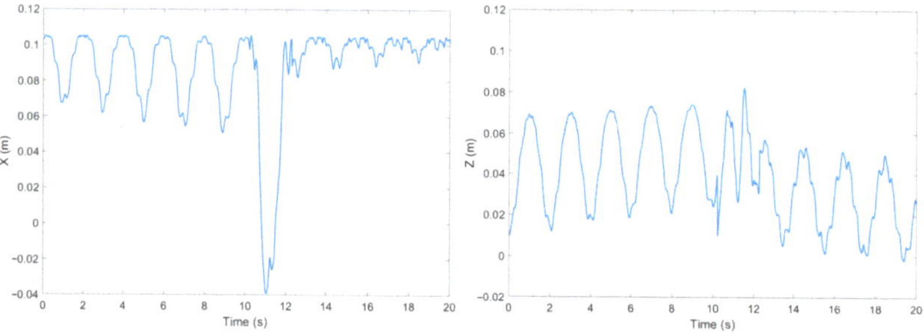

Figure 12. PneuNet displacement on for previous driving force (X-left and Z-right).

The displayed images confirm the supposition by the displacements presented on both axis evolutions. It is important to note that the force amplitude driving the PneuNet decreased compared to the previous experiments. The consequence was the decreased Z displacement and damping of the oscillations.

In the current experiment, the relation between the SMA-generated force directly influencing the PneuNet behavior and the oscillatory dynamics produced by the elastic characteristic. Each force interval described above generates a specific outcome, combining the actuation speed of the SMA and the PneuNet shape recovery propriety.

Corroborating the experiments described above, important ideas can be extracted. One essential aspect is the generated actuation force from the SMA, presented as the PneuNet driving signal acting antagonistically to the elastic shape recovery propriety and towards a damped, smoothed displacement. In the presented experiments, we validated that the hybrid architecture benefits from a superposition behavior regarding the advantages of each component. This result allows for future development in both modeling and practical applications.

6. Conclusions and Future Work

The present work proves that soft robotics is at the beginning of the exploratory phase and still has vast areas to uncover and understand. The proposed architecture shows the potential of hybrid soft actuation in the performance enhancement of the individual components. The concept proves that more complex structures could be tested and verified and that the numerical representations of soft components are inter-connectable using slight compatibility regarding scales and approximations.

In the detailed analysis, the hybrid architecture uses mathematical models that are approximations of ideal physical components. One critical remark is that the numerical data (namely the force signal) transferred from one actuator to the other are suitable for polynomial approximation. Based on this, the resulting dynamical behavior shows enhanced performance compared to individual components' performance.

The experimental results highlight that the behavior of the force generated by the SMA wire model can drive the PneuNet integrated model. The resulting displacements contain a combined influence of the component proprieties. The elastic characteristics dampen the SMA wire's fast responses, while the oscillatory dynamic of the PneuNet is reduced by the constant force maintained by the SMA.

The presented procedure proves functional for the test case but can be extrapolated and used for different hybrid structures. The only required consideration is the conversion between the physical data, creating physical quantity compatibility and, if necessary, converting the amplitude from one actuator to the other, representing a transfer element in the real system.

Regarding future work, one of the main goals will be to integrate elements from fluid mechanics and thermodynamics in a hybrid architecture. Also, sensing elements can be used to widen the application areas. The resulting schemes would allow for the simulation of multiple case studies with potential physical implementation. This can provide novel actuation types, hybrid structures, and more complex testing scenarios.

Last, but not least, the required work to firmly conclude the hybrid performance is to physically implement and test the simulated structure while measuring and evaluating data and outcomes. Although the real testing plant could be influenced by material imperfections and various surrounding thermal or positioning conditions, the continuous phenomena at the material level of the SMA and PneuNet, the effect discretized in modeling, could prove to be a more fluent combined behavior.

Author Contributions: Conceptualization, both; methodology, both; software and validation, F.-A.B.; formal analysis, A.B.; writing—original draft preparation, F.-A.B.; writing—review and editing, both; supervision A.B. All authors have read and agreed to the published version of the manuscript.

Funding: This research received no external funding

Data Availability Statement: The data supporting this study's findings are available from the first author or corresponding author upon reasonable request.

Conflicts of Interest: The authors declare no conflicts of interest.

References

1. Troccaz, J.; Dagnino, G.; Yang, G.Z. Frontiers of medical robotics: From concept to systems to clinical translation. *Annu. Rev. Biomed. Eng.* **2019**, *21*, 193–218. [CrossRef] [PubMed]
2. Kheirikhah, M.M.; Rabiee, S.; Edalat, M.E. A review of shape memory alloy actuators in robotics. In Proceedings of the RoboCup 2010: Robot Soccer World Cup XIV 14, Singapore, 25 June 2011; pp. 206–217.
3. El-Atab, N.; Mishra, R.B.; Al-Modaf, F.; Joharji, L.; Alsharif, A.A.; Alamoudi, H.; Diaz, M.; Qaiser, N.; Hussain, M.M. Soft actuators for soft robotic applications: A review. *Adv. Intell. Syst.* **2020**, *2*, 2000128. [CrossRef]
4. Xavier, M.S.; Fleming, A.J.; Yong, Y.K. Finite element modeling of soft fluidic actuators: Overview and recent developments. *Adv. Intell. Syst.* **2021**, *3*, 2000187. [CrossRef]
5. da Veiga, T.; Chandler, J.H.; Lloyd, P.; Pittiglio, G.; Wilkinson, N.J.; Hoshiar, A.K.; Harris, R.A.; Valdastri, P. Challenges of continuum robots in clinical context: A review. *Prog. Biomed. Eng.* **2020**, *2*, 032003. [CrossRef]
6. Park, Y.L.; Santos, J.; Galloway, K.G.; Goldfield, E.C.; Wood, R.J. A soft wearable robotic device for active knee motions using flat pneumatic artificial muscles. In Proceedings of the 2014 IEEE International Conference on Robotics and Automation (ICRA), Hong Kong, China, 31 May–7 June 2014; pp. 4805–4810.
7. Tabasi, S.F.; Banihashemi, S. Design and mechanism of building responsive skins: State-of-the-art and systematic analysis. *Front. Archit. Res.* **2022**, *11*, 1151–1176. [CrossRef]
8. Heng, W.; Solomon, S.; Gao, W. Flexible electronics and devices as human–machine interfaces for medical robotics. *Adv. Mater.* **2022**, *34*, 2107902. [CrossRef] [PubMed]
9. Pan, M.; Yuan, C.; Liang, X.; Dong, T.; Liu, T.; Zhang, J.; Zou, J.; Yang, H.; Bowen, C. Soft actuators and robotic devices for rehabilitation and assistance. *Adv. Intell. Syst.* **2022**, *4*, 2100140. [CrossRef]
10. Stokes, A.A.; Shepherd, R.F.; Morin, S.A.; Ilievski, F.; Whitesides, G.M. A hybrid combining hard and soft robots. *Soft Robot.* **2014**, *1*, 70–74. [CrossRef]
11. Golgouneh, A.; Beaudette, E.; Woelfle, H.; Li, B.; Subash, N.; Redhouse, A.J.; Jones, M.; Martin, T.; Lobo, M.A.; Holschuh, B.; et al. Design of a hybrid SMA-pneumatic based wearable upper limb exoskeleton. In Proceedings of the 2021 ACM International Symposium on Wearable Computers, Virtual, 21–26 September 2021; pp. 179–183.
12. Allen, E.A.; Swensen, J.P. Design of a Highly-Maneuverable Pneumatic Soft Actuator Driven by Intrinsic SMA Coils (PneuSMA Actuator). In Proceedings of the 2020 IEEE/RSJ International Conference on Intelligent Robots and Systems (IROS), Las Vegas, NV, USA, 24 October–24 January 2020; pp. 8667–8672.
13. Shibly, H.; Söffker, D. Mathematical models of shape memory alloy behavior for online and fast prediction of the hysteretic behavior. *Nonlinear Dyn.* **2010**, *62*, 53–66. [CrossRef]
14. Walker, J.; Zidek, T.; Harbel, C.; Yoon, S.; Strickland, F.S.; Kumar, S.; Shin, M. Soft robotics: A review of recent developments of pneumatic soft actuators. *Actuators* **2020**, *9*, 3. [CrossRef]
15. Brojan, M.; Bombač, D.; Kosel, F.; Videnič, T. Shape memory alloys in medicine Materiali z oblikovnim spominom v medicini. *RMZ-Mater. Geoenviron.* **2008**, *55*, 173–189.
16. Crews, J.H.; Smith, R.C.; Pender, K.M.; Hannen, J.C.; Buckner, G.D. Data-driven techniques to estimate parameters in the homogenized energy model for shape memory alloys. *J. Intell. Mater. Syst. Struct.* **2012**, *23*, 1897–1920. [CrossRef]
17. Zhuo, M. Timescale competition dictates thermo-mechanical responses of NiTi shape memory alloy bars. *Int. J. Solids Struct.* **2020**, *193*, 601–617. [CrossRef]
18. DYNALLOY Inc. Technical Characteristics of FLEXINOL® Actuator Wires. 2023. Available online: https://www.dynalloy.com/pdfs/TCF1140.pdf (accessed on 3 October 2023).
19. Bengisu, M.; Ferrara, M. *Materials That Move: Smart Materials, Intelligent Design*; Springer: Cham, Switzerland, 2018.
20. Abdullah, E.J.; Soriano, J.; Fernández de Bastida Garrido, I.; Abdul Majid, D.L. Accurate position control of shape memory alloy actuation using displacement feedback and self-sensing system. *Microsyst. Technol.* **2021**, *27*, 2553–2566. [CrossRef]
21. Kim, M.S.; Heo, J.K.; Rodrigue, H.; Lee, H.T.; Pané, S.; Han, M.W.; Ahn, S.H. Shape memory alloy (SMA) actuators: The role of material, form, and scaling effects. *Adv. Mater.* **2023**, *35*, 2208517. [CrossRef] [PubMed]
22. Gariya, N.; Kumar, P. A comparison of plane, slow pneu-net, and fast pneu-net designs of soft pneumatic actuators based on bending behavior. *Mater. Today Proc.* **2022**, *65*, 3799–3805. [CrossRef]
23. Brașoveanu, F.A. Soft Robotics Modeling and Simulation. In Proceedings of the IFToMM International Symposium on Science of Mechanisms and Machines (SYROM), Iasi, Romania, 17–18 November 2022; Springer: Cham, Switzerland, 2022; pp. 371–380.
24. Mathew, A.T.; Hmida, I.M.B.; Armanini, C.; Boyer, F.; Renda, F. Sorosim: A Matlab toolbox for hybrid rigid-soft robots based on the geometric variable-strain approach. *IEEE Robot. Autom. Mag.* **2022**, *30*, 106–122. [CrossRef]
25. Brașoveanu, F.A.; Burlacu, A. Soft Robotics: A Numerical Evaluation of Model-Based PneuNet Simulation. In Proceedings of the International Symposium on Measurements and Control in Robotics (ISMCR), Iasi, Romania, 21–22 September 2023; Springer: Cham, Switzerland, 2023.

Article

Hybrid Nursing Robot Based on Humanoid Pick-Up Action: Safe Transfers in Living Environments for Lower Limb Disabilities

Jiabao Li [1,2], Chengjun Wang [1,2,*] and Hailong Deng [3]

1   School of Artificial Intelligence, Anhui University of Science & Technology, Huainan 232001, China; 2022100093@aust.edu.cn
2   Anhui Artificial Intelligence Laboratory, Artificial Intelligence Research Institute of Hefei Comprehensive National Science Center, Hefei 230022, China
3   School of Mechanical Engineering, Anhui University of Science & Technology, Huainan 232001, China; 2021200068@aust.edu.cn
*   Correspondence: cjwang@aust.edu.cn; Tel.: +86-17355477119

Abstract: This research paper outlines the development of a modular and adjustable transfer care robot aimed at enhancing safe and comfortable transfers for individuals with lower limb disabilities. To design this robot, we utilized a 3D motion capture system to analyze the movements of a person assisting another person and to determine the necessary range of motion and workspace for the robot. Based on this analysis, we developed a 3-UPS + UPR parallel spreader to transfer the person receiving care. We also conducted kinematic and dynamic analyses of the parallel spreader to validate its operational space and to obtain the force change curve for the drive. To evaluate the robot's performance, we enlisted the help of ten volunteers with varying heights and weights. Our findings indicate that the pressure distribution during transfers remained largely consistent. Additionally, the surveys revealed that those receiving care perceived the robot as being capable of securely and comfortably transferring individuals between different assistive devices. This modular and adaptable transfer assistance robot presents a promising solution to the challenges encountered in caregiving.

Keywords: hybrid nursing robot; humanoid pick-up action; workspace; dynamical model; comfort analysis

Citation: Li, J.; Wang, C.; Deng, H. Hybrid Nursing Robot Based on Humanoid Pick-Up Action: Safe Transfers in Living Environments for Lower Limb Disabilities. *Actuators* **2023**, *12*, 438. https://doi.org/10.3390/act12120438

Academic Editors: Ioan Doroftei and Bing Li

Received: 9 October 2023
Revised: 20 November 2023
Accepted: 23 November 2023
Published: 24 November 2023

1. Introduction

With the aggravated aging of the population, the number of disabled/semi-disabled persons is increasing, which causes many social problems, such as the lack of care and increased bodily injuries to nursing persons [1–3]. For disabled/semi-disabled persons, nursing services include transfer support, excretion support, guardian support, and spiritual consolation, wherein transfer support poses the most serious risk of bodily injury to the nursing person [4,5]. Transfer support refers to the heavy physical work of assisting or carrying a disabled person to transfer them between living appliances such as a bed, wheelchair, a close stool, or sofa. It is a key factor that causes home nursing difficulty, disharmonious family relationships, and other social problems [6].

The common transfer carriers are divided into direct and indirect carriers [7]. Direct carriers adopt a bed–wheelchair integrated design. Based on the Internet of Things (IoT), Muangmeesri and Wisaeng [8] designed an intelligent wheelchair with a health monitoring function. Zhou et al. [9] designed an intelligent nursing bed that integrates the functions of a traditional nursing bed and a powered wheelchair. Lang et al. [10] developed a smart nursing bed designed to facilitate automatic defecation for patients. Nevertheless, the expensive and non-transferable equipment transfer approach poses limitations, as it does not enable transfers between various devices. At present, the indirect carrier is the popular transfer carrier. Wang et al. [11] designed a power-driven transfer device that uses a sling to lift and move a nursed person. It is simple and cheap, but it is less comfortable and can easily cause secondary injury. The RI-MAN humanoid bilateral-arm rehabilitation robot [12] designed by the RIkagaku KENkyusho/Institute of Physical and Chemical Research has 19 degrees of freedom (DOFs) and contains two coupled six-DOF robotic arms and multiple sensors for the rapid perception of the external environment. The second generation of the robot nursing assistant (RoNA) launched by HStar [13,14] adopts a tandem elastic drive system that effectively guarantees the safety and flexibility of two arms and has a loading capacity of 227 kg. It is also equipped with a three-dimensional (3D) intelligent navigation system. Its shortcoming is that it requires the assistance of two medical care personnel and is thus less efficient. Zhang et al. [15] created a nursing robot with an aluminum alloy structure by focusing on the center of gravity. This robot can move safely and efficiently along a predefined path between two work points. However, the use of motor-driven arm joints proves to be expensive and unsuitable for widespread adoption. Liu et al. [16] developed a humanoid back-holding transfer rehabilitation robot that uses a three-DOF chest support to support the chest of a nursing person while it carries the nursed person on its back for transfer. However, this robot is complex and large. To sum up, the existing transfer equipment, aimed at ensuring the safety and comfort of care, comes with elevated costs and equipment complexity. This high price tag poses challenges for the broader population to afford such equipment [17–19].

Simultaneously, the emotional well-being of an individual receiving care and their comfort during the transfer process play significant roles in caregiving. Previous studies [12–14] have employed humanoid robots, demonstrating that human-like movements can enhance the emotional connection between the robot and the individual while improving comfort. However, none of the aforementioned studies analyze the movements of a person embracing another individual, instead opting for a costly robotic arm with multiple motors.

To create a transfer care robot that is safe, comfortable, and cost-effective while mimicking the anthropomorphic hugging and transferring motions, our approach involved using a 3D motion capture system to analyze human hugging movements and determine the necessary range of motion and workspace for the robot. Building on this analysis, we developed a hybrid transfer robot equipped with a 3-UPS + UPR parallel spreader designed for optimal structural stability. To ensure the functionality and safety of the designed robot, we conducted kinematic and dynamic analyses. Subsequently, we fabricated a prototype and conducted experiments with recruited volunteers. The results demonstrated the robot's ability to comfortably transfer caregivers between chairs, beds, sofas, wheelchairs, and other everyday living appliances. This success suggests the potential for the widespread application of this innovative robot.

2. Analysis of the Pick-Up Action

In this study, we employed a 3D motion capture system to determine the necessary degrees of freedom for our mechanism's design and its range of motion. Additionally, we conducted tests involving several individuals who require care to establish the required workspace, tailored to the needs of a wide range of individuals. The specific body parameters for individuals who can benefit from this robot's assistance are outlined based on the inertial parameters defined by the General Administration of Quality Supervision, Inspection and Quarantine of the People's Republic of China, China National Committee for Standardization [20]. These parameters include a height range of 1500 mm to 1880 mm and a weight range of 50 KG to 90 KG.

2.1. Design of the Pick-Up Test

In this test, a NOKOV optical 3D motion capture system was used to capture the actions of the nursed person in the pick-up process [21,22]. According to the standard human skeleton model [23] given in the ISO/IEC19774-Human animation (H-Anim) standard, "r_shoulder", "l_shoulder", "vl5", and "sacroiliac" in the model were selected as the markers (Figure 1). As shown in the figure, a nursed person sat on a chair, and a nursing person picked up the nursed person and moved them by walking.

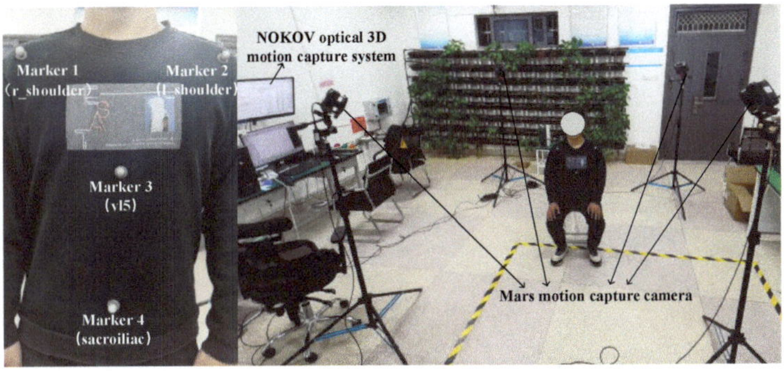

Figure 1. Test arrangement for the pick-up action.

The nursed persons were five volunteers with different heights and weights, as shown in Table 1. The first nursing person was a male, aged 23, 1750 mm in height, and 85 kg in weight. To avoid daytime sunshine from affecting the motion capture, the test was conducted at night. Before the test, the nursed persons and the nursing person did not drink, but they had a good rest. In addition, to prevent the physical fatigue caused by the continuous tests from affecting the pick-up action, the nursed persons and the nursing person were asked to take a 10 min break before participating in the next set of tests.

Table 1. Information of the nursed persons.

	Height, mm	Weight, kg	Sex	Age
Nursed person 1	1520	53	Female	23
Nursed person 2	1650	70	Male	24
Nursed person 3	1700	72	Male	24
Nursed person 4	1750	78	Male	29
Nursed person 5	1810	85	Male	35

2.2. Analysis of the Pick-Up Action

The pick-up and transfer actions, illustrated in Figure 2, included four stages: approaching, picking up, steady movement, and lowering. First, the nursed person sat on a chair, and the nursing person approached the nursed person and bent to an appropriate position. Second, the nursing person maintained their leg posture when approaching the nursed person, put their arms on the back and thighs of the nursed person, and then stood up. In this process, the back angle of the nursed person changed a little. Third, the nursing person adjusted the posture of the nursed person to a "comfortable" state and leaned the body of the nursed person backward. Fourth, the nursing person leaned the nursed person's body frontward and slowly lowered them down into the chair.

A coordinate system was established by considering the body of the nursed person as a rigid rod and considering the projection center of the chair on the ground as the origin (Figure 3). Using an example of a person being cared for who has a height of 1700 mm and a weight of 72 kg, the entire lifting process takes 18 s. During this process, we observed the change in the angle between the cared-for person's body and the X-axis, as illustrated in Figure 4. The nursed person's subjective feeling in the pick-up process was affected by their posture and the change in the interaction between the two persons due to the posture variation; when the said angle was within $10°–25°$, the nursed person felt "comfortable" as the forces on their back, legs, and other contact parts were acceptable [24].

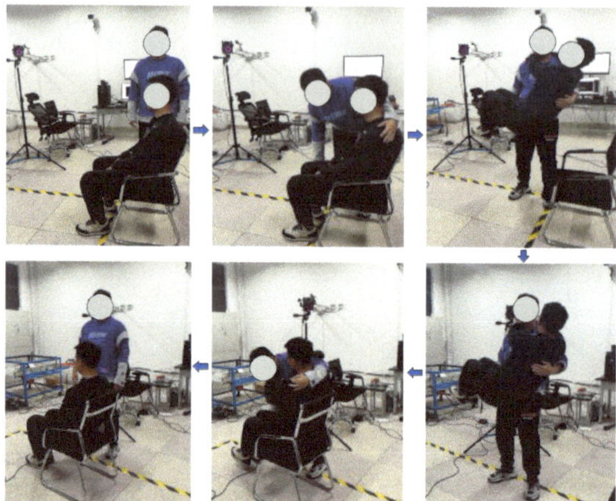

Figure 2. Pick-up actions performed in the pick-up test.

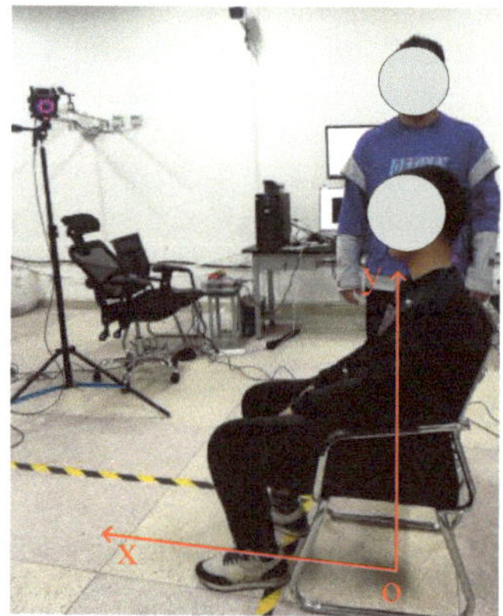

Figure 3. Coordinate system of the pick-up action.

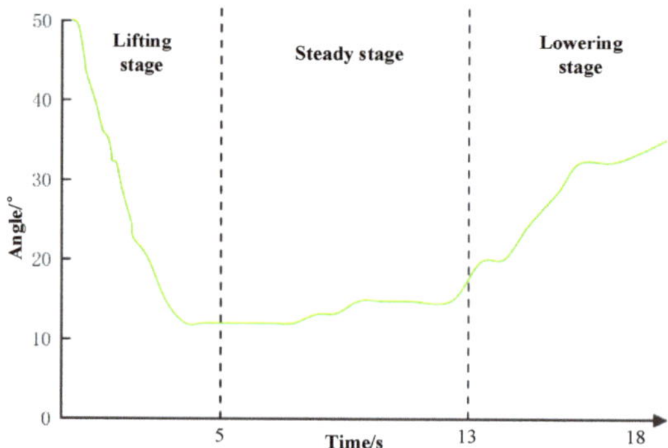

Figure 4. Variation of the body angle.

Figure 5 provides the minimum trajectory ranges of the markers on the five nursed persons. The vertical displacement of each marker was nondimensionalized by the corresponding nursed person's height as the denominator. The smallest trajectory range covers

640 mm horizontally and 280 mm vertically. These minimum trajectory ranges are crucial for designing the workspace of the robot. A requirement is that the workspace of the sling used in this innovative design should be greater than the minimum trajectory range.

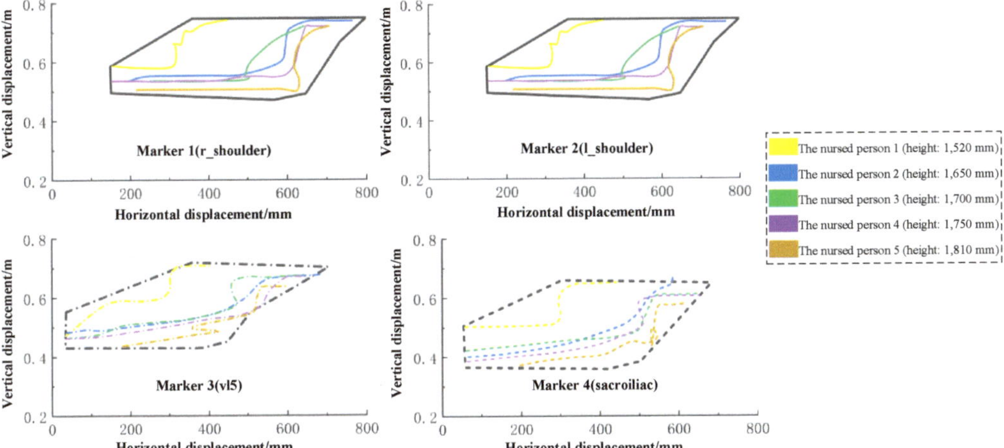

Figure 5. Minimum trajectory ranges of the markers on the nursed persons.

The test results revealed that the forces on the nursed person changed with the change in the nursing person's posture, which directly affected their comfort. Therefore, based on the features of the pick-up action, the robotic pick-up sling should have three DOFs (for horizontal movement, vertical movement, and rotation), as shown in Figure 6.

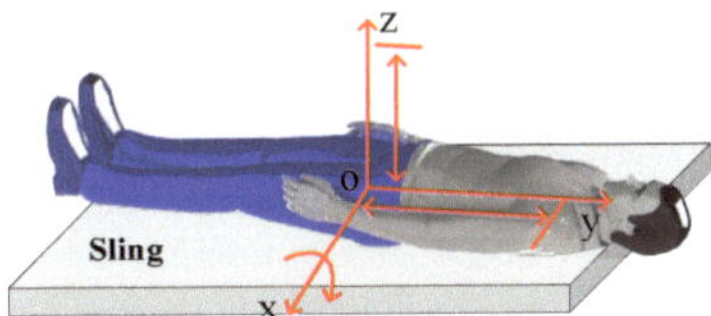

Figure 6. Concept map of the robotic sling.

3. Hybrid Robot: Structure and Functionality

As illustrated in Figure 7, The modular hybrid robot, described in this paper, incorporates several components including a mobile chassis, a 3UPS + UPR parallel spreader, a linear skid module, a binocular vision system, a protective hood, a slewing device, and a lifting device. Its dimensions are 1000 mm × 700 mm × 1750 mm, supporting a load capacity of 50 KG to 90 KG. These specifications stem from an in-depth analysis of the physical characteristics of the intended user group, detailed in Section 2. The focus of this research centers on the 3-UPS + UPR parallel spreader.

The 3-UPS + UPR's three-dimensional movement is facilitated by adjusting the lengths of the four electrically driven rods. These rods offer a motion range of 0–350 mm, a maximum thrust of 2500 N, and a telescopic speed of 7 mm/s. Coupled with the horizontal slide movement, this mechanism allows for precise positioning relative to the caregiver. The module's design prioritizes structural integrity and stability, effectively replacing traditional multi-joint robotic arms. This aspect is crucial in minimizing the risk of mechanical failures during operation, thereby safeguarding the user's safety during transfer processes.

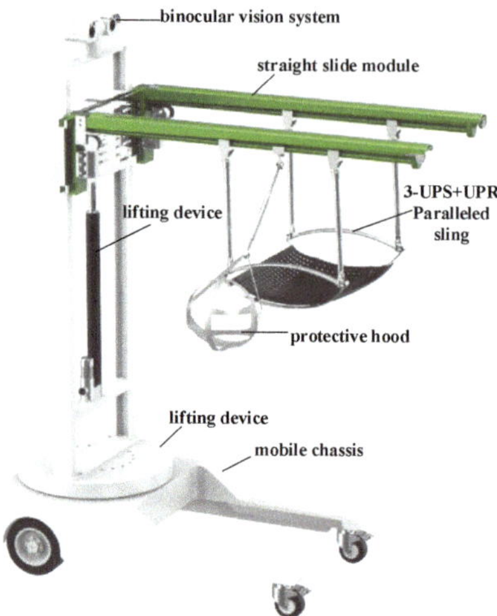

Figure 7. The series-parallel transfer nursing robot.

The work process of this robot is as follows: when detecting a nursed person, the binocular vision system moves the mobile chassis to a designated position and adjusts the 3-UPS + UPR paralleled sling to a position that is wearable for the nursed person through the rotatory device, lifting device, and straight pulley block modules; when the nursed person wears the sling well, the paralleled sling can pick up the nursed person and adjust the postures until the transfer nursing process is finished.

In response to the requirements identified during the analysis of pick-up and carry movements, the 3-UPS + UPR parallel spreader is designed to offer three essential degrees of freedom: horizontal movements, vertical movements, and rotational capability. This design specifically addresses the diverse mobility needs of individuals with lower limb disabilities.

As illustrated in Figure 8, the 3-UPS + UPR paralleled mechanism is composed of a fixed platform, a mobile platform, three identical UPS branches (hooke joint (U), sliding pair (P), and spherical joint (S)), and a UPR branch. The four branches are distributed like a square on the fixed platform.

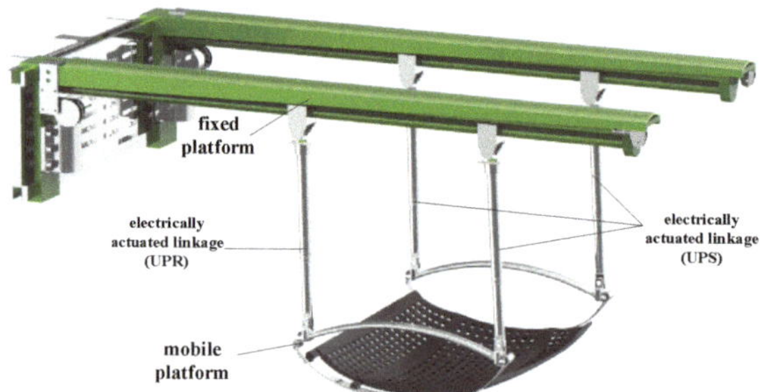

Figure 8. The 3-UPS + UPR paralleled sling.

4. Kinematic Analysis

To confirm whether there is a sudden change in the driving force of the parallel mechanism during the holding motion, which could be potentially hazardous, a kinematic analysis is required. This analysis will help to derive the position inverse solution and the Jacobian matrix for the parallel mechanism, establishing the groundwork for a subsequent dynamic analysis. Simultaneously, it is essential to determine the workspace to ensure that it exceeds the minimum trajectory range required for the cared-for person.

4.1. Inverse Solution for Positions

To pick up a person with the robotic 3-UPS + UPR paralleled sling, the kinematic parameters of the sling were calculated based on the trajectory of the humanoid pick-up motion. Further, the topological structure of the 3-UPS + UPR paralleled sling was established, as shown in Figure 9.

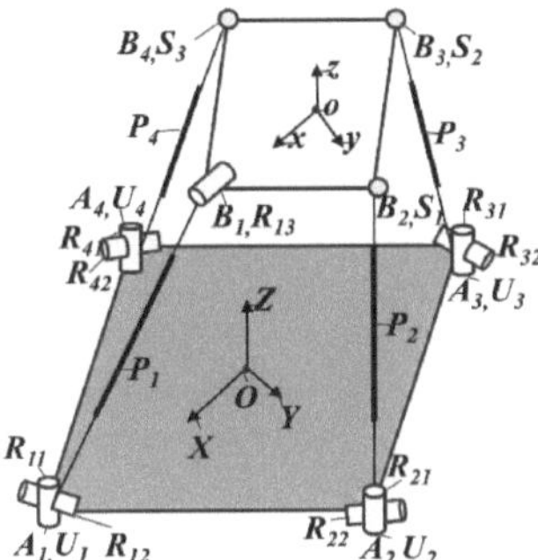

Figure 9. The topological structure of the 3-UPS + UPR paralleled sling mechanism.

A base coordinate system (O) is established at the center of the fixed platform, where the center of the square is the origin O, the X-axis is parallel to line A_1A_3, the Y-axis is parallel to line A_2A_4, and the Z-axis is vertical to face $A_1A_2A_3A_4$; a moving coordinate system (o) is established at the center of the mobile platform, with the origin o at the center of the square mobile platform, the x-axis parallel to line B_1B_3, the y-axis parallel to line B_2B_4, and the z-axis vertical to face $B_1B_2B_3B_4$. In this paper, the first shaft $R_{i1}(i = 1,2,3,4)$ of each hooke joint is along the Z-axis direction and vertical to the second shaft $R_{i2}(i = 1,2,3,4)$, the second shaft R_{i2} of each hooke joint is vertical to the driving pair Pi of each branch, the second shaft R_{12} of the hooke joint in the UPR branch is vertical to the R_{13}, and the rotating axis of R_{13} is parallel to the x-axis of the mobile platform.

As for the inverse kinematic of the position of the 3-UPS + UPR mechanism, the length vector l_i of the drive rod in each branch of the mechanism was determined by using a rotation matrix and constraint equation, given that the length from the center to the vertex of the fixed platform is L, the length from the center to each vertex of the mobile platform is l, and the pose of the center o of the mobile platform relative to the fixed coordinate system is (x, y, z, α) For the 3-UPS + UPR paralleled mechanism, the constraint equation is determined, as shown below, based on the geometrical relationships of the revolving pairs $R_{12}\perp A_1B_1$, $R_{12}\perp R_{13}$, $R_{11}\perp R_{12}$, and $R_{13}\perp A_1B_1$:

$$\begin{cases} R_{12} \times A_1B_1 = 0 \\ R_{12} \times R_{13} = 0 \\ R_{13} \times A_1B_1 = 0 \\ R_{12} \times R_{11} = 0 \end{cases} \tag{1}$$

The variables k_i, h_i, l_i, and $\overrightarrow{r}$ are set, representing the unit direction vectors of oA_i, oB_i, A_iB_i, and oo in the base coordinate system, and e_i is set as the length of the revolving pair, where

$$\begin{cases} k_1 = (L,0,0)' \\ k_2 = (0,L,0)' \\ k_3 = (-L,0,0)' \\ k_4 = (0,-L,0)' \end{cases} \tag{2}$$

The pose transformation matrix of each vertex Bi on the mobile platform in the base coordinate system is

$$R = \begin{bmatrix} \cos\theta & -\sin\theta & 0 \\ \sin\theta & \cos\theta & 0 \\ 0 & 0 & 1 \end{bmatrix} \tag{3}$$

According to the position transformation relation, the direction vector of the vector h_i in the base coordinate system (O) can be expressed as

$$\begin{cases} h_1 = {}^{o}_{o_1}R(l,0,0)' \\ h_2 = {}^{o}_{o_1}R(0,l,0)' \\ h_3 = {}^{o}_{o_1}R(-l,0,0)' \\ h_4 = {}^{o}_{o_1}R(0,-l,0)' \end{cases} \tag{4}$$

Hence, the coordinates of each vertex on the mobile platform in the base coordinate system are obtained as follows: D is the coordinate of a vertex on the mobile platform in the base coordinate system, R is the pose transformation matrix of each vertex B_i on the mobile platform in the base coordinate system, m is the coordinate of the vertex on the mobile platform in the moving coordinate system, and P is the coordinate of the vertex on the mobile platform in the base coordinate system.

$$^{o}D_i = Rmm_i + P \tag{5}$$

Thus, the vector equation of each branched mobile chain in the 3-UPS + UPR paralleled mechanism is established as shown below:

$$l_i = l_i e_i = \vec{r} + h_i - k_i \tag{6}$$

In combination with the constraint equation, the rotation matrix, and the vector equations of the branches, the position inverse kinematic expression of the 3UPS + UPR paralleled mechanism is obtained as shown below:

$$\begin{cases} l_1 = \pm\sqrt{L^2 - (y + l\sin\theta - a) - z^2} + (x + l\cos\theta) \\ l_2 = \pm\sqrt{L^2 - (x - l\sin\theta - a) - z^2} + (y + l\cos\theta) \\ l_3 = \pm\sqrt{L^2 - (y - l\sin\theta - a) - z^2} + (-x - l\cos\theta) \\ l_4 = \pm\sqrt{L^2 - (x + l\sin\theta + a) - z^2} + (-y - l\cos\theta) \end{cases} \tag{7}$$

4.2. Solution of the Jacobian Matrix

The first-order influential factor matrix of the mechanism was calculated using the differential transform method based on the symbolic operation. Then, the first-order motion influential factors were used to solve the velocity Jacobian matrix of the 3-UPS + UPR paralleled mechanism. According to the motion feature of the mechanism, the length of the drive rod can be expressed as

$$l_i = \sqrt{(l_i e_i)^T (l_i e_i)} = f_i(z, \alpha, \beta) \tag{8}$$

By calculating the derivative of time based on the above equation, the velocity of the drive rod in each branched chain is obtained as follows:

$$\dot{l_i} = \frac{\partial f_i(z, \alpha, \beta)}{\partial z}\dot{z} + \frac{\partial f_i(z, \alpha, \beta)}{\partial \alpha}\dot{\alpha} + \frac{\partial f_i(z, \alpha, \beta)}{\partial \beta}\dot{\beta} \tag{9}$$

Hence, the mapping relationship between the velocity of the mobile platform and the driving velocity can be expressed as

$$
\begin{bmatrix} \dot{l}_1 \\ \dot{l}_2 \\ \dot{l}_3 \\ \dot{l}_3 \\ \dot{l}_4 \end{bmatrix} = \begin{bmatrix} \frac{\partial f_1}{\partial z} & \frac{\partial f_1}{\partial \alpha} & \frac{\partial f_1}{\partial \beta} \\ \frac{\partial f_2}{\partial z} & \frac{\partial f_2}{\partial \alpha} & \frac{\partial f_1}{\partial \beta} \\ \frac{\partial f_3}{\partial z} & \frac{\partial f_3}{\partial \alpha} & \frac{\partial f_1}{\partial \beta} \\ \frac{\partial f_4}{\partial z} & \frac{\partial f_4}{\partial \alpha} & \frac{\partial f_1}{\partial \beta} \end{bmatrix} \begin{bmatrix} \dot{z} \\ \dot{\alpha} \\ \dot{\beta} \end{bmatrix} \tag{10}
$$

Let $\dot{l} = \begin{bmatrix} \dot{l}_1 & \dot{l}_2 & \dot{l}_3 & \dot{l}_4 \end{bmatrix}^T$ and $V_o = \begin{bmatrix} \dot{z} & \dot{\alpha} & \dot{\beta} \end{bmatrix}^T$. Then,

$$
\dot{l} = JV_o \tag{11}
$$

where $\dot{l}$ is the velocity of the drive rod, J is the 1-order influential factor matrix of the mechanism, which is written as $J = [J_1 \, J_2 \, J_3 \, J_4]^T$, and V_o is the motion velocity of the mobile platform. The Jacobian matrix of the 3-UPS + UPR paralleled mechanism has four rows and three columns.

4.3. Analysis of the Workspace

The pose adjustability of the paralleled sling depends on the workspace of the paralleled mechanism. To realize "global" pose adjustment for a nursed person, the reachable workspace of the mechanism can be calculated with a numerical method and analytical method. However, the latter method requires an explicit expression of the position solution of the mechanism and has a complicated calculation process and high cost. Hence, the numerical method was used to calculate the flexible workspace of this mechanism.

The workspace of the 3-UPS + UPR paralleled sling is affected by the structural dimension of the mechanism, the rotation angle limit of the revolving pair, and the interference of the connecting rods. The dimensions of the platforms in the mechanism are listed in Table 2.

Table 2. Geometric and inertial parameters of the mechanism.

Simulated Parameter	Value	Unit
L, l, z	0.35, 0.30, 0.32	m
l_{mi} $(i = 1, 2, 3, 4)$	0.32	m
m_o, m_{li} $(i = 1,2,3,4)$	4.035, 1.524	kg
I_o	diag [0.0801 0.0425 0.1225]	kg·m^2
I_{li}	diag [0.0829 0.0639 0.0211]	kg·m^2
f_o	$[0\ 0\ {-}120]^T$	N
n_o	$[0\ 50\ 0]^T$	N·m

The constraint conditions are as follows: (a) the travel of the driving pair ranges within 320 mm–600 mm; (b) the rotation angles (α, β, and λ) of the hooke joint and spherical joint are limited to $\pm 25°$.

The following displays the process flow in calculating the workspace of the paralleled mechanism with the numerical method:

1. Set the pre-workspace of the mechanism and divide the uniform step size equally.
2. Substitute the position parameters and posture rotation angle in the workspace into the constraint equation and establish a system of nonlinear equations.
3. Use the homotropy method to obtain multiple groups of numerical solutions, then compare this set of solutions with the preset constraint conditions, and screen out the correct solution.

4. All solutions satisfying the constraints in the pre-workspace are reachable workspaces, as presented in Figure 10.

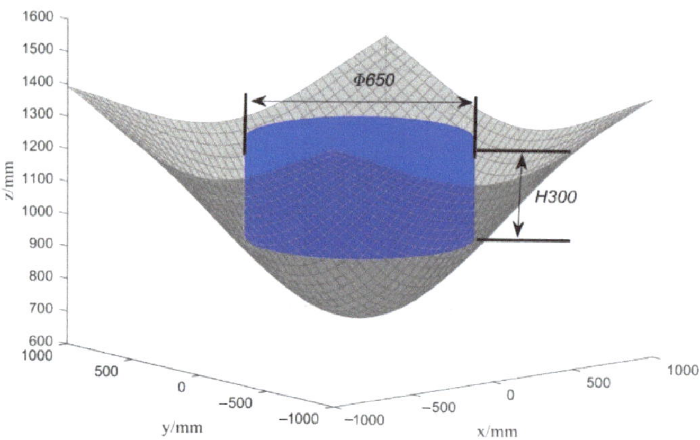

Figure 10. The workspace of the 3-UPS + UPR paralleled mechanism.

Meanwhile, the SoildWorks software was used to verify the correctness of the workspace of the paralleled mechanism based on the numerical solution, with the following major process flow:

1. Establish a 3-UPS + UPR paralleled mechanism model in the SoildWorks software.
2. Read and tabulate the equal division points in step 1 of the numerical solution, and then use Visual Studio to compile an EXE file, and read the elongation of each rod and the angle of the shafts of the revolving pair while inputting all points in the table into the center of the mobile platform of the mechanism successively.
3. Delete the point locations that cannot be generated by SoildWorks, screen out the points satisfying the constraint equation, and reserve the remaining points.
4. Generate the remaining points into curves in SoildWorks and seam the curved surfaces, obtaining the workspace, as provided in Figure 11, and the volume parameter.

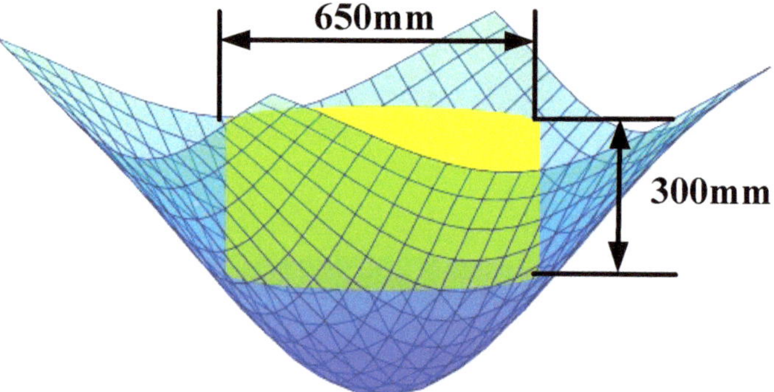

Figure 11. Simulated diagram of the workspace of the 3-UPS + UPR paralleled mechanism.

As discovered by comparing the workspace figure obtained based on the numerical solution with the simulated diagram, the simulated workspace is identical to the calculated workspace, which proves that the conducted work is correct. Further, the inscribed cylinder in the workspace, as shown in the reachable workspace figure, was calculated, obtaining a task workspace with a diameter of 650 mm and a height of 300 mm. The workspace of the robot exceeds the minimum trajectory required for the individual receiving care. This result verifies that the designed robot sling satisfies the transfer demands of most nursed persons.

5. Dynamical Analysis and Simulation

To ensure the safety of the transfer process for the individual receiving care and to avoid abrupt changes in the driving force, a dynamic analysis of the 3-UPS + UPR parallel mechanism is conducted using the principle of virtual work. The parameters for the moving platform position are determined based on the humanoid hold-and-take action, enabling us to derive the curve depicting changes in its driving force.

5.1. Dynamical Analysis

Performing a dynamical analysis of a paralleled mechanism based on the principle of virtual work is one of the analysis methods for such systems [25–27]. This method involves a high calculation speed and low complexity, without calculating the constraining force and moment. Thus, the principle of virtual work has been used to analyze the dynamics of the 3-UPS + UPR paralleled mechanism.

The mass, m_o, of the mobile platform; the external force, f_o, and the moment, n_o, on the barycenter of the mobile platform; and their references, f_o and n_o, in the fixed coordinate system are given as follows: $f_o = Rf_o$ and $n_o = Rn_o$.

Taking I_o as the inertial matrix of the mobile platform in the mobile coordinate system, the inertial matrix of the mobile platform in the fixed coordinate system is expressed as $I_o = RI_oR^T$.

The relationship between the force and moment on the barycenter of the mobile platform is expressed as

$$F_o = \begin{bmatrix} f_o + m_o g - m_o \dot{v}_o \\ n_o - I_o w_o - w_o \times (I_o w_o) \end{bmatrix} \tag{12}$$

When each branched chain only bears gravity, given the mass m_{li} of the drive rod in a branched chain, its inertial matrix in the branched system is represented as I_{li}. The force and moment on the branched chain are indicated as

$$F_u = \begin{bmatrix} m_u g - m_{ii} a_{mi} \\ -{}^o I_v w_v - w_v \times ({}^o I_v w_v) \end{bmatrix} \tag{13}$$

Based on Equations (12) and (13), the dynamical equation of the mechanism is established based on the principle of virtual work as follows:

$$\delta x_o^T F_o + \delta q^T F + \sum_{i=1}^{4} \delta x_u^T F_u = 0 \tag{14}$$

where δx_o is the virtual displacement of the mobile platform, F is the driving force, δq is the virtual displacement of the driving force, and δx_{li} is the virtual displacement corresponding to F_{li}.

The virtual displacement shown in Equation (14) and the generalized virtual displacement of the mechanism satisfy the geometric constraints of the 3-UPS + UPR mechanism itself, and their relationships can be denoted as $\delta q = J \delta x_o$ and $\delta x_{li} = J_{li} \delta x_o$. Equation (14) can be written as

$$\delta x_o^T \left(F_o + J^T F + \sum_{i=1}^{4} J_i^T F_u \right) = 0 \tag{15}$$

Equation (15) is satisfied for any δx_o. Hence,

$$F_o + J^T F + \sum_{i=1}^{4} J_{ii}^T F_{ii} = 0 \tag{16}$$

In Equation (16),

$$F = -(J^T) + (F_o + \sum_{i=1}^{4} J_{ii}^T F_{ii}) + F_o K \tag{17}$$

$(J^T)^+$, F_o, and K are the pseudo-inverse matrix of J^T, the base vector of the null space, and the base vector coefficient of the null space, respectively. In other words, Equation (17) can be treated as the dynamic model of this mechanism. For the 3-UPS + UPR redundancy parallel mechanism, Equation (17) comprises three linear equations, which, however, have four unknown numbers. Hence, the solutions of these equations are not unique. Based on the two-norm solution of the driving force, Equation (17) can be denoted as

$$F = -(J^T) + (F_o + \sum_{i=1}^{4} J_{ii}^T F_{ii}) \tag{18}$$

5.2. Numerical Simulation for Verification

The simulation performed using MATLAB and the ADAMS software was carried out to verify the validity of the kinematic and dynamical models established for the paralleled mechanism. The parameters of the 3-UPS + UPR paralleled mechanism (Table 2) were determined as per the work requirement and workspace of the transfer nursing robot.

Utilizing the trajectory of the humanoid pick-up motion, we made an assumption regarding the pose parameters of the mobile platform.

$$\begin{cases} \alpha(t) = 0.1\sin(\pi t/10)(\pi/50)\cos(\pi t/10) + \pi/60 \\ \beta(t) = 0.1\cos(\pi t/10) + 0.15 \\ z(t) = 0.32 + 0.5\cos(\pi t/10) \end{cases}$$

The driving force curves of the drive rods in the 3-UPS + UPR paralleled mechanism were calculated based on the motion trajectory of the mobile platform and the aforementioned kinematic and dynamical equations. Figure 12 shows the driving force curves calculated using Matlab, and Figure 13 shows the driving force curves simulated using ADAMS, where the step size was 0.2 s and the calculation time was 20 s. The results reveal that the driving force varied continuously and steadily such that the nursed person could be transferred steadily. The theoretically calculated results were almost equal to the simulated results, with a maximum error of 3%, which is allowable. Hence, it was proven that the dynamical model of the mechanism is valid.

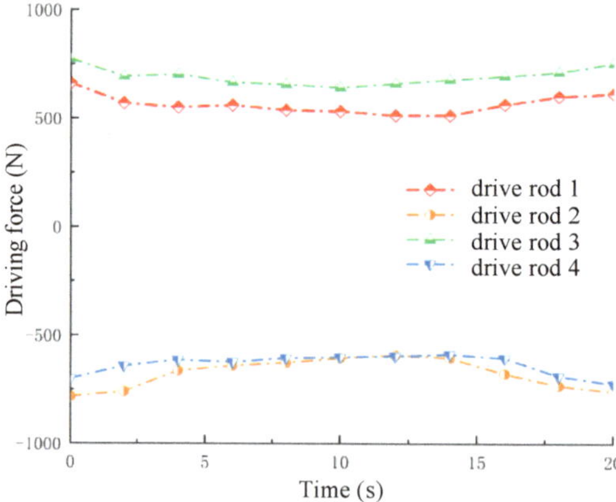

Figure 12. Theoretical variation curve of the driving force.

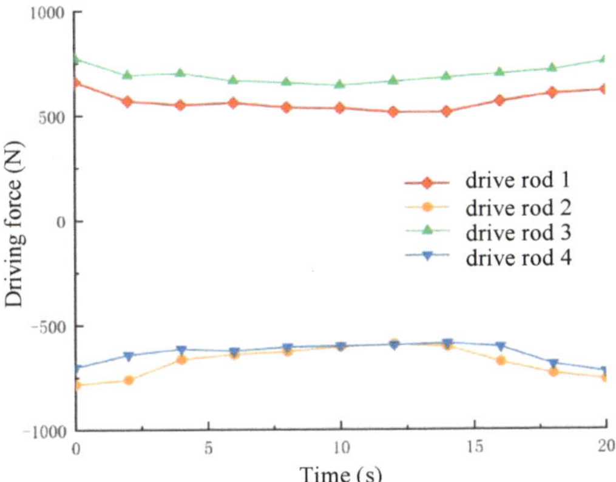

Figure 13. Simulated variation curve of the driving force.

6. Functional Test and Analysis of the Prototype

6.1. Functional Test

The series-parallel transfer nursing robot has been designed mainly for transferring a person with healthy upper limbs and disabled lower limbs between different living appliances. In this test, we selected ten young adults aged 20–35 with different heights and weights (Table 3), but with healthy upper limbs and disabled lower limbs, to experience the transfer function of the robot designed in this study.

Table 3. Information about the test subjects.

	Height, mm	Weight, kg	Sex	Age
Nursed person 1	1550	60	Male	23
Nursed person 2	1670	70	Male	24
Nursed person 3	1710	73	Male	24
Nursed person 4	1750	78	Male	29
Nursed person 5	1800	85	Male	35
Nursed person 6	1850	80	Male	26
Nursed person 7	1700	56	Female	26
Nursed person 8	1650	55	Female	25
Nursed person 9	1780	78	Male	24
Nursed person 10	1790	80	Male	25

The study was approved by the Biomedical Research Ethics Committee of Anhui University of Science and Technology (Anhui, China) in accordance with the Declaration of Helsinki. All methods were carried out in accordance with relevant guidelines and regulations. Written informed consent was obtained from all individual patients included in the study.

The transfer nursing process of a test subject with a height of 1750 mm and a 70 kg weight (the nursed person), as an example, is illustrated in Figure 14. At the beginning, the nursed person adjusted his sitting posture and the chair height as per his subjective feeling. Once the preparatory work was finished, in the course of nursing, the parallel spreader was shifted to the individual being cared for using the lifting device and the linear slide module. This movement was carried out in accordance with the position described in the previous section for adjusting the humanoid holding action mechanism. The nursing staff assisted in placing the individual on the spreader. The parallel spreader adapted the length of the individual drive rods based on the posture parameters of the humanoid lifting action, simulating the lifting action to raise the cared-for person. After that, the nursed person moved his head to the protective hood and laid flat, entering a steady convey stage

(Figure 14a–c). After approaching the ward bed, the sling was used to readjust the lengths of the drive rods and lower the nursed person down onto the bed slowly (Figure 14d–e). At this time, the transfer nursing was finished.

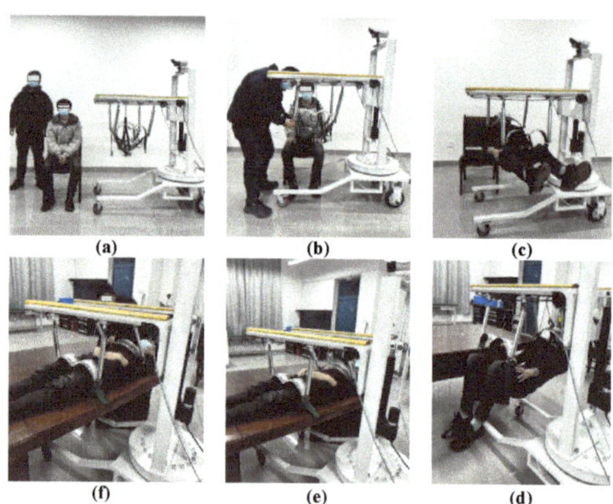

Figure 14. Transfer test on the robot. (**a**) starting transport (**b**) preparatory stage (**c**) Initial stage of transport (**d**) Medium-term transport (**e**) Post-transport (**f**) End of transport.

In this functional test, the robot completed the transfer of ten test objects successfully and the test objects also gave good feedback without adverse reactions.

6.2. "Comfort" Test

In the transfer process, the forces on the parts of the nursed person contacting the robot were the direct factors affecting their comfort. The pressure on their back changed with the change in the posture of the 3-UPS + UPR paralleled sling. The vibration in the movement also contributed to the pressure change. Therefore, arrayed flexible tactile sensors were provided on the contact parts (back) between the sling and the nursed person, and a pressure acquisition system was used to record the change in the pressure distribution on the backs of the ten test subjects during the transfer.

At the beginning of the test, the test subjects adjusted the chair height and position as per their specific heights and subjective feelings. Then, the robot's motion trajectory was designed based on each test subject's subjective feeling, respectively. According to the motion modes of the robot, the transfer was divided into four stages: the starting stage (the nursed person leaves the chair), the lifting stage, the steady stage, and the lowering stage. Figure 15 exhibits the pressure changes on the ten test objects in the four stages, where the maximum pressure was 26.6 kPa, the minimum pressure was 4.5 kPa, and the maximum variation was 24.1 kPa. A person can noticeably feel a pressure change of 24.1 kPa, which falls within the normal acceptable range. In the test, the test subjects' pressure distributions changed less in four states, with an average pressure variation of 5.2 kPa/s. Therefore, their pressure distributions wereless affected by the change in the posture of the 4-UPR sling and the vibration during transferring. This result verifies that this prototype is stable and can transfer a nursed person comfortably.

As pain or comfort is based on the subjective judgment of individuals, a questionnaire survey and analysis was also made on whether the nursed persons felt "comfortable" or not in the test stages. In the subjective comments, the nursed persons' comfort degrees were divided into four levels: "uncomfortable", "acceptable, "good", and "comfortable". The comfort degree comments of the nursed persons in the stages are exhibited in Figure 16. As can be seen in Figure 16, the nursed persons did not feel uncomfortable in the starting stage and steady stage; one nursed person felt uncomfortable in the lifting stage, and two nursed persons felt uncomfortable in the lowering stage; and no nursed person felt uncomfortable in the whole transfer process, and those who felt good and comfortable accounted for 80%. The pressure distribution test and subjective comments all reveal that most nursed persons feel comfortable in the test process. Hence, the designed robot can transfer a nursed person comfortably.

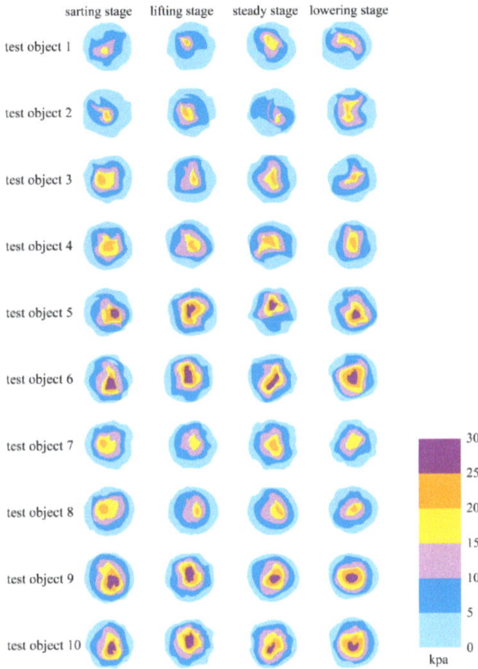

Figure 15. Contact pressure distribution on the backs of different test subjects.

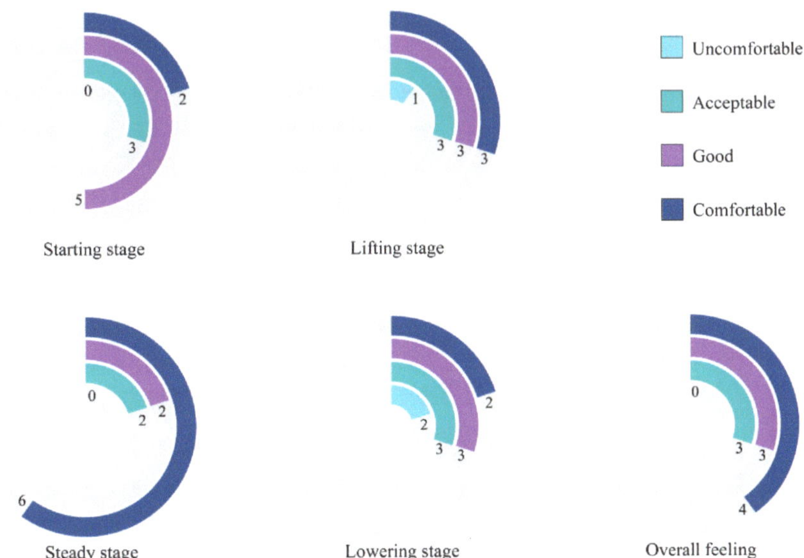

Figure 16. Comments of the nursed persons' comfort degree in the test stages.

7. Conclusions

This paper examines the action of humanoid hugging and employs this analysis to design a hybrid shift-riding care robot. To begin, a 3D motion capture system is utilized to record the human hugging action, allowing for an analysis of the movement path of the person being cared for. This analysis yields information on the degrees of freedom and workspace relevant to this action. Subsequently, a transfer-riding nursing robot with a 3-UPS + UPR parallel mechanism is devised. Kinematics and dynamics are employed to assess the workspace and stability of the robot. The results indicate that the robot's workspace exceeds the minimum trajectory range required for caring for the individual, meeting the necessary nursing care standards. Furthermore, during the humanoid-carrying motion, the driving force in each branch of the parallel mechanism transitions smoothly, without sudden changes, thus fulfilling stability requirements. In a prototype experiment, all volunteers report positive experiences when the robot performs humanoid holding care, and the change in the back pressure remains within an acceptable range. This robot offers convenience and reduces the workload of caregivers.

In contrast to the dual-arm humanoid care robot, this robot utilizes a hybrid plus modular structural design, providing improved structural stability at a lower cost. In the future, we plan to enhance the robot's structure, develop more advanced control algorithms, and undertake clinical validation to better address caregivers' requirements.

Author Contributions: J.L.: conceptualization, writing—original draft, writing—review and editing, data curation, formal analysis, and visualization. C.W.: conceptualization, methodology, writing—review and rditing, supervision, and resources. H.D.: writing—review and editing, data curation, and investigation. All authors have read and agreed to the published version of the manuscript.

Funding: This work was supported by the University Synergy Innovation Program of Anhui Province (grant numbers GXXT-2022-053); Anhui New Era Education Quality Project (Graduate Education); Provincial Graduate students "Innovation and Entrepreneurship Star" (grant number 2022cxcyzx127); and Huainan City science and technology plan project (grant number 2021A242).

Institutional Review Board Statement: This study was approved by the Biomedical Research Ethics Committee of Anhui University of Science and Technology (Anhui, China) in accordance with the Declaration of Helsinki. All methods were carried out in accordance with relevant guidelines and regulations. Written informed consent was obtained from all individual patients included in the study.

Data Availability Statement: The datasets used and analyzed during the current study are not available for public disclosure and are available from the corresponding authors upon reasonable request.

Acknowledgments: The authors thank the volunteers who contributed to the experiment.

Conflicts of Interest: The authors declare no conflict of interest. The funders had no role in the design of the study; in the collection, analyses, or interpretation of data; in the writing of the manuscript; or in the decision to publish the results.

References

1. van Dam, K.; Gielissen, M.; Reijnders, R.; van der Poel, A.; Boon, B. Experiences of Persons With Executive Dysfunction in Disability Care Using a Social Robot to Execute Daily Tasks and Increase the Feeling of Independence: Multiple-Case Study. *JMIR Rehabil. Assist. Technol.* **2022**, *9*, e41313. [CrossRef]
2. Pedace, C.; Rosa, A.; Francesconi, P.; Acampora, A.; Ricciardi, W.; Damiani, G. Governance in a project addressing care of disabled elderly persons within the regional healthcare system of Tuscany, Italy. *Ig. Sanita Pubbl.* **2017**, *73*, 351–372. [PubMed]
3. Moon, C.H.; Groman, R.; Jasak, R.S.; Burnetta, E.C.; Gonzalez-Fernandez, M.; Annaswamy, T.; Jayabalan, P.; Venesy, D.A.; Sereiko, T.J.; Flanagan, S.R. PM&R BOLD: The American Academy of Physical Medicine and Rehabilitation's strategic initiative to envision—And effectuate—The future of care across the rehabilitation care continuum. *PM&R J. Inj. Funct. Rehabil.* **2022**, *14*, 1497–1508. [CrossRef]
4. van Rooijen, M.; Lenzen, S.; Dalemans, R.; Beurskens, A.; Moser, A. Stakeholder engagement from problem analysis to implementation strategies for a patient-reported experience measure in disability care: A qualitative study on the process and experiences. *Health Expect.* **2021**, *24*, 53–65. [CrossRef] [PubMed]
5. Yang, K.; Lan, L.; Shen, P.; Wen, X.; Yue, P.; Luo, L. Application of APN model based on nursing studio in home rehabilitation of patients with ischemic stroke. *Chin. Nurs. Res.* **2022**, *36*, 3906–3911.

6. Garg, A.; Lobner, K.; Mitchell, R.; Song, J.; Egbunine, A.; Kudchadkar, S.R. PP324 [Critical Care Rehabilitation» Rehabilitation]: Health disparities in pediatric rehabilitation: A systematic review. *Pediatr. Crit. Care Med.* **2022**, *23*, 11S. [CrossRef]
7. Mireles, C.; Sanchez, M.; Cruz-Ortiz, D.; Salgado, I.; Chairez, I. Home-care nursing controlled mobile robot with vital signal monitoring. *Med. Biol. Eng. Comput.* **2023**, *61*, 399–420. [CrossRef]
8. Muangmeesri, B.; Wisaeng, K. IoT-Based Discomfort Monitoring and a Precise Point Positioning Technique System for Smart Wheelchairs. *Appl. Syst. Innov.* **2022**, *5*, 103. [CrossRef]
9. Zhou, S. Research on Design and Key Technologies of Compound Intelligent Nursing Bed. Master's Thesis, Jiangnan University, Wuxi, China, 2018.
10. Lang, J.; Zhang, Z.; Zhao, J. Mechanical analysis and simulation of intelligent nursing bed. *IOP Conf. Ser. Mater. Sci. Eng.* **2019**, *677*, 022069. [CrossRef]
11. Wang, C.; Hu, K.; Wang, X.; Dai, R. Design and kinematics simulation analysis of disabled shifter. *Mach. Design Manuf.* **2020**, *11*, 279–283. [CrossRef]
12. Qassim, H.M.; Wan Hasan, W.Z. A Review on Upper Limb Rehabilitation Robots. *Appl. Sci.* **2020**, *10*, 6976. [CrossRef]
13. Narayanan, K.L.; Krishnan, R.S.; Son, L.H.; Tung, N.T.; Julie, E.G.; Robinson, Y.H.; Kumar, R.; Gerogiannis, V.C. Fuzzy Guided Autonomous Nursing Robot through Wireless Beacon Network. *Multimed. Tools Appl.* **2021**, *81*, 3297–3325. [CrossRef] [PubMed]
14. Cao, S.; Kazuki, N.; Quan, C.; Luo, Z. On robotic rehabilitation of human dual arms' coordinative function. *Int. J. Appl. Electromagn. Mech.* **2016**, *52*, 943–950. [CrossRef]
15. Zhang, L.; Gao, H.; Xu, H.; Song, J. Trajectory planning of the nursing robot based on the center of gravity for aluminum alloy structure. *Rev. Adv. Mater. Sci.* **2021**, *60*, 731–743. [CrossRef]
16. Liu, Y.; Guo, S.; Chen, G.; Liu, J.; Gan, Z. Bionic motion planning and the analysis for human comfort of a piggyback nursing-care robot for transfer tasks. *J. Mech. Eng.* **2020**, *56*, 147–156.
17. Nielsen, S.; Langensiepen, S.; Madi, M.; Elissen, M.; Stephan, A.; Meyer, G. Implementing ethical aspects in the development of a robotic system for nursing care: A qualitative approach. *BMC Nurs.* **2022**, *21*, 180. [CrossRef] [PubMed]
18. Lim, M.J.; Song, W.K.; Kweon, H.; Ro, E.R. Care robot research and development plan for disability and aged care in Korea: A mixed-methods user participation study. *Assist. Technol.* **2022**, *35*, 292–301. [CrossRef]
19. Ohneberg, C.; Stöbich, N.; Warmbein, A.; Rathgeber, I.; Fischer, U.; Eberl, I. Potential Uses of Assistive Robotic Systems in Acute Inpatient Care. *Stud. Health Technol. Inform.* **2022**, *294*, 801–802. [CrossRef]
20. *GB/T 17245-2004*; Inertial Parameters of Adult Human Body. Standardization Administration of China: Beijing, China, 2004.
21. Xu, C.; Wang, X.; Duan, S.; Wan, J. Spatial-temporal constrained particle filter for cooperative target tracking. *J. Network Comput. Appl.* **2021**, *176*, 102913. [CrossRef]
22. Chen, W.; Li, Z.; Cui, X.; Zhang, J.; Bai, S. Mechanical design and kinematic modeling of a cable-driven arm exoskeleton incorporating inaccurate human limb anthropomorphic parameters. *Sensors* **2019**, *19*, 4461. [CrossRef]
23. *ISO/IEC FCD 19774: 200x*; Humanoid Animation, Annex B, Feature Points for the Human Body. ISO: Geneva, Switzerland, 2003.
24. Huang, R.; Li, H.; Suomi, R.; Li, C.; Peltoniemi, T. Intelligent physical robots in health care: Systematic literature review. *J. Med. Internet Res.* **2023**, *25*, e39786. [CrossRef] [PubMed]
25. Chen, X.; Yang, W. Dynamic modeling and analysis of spatial parallel mechanism with revolute joints considering radial and axial clearances. *Nonlinear Dynam.* **2021**, *106*, 1929–1953. [CrossRef]
26. Hess-Coelho, T.A.; Orsino, R.M.M.; Malvezzi, F. Modular modelling methodology applied to the dynamic analysis of parallel mechanisms. *Mech. Mach. Theory* **2021**, *161*, 104332. [CrossRef]
27. Chen, X.; Guo, J. Effects of spherical clearance joint on dynamics of redundant driving spatial parallel mechanism. *Robotica* **2021**, *39*, 1064–1080. [CrossRef]

Article

Adaptive Super-Twisting Sliding Mode Control of Underwater Mechanical Leg with Extended State Observer

Lihui Liao, Luping Gao *, Mboulé Ngwa , Dijia Zhang, Jingmin Du * and Baoren Li

Institute of Marine Mechatronics and Equipment, School of Mechanical Science and Engineering, Huazhong University of Science and Technology, Wuhan 430074, China; lhliao@hust.edu.cn (L.L.); ngwa_m@yahoo.fr (M.N.); zhangdijia323@163.com (D.Z.)
* Correspondence: m202170603@hust.edu.cn (L.G.); hustdjm@hust.edu.cn (J.D.)

Abstract: Underwater manipulation is one of the most significant functions of the deep-sea crawling and swimming robot (DCSR), which relies on the high-accuracy control of the body posture. As the actuator of body posture control, the position control performance of the underwater mechanical leg (UWML) thus determines the performance of the underwater manipulation. An adaptive super-twisting sliding mode control method based on the extended state observer (ASTSMC-ESO) is proposed to enhance the position control performance of the UWML by taking into account the system's inherent nonlinear dynamics, uncertainties, and the external disturbances from hydrodynamics, dynamic seal resistance, and compensation oil viscous resistance. This newly designed controller incorporates sliding mode (SMC) feedback control with feedforward compensation of the system uncertainties estimated by the ESO, and the external disturbances of the hydrodynamics by fitting the parameters, the dynamic seal resistance, and the compensation oil viscous resistance to the tested results. Additionally, an adaptive super-twisting algorithm (AST) with integral action is introduced to eliminate the SMC's chattering phenomenon and reduce the system's steady-state error. The stability of the proposed controller is proved via the Lyapunov method, and the effectiveness is verified via simulation and comparative experimental studies with SMC and the adaptive fuzzy sliding mode control method (AFSMC).

Keywords: deep-sea crawling and swimming robot; underwater mechanical leg; extended state observer; adaptive super twisting control; sliding mode control; hydrodynamics; dynamic seal resistance; compensation oil viscous resistance

Citation: Liao, L.; Gao, L.; Ngwa, M.; Zhang, D.; Du, J.; Li, B. Adaptive Super-Twisting Sliding Mode Control of Underwater Mechanical Leg with Extended State Observer. *Actuators* **2023**, *12*, 373. https://doi.org/10.3390/act12100373

Academic Editor: Ioan Doroftei

Received: 3 September 2023
Revised: 22 September 2023
Accepted: 25 September 2023
Published: 27 September 2023

1. Introduction

With the increasingly continuous consumption of land resources, the pace of human exploration and the development of ocean resources is accelerating and gradually moving from shallow waters to the deep seas. Recently, with the development of robot technology, numerous robots have been increasingly applied in marine engineering. Therefore, deep-sea robots have attracted more and more attention to research. Compared with wheeled robots and tracked robots, legged robots have excellent motion flexibility, strong obstacle avoidance ability, and excellent terrain adaptation [1], making them a hot research topic in recent years [2–4]. As a special type of legged robots, the DCSR combines swimming, walking, and underwater manipulating functions, therefore being an important equipment for marine engineering.

Since the robot body is the base of the manipulator equipped on the robot, high-accuracy position control of the UWML is of great significance to improve the robot's posture control performance, thus enhancing the manipulating performance. Currently, model-based control is the primary method of the high-performance control of dynamic systems, but an accurate model of the control object is required. However, as an inherent nonlinear and strongly coupled dynamic system, the UWML not only has complex nonlinear characteristics but is also influenced by system uncertainties (i.e., parametric uncertainties and unmodeled uncertainties) and external disturbances caused by the deep-sea environment (i.e., hydrodynamic, dynamic seal resistance, and compensation oil viscous resistance). The combined effect of these factors makes it difficult to establish an accurate dynamic model of the UWML, thus exacerbating the difficulty of control.

To overcome nonlinearity and the uncertainties of manipulators, many elegant control techniques [5–11] were proposed. Among them, the SMC-based methods are the most widely used due to their simple design and robustness against uncertainties and disturbances [12]. However, these methods require an infinite switching control action to handle system uncertainties, which would aggravate the undesired chattering phenomenon. Moreover, the upper bounds of the external disturbances and the system uncertainties must be well known to obtain a stable closed-loop control law. Another issue in SMC-based methods is that of asymptotic convergence, which may not meet the high-performance control of the robotic systems [13].

To restrain chattering, some momentous methods, such as the boundary layer method, the modified switching signal with saturation, etc., have been introduced; however, the presence of the finite steady-state error may cause degradation of the tracking ability [14]. Therefore, high-order sliding mode control (HOSMC) approaches, which provide effective solutions to the chattering problem without sacrificing the control performance, were developed to improve the performance of the classical SMC [15]. In particular, the super-twisting control (STC) has been a widely applied HOSMC since its inception [16].

As a prototype of the HOSMC, STC can effectively suppress the chattering phenomenon since the discontinuous signum function is concealed in the time derivative of the sliding variable [15,17]. However, if the states are far from the sliding surface, the bounded correction terms will result in a very slow convergence of the sliding variable [18]. Moreover, although STC yields a smooth control signal, chattering is still possible. For example, in situations where the upper bound of system uncertainties is difficult to obtain accurately, the largest possible parameters are often selected to ensure the finite-time stability condition, which may result in system oscillation [19]. Therefore, various new STC algorithms have been proposed to enhance the performance of traditional STC. A modified STC was proposed to improve the convergence speed and robustness, in which two linear correction terms were added to the traditional STC to form a double-closed-loop feedback structure [18]. An adaptive STC (ASTC) was synthesized to avoid the problem of difficulty in determining the upper bound of system uncertainties in practical applications, where a double-layer gain-adaptation function was introduced into the traditional STC to determine the two control parameters [20].

In addition, many researchers also conducted a series of observer-based control methods to deal with the uncertainties and disturbances of the dynamic systems, in which observers were applied to estimate the total system uncertainties and disturbances to synthesize a feedforward compensation. Such observers include the disturbance observer (DOB) [21], the sliding mode disturbance observer (SMDO) [22,23], the extended state observer (ESO) [24–27], etc. An adaptive fuzzy sliding mode control method based on a disturbance observer was proposed to control a manipulator, in which a fuzzy system was adopted to approximate the modeling uncertainties while the disturbances were estimated via a DOB online, respectively, and both were then compensated in SMC [21]. Due to the advantages of estimating the system's states and uncertainties simultaneously with little model information and a simple design process, ESO was widely applied in the control of manipulators. An adaptive extended state observer (ESO) is developed to estimate the unmeasured states and eliminate the impact of the unknown disturbances and parameter uncertainties for the control of an aircraft skin inspection robot [26]. To deal with the time-varying output constraints and external disturbances of a manipulator, an ESO was applied to estimate the unmeasurable states and total disturbances, which were then incorporated into the controller design [25].

Based on the above analysis and the UWML's system characteristics, by borrowing ideas from ESO and AST and integrating them via an SMC control action, a novel adaptive super-twisting sliding mode control method based on the extended state observer (ASTSMC-ESO) is proposed for the high-accuracy position control of the UWML. The introduction of AST can effectively suppress the chattering effect and enhance the steady-state control accuracy while ensuring the robustness of the feedback controller SMC. The integration of ESO and the experimentally tested data can make a feedforward compensation for the system uncertainties, and the external disturbances of the hydrodynamic force, dynamic seal resistance, and compensation oil viscous resistance.

The rest of this paper is organized as follows. Section 2 establishes the models of the UWML and gives the problem formulation. Section 3 presents the ESO design procedure, while Section 4 carries out the ASTSMC-ESO design procedure. Sections 5 and 6 describe the simulation and experimental verification results, respectively, and some conclusions are made in Section 7.

2. Models and Problem Formulation

2.1. Kinematic Model

The UWML is a 3-DOF serial mechanism shown in Figure 1. The D-H coordinate systems and the corresponding parameters used to develop the kinematic model of the UWML are shown in Figure 2 and Table 1, respectively.

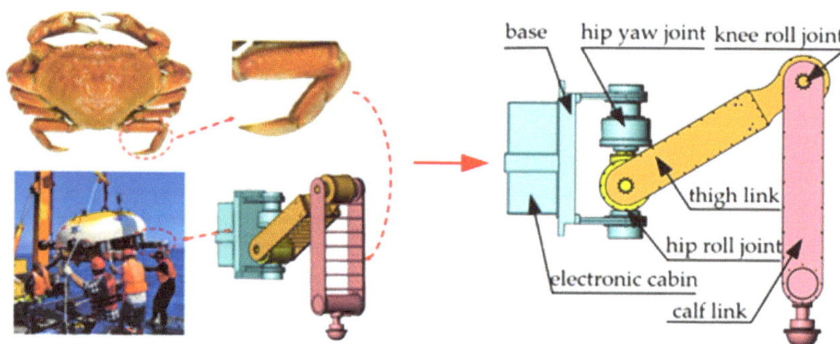

Figure 1. The structure of the UWML.

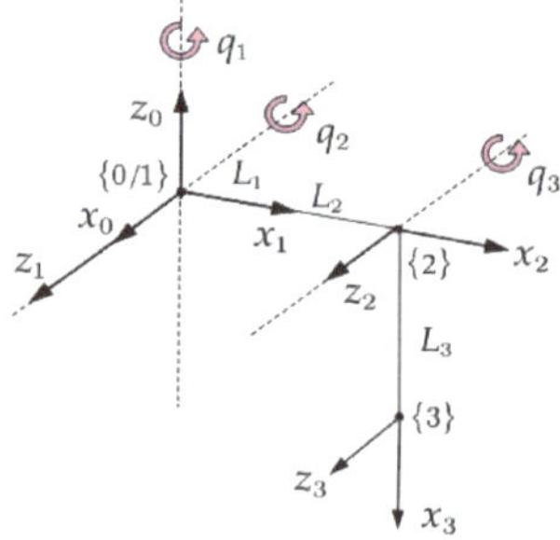

Figure 2. D-H coordinate systems of the UWML.

Table 1. D-H parameters of the UWML.

#	q	d	a	α
0–1	q_1	0	L_1 (0 m)	90°
1–2	q_2	0	L_2 (0.66 m)	0°
2–3	q_3	0	L_3 (0.85 m)	0°

The coordinates of the foot tip of the UWML with respect to the base frame {0} can be determined as:

$$^{0}p_3 = \begin{bmatrix} x \\ y \\ z \end{bmatrix} = \begin{bmatrix} -s_1(L_1 + L_2c_2 + L_3s_{23}) \\ c_1(L_1 + L_2c_2 + L_3s_{23}) \\ L_2s_2 - L_3c_{23} \end{bmatrix} \tag{1}$$

Then, the joint angles can be derived using the inverse kinematics, as follows:

$$\begin{cases} q_1 = \arctan2(-x,\ y) \\ q_2 = \arccos\left(\dfrac{-z}{\sqrt{M^2 + N^2}}\right) - \varphi \\ q_3 = \arcsin\left(\dfrac{A^2 + z^2 - L_2^2 - L_3^2}{2L_2L_3}\right) \end{cases} \tag{2}$$

where x, y, and z are the coordinates of the foot tip, $s_i = \sin(q_i), c_i = \cos(q_i), s_{ij} = \sin(q_i + q_j)$, $c_{ij} = \cos(q_i + q_j), A = y/c_1 - L_1, M = L_2 + L_3s_3 = \rho\sin(\varphi), N = L_3c_3 = \rho\cos(\varphi)$, and $\varphi = \arctan2(M,\ N)$.

2.2. Dynamic Model

When taking into account the external disturbances of hydrodynamics τ_w, dynamic seal resistance τ_f, compensation oil viscous resistance τ_s, and the system uncertainty d, the dynamic model of the UWML is:

$$\tau = M\ddot{q} + C\dot{q} + G + \tau_f + \tau_s - \tau_w + d \tag{3}$$

where $\dot{q}$ and $\ddot{q}$ are the joint velocity and acceleration, respectively. M, C, and G are the inertia matrix, the coriolis/centrifugal matrix, and the gravity vector, respectively.

In general, due to parameter perturbation, testing errors, and unmodeled dynamics, the model (3) differs from the actual model of the UWML, and the differences become uncertainties of the system. When considering the impact of system uncertainties, the system parameters M, C, and G and the external disturbances τ_w, τ_f, and τ_s can be represented as:

$$\begin{cases} M = M_n + \Delta M \\ C = C_n + \Delta C \\ G = G_n + \Delta G \\ \tau_f = \tau_{fn} + \Delta\tau_f \\ \tau_s = \tau_{sn} + \Delta\tau_s \\ \tau_w = \tau_{wn} + \Delta\tau_w \end{cases} \tag{4}$$

where M_n, C_n and G_n are the nominal values of the inertia matrix, the coriolis/centrifugal matrix, and the gravity vector of the UWML, respectively, which can be found in the author's published article [28]. $\Delta M, \Delta C$, and ΔG are the modeling uncertainties. τ_{fn} and τ_{sn} are the tested results of the dynamic seal resistance and compensation oil viscous resistance, respectively, which are shown in Figures 3 and 4; τ_{wn} is the hydrodynamics calculated with the fitted parameters, of which the detailed analysis can be found in the author's recently published article [29]. $\Delta\tau_f, \Delta\tau_s$, and $\Delta\tau_w$ represent the uncertainties caused by the testing errors.

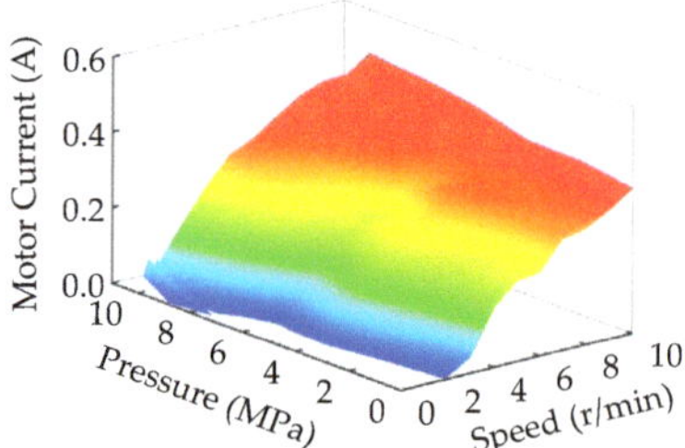

Figure 3. Compensation oil viscous resistance with different rotation speed and pressure.

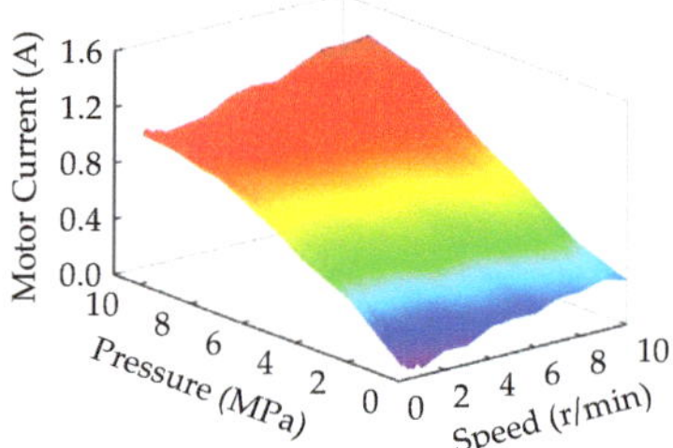

Figure 4. Dynamic seal resistance with respect to rotation speeds and pressure.

Substituting (4) into (3) yields the nonlinear dynamic model of the UWML:

$$\tau = M_n \ddot{q} + C_n \dot{q} + G_n + \tau_{fn} + \tau_{sn} - \tau_{wn} + \Delta \tag{5}$$

where $\Delta = \Delta M \ddot{q} + \Delta C \dot{q} + \Delta G + \Delta \tau_f + \Delta \tau_s - \Delta \tau_w + d$ represents the total dynamic uncertainties caused by the parameter uncertainties, testing errors, and unmodeled dynamics. According to the robotic theory, the inertia matrix M_n is a bounded symmetric positive definite matrix, and $M_n - 2C_n$ is a skew-symmetric matrix, satisfying $x^T(M_n - 2C_n)x = 0$ for any vector x.

To simplify the design and analysis, define the state variable of the UWML as $x = [x_1, x_2]^T = [q, \dot{q}]^T$, then the dynamic model (5) can be rewritten in a state-space form:

$$\begin{cases} \dot{x}_1 = x_2 \\ \dot{x}_2 = M_n^{-1}\tau - M_n^{-1}\left(C_n x_2 + G_n + \tau_{fn} + \tau_{sn} - \tau_{wn}\right) - M_n^{-1}\Delta \end{cases} \tag{6}$$

2.3. Control Objective

The control objective is to design a high-performance controller for the UWML with strong nonlinearities (i.e., structure nonlinearity, friction resistance, and oil resistance nonlinearity) and uncertainties (i.e., parameter uncertainties, unmodeled dynamics) to ensure the actual foot trajectory $y(t)$ tracking the planned trajectory $y_d(t)$ as close as possible.

3. Observer Design

To reduce the impact of uncertainties on the UWML's control performance, the total uncertainties of the dynamic system (6) have to be compensated in the controller, which can be extended to a new state variable and estimated via an ESO. Based on this idea, the uncertainties $-M_n^{-1}\Delta$ is extended to a state variable x_3, and its time derivative is set to be $h(t)$. Thus, the extended dynamic model of the UWML can be expressed as:

$$\begin{cases} \dot{x}_1 = x_2 \\ \dot{x}_2 = M_n^{-1}\tau - M_n^{-1}\left(C_n x_2 + G_n + \tau_{fn} + \tau_{sn} - \tau_{wn}\right) + x_3 \\ \dot{x}_3 = h(t) \end{cases} \tag{7}$$

Assumption 1. *$h(t)$ is bounded, and there exists a constant $h_0 > 0$ such that $\|h(t)\| \leq h_0$.*

Based on the extended system model (7), the ESO can be designed as:

$$\begin{cases} \dot{\hat{x}}_1 = \hat{x}_2 + l_1\omega_0(x_1 - \hat{x}_1) \\ \dot{\hat{x}}_2 = M_n^{-1}\tau - M_n^{-1}(C_n x_2 + G_n + \tau_{fn} + \tau_{sn} - \tau_{wn}) + \hat{x}_3 + l_2\omega_0^2(x_1 - \hat{x}_1) \\ \dot{\hat{x}}_3 = l_3\omega_0^3(x_1 - \hat{x}_1) \end{cases} \tag{8}$$

where $\hat{x} = [\hat{x}_1, \hat{x}_2, \hat{x}_3]^T$ is the estimated state, $L = [l_1, l_2, l_3]^T$ is the observer gain, and $\omega_0 > 0$ is the observer bandwidth.

To ensure the stability of ESO, the observer gain is designed to satisfy the following polynomial:

$$p(s) = s^3 + l_1 s^2 + l_2 s + l_3 = (s+1)^3 \tag{9}$$

Let $\tilde{x} = x - \hat{x} = [\tilde{x}_1, \tilde{x}_2, \tilde{x}_3]^T$ denote the observation error, then the dynamic of the observation error can be obtained from (7) and (8):

$$\begin{cases} \dot{\tilde{x}}_1 = \tilde{x}_2 - l_1\omega_0\tilde{x}_1 \\ \dot{\tilde{x}}_2 = \tilde{x}_3 - l_2\omega_0^2\tilde{x}_1 \\ \dot{\tilde{x}}_3 = h(t) - l_3\omega_0^3\tilde{x}_1 \end{cases} \tag{10}$$

Define the proportional observation error $\varepsilon_i = \tilde{x}_i/\omega_0^{i-1}$, $i = 1, 2, 3$. Then, the dynamic of the proportional observation error can be obtained according to (10):

$$\dot{\varepsilon} = \omega_0 A\varepsilon + B\frac{h(t)}{\omega_0^2} \tag{11}$$

where $\varepsilon = [\varepsilon_1, \varepsilon_2, \varepsilon_3]^T$, $A = \begin{bmatrix} -3I_{3\times3} & I_{3\times3} & 0_{3\times3} \\ -3I_{3\times3} & 0_{3\times3} & I_{3\times3} \\ -I_{3\times3} & 0_{3\times3} & 0_{3\times3} \end{bmatrix}$, $B = \begin{bmatrix} 0_3 \\ 0_3 \\ I_3 \end{bmatrix}$; $0_{3\times3}$ and $I_{3\times3}$ are the 3×3 zero matrix and the identity matrix, respectively; and 0_3 and I_3 are the 3-dimensional zero vector and the unit vector, respectively.

Theorem 1 [27,30]. *Under Assumption 1, according to the dynamic Equation (6), when designing an ESO (8), there exist constants $\sigma_{ij} > 0$, $c > 0$, and finite time $T_1 > 0$, such that the state estimation error $\tilde{x}_{ij}$ is bounded, and its value of the state estimation error can be adjusted by changing the bandwidth ω_0:*

$$|\tilde{x}_{ij}| \leq \sigma_{ij}, \quad \sigma_{ij} = O\left(\frac{1}{\omega_0^c}\right); \quad i = j = 1, 2, 3 \quad \forall t \geq T_1 \tag{12}$$

4. Controller Design

4.1. Design of SMC-ESO Controller

Since the UWML is subject to strong external disturbances and uncertainties when working in a complex underwater environment, an SMC-ESO control method, possessing strong robustness, is adopted to deal with the trajectory tracking problem of the UWML intuitively.

The tracking errors of the joint position and velocity of the UWML are defined as e_1 and e_2, respectively:

$$e_1 = x_{1d} - x_1 \tag{13}$$

$$e_2 = \dot{x}_{1d} - x_2 \tag{14}$$

where x_{1d} and $\dot{x}_{1d}$ are the joint position and velocity command obtained from the reference foot trajectory via inverse kinematics (6).

Then, the sliding mode surface function s can be defined as:

$$s = e_2 + \lambda_1 e_1 \tag{15}$$

where $\lambda_1 \in R^{3\times3}$ is a positive definite diagonal parameter matrix.

Based on Equations (7) and (13)–(15), the time derivative of s can be expressed as:

$$\begin{aligned} \dot{s} &= \dot{e}_2 + \lambda_1 \dot{e}_1 \\ &= \ddot{x}_{1d} + \lambda_1 e_2 - M_n^{-1}\tau + M_n^{-1}\left(C_n x_2 + G_n + \tau_{fn} + \tau_{sn} - \tau_{wn}\right) - x_3 \end{aligned} \tag{16}$$

From (16), the SMC-ESO with an exponential convergence rate can be designed as:

$$\tau = \tau_{eq} + \tau_{sw} \tag{17}$$

$$\tau_{eq} = M_n\left(\ddot{x}_{1d} + \lambda_1 e_2\right) - M_n \hat{x}_3 + C_n x_2 + G_n + \tau_{fn} + \tau_{sn} - \tau_{wn} \tag{18}$$

$$\tau_{sw} = M_n\left(\lambda_1 s + \lambda_r \mathrm{sign}(s)\right) \tag{19}$$

where $-M_n \hat{x}_3$ is the estimated value of the system uncertainties Δ, which can be calculated from the extended state of the ESO; λ_2 and $\lambda_r \in R^{3\times3}$ are positive definite diagonal parameter matrices; and $\mathrm{sign}(s)$ is the signum function vector.

Although the above SMC-ESO can ensure the stability of the closed-loop system, it has inherent defects of conventional SMC. On the one hand, the sliding mode switching control action τ_{sw} contains a discontinuous signum function $\mathrm{sign}(s)$ leading to the "chattering" phenomenon. On the other hand, the gain λ_r of τ_{sw} must be selected based on the upper bound of the system uncertainties, which are usually unknown. If the value is too conservative to be large, it will exacerbate the system chattering. However, if a smaller value is selected, it may cause the system to be unstable. To avoid sacrificing the control accuracy or robustness by using traditional methods such as the boundary layer method and the modified switching signal with saturation to suppress chattering, this paper introduces the AST into the above SMC-ESO to eliminate the chattering and improve the control performance of the system.

4.2. Design of ASTSMC-ESO Controller

To eliminate the chattering effect of the conventional SMC, an ASTSMC-ESO is proposed, in which a super-twisting control action [16,31] is applied to replace the discontinuous sliding mode control law τ_{sw}:

$$\begin{cases} \tau_{sw} = M_n\left[\alpha(t)|s|^{\frac{1}{2}}\mathrm{sign}(s) + k_1(t)s - \xi + \phi(s,\, L(t))\right] \\ \dot{\xi} = -\beta(t)\mathrm{sign}(s) - k_2(t)s \end{cases} \tag{20}$$

where ξ is an intermediate variable; $L(t)$ is the parameter adaptive function to be designed later; and $\alpha(t)$, $\beta(t)$, $k_1(t)$, and $k_2(t)$ are positive definite parameter matrices; $\phi(s,\, L(t)) = -\dot{L}(t)s/L(t)$.

In the above super-twisting algorithm, the values of $\alpha(t)$ and $\beta(t)$ should be determined based on the upper bound of system uncertainties and its rate of change. For the UWML, since the system uncertainties have been estimated via the expanded state observer,

$\alpha(t)$ and $\beta(t)$ should be determined based on the upper bound of the estimation error and its rate of change. Since these two values are both difficult to determine, the adaptive laws are designed for the parameters of (20), which are shown in (21):

$$\begin{cases} \alpha(t) = \sqrt{L(t)}\alpha_0 \\ \beta(t) = L(t)\beta_0 \\ k_1(t) = L(t)k_{10} \\ k_2(t) = L^2(t)k_{20} \end{cases} \tag{21}$$

where α_0, β_0, k_{10}, and k_{20} are the constants, and $L(t)$ are adaptive functions that will be designed using a double-layer gain-adaptation algorithm in Section 4.4.

Substituting (17)–(20) into (16) yields:

$$\dot{s} = -\tilde{x}_3 - \alpha|s|^{\frac{1}{2}}\operatorname{sign}(s) - k_1 s - \int_0^t (\beta\operatorname{sign}(s) + k_2 s)dt - \phi(s, L) \tag{22}$$

Define $z = -\int_0^t (\beta\operatorname{sign}(s) + k_2 s)dt - \tilde{x}_3$, $f(t) = -\dot{\tilde{x}}_3$, then (22) can be rewritten as:

$$\begin{cases} \dot{s} = -\alpha|s|^{\frac{1}{2}}sign(s) - k_1 s + z - \phi(s,\ L) \\ \dot{z} = -\beta sign(s) - k_2 s + f(t) \end{cases} \tag{23}$$

According to Theorem 1 and Equation (10), it can be seen that $f(t) = -\dot{\tilde{x}}_3 = -h(t) + l_3\omega_0^3\tilde{x}_1$ is bounded. Let its upper bound be denoted as σ_0, thus $\|f(t)\| \le \sigma_0$.

4.3. Controller Stability Analysis

The stability proof of the ASTSMC-ESO is relatively complex. Here, only the following stability theorem is presented. The detailed proof process can be found in Appendix A.

Theorem 2. *Based on Assumption 1, Lemma 1, and Theorem 1, the ASTSMC-ESO controller (17), (18), (20), and the parameter adaptive function (21) are designed for the dynamic system (7) of the UWML. When $L_i(t) > \sigma_{0i} \ge |f_i(t)|$, and the appropriate parameters α_{0i}, β_{0i}, k_{10i} and k_{20i} are selected to make the matrix P_i, B_i, and Q_i (which was expressed as (24)–(26)) positive definite, then the closed-loop system of the UWML is stable, and the joint position error e_{1i} will converge to the origin in finite time, where $i = 1, 2, 3$.*

$$P_i = \frac{1}{2}\begin{bmatrix} 4\beta_{0i} + \alpha_{0i}^2 & \alpha_{0i}k_{10i} & -\alpha_{0i} \\ \alpha_{0i}k_{10i} & 2k_{20i} + k_{10i}^2 & -k_{10i} \\ -\alpha_0 & -k_{10i} & 2 \end{bmatrix} \tag{24}$$

$$B_i = k_{10i}\begin{bmatrix} \beta_{0i} + 2\alpha_{0i}^2 & 0 & 0 \\ 0 & k_{20i} + k_{10i}^2 & -k_{10i} \\ 0 & -k_{10i} & 1 \end{bmatrix} \tag{25}$$

$$Q_i = A_i - C_i \tag{26}$$

$$A_i = \frac{\alpha_{0i}}{2}\begin{bmatrix} 2\beta_{0i} + \alpha_{0i}^2 & 0 & -\alpha_{0i} \\ 0 & 2k_{20i} + 5k_{10i}^2 & -3k_{10i} \\ -\alpha_{0i} & -3k_{10i} & 1 \end{bmatrix} \tag{27}$$

$$C_i = \begin{bmatrix} -\alpha_{0i}\vartheta_i & -\frac{1}{2}k_{10i}\vartheta_i & \vartheta_i \\ -\frac{1}{2}k_{10i}\vartheta_i & 0 & 0 \\ \vartheta_i & 0 & 0 \end{bmatrix} \tag{28}$$

$$\vartheta_i = \frac{f_i(t)sign(s_i)}{L_i(t)} \tag{29}$$

Compared to the conventional ASTSMC-ESO algorithm applied in other fields [32], The main improvement of the method proposed in this paper is to increase the convergence speed of the system states. It can be seen from the proof results (A16) of the ASTSMC-ESO in this paper, when the Lyapunov function V is relatively close to the equilibrium, the nonlinear term $\lambda_2 V_i^{1/2}$ is much bigger than the linear term $\lambda_1 V_i$, so the nonlinear term $\lambda_2 V_i^{1/2}$ mainly determines the convergence speed. When the Lyapunov function V is far from the equilibrium, the linear term $\lambda_1 V_i$ is much bigger than the nonlinear term $\lambda_2 V_i^{1/2}$, so the linear term $\lambda_1 V_i$ mainly determines the convergence speed. Nevertheless, for conventional ASTSMC-ESO, no matter where the Lyapunov function is, the convergence speed is determined only by the nonlinear term $\lambda_2 V_i^{1/2}$, so when the Lyapunov function V is far from the equilibrium, slow convergence speed will present [20]. The above analysis shows that by adding linear terms to the conventional AST, the ASTSMC-ESO designed in this paper improves the convergence characteristics. From the viewpoint of the controller, the newly designed ASTSMC-ESO is equivalent to adding a proportional and integral sliding mode control action than the conventional one. Therefore, the convergence speed and control accuracy are improved.

4.4. Design of the Adaptive Function $L_i(t)$

As mentioned in the previous design process, the parameters $\alpha_i(t)$, $\beta_i(t)$, $k_{1i}(t)$, and $k_{2i}(t)$ of (20) must be adjusted via the adaptive function $L_i(t)$. To ensure stability, $L_i(t)$ should satisfy the condition $L_i(t) > |f_i(t)|$, where a dual-layer adaption algorithm is adopted to design $L_i(t)$.

According to the concept of "equivalent control", the system (23) enters the sliding mode surface when $s_i = \dot{s}_i = 0$ and $z_i = \dot{z}_i = 0$. To maintain the system trajectory on the sliding surface, the equivalent effect of the discontinuous switching term $\beta_i sign(s_i)$, denoted as u_{eqi}, can be used to compensate for $f_i(t)$:

$$u_{eqi} = \beta_i sign(s_i)\big|_{eq} = f_i(t) \tag{30}$$

In practical applications, u_{eqi} can be estimated from $\beta_i sign(s_i)$ online via a low-pass filter:

$$\dot{u}_{eqi} = \frac{1}{\tau}(\beta_i sign(s_i) - u_{eqi}) \tag{31}$$

where τ is the time constant of the filter.

From (30) and (31), an online estimation of $f_i(t)$ can be achieved by filtering the discontinuous switch term. This estimation value can be used to design $L_i(t)$:

$$L_i(t) = l_{0i} + l_i(t) \tag{32}$$

where l_{0i} is a small positive design parameter, $l_i(t)$ is a time-varying parameter whose rate of change is defined as the first-layer adaptive algorithm:

$$\dot{l}_i(t) = -(\rho_{0i} + \rho_i(t))sign(\delta_i) \tag{33}$$

where ρ_{0i} is a constant positive parameter, $\rho_i(t)$ is a time-varying parameter, and its rate of change is defined as the second-layer adaptive algorithm:

$$\dot{\rho}_i(t) = \gamma_i|\delta_i| \tag{34}$$

where γ_i is a positive design parameter.

The variable δ_i in the above two-layer adaptive rate is:

$$\delta_i = L_i(t) - \frac{1}{a_i \beta_{0i}} |u_{eqi}| - \epsilon_i \tag{35}$$

where $0 < a_i < 1/\beta_{0i} < 1$, and ϵ_i is a small positive parameter.

Using the above dual-layer adaption algorithm, the condition $L_i(t) > |f_i(t)|$ can be guaranteed within a finite time, which can be described as the following theorem.

Theorem 3 [20]. *Consider the system in (23) is subject to uncertainty $f(t)$, which satisfies the two constraints $|f_i(t)| < \sigma_{0i}$ and $|\dot{f}_i(t)| < \sigma_{1i}$, where the positive scalars a_0 and a_1 are finite but unknown. Then, the dual-layer adaption algorithm in (32)–(34) ensures $L_i(t) > |f_i(t)|$ in finite time.*

Theorem 3 is a supplement to Theorem 2, which resolves the parameter selecting-problem of the ASTSMC-ESO. In practical applications, in order to reduce the influence of noise and disturbances, $\dot{\rho}_i(t)$ can be obtained with a dead zone δ_{i0}:

$$\dot{\rho}_i(t) = \begin{cases} \gamma_i |\delta_i|, & \text{if } |\delta_i| > \delta_{i0} \\ 0, & \text{otherwise} \end{cases} \tag{36}$$

From Equations (17), (18), (20), and (21), it can be seen that the ASTSMC-ESO mainly consists of two parts: τ_{eq} is the model compensation term, which enables the system output to quickly track the desired trajectory, while τ_{sw} is the stabilizing feedback term, which ensures that the tracking error of the system is stable. Further analysis of the composition of τ_{sw} reveals that, in addition to the conventional approaching term $|s|^{\frac{1}{2}}\text{sign}(s)$ and the exponential approaching term $k_1 s$, it also includes an integral term of the signum function of the sliding mode variable $\int_0^t \beta \text{sign}(s) dt$ and an integral term of the sliding mode variable itself $\int_0^t k_2 s dt$. The former compensates for the estimation error of the system uncertainties, while the latter reduces the steady-state error of the system, thereby improving both the dynamic convergence speed and the steady-state control accuracy of the system. In addition, an adaptive function $L_i(t)$ is designed to adjust the parameters $\alpha_i(t)$, $\beta_i(t)$, $k_{1i}(t)$, and $k_{2i}(t)$ of the controller, which effectively eliminates the chattering phenomenon of the SMC. The principle of the ASTSMC-ESO control system for the UWML is shown in Figure 5.

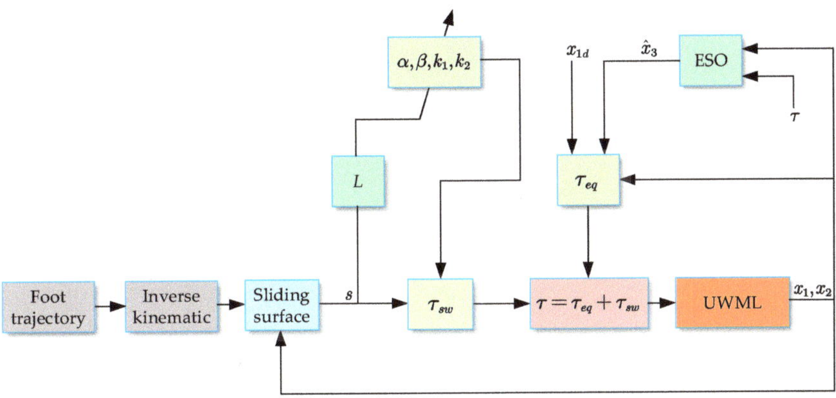

Figure 5. The principle of the ASTSMC-ESO controller for the UWML.

5. Simulation Results

To verify the effectiveness of the proposed ASTSMC-ESO, the trajectory tracking performance of the UWML is studied via simulation.

5.1. Simulation Model

The simulation model of the UWML is shown in Figure 6, which is built using Simulink/Simscape. The parameters of the UWML are listed in Table 2. To simulate the effects of the dynamic seal resistance τ_f, compensation oil viscous resistance τ_s, and the hydrodynamic τ_w, a tabular form of τ_s and τ_w from the tested results in Figures 3 and 4, which are added to the dynamic model of the UWML, while τ_w is added according to the calculation formulas with the fitted hydrodynamic parameters. The initial values of the system states are $x_1 = [0.294, 0.158, -0.105]^{\mathrm{T}}$, and $x_2 = [0, 0, 0]^{\mathrm{T}}$; the initial values of the states of ESO are $\hat{x}_1 = [0, 0, 0]^{\mathrm{T}}$, $\hat{x}_2 = [0, 0, 0]^{\mathrm{T}}$, and $\hat{x}_3 = [0, 0, 0]^{\mathrm{T}}$, while the bandwidth is $\omega_0 = 60$. The controller parameters are presented in Table 3. For convenience, a cycloidal foot trajectory [28] is planned for the UWML, which is shown in (37), with the step length $S_0 = 0.4$ m, step height $H_0 = 0.2$ m, step period $T = 8$ s, translational phase period $T_m = 4$ s, and initial height of the foot tip $z_0 = -0.61$.

$$
\begin{cases}
x = \begin{cases} \frac{S_0}{2\pi}\left(\frac{2\pi t}{T_m} - \sin\left(\frac{2\pi t}{T_m}\right)\right) - \frac{S_0}{2}, & (0 \le t \le T_m) \\ S_0 - \frac{S_0}{2\pi}\left(\frac{2\pi t}{T_m} - \sin\left(\frac{2\pi t}{T_m}\right)\right) - \frac{S_0}{2}, & (T_m < t \le T) \end{cases} \\
y = L_2, \, (0 \le t \le T_m) \\
z = \begin{cases} 2H_0\left(\frac{t}{T_m} - \frac{1}{4\pi}\sin\left(\frac{4\pi t}{T_m}\right)\right) + z_0, & (0 \le t \le T_m/2) \\ -2H_0\left(\frac{t}{T_m} - \frac{1}{4\pi}\sin\left(\frac{4\pi t}{T_m}\right)\right) + 2H_0 + z_0, & (T_m/2 < t \le T_m) \\ z_0, & (T_m \le t \le T) \end{cases}
\end{cases}
\tag{37}
$$

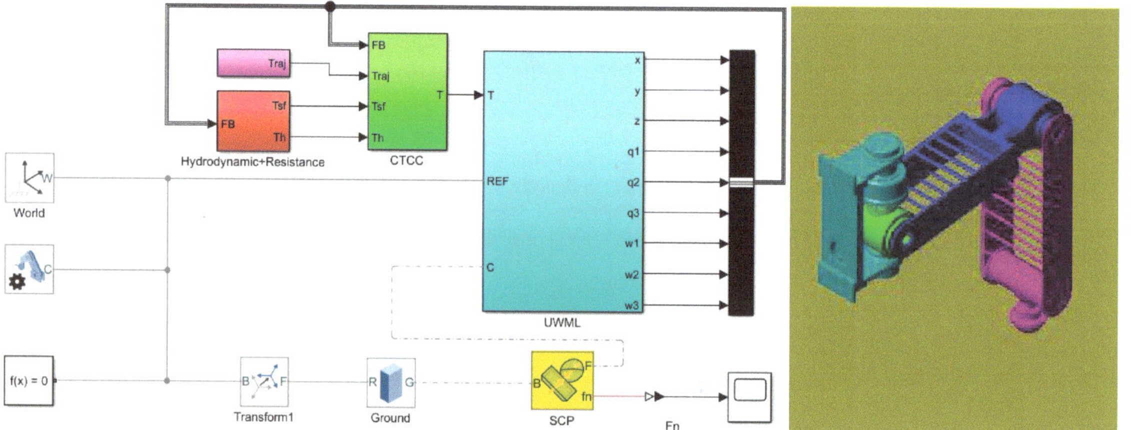

Figure 6. Simulation model of the UWML.

Table 2. Parameters of the UWML.

Link	Mass (kg)	Center of Mass (m)	Inertia Matrix (kg·m^2)
1	10.758	$[0, 0.001, -0.017]^T$	diag{0.044, 0.039, 0.032}
2	19.261	$[-0.154, 0, -0.014]^T$	diag{0.135, 1.455, 1.375}
3	10.375	$[-0.391, 0, -0.045]^T$	diag{0.138, 2.437, 2.327}

Table 3. Parameters of the simulation controller.

Controller	Parameters
ASTSMC-ESO	$\lambda_1 = \text{diag}\{400, 800, 800\}$, $k_{10} = \text{diag}\{12, 20, 4\}$, $k_{20} = \text{diag}\{200, 200, 200\}$, $\alpha_0 = \text{diag}\{6.32, 6.32, 8.94\}$, $\beta_0 = \text{diag}\{5, 5, 10\}$, $\tau = \text{diag}\{0.01, 0.01, 0.01\}$, $a = \text{diag}\{0.86, 0.86, 0.86\}$, $l_0 = \text{diag}\{0.5, 0.5, 0.5\}$, $\rho_0 = \text{diag}\{0.5, 0.5, 0.5\}$, $\gamma = \text{diag}\{0.002, 0.002, 0.002\}$, $\delta_0 = \text{diag}\{50, 50, 50\}$, $\epsilon = \text{diag}\{0.1, 0.1, 0.1\}$

5.2. Simulation Study

Three working conditions were studied with the ASTSMC-ESO. Condition 1 is the basic condition, where no disturbances are added to the UWML, and the action time is $0\,\text{s} \leq t \leq 20\,\text{s}$, $40\,\text{s} \leq t \leq 60\,\text{s}$, and $80\,\text{s} \leq t \leq 100\,\text{s}$, respectively. Condition 2 is the condition where the modeling uncertainty disturbances are added, of which the parameters M, C, and G of the UWML are set with an amplitude of 50% variation during the time period of $20\,\text{s} \leq t \leq 40\,\text{s}$, and the changing law is shown in Equation (38). Condition 3 is the condition where the external disturbances are added, of which the system uncertainty d in the dynamic equation of the UWML has a fluctuation of 50 Nm during the time period of $60\,\text{s} \leq t \leq 80\,\text{s}$, and the variation law is shown in Equation (39). The simulation results are shown in Figures 7–10.

$$\begin{cases} M = M_n[1 + 0.5\sin(0.25\pi t)](20 \leq t \leq 40) \\ C = C_n[1 + 0.5\sin(0.25\pi t)](20 \leq t \leq 40) \\ G = G_n[1 + 0.5\sin(0.25\pi t)](20 \leq t \leq 40) \end{cases} \tag{38}$$

$$d = 50\sin(0.25\pi t)(60 \leq t \leq 80) \tag{39}$$

From the foot trajectory tracking results in Figure 7, it can be seen that the ASTSMC-ESO can accurately control the UWML to move along the planned trajectory under all the three working conditions with tracking errors within 10^{-2} mm in all three directions of the workspace, which illustrates the effectiveness of the ASTSMC-ESO. In addition, from Figure 7, we can also see that the foot trajectory tracking errors under two external disturbances are not significantly different from those without disturbances, indicating that the ASTSMC-ESO has excellent robustness. Further analysis of the joint position control results in Figure 8 shows that the actual joint position almost coincides with the reference joint position. Even under the time-varying modeling uncertainties with an amplitude up to 50% of the nominal value of the system parameters in working condition 2 and the external disturbances with an amplitude up to 50 Nm in working condition 3, the position tracking error of each joint can still be maintained within 10^{-5} rad and is hardly affected by external disturbances.

The estimated uncertainties in Figure 9 show that no matter which working conditions, the ESO can accurately estimate the system uncertainties. This indicates that the ESO has good estimation ability for the system uncertainties. As can be seen, with the help of the ESO, the uncertainties of different working conditions can be effectively compensated via ASTSMC-ESO, so that the joint control torque can respond quickly and accurately to reduce the impact of these uncertainties on the system performance. In addition, it can be seen from Figure 10 that the parameter's adaptive function $L_i(t)$ has good convergence performance and can be adjusted online periodically according to the motion of UWML, thereby eliminating the chattering of SMC.

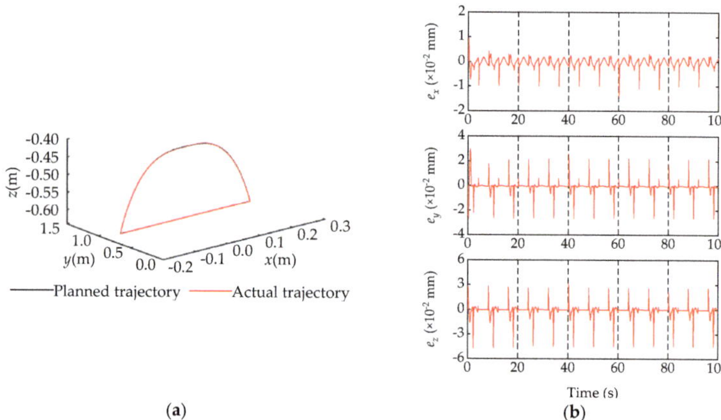

Figure 7. Foot trajectory tracking results. (**a**) Foot trajectory; (**b**) Foot trajectory tracking error.

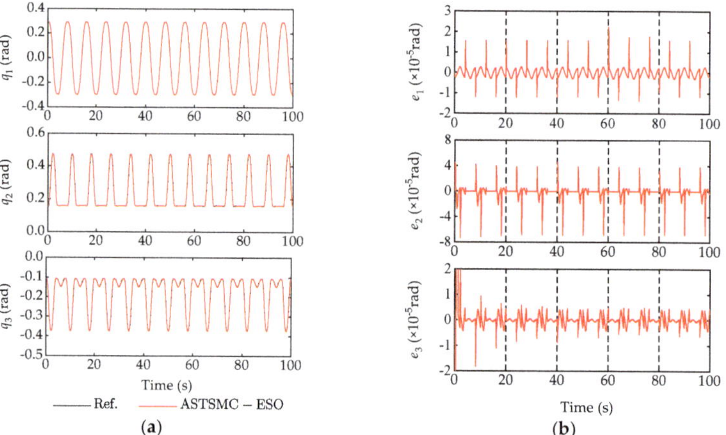

Figure 8. Joint position tracking results. (**a**) Joint position; (**b**) Joint position tracking error.

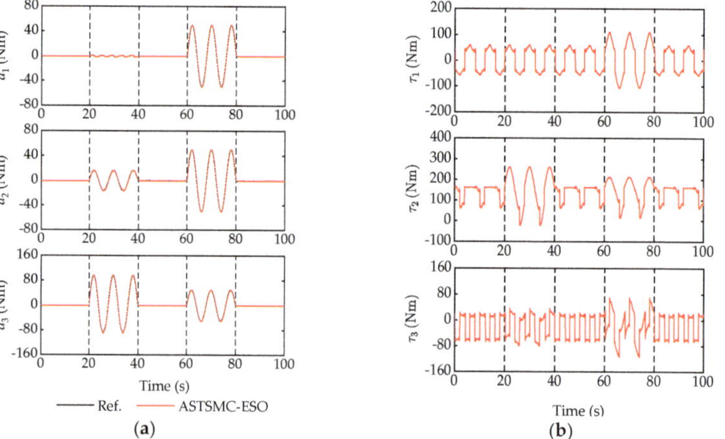

Figure 9. System uncertainty estimation and joint control torque. (**a**) Uncertainties estimation; (**b**) Joint control torque.

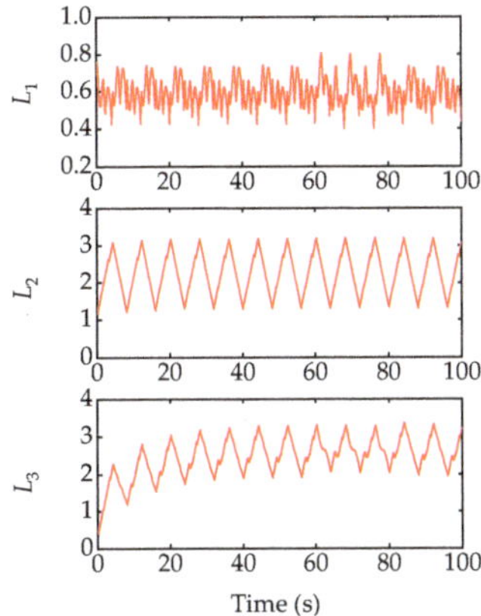

Figure 10. Parameter adaptive function $L_i(t)$.

6. Comparative Experimental Results

To further test the performance of the ASTSMC-ESO, experimental research was conducted with the UWML, of which the test platform is presented in Figure 11, and its main components are listed in Table 4. To illustrate the control performance, comparative studies were carried out on the ASTSMC-ESO, SMC [33], and AFSMC [34].

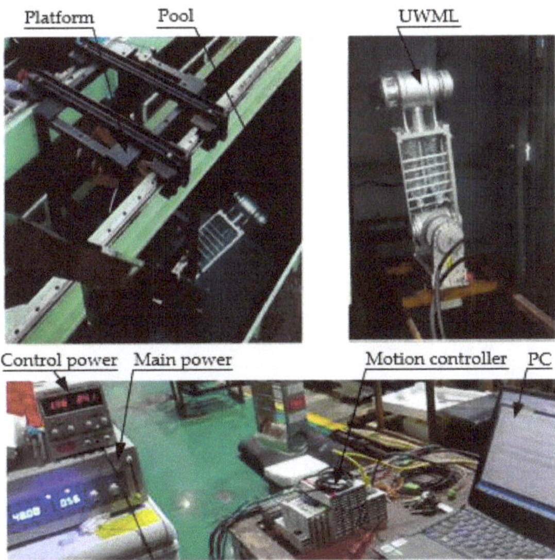

Figure 11. The experimental platform of the UWML.

Table 4. The main components of the experimental platform of the UWML.

Name	Specification	Name	Specification
Motor for joint 1/2/3	Kollmorgen TBMS-7646-A	Velocity sensor of joint 2	Tamagawa S2620N271E14
Reducer for joint 1/2/3	Benrun BHS32-160	Position sensor of joint 3	Tamagawa S2640N321E64
Motor driver for joint 1/2/3	Elmo G-SOLTWI15/100ER1	Velocity sensor of joint 3	Tamagawa S2620N271E14
Position sensor of joint 1	Tamagawa TS2660N31E148	Motion controller	Beckhoff CX5140 PLC
Velocity sensor of joint 1	Tamagawa TS2620N271E14	Main power	DC48V
Position sensor of joint 2	Tamagawa TS2620N271E14	Control power	DC24V

6.1. Comparison Controllers

(1) SMC: This is the traditional sliding mode controller with the compensation of external disturbances of τ_f, τ_s, and τ_w:

$$\tau = M_n(\ddot{x}_{1d} + \lambda_1 e_2) + C_n x_2 + G_n + \tau_{fn} + \tau_{sn} - \tau_{wn} + \lambda_2 s + \lambda_r \text{sign}(s) \tag{40}$$

where λ_1, λ_2, and λ_r are positive definite diagonal parameter matrices. In order to suppress the chattering phenomenon caused by the signum function sign(), the saturation function sat() with a boundary layer thickness of δ is used instead of the sign() function in the application.

(2) AFSMC: This is the adaptive fuzzy sliding mode controller with the compensation of disturbances of τ_f, τ_s, and τ_w:

$$\tau = M_n(\ddot{x}_{1d} + \lambda_1 e_2) + C_n x_2 + G_n + \tau_{fn} + \tau_{sn} - \tau_{wn} + \lambda_2 s + \lambda_r \tag{41}$$

$$\hat{\lambda}_{ri} = \hat{\theta}_i^T \psi_i(s_i) \tag{42}$$

$$\dot{\hat{\theta}}_i = -\Gamma s_i \psi_i(s_i) \tag{43}$$

where λ_1, λ_2, and Γ are positive definite diagonal parameter matrices. The elements λ_{ri} in λ_r are approximated by an adaptive fuzzy logic system (AFLS), as shown in (42), with a weight adaptive rate $\dot{\hat{\theta}}_i$ given by (43). The fuzzy rules are shown in Table 5, and the Gaussian functions (44) are chosen as the membership functions, where $i = 1, 2, 3$ and $j = 1, 2, 3, 4, 5$.

$$\begin{cases} \psi_j(s_1) = \exp\left(-\left[\dfrac{s_1 + \pi/12 - (j-1)\pi/24}{3\pi/48}\right]^2\right) \\[2ex] \psi_j(s_2) = \exp\left(-\left[\dfrac{s_2 + \pi/12 - (j-1)\pi/24}{20\pi/48}\right]^2\right) \\[2ex] \psi_j(s_3) = \exp\left(-\left[\dfrac{s_3 + \pi/12 - (j-1)\pi/24}{3\pi/48}\right]^2\right) \end{cases} \tag{44}$$

Table 5. Fuzzy rules of λ_{ri}.

Condition	Conclusion
IF s_i is NB	THEN λ_{ri} is NB
IF s_i is NS	THEN λ_{ri} is NS
IF s_i is ZE	THEN λ_{ri} is ZE
IF s_i is PS	THEN λ_{ri} is PS
IF s_i is PB	THEN λ_{ri} is PB

6.2. Experimental Study

The controller parameters of the experimental study are shown in Table 6. For fairness of comparison, some parameters of ASTSMC-ESO and AFSMC are inherited from the SMC. To evaluate the performance of the three control algorithms, the maximum error M_e, average error μ_e, and standard deviation error σ_e of the foot trajectory were used as indicators [35]. The results are shown in Figures 12–14, and the performance evaluation results are listed in Table 7.

Table 6. Controller parameters of the experimental study.

Controller	Parameters
SMC	$\lambda = \text{diag}\{400,\ 600,\ 600\}$, $k_1 = \text{diag}\{10,\ 20,\ 8\}$, $k_2 = \text{diag}\{150,\ 180,\ 100\}$, $\delta = \text{diag}\{0.1,\ 0.1,\ 0.1\}$
AFSMC	$\lambda = \text{diag}\{400,\ 600,\ 600\}$, $k_1 = \text{diag}\{10,\ 20,\ 8\}$, $\Gamma = \text{diag}\{1500,\ 1500,\ 1500\}$
ASTSMC-ESO	$\lambda = \text{diag}\{400,\ 600,\ 600\}$, $k_{10} = \text{diag}\{10,\ 20,\ 8\}$, $k_{20} = \text{diag}\{100,\ 100,\ 100\}$, $\alpha_0 = \text{diag}\{6.32,\ 6.32,\ 8.94\}$, $\beta_0 = \text{diag}\{5,\ 5,\ 10\}$, $\tau = \text{diag}\{0.01,\ 0.01,\ 0.01\}$, $a = \text{diag}\{0.19,\ 0.19,\ 0.38\}$, $l_0 = \text{diag}\{0.5,\ 0.5,\ 0.5\}$, $\rho_0 = \text{diag}\{0.5,\ 0.5,\ 0.5\}$, $\gamma = \text{diag}\{0.005,\ 0.005,\ 0.005\}$, $\delta_0 = \text{diag}\{80,\ 80,\ 80\}$, $\epsilon = \text{diag}\{0.1,\ 0.1,\ 0.1\}$

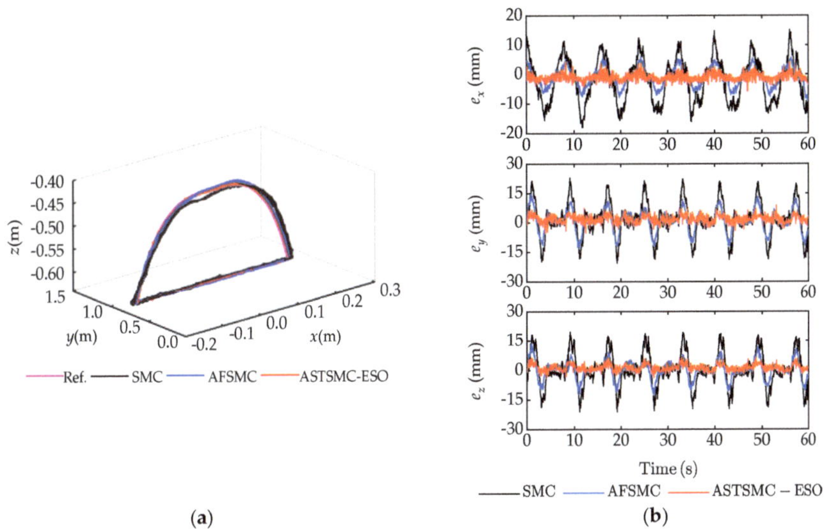

Figure 12. Foot trajectory tracking results with different controllers. (**a**) Foot trajectory; (**b**) Foot trajectory tracking error.

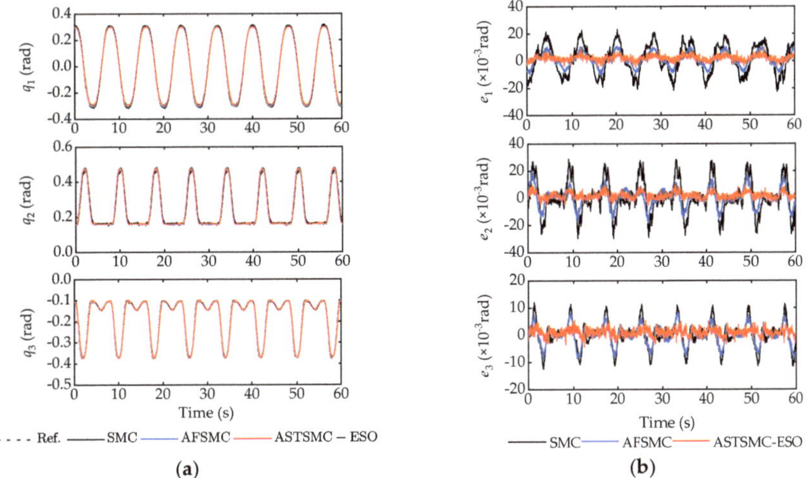

Figure 13. Joint position tracking results with different controllers. (**a**) Joint position; (**b**) Joint position tracking error.

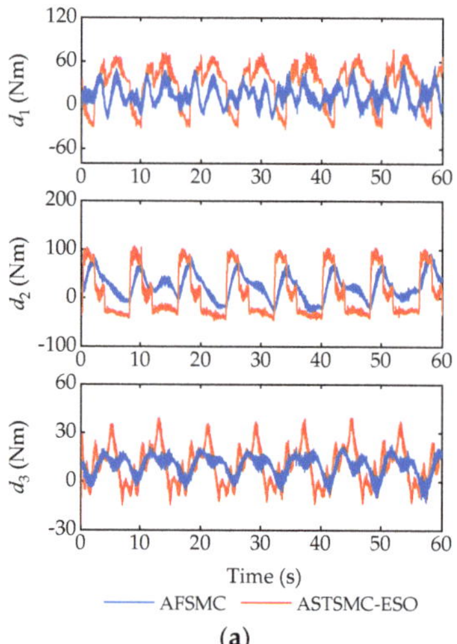
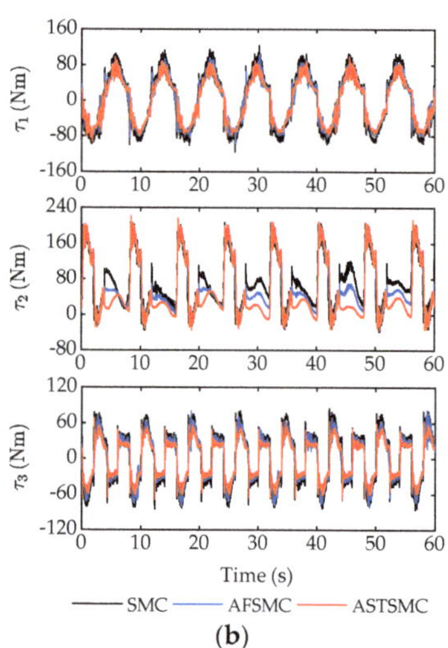

Figure 14. System uncertainty estimation and joint control torque with different controllers. (**a**) Uncertainty estimation; (**b**) Joint control torque.

Table 7. Evaluation results of the trajectory tracking performance with different controllers.

Controller	M_e (mm)	μ_e (mm)	σ_e (mm)
SMC	[17.90, 22.76, 21.15]	[6.42, 6.52, 6.17]	[4.25, 6.05, 5.78]
AFSMC	[8.94, 14.67, 13.73]	[2.79, 4.32, 3.44]	[2.01, 3.91, 3.04]
ASTSMC-ESO	[5.92, 8.28, 6.26]	[1.42, 1.96, 1.54]	[0.91, 1.49, 1.28]

The desired foot trajectory and corresponding tracking performance under the three controllers are shown in Figure 12. As seen, the proposed ASTSMC-ESO controller performs better than the other two controllers throughout the movement. From the performance evaluation results in Table 7, it can be seen that all the performance indicators of ASTSMC-ESO are better than SMC and AFSMC. Comparing the joint positions in Figure 13, it can be seen that through uncertainty compensation, the joint position control accuracy can be effectively improved. Further comparing of the joint position errors of ASTSMC-ESO and AFSMC indicates that the former has better robustness than the latter, which is mainly due to the better control mechanism of ASTSMC-ESO than AFSMC.

From Figure 14, we can conclude that the differences in the trajectory tracking performance are mainly due to the differences in the joint control torques of each control method, even though they seem minor. Furthermore, we can see that the torque differences of joint 2 are more significant than those of the other two joints, which are mainly caused by the accuracy of uncertainty compensation, as shown in Figure 14a. AFSMC and ASTSMC-ESO estimate and compensate for the system uncertainties with different methods, so their output control torques can effectively improve the trajectory tracking performance compared with SMC. However, due to the difference in estimation performance, ASTSMC-ESO can compensate for the system uncertainties more precisely than AFSMC, thus further improving the system performance. In addition, it can be seen that the chattering phenomenon of the joint control torques of AFSMC and ASTSMC-ESO is effectively eliminated, which indicates that the AFLS and AST can effectively suppress the chattering effect of SMC while ensuring the control performance of the system.

7. Conclusions

In this paper, a novel adaptive super-twisting sliding mode control method with an extended state observer is proposed for the high-accuracy position control of the UWML in the presence of both uncertainties and external disturbances, which considers the system uncertainties and the external disturbances from the underwater working environment. On the one hand, an accurate model compensation is made with SMC feedback control based on the model information of the UWML to ensure a quick response performance. On the other hand, a feedforward compensation of the system uncertainties is achieved with the estimated uncertainties by ESO to reduce their impact on the control performance. Finally, the AST is introduced to eliminate the chattering phenomenon and further improve the steady-state control accuracy. Simulation and comparative experimental studies were conducted to illustrate the effectiveness of this proposed control scheme, which shows that the proposed controller can effectively compensate for system uncertainties and disturbances and significantly enhance system control accuracy and robustness without the steady-state chattering effect.

Author Contributions: Conceptualization, resources, and writing—original draft, L.L. and M.N.; simulation analysis and data curation, L.G.; experimental analysis and visualization, L.L., M.N. and D.Z.; formal analysis, experimental guidance, and writing—review and editing, J.D.; supervision and methodology, B.L. All authors have read and agreed to the published version of the manuscript.

Funding: This research received no external funding.

Data Availability Statement: Not applicable.

Acknowledgments: The authors would like to thank Xinliang Wang, Dingfeng Liu, Kang Zhang, etc., researchers of the Second Ship Design and Research Institute of Wuhan, for their technical support.

Conflicts of Interest: The authors declare no conflict of interest.

Appendix A. Proof of Theorem 2

Before the proof of Theorem 2, the following finite-time convergence lemma is introduced:

Lemme A1 [36]. *For nonlinear system:*

$$x(t) = f(x(t)), \; f(0) = 0, \; x \in D \tag{A1}$$

if there exists a continuously differentiable positive definite function $V(x)$ and parameters $\lambda_1 > 0$, $\lambda_2 > 0, 0 < \theta < 1$, such that the following inequality holds:

$$\dot{V}(x) \leq -\lambda_1 V(x) - \lambda_2 V^\theta(x) \tag{A2}$$

then the state of the system (A1) will converge to the origin in finite time t_r, which satisfies:

$$t_r \leq \frac{1}{\lambda_1(1-\theta)} \ln \frac{\lambda_2 + \lambda_2 V^{1-\theta}(0)}{\lambda_2} \tag{A3}$$

To simplify the proof, system (20) is decomposed into three subsystems, where the *ith* subsystem is:

$$\begin{cases} \dot{s}_i = -\alpha_i |s_i|^{1/2} sign(s_i) - k_{1i} s_i + z_i + \phi_i(s_i, L_i) \\ \dot{z}_i = -\beta_i sign(s_i) - k_{2i} s_i + f_i(t) \end{cases} \tag{A4}$$

Define the following Lyapunov function for the subsystem (A4):

$$V_i(s_i, z_i, L_i) = 2\beta_i |s_i| + k_{2i} s_i^2 + \frac{1}{2} z_i^2 + \frac{1}{2} \zeta_i^2 \tag{A5}$$

where $\zeta_i = \alpha_i |s_i|^{1/2} sign(s_i) + k_{1i} s_i - z_i$.

The time derivative of $V_i(s_i,\ z_i,\ L_i)$ is:

$$\dot{V}_i(s_i,\ z_i,\ L_i) \;=\; \frac{\partial V_i}{\partial s_i}\dot{s}_i + \frac{\partial V_i}{\partial z_i}\dot{z}_i + \frac{\partial V_i}{\partial L_i}\dot{L}_i \;=\; \dot{V}_{i1} + \dot{V}_{i2} \tag{A6}$$

where $\dot{V}_{i1}$ and $\dot{V}_{i2}$ are given below:

$$\begin{aligned}
\dot{V}_{i1} &= \frac{\partial V_i}{\partial s_i}\dot{s}_i + \frac{\partial V_i}{\partial z_i}\dot{z}_i \;=\; \beta_i \mathrm{sgn}(s_i)\dot{s}_{i0} + 2k_{2i}s_i\dot{s}_{i0} \\
&\quad + z_i\dot{z}_i\left(\alpha_i|s_i|^{1/2}\mathrm{sign}(s_i) + k_{1i}s_i - z_i\right)\left(\frac{\alpha_i}{2|s_i|^{1/2}}\dot{s}_{i0} + k_{1i}\dot{s}_{i0} - \dot{z}_i\right) \\
&\quad + 2\beta_i\mathrm{sgn}(s_i)\phi_i + 2k_{2i}s_i\phi_i + \left(\alpha_i|s_i|^{1/2}\mathrm{sign}(s_i) + k_{1i}s_i - z_i\right)\left(\frac{\alpha_i}{2|s_i|^{1/2}}\phi_i + k_{1i}\phi_i\right)
\end{aligned} \tag{A7}$$

$$\begin{aligned}
\dot{V}_{i2} &= \frac{\partial V_i}{\partial L_i}\dot{L}_i \;=\; 2\dot{L}_i\beta_{0i}\mathrm{sgn}(s_i)s_i + 2L_i\dot{L}_ik_{20i}s_i^2 \\
&\quad + \left(\alpha_i|s_i|^{1/2}\mathrm{sign}(s_i) + k_{1i}s_i - z_i\right)\left(\frac{1}{2\sqrt{L_i}}\dot{L}_i\alpha_{i0}|s_i|^{1/2}\mathrm{sign}(s_i) + \dot{L}_ik_{1i0}s_i\right) \\
&= -2\beta_i\mathrm{sgn}(s_i)\phi_i - 2k_{2i}s_i\phi_i - \left(\alpha_i|s_i|^{1/2}\mathrm{sign}(s_i) + k_{1i}s_i - z_i\right)\left(\frac{\alpha_i}{2|s_i|^{1/2}}\phi_i + k_{1i}\phi_i\right)
\end{aligned} \tag{A8}$$

where $\dot{s}_{i0} = -\alpha_i|s_i|^{1/2}\mathrm{sign}(s_i) - k_{1i}s_i + z_i$.

Substituting (A7) and (A8) into (A6) yields:

$$\begin{aligned}
\dot{V}_i(s_i,\ z_i,\ L_i) &= \dot{V}_{i1} + \dot{V}_{i2} \\
&= 2\beta_i\mathrm{sgn}(s_i)\dot{s}_{i0} + 2k_{2i}s_i\dot{s}_{i0} + z_i\dot{z}_i + \left(\alpha_i|s_i|^{1/2}\mathrm{sign}(s_i)\right. \\
&\quad \left. + k_{1i}s_i - z_i\right)\left(\frac{\alpha_i}{2|s_i|^{1/2}}\dot{s}_{i0} + k_{1i}\dot{s}_{i0} - \dot{z}_i\right)
\end{aligned} \tag{A9}$$

Applying $\dot{s}_{i0} = -\alpha_i|s_i|^{1/2}\mathrm{sign}(s_i) - k_{1i}s_i + z_i$ and $\dot{z}_i = -\beta_i\mathrm{sign}(s_i) - k_{2i}s_i + f_i(t)$ to (A9), we can obtain:

$$\dot{V}_i(s_i,\ z_i,\ L_i) \;=\; -\frac{\sqrt{L_i}}{|s_i|^{1/2}}\xi_i^T(A_i - C_i)\xi_i - L_i\xi_i^T B_i\xi_i \tag{A10}$$

where $\xi_i = \left[\sqrt{L_i}|s_i|^{1/2}\mathrm{sign}(s_i)f_i,\ L_is_i,\ z_i\right]^T$, A_i, B_i and C_i are shown in (25), (27), and (28), respectively.

From (A10) we can see that by selecting the appropriate parameters α_{0i}, β_{0i}, k_{10i}, and k_{20i} such that the matrices $Q_i = A_i - C_i$ and B_i are positive definite, then $\dot{V}_i(s_i,\ z_i,\ L_i)$ can be transformed into:

$$\dot{V}_i \leq -\frac{\sqrt{L_i}}{|s_i|^{1/2}}\lambda_{min}(Q_i)\|\xi_i\|^2 - L_i\lambda_{min}(B_i)\|\xi_i\|^2 \tag{A11}$$

From the definition of $V_i(s_i,\ z_i,\ L_i)$ and ξ_i, $V_i(s_i,\ z_i,\ L_i)$, it can also be expressed as:

$$V_i(s_i,\ z_i,\ L_i) \;=\; \xi_i^T P_i\xi_i \tag{A12}$$

Therefore, by selecting appropriate parameters α_{0i}, β_{0i}, k_{10i} and k_{20i} such that the matrix P_i is positive definite, then $V_i(s_i,\ z_i,\ L_i)$ can be transformed into:

$$V_i(s_i,\ z_i,\ L_i) \leq \lambda_{max}(P_i)\|\xi_i\|^2 \tag{A13}$$

From (A13), we can obtain:

$$\|\xi_i\|^2 \leq \frac{V_i}{\lambda_{max}(P_i)} \tag{A14}$$

$$|s_i|^{1/2} = |\xi_{i1}| \leq \|\xi_i\| \leq \frac{V_i^{1/2}}{\lambda_{\max}^{1/2}(P_i)} \tag{A15}$$

Substituting (A14) and (A15) into (A11), we can obtain:

$$\dot{V}_i \leq -\frac{L_i \lambda_{min}(Q_i)\lambda_{min}^{1/2}(P_i)}{\lambda_{max}(P_i)} V_i^{1/2} - L_i \frac{\lambda_{min}(B_i)}{\lambda_{max}(P_i)} V_i = -\lambda_1 V_i - \lambda_2 V_i^{1/2} \tag{A16}$$

where $\lambda_1 = L_i\lambda_{\min}(B_i)/\lambda_{\max}(P_i)$, $\lambda_2 = L_i\lambda_{\min}(Q_i)\lambda_{\min}^{1/2}(P_i)/\lambda_{\max}(P_i)$.

According to the conclusions in [31,37], when $L_i(t) > \sigma_{0i} \geq |f_i(t)|$, $|\vartheta_i| = \left|\frac{f_i(t)\mathrm{sign}(s_i)}{L_i(t)}\right| < 1$ and the selecting parameters are α_{0i}, β_{0i}, k_{10i}, and k_{20i} to satisfy the following conditions in (A17), it can be guaranteed that B_i, P_i, and Q_i are positive definite matrices.

$$\alpha_{0i} > 5^{0.25}, \ \beta_{0i} > 1, \ k_{10i} > 0, \ k_{20i} > \frac{8k_{10i}^2\beta_{0i} + 22k_{10i}^2 + 9\alpha_{0i}^2 k_{10i}^2}{4(\beta_{0i} - 1)} \tag{A17}$$

when B_i, P_i, and Q_i are positive definite matrices, then $\lambda_1 > 0$ and $\lambda_2 > 0$. According to Lemma 1, the result of (A16) shows that the state of subsystem (A4) can converge to the origin within a finite time t_r, which completes the proof of Theorem 2.

References

1. Xu, S.; He, B.; Hu, H. Research on Kinematics and Stability of a Bionic Wall-Climbing Hexapod Robot. *Appl. Bionics Biomech.* **2019**, *2019*, 6146214. [CrossRef]
2. Guizzo, E. By leaps and bounds: An exclusive look at how Boston dynamics is redefining robot agility. *IEEE Spectr.* **2019**, *56*, 34–39. [CrossRef]
3. Carlo, J.D.; Wensing, P.M.; Katz, B.; Bledt, G.; Kim, S. Dynamic Locomotion in the MIT Cheetah 3 through Convex Model-Predictive Control. In Proceedings of the 2018 IEEE/RSJ International Conference on Intelligent Robots and Systems (IROS), Madrid, Spain, 1–5 October 2018; pp. 1–9.
4. Shim, H.; Yoo, S.Y.; Ju, B.H.; Kang, H. Development of arm and leg for seabed walking robot CRABSTER200. *Ocean Eng.* **2016**, *116*, 55–67. [CrossRef]
5. Yang, C.; Yao, F.; Zhang, M.J.; Zhang, Z.Q.; Wu, Z.Z.; Dan, P.J. Adaptive Sliding Mode PID Control for Underwater Manipulator Based on Legendre Polynomial Function Approximation and Its Experimental Evaluation. *Appl. Sci.* **2020**, *10*, 1728. [CrossRef]
6. Yao, J.J.; Wang, C.J. Model reference adaptive control for a hydraulic underwater manipulator. *J. Vib. Control* **2011**, *18*, 893–902. [CrossRef]
7. Lee, M.; Choi, H.S. A Robust Neural Controller for Underwater Robot Manipulators. *IEEE Trans. Neural Netw.* **2000**, *11*, 1465–1470.
8. Zhou, Z.C.; Tang, G.Y.; Huang, H.; Han, L.J.; Xu, R.K. Adaptive nonsingular fast terminal sliding mode control for underwater manipulator robotics with asymmetric saturation actuators. *Control Theory Technol.* **2020**, *18*, 81–91. [CrossRef]
9. Zhou, Z.C.; Tang, G.Y.; Xu, R.K.; HAN, L.J.; Cheng, M.L. A Novel Continuous Nonsingular Finite-Time Control for Underwater Robot Manipulators. *J. Mar. Sci. Eng.* **2021**, *9*, 269. [CrossRef]
10. Chatchanayuenyong, T.; Parnichkun, M. Neural network based-time optimal sliding mode control for an autonomous underwater robot. *Mechatronics* **2006**, *16*, 41–87. [CrossRef]
11. Yao, B.; Tomizuka, M. Adaptive robust motion and force tracking control of robot manipulators in contact with compliant surfaces with unknown stiffness. *J. Dyn. Syst. Meas. Control* **1998**, *120*, 232–240. [CrossRef]
12. Sabanovic, A. Variable structure systems with sliding modes in motion control—A survey. *IEEE Trans. Ind. Inform.* **2011**, *2*, 212–223. [CrossRef]
13. Doulgeri, Z. Sliding regime of a nonlinear robust controller for robot manipulators. *IEE Proc. Control Theory Appl.* **1999**, *146*, 493–498. [CrossRef]
14. Goel, A.; Swarup, A. Adaptive fuzzy high-order super-twisting sliding mode controller for uncertain robotic manipulator. *J. Intell. Syst.* **2017**, *26*, 697–715. [CrossRef]
15. Levant, A. Higher-order sliding modes, differentiation and output-feedback control. *Int. J. Control* **2003**, *76*, 924–941. [CrossRef]
16. Nagesh, I.; Edwards, C. A multivariable super-twisting sliding mode approach. *Automatica* **2014**, *50*, 984–988. [CrossRef]
17. Polyakov, A.; Poznyak, A. Reaching time estimation for "super-twisting" second order sliding mode controller via Lyapunov function designing. *IEEE Trans. Autom. Control* **2009**, *54*, 1951–1955. [CrossRef]
18. Yang, Y.; Qin, S. A new modified super-twisting algorithm with double closed-loop feedback regulation. *Trans. Inst. Meas. Control* **2017**, *39*, 1603–1612. [CrossRef]

19. Boiko, I.; Fridman, L. Analysis of chattering in continuous sliding-mode controllers. *IEEE Trans. Autom. Control* **2005**, *50*, 1442–1446. [CrossRef]

20. Edwards, C.; Shtessel, Y. Adaptive dual-layer super-twisting control and observation. *Int. J. Control* **2016**, *89*, 1759–1766. [CrossRef]

21. Van, M.; Ge, S.S. Adaptive Fuzzy Integral Sliding-Mode Control for Robust Fault-Tolerant Control of Robot Manipulators with Disturbances Observer. *IEEE Trans. Fuzzy Syst.* **2021**, *29*, 284–1296. [CrossRef]

22. Zhu, Y.K.; Qiao, J.Z.; Guo, L. Adaptive Sliding Mode Disturbances Observer-Based Composite Control with Prescribed Performance of Space Manipulators for Target Capturing. *IEEE Trans. Ind. Electron.* **2019**, *66*, 1973–1983. [CrossRef]

23. Xi, R.; Xiao, X.; Ma, T.; Yang, Z. Adaptive Sliding Mode Disturbances Observer Based Robust Control for Robot Manipulators Towards Assembly Assistance. *IEEE Robot. Autom. Lett.* **2022**, *7*, 6139–6146. [CrossRef]

24. Xu, B.; Ji, S.; Zhang, C.; Chen, C.; Ni, H.; Wu, X. Linear-extended-state-observer-based prescribed performance control for trajectory tracking of a robotic manipulator. *Ind. Robot Int. J. Robot. Res. Appl.* **2021**, *48*, 544–555. [CrossRef]

25. Tran, D.T.; Jin, M.; Ahn, K.K. Nonlinear Extended State Observer Based on Output Feedback Control for a Manipulator with Time-Varying Output Constraints and External Disturbances. *IEEE Access* **2019**, *7*, 156860–156870. [CrossRef]

26. Wu, X.; Wang, C.; Hua, S. Adaptive extended state observer-based nonsingular terminal sliding mode control for the aircraft skin inspection robot. *J. Intell. Robot. Syst.* **2020**, *98*, 721–732. [CrossRef]

27. Yao, J.Y.; Jiao, Z.X.; Ma, D.W. Adaptive robust control of DC motors with extended state observer. *IEEE Trans. Ind. Electron.* **2013**, *61*, 3630–3637. [CrossRef]

28. Liao, L.H.; Li, B.R.; Wang, Y.Y.; Xi, Y.; Zhang, D.J.; Gao, L.L. Adaptive fuzzy robust control of a bionic mechanical leg with a high gain observer. *IEEE Access* **2021**, *9*, 134037–134051. [CrossRef]

29. Liao, L.H.; Li, B.R.; Zhang, D.J.; Ngwa, M.; Gao, L.P.; Du, J.M. Research on the Influence of Underwater Environment on the Dynamic Performance of the Mechanical Leg of a Deep-sea Crawling and Swimming Robot. *arXiv* **2023**, arXiv:2308.14393.

30. Zheng, Q.; Gao, L.; Gao, Z. On stability analysis of active disturbances rejection control for nonlinear time-varying plants with unknown dynamics. In Proceedings of the IEEE Conference on Decision and Control, New Orleans, LA, USA, 12–14 December 2007; pp. 3501–3506.

31. Yang, Y.J.; Liao, Y.; Yin, D.W.; Zheng, Y.X. Adaptive dual layer fast super twisting control algorithm. *Control Theory Appl.* **2016**, *33*, 1119–1127.

32. Zhang, M.; Guan, Y.; Zhao, W. Adaptive super-twisting sliding mode control for stabilization platform of laser seeker based on extended state observer. *Optik* **2019**, *199*, 163337. [CrossRef]

33. Luo, G.S. Research on Subsea 7 Function Maser-Slave Hydraulic Manipulator and Its Nonlinear Robust Control. Ph.D. Thesis, Zhejiang University, Hangzhou, China, 2013.

34. Zhong, Y.G.; Yang, F. Dynamic Modeling and adaptive fuzzy sliding mode control for multi-link underwater manipulators. *Ocean Eng.* **2019**, *187*, 106202.

35. Yao, J.Y.; Deng, W.X.; Jiao, Z.X. Adaptive Control of Hydraulic Actuators with LuGre Model-Based Friction Compensation. *IEEE Trans. Ind. Electron.* **2015**, *62*, 6469–6477. [CrossRef]

36. Wang, Z. Adaptive smooth second-order sliding mode control method with application to missile guidance. *Trans. Inst. Meas. Control* **2017**, *39*, 848–860. [CrossRef]

37. Tan, J.; Zhou, Z.; Zhu, X.P.; Xu, X.P. Fast super twisting algorithm and its application to attitude control of flying wing UAV. *Control Decis.* **2016**, *31*, 143–148.

Article

Design and Simulation of a Seven-Degree-of-Freedom Hydraulic Robot Arm

Jun Zhong *, Wenjun Jiang, Qianzhuang Zhang and Wenhao Zhang

College of Mechanical & Electrical Engineering, Hohai University, Changzhou 213022, China;
jnate@hhu.edu.cn (W.J.); zqz@hhu.edu.cn (Q.Z.); 211319010035@hhu.edu.cn (W.Z.)
* Correspondence: zhongjun@hhu.edu.cn; Tel.: +86-18651991185

Abstract: The current reliance on manual rescue is inefficient, and lightweight, highly flexible, and intelligent robots need to be investigated. Global seismic disasters occur often, and rescue jobs are defined by tight timetables and high functional and intellectual requirements. This study develops a hydraulically powered redundant robotic arm with seven degrees of freedom. To determine the force situation of the robotic arm in various positions, the common digging and handling conditions of the robotic arm are dynamically simulated in ADAMS. A finite element analysis is then performed for the dangerous force situation to confirm the structural strength of the robotic arm. The hydraulic manipulator prototype is manufactured, and stress–strain experiments are conducted on the robotic arm to verify the finite element simulation's reliability.

Keywords: hydraulic robotic arm; redundant manipulator; manipulator dynamics simulation; finite element analysis; stress–strain experiments

1. Introduction

Efficient and quick rescue operations in the aftermath of natural disasters such as earthquakes and mudslides are a complex issue. Rescue equipment can be broadly categorized into two types: search and rescue. While the search type is used to locate trapped individuals, the rescue type is designed to perform tasks such as grasping, handling, and shearing that would otherwise require manual labor. Robots are increasingly being used for rescue operations to improve work flexibility and expand the range of operations. Over the past decades, different types of rescue robots have been developed, such as snake-shaped search and rescue robots [1–3], bionic crawling search and rescue robots [4,5], wheel-footed robots with enhanced obstacle-crossing capabilities [6–8], aircraft-based rescue robots [9,10], rope-assisted climbing robots for applications in mountainous environments [11], and deformable robots for water rescue [12]. However, few of these robots have significant excavation and handling capabilities. Mitsubishi Heavy Industries developed a robot that could load 5 kg heavy objects and a robot that could load 10 kg heavy objects by adapting to a variety of environments, which are compact, flexible, and can enter narrow spaces [13]. Japan's TMSUK developed the "Enryu T-53" [14]—a crawler-type double-robotic-arm robot with high load capacity and suitable for handling, capable of lifting 100 kg of weight with a single arm. Wolf et al. [15] of Carnegie Mellon University developed a rescue robot with a wheeled motion platform and a multi-degree-of-freedom extension robotic arm, and the bottom moving platform expands the operating range of the robot. Shandong University, in cooperation with Luban Machinery Technology Corporation, developed a lightweight multifunctional dual-arm rescue robot with an overall weight of less than 5 tons, which can realize a variety of remote rescue functions [16].

The weight and load capacity of the rescue robot are the most important performance indicators, and the robotic arm, as the main actuator of the robot, plays an important role in the rescue process. Optimizing the structure and size of the robotic arm, reducing its own weight, and improving its load to weight ratio under the premise of ensuring

Citation: Zhong, J.; Jiang, W.; Zhang, Q.; Zhang, W. Design and Simulation of a Seven-Degree-of-Freedom Hydraulic Robot Arm. *Actuators* **2023**, *12*, 362. https://doi.org/10.3390/act12090362

Academic Editor: Ioan Doroftei

Received: 15 August 2023
Revised: 9 September 2023
Accepted: 12 September 2023
Published: 14 September 2023

its load capacity can effectively reduce consumables and control production costs on the one hand, and reduce energy consumption and improve work efficiency on the other. There are various ways to optimize the robotic arm, which can be optimized and improved in terms of structure [17–20], material [21–24], and process [25–27], in particular, combining topology optimization with manufacturing processes [28–30]. Rout et al. [31] used evolutionary optimization methods for two-degree-of-freedom planar manipulators and four-degree-of-freedom SCARA manipulators to design dimensional parameters and weights, and the method minimized the sensitivity of disturbing factors affecting accuracy and repeatability to minimum. Zhou L et al. [17] from Aalborg University combined finite element strength analysis based on kinematic performance, dynamics requirements, and lightweight design of a five-degree-of-freedom robotic arm in terms of both drive chain and structural dimensions. Liu W [32] used the non-inferiority ranking genetic algorithm NSGA-II (non-dominated sorting genetic algorithm II) with bit matrix representation to optimize the topology for multiple objectives such as mass and load capacity for lightweighting.

In this study, a seven-degree-of-freedom hydraulic robotic arm for emergency rescue is proposed. A model of the robotic arm is built, and in order to reduce the weight of the robotic arm, 7075 aluminum alloy is used as the main body material. Milling is then used to remove any excess material from the robotic arm while still maintaining the strength of the device and ensuring that it meets the load premise of the ideal weight. This study builds the dynamics model under the two working scenarios of excavation and heavy lifting and explores the finite element statics under these two working scenarios in order to test the strength of the robotic arm in the actual operation. The accuracy of the results of the finite element analysis was then confirmed by conducting stress–strain testing on a robotic arm prototype. This paper is organized as follows: the second part introduces the design and basic parameters of the robotic arm; the third part introduces the dynamic simulation of excavation and the handling of heavy objects; the fourth part introduces the finite element analysis of these conditions; and the fifth part introduces the prototype and finite element experiments carried out.

2. Structural Design of 7-DOFs Rescue Robotic Arm

The main challenges encountered during the process of its operation and performance requirements are as follows for earthquakes and other disasters that need to be specifically implemented after the rescue task to carry out the structural design of a rescue robotic arm, mainly including handling heavy objects, digging the ground, supporting the wall, and shear crushing and other conditions.

(1)	Because the homes of the current population are mostly made of reinforced concrete, earthquakes cause the walls to come off and crack. As a result, the robotic arm should be able to support and transport the collapsed wall with enough weight capacity.

(2)	The post-earthquake debris formation is largely random, and the situation for rescue operations is very unstable. The robotic arm must have a high degree of flexibility as well as strong adaptive and movement capabilities for the complex road surface in order to respond to the complex and changing rescue needs. In order to rescue the trapped people from the small space, the robotic arm must be able to quickly adjust its position to the best rescue state.

(3)	After the earthquake, the rescue road is severely blocked, making it impossible for rescue supplies to be transported to the disaster site by land at first. In order to meet helicopter lifting requirements, the weight of the entire robot arm should be kept within a reasonable range.

Table 1 below displays the primary design criteria for the hydraulic rescue robot arm.

Table 1. Mechanical arm design parameters table.

Performance Indicators	Parameters
Number of degrees of freedom	7
Weight	500 kg
Working radius	4 m
Maximum load capacity	300 kg
Power type	Hydraulic

Contrarily, industrial robots have greater flexibility and finer operations, but their drawbacks are insufficient load capacity and more stringent environmental requirements. Conventional construction machinery and equipment have strong environmental adaptability and large load capacities, but their disadvantages are relatively bulky and inflexible activities. A hydraulic machinery arm with seven degrees of freedom is thus constructed as follows Figure 1 after combining the benefits and drawbacks of industrial machinery equipment and industrial robots. The rotary base, boom, two arms, three arms, rotary hammer joint, swing, and end rotary joint are the key parts.

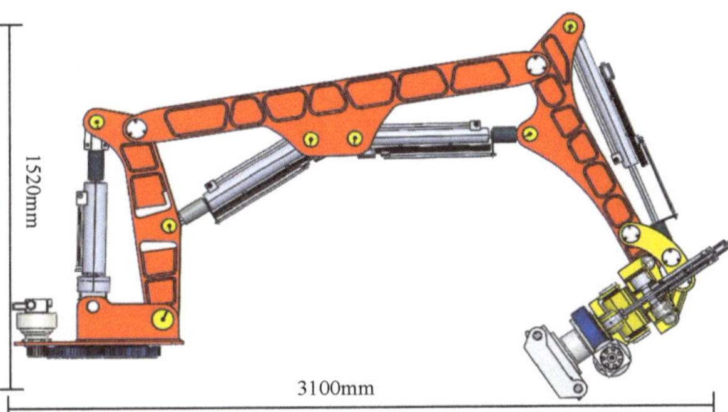

Figure 1. Mechanical arm structure schematic.

Aluminum alloy 7075 is the primary material utilized to meet the robotic arm's high strength and lightweight criteria. However, compared with steel, aluminum alloy material performs far poorer when it comes to welding. As a result, none of the robotic arm's arms are machined and no welding technology is used. The arm makes extensive use of carving and excavation to ensure lightweight. Figure 2 depicts the key parts of the robotic arm, with (a)–(e) standing for the base, primary arm, second arm, third arm, and end wrist.

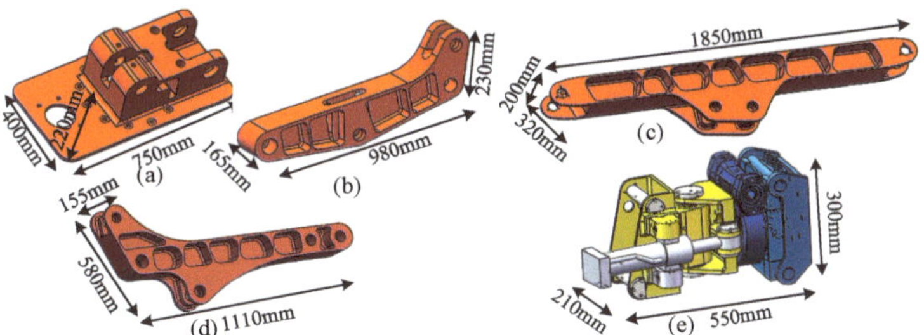

Figure 2. Models of the main parts of the robot arm: (**a**) Structural diagram of base; (**b**) Structural diagram of the first arm; (**c**) Structural diagram of the second arm; (**d**) Structural diagram of the third arm; (**e**) Structural diagram of the wrist joint.

3. Mechanical Performance Analysis of Robotic Arm under Different Working Conditions

In the actual post-earthquake rescue process, the most typical working conditions of the rescue robotic arm are excavating and damaging the road and lifting heavy objects, and the robotic arm is prone to structural failure due to insufficient strength or excessive deformation under these two dangerous working conditions. In this section, the virtual prototype of the robotic arm is built, and the mechanical performance analysis of the working device is completed by using ADAMS (version 2019) simulation software for these two typical working conditions.

3.1. Dynamic Simulation of Robotic Arm under Excavation Conditions

When the robotic arm performs the task of digging and damaging the road, it will adjust its digging posture according to the changes of the digging target, and it needs to carry out simulation analysis for the most typical and extreme working conditions of the force on the robotic arm during the digging operation. If the arm can operate normally under this condition and there is no safety hazard, then other digging conditions will also meet the requirements of safe operation, so this subsection will simulate the most dangerous and extreme conditions of the force on the arm during digging.

When the mechanical arm is digging at the maximum depth, the force on the big arm and the second arm is the greatest, and it is at the weakest state of the working arm. At this time, the hydraulic cylinders of the big arm and the third arm are fully extended, the hydraulic cylinders of the second arm are fully retracted, and the bucket tip is vertically downward, as shown in Figure 3a. When the line between the tip of the bucket tooth and the articulation point of the third arm and the swing frame is perpendicular to the swing-frame cylinder, the torque on the swing-frame cylinder is the greatest at this time, as shown in Figure 3b. Special attention is needed, when the bucket tooth tip is in common line with the swing frame, swing-frame cylinder, and three arms, as shown in Figure 3c. At this time, the mechanical arm is in the posture of digging radius, and the digging depth is large; the form of force is more dangerous and an extreme situation, which needs to be taken into account during simulation.

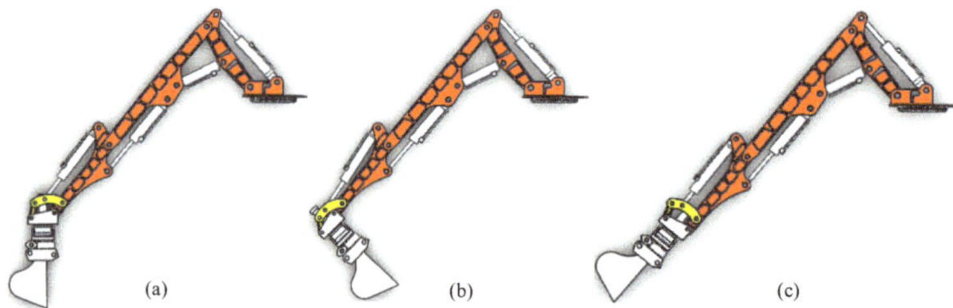

Figure 3. Dangerous posture of the robot arm under digging conditions. (**a**) Maximum depth attitude, (**b**) Maximum bending moment attitude, (**c**) Tooth tip and working arm co-linear attitude.

According to the actual action requirements of excavation, and at the same time to include the above three extreme postures, the design of the excavation working conditions shown in Figure 4, first of all, the big arm and three-arm cylinders in turn extended to the state of full stroke, the second arm hydraulic cylinder fully retracted and maintained the state of no movement with drive, and, then, the swing-frame hydraulic cylinder from being fully retracted gradually extended, driving the bucket around the rotation point to dig, while the mechanical arm, in turn, went through the above three extreme postures and finally the three arms, big arm, and swing-frame hydraulic cylinder in turn retracted. The mechanical arm returned back to the original posture.

Figure 4. Bucket movement trajectory under digging conditions of the robot arm.

The main purpose of this simulation is to measure the force at the articulation point of each working arm, and the calculation and analysis found that the internal force of the wrist joint has almost no effect on the simulation results. Therefore, in order to simplify the operation, based on the ADAMS virtual prototype simulation model of the arm digging conditions, verified in the previous subsection, we use Boolean operations to combine the wrist joints into a whole. The tangential resistance gradually increases from 0 at 6.6 s, reaches the peak value of 8500 N after 4 s, and decreases to 0 after 2 s; the lifting resistance gradually increases from 0 at 6.6 s, reaches the maximum value of 2460 N after 6 s, and decreases rapidly at 20 s, and decreases to 0 after 1 s. Firstly, marker points are established at the position of the center of gravity of the bucket and the center point of the tooth tip, respectively, and the corresponding one-way force is added. The displacement driving function of each hydraulic cylinder and the change curve of the digging resistance are set by using STEP function, as shown in Table 2 and Figure 5.

Table 2. The displacement drive function of each hydraulic cylinder of the robot arm under digging conditions.

Drivers	Motion Functions
The first arm hydraulic cylinder driver	$\mathrm{STEP}(\mathrm{TIME},0,0,2.7,270)+\mathrm{STEP}(\mathrm{TIME},16.6,0,19,-270)$
The second arm hydraulic cylinder driver	0
The third arm hydraulic cylinder driver	$\mathrm{STEP}(\mathrm{TIME},2.7,0,6.6,390)+\mathrm{STEP}(\mathrm{TIME},12.6,0,16.6,-390)$
Rotary hammer hydraulic cylinder driver	$\mathrm{STEP}(\mathrm{TIME},6.6,0,12.6,295)+\mathrm{STEP}(\mathrm{TIME},19,0,22,-295)$
Tangential resistance	$\mathrm{STEP}(\mathrm{TIME},6.6,0,10.6,8500)+\mathrm{STEP}(\mathrm{TIME},10.6,0,12.6,-8500)$
Normal resistance	$\mathrm{STEP}(\mathrm{TIME},6.6,0,8.6,-1700)+\mathrm{STEP}(\mathrm{TIME},8.6,0,12.6,1700)$
Lifting resistance	$\mathrm{STEP}(\mathrm{TIME},6.6,0,12.6,-2460)+\mathrm{STEP}(\mathrm{TIME},20,0,21,2460)$

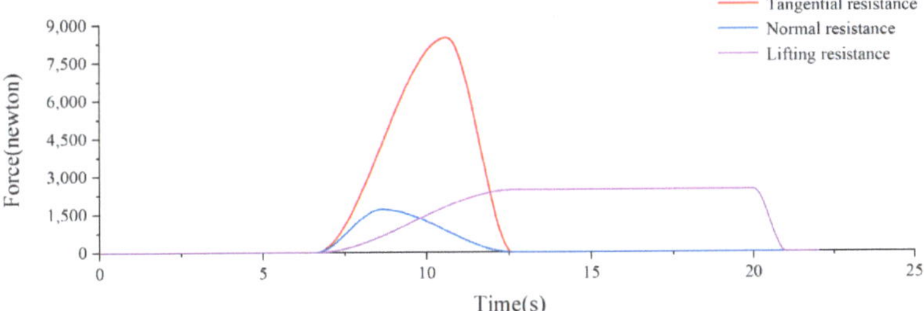

Figure 5. Digging resistance change curve.

Before running the simulation, in order to quickly identify and extract the force change curves at the target articulation points in the "ADAMS/Postprocessor" module and to make the following clearer, the articulation points of the robot arm are numbered and named as shown in Figure 6.

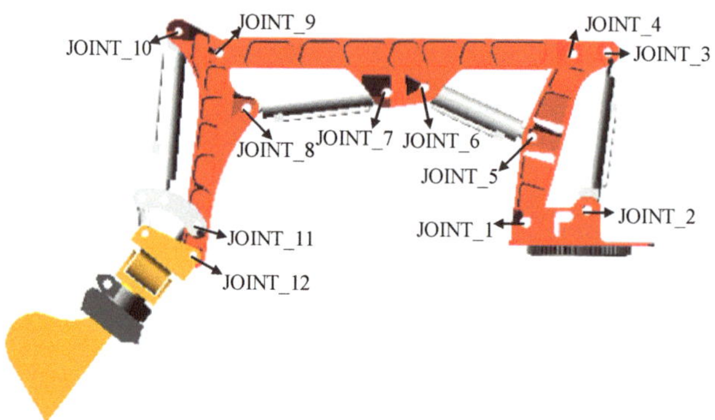

Figure 6. Diagram of the articulation point of each working arm of the rescue robot arm.

We set the simulation time to 22 s, step size 0.01, and ran the simulation. After the simulation, the force situation of each articulation point of the robot arm is viewed by the post-processing module. The load components of each joint's articulation point in *X*-axis and *Y*-axis are shown in Figure 7, and (a) to (d) indicate the force of the articulation points of the base, large arm, second arm, and third arm joints, respectively.

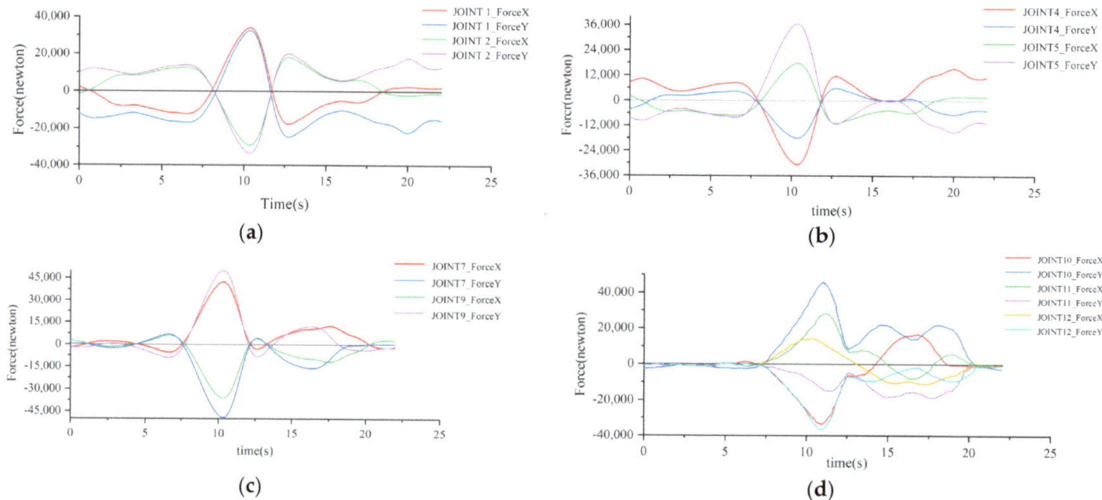

Figure 7. Force diagram of arm joint articulation point under digging conditions: (**a**) Force at the articulation point of the base; (**b**) Force at the articulation point of the first arm; (**c**) Force at the articulation point of the second arm; (**d**) Force at the articulation point of the third arm.

Analysis of Figure 7 shows that the trend of the force at the articulation points of the base, large arm, and two arm joints is basically the same. Among them, the positive and negative force values only indicate the direction of the force. For the first 6.6 s, the big arm and the three-arm hydraulic cylinders were extended to the state of full stroke in turn, and the mechanical arm under the influence of the self-weight force is small. As the digging depth increases, the strength slowly rises. From the 6.6 s to start digging, due to the bucket contact with the ground at 7.2 s, the ground support force on the bucket offsets the influence of the mechanical arm self-weight, so that the force component at each articulation point gradually decreases between 6.6 s and 7.2 s to 0. With the increase of digging resistance, the force of each articulation point also gradually increases, and at 10.6 s, the digging depth and resistance reach the maximum, while the component of each joint articulation point in the X and Y axes is also close to the maximum. After the moment of 10.6 s, the digging resistance starts to decrease, and at the moment of 11.5 s, the force at each articulation point is again close to 0 under the compound action of cutting resistance and self-weight of the arm, and stops digging at 12.6 s, when the amount of material in the bucket reaches the maximum and the lifting resistance increases to another peak, and the force at each articulation point reaches the maximum at this time. After 12.6 s, hydraulic cylinders of the three arms start to shrink while driving the full-load bucket up, and the force on the articulation point of each joint is gradually reduced by the reduction of the force arm. At 20 s, the material is unloaded, the lifting resistance starts to decrease rapidly, and the force at each articulation point also decreases rapidly. At 21 s, the material is unloaded, and the swing hydraulic cylinder gradually retracts to the shortest state and drives the unloaded bucket to the initial attitude of the arm at 22 s, during which the force component at each articulation point remains basically unchanged.

3.2. Dynamic Simulation of Robotic Arm Handling Heavy Objects

Firstly, a workpiece with a weight of 200 kg is created in Solidworks and assembled with the end appliance gripper. Then, the end appliance bucket of the robot arm is deleted and replaced with the end appliance gripper, and finally a virtual simulation model of the robot arm under handling conditions is created in ADAMS according to the method described in the previous section. In order to accurately measure the force at each articulation point of the robot arm in the process of lifting the workpiece, the end gripper is set to carry the workpiece in the whole working range of the robot arm and to go through three extreme cases of highest position, horizontal position, and lowest position in turn. The initial position and the movement trajectory of the gripper finger end are shown in Figure 8. STEP function is used to set the displacement drive function of each hydraulic cylinder, as shown in Table 3 below. Firstly, the two-arm cylinder and the three-arm cylinder are set to extend to the full-stroke posture from the shortest state, and the swing-frame cylinder is retracted to the shortest state from the full-stroke state, at which time the mechanical arm reaches the highest point, then the mechanical arm passes through the horizontal position during the extension of the large-arm cylinder and swing-frame cylinder, and the mechanical arm passes through the lowest position when the two-arm cylinder retracts to the shortest posture, and finally the large-arm cylinder and the three-arm cylinder retract to the shortest state in turn, and the mechanical arm returns to the original posture.

We set the simulation time to 118 s, and got the force change curve of each articulation point of the robot arm under the handling condition, where the load components of each joint's articulation point in X-axis and Y-axis are shown in Figure 9, and (a) to (d) indicate the force of the articulation points of the base, large arm, second arm, and third arm joints, respectively.

Table 3. The displacement drive function of each hydraulic cylinder of the robot arm under handling conditions.

Drivers	Motion Functions
The first arm hydraulic cylinder driver	$STEP(TIME, 49, 0, 59, 270) + STEP(TIME, 108, 0, 118, -270)$
The second arm hydraulic cylinder driver	$STEP(TIME, 0, 0, 10, 378) + STEP(TIME, 82, 0, 92, -378)$
The third arm hydraulic cylinder driver	$STEP(TIME, 10, 0, 26, 390) + STEP(TIME, 92, 0, 108, -390)$
Rotary hammer hydraulic cylinder driver	$STEP(TIME, 26, 0, 49, -295) + STEP(TIME, 59, 0, 82, 295)$

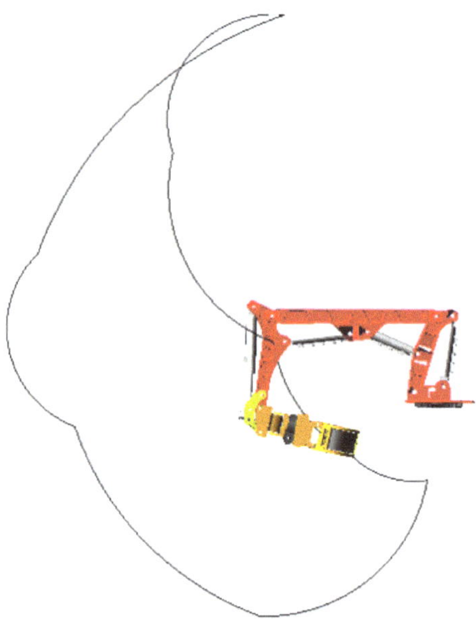

Figure 8. Robotic arm trajectory under handling conditions.

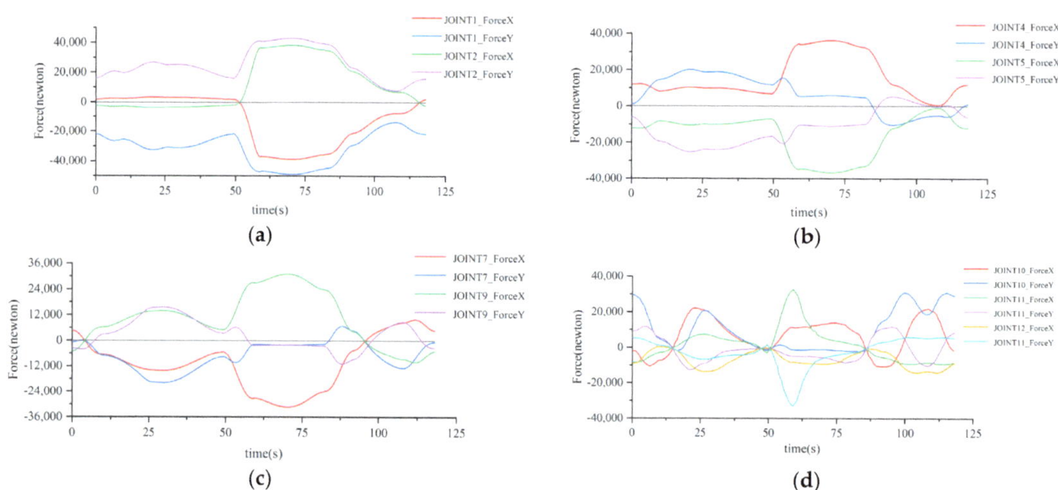

Figure 9. Force diagram of joint articulation point of robot arm under handling conditions: (**a**) Force at the articulation point of the base; (**b**) Force at the articulation point of the first arm; (**c**) Force at the articulation point of the second arm; (**d**) Force at the articulation point of the third arm.

As can be seen from Figure 9, the change in patterns and trends of the base, big arm, and second arm amplitude curves are basically the same, and the force at the hinge point of the third arm is slightly more complicated. Before the first 26 s, the second arm cylinder and the third arm cylinder are extended in turn while making the workpiece rise slowly. As the center of gravity of the workpiece is farther and farther from the base, the moment of action on the base hinge point becomes bigger and bigger, and the force at each hinge point also rises slowly. At the moment of 26 s, the swing-frame cylinder starts to shrink while driving the workpiece close to the base, the moment gradually decreases so that the

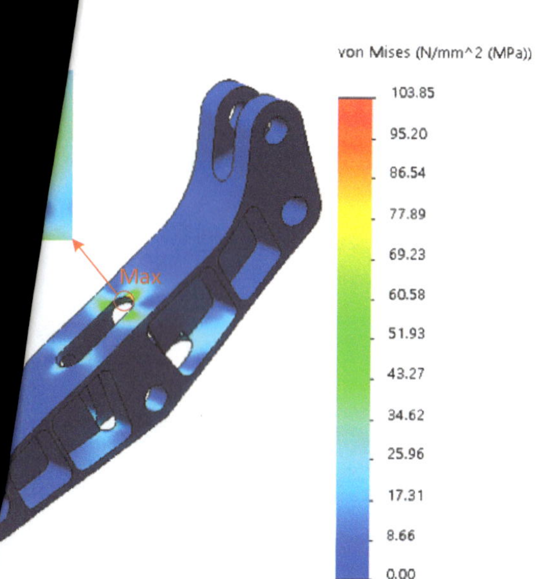

Stress cloud of first arm under excavation conditions.

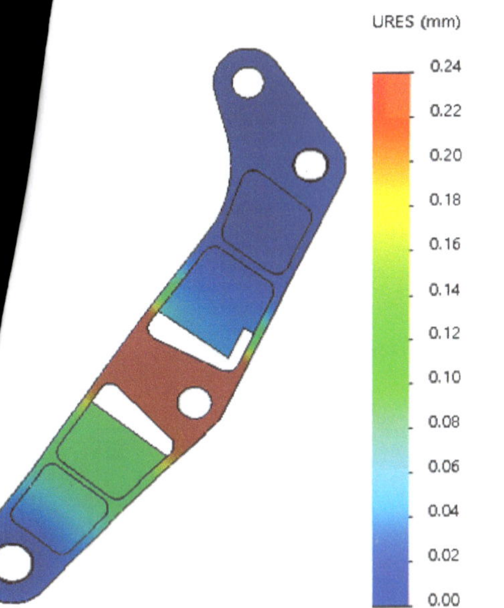

Figure 15. Strain cloud of first arm under excavation conditions.

The parts of the second arm where the stress changes more and the stress maximum are shown in Figure 16. The maximum stress appears at the articulation hole linked with the big arm. The stress maximum is 47.82 MPa, much less than the stress limit 505 MPa. According to the stress cloud diagram, the stress at the middle reinforcement of the second arm, the articulation hole, and the stress at the back is larger. The maximum position of deformation appears at the end of the second arm and the three arms' articulated shaft hole ear plate, as shown in Figure 17. The maximum value of deformation is 0.96 mm, and close to the robot arm base direction, the deformation is relatively small.

force at each articulation point slowly decreases until 49 s, and the robot arm moves to the highest position. After 49 s, the big arm cylinder starts to elongate, driving the workpiece away from the base while making the robot arm posture gradually approach the horizontal position; the workpiece and the wrist joint gravity on each articulation point produces the moment gradually increases, so that the force at the articulation point continuously increases. At the instance of 70 s, the end wrist joint is in common with the three arms and the second arm; at this time, the arm is in horizontal posture, the force component and amplitude of each articulation point reach the maximum value, and, after the horizontal position, the force at each articulation point gradually decreases. At 92 s, the hydraulic cylinder of the second arm is completely retracted; at this time, the arm is in the lowest position. After 92 s, the three-arm hydraulic cylinder starts to retract while driving the workpiece closer to the base. Due to the reduction of the action moment, the force at each articulation point slowly decreases, reaching 108 s, and the three-arm hydraulic cylinder completely retracts. The force at each articulation point reaches the minimum value, and then the big arm hydraulic cylinder starts to retract while driving the workpiece away from the base. Due to the increase of the action moment, the combined force at each articulation point increases, and, by 118 s, the mechanical arm returns to the original posture.

4. Finite Element Static Analysis of Key Components

4.1. Pre-Processing of Robot Arm Finite Element Model

Although the seven-degrees-of-freedom redundant rescue robot arm designed in this topic is completed in Solidworks, the overall structure is more complex. Considering the aesthetic appearance of the robot arm and the requirements of some parts' processing characteristics, many features such as rounded corners, small holes, chamfers, and tabs are retained at the early stage of structural design, and if these features are not simplified, stress singularities will be easily generated during analysis, which will lead to a large difference between the analysis results and the real value, and even the failure of the mesh division and the crash of the solver. Therefore, before the analysis, the analysis object should be reasonably simplified, the insignificant geometric features compressed or removed, and the reinforcement retained. Rounded corners and other geometric features have a greater impact on the calculation results, and special attention must be paid to the location where the analysis object is very likely to produce stress concentration to reduce the impact of stress concentration. The stress results of different nodes of the base at different mesh densities are shown in Figure 10. Therefore, when carrying out the mesh division, the region with greater influence on the analysis results uses the mesh control function to refine the mesh, and the quality should be high. When carrying out pre-processing work, it is necessary to ensure that the stiffness of the finite element analysis model before and after simplification is consistent; otherwise, it is likely to lead to a large difference between the analysis results and the actual value. The results of the mesh division of the three arms, two arms, large arm, and base are shown in Figure 11a–d, respectively.

In order to limit the degrees of freedom of each rigid body, the boundary conditions of the model need to be set before the analysis; according to the actual working conditions of the fixed end and free end of the robot arm, to add constraints and loads, the process is as follows:

(1) Add reasonable restraint: The robot arm near the base member can be regarded as the fixed end, and near the end apparatus as the free end. Using the fixture function, a fixed hinge constraint is set on the selected axis hole surface to restrict the hinge point.

(2) Applying load: Therefore, the bearing load is added to the pin hole at the free end. When assigning the load size, considering that each hole has two bearing surfaces, half of the combined force obtained from the motion simulation should be added to each bearing surface. Since the force at the hinge point is not constant, but fluctuates according to a certain rule, according to engineering experience, the fluctuation is in line with the parabolic distribution, so the distribution of load is more compounded with the actual working condition by choosing parabolic distribution. The mesh parameters for each component are shown in Table 4.

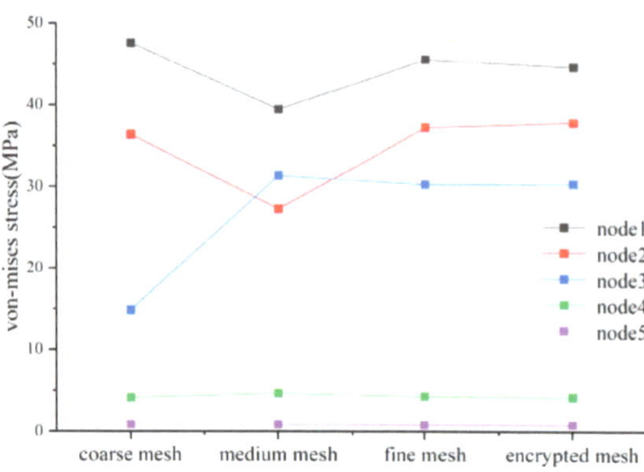

Figure 10. Results of grid independence analysis.

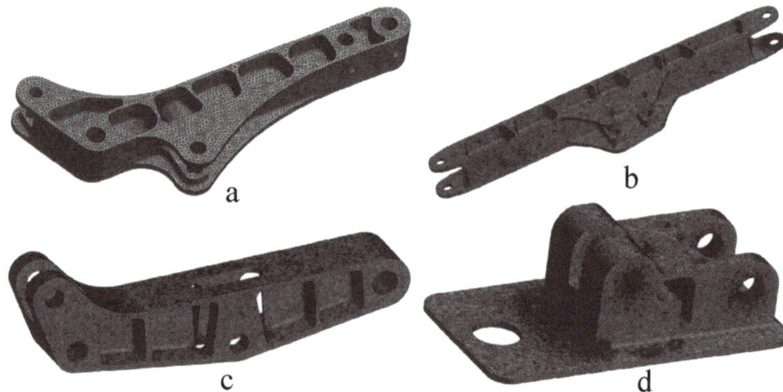

Figure 11. Schematic diagram of components meshing: (**a**) Schematic diagram of the third arm meshing; (**b**) Schematic diagram of the second arm meshing; (**c**) Schematic diagram of the first arm meshing; (**d**) Schematic diagram of base meshing.

Table 4. Grid parameters.

Component	Number of Nodes	Number of Cells	Aspect Ratio < 3 (%)	Flexure Unit (%)
Base	415,702	275,829	99.4	0
The first arm	453,166	296,149	99.5	0
The second arm	1,248,279	827,042	99.2	0
The third arm	774,436	510,797	99.1	0

4.2. Results of the Static Analysis of the Robot Arm under Excavation Conditions

According to the dynamic simulation results of ADAMS under the digging condition of the arm, the working arm is adjusted to the attitude corresponding to the maximum force at the articulation point, and the pre-processing of the finite element analysis model of the working arm and the displacement boundary conditions are completed according to the static force analysis process under the handling condition in the previous section, and finally the X-axis and Y-axis load components corresponding to the maximum force at the articulation point are added to the corresponding articulated shaft holes in the form of bearing load with the local coordinate system as the reference coordinate system to complete the application of the load boundary conditions.

Figure 14

Figure .

Figure 13. Strain cloud of

The parts of the big
Figure 14. The maximum s
is 103.85 MPa, which is muc
cloud diagram, it is seen that
From the deformation cloud d
the articulated shaft hole of the
the maximum deformation is 0.

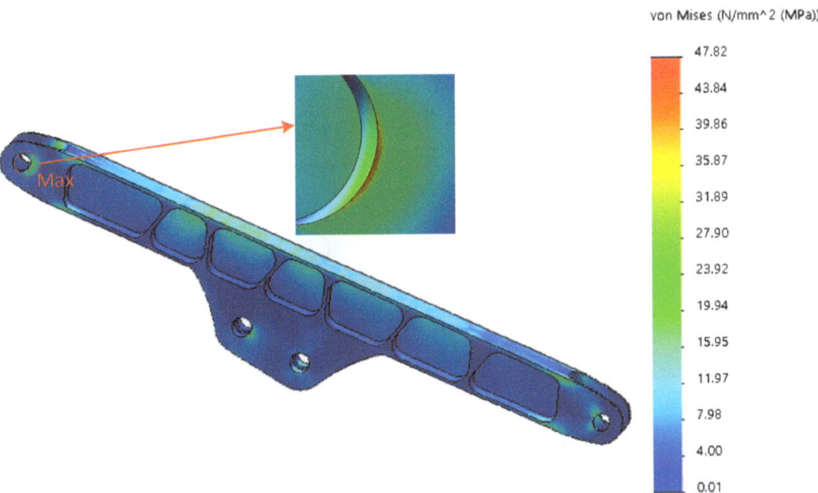

Figure 16. Stress cloud of second arm under excavation conditions.

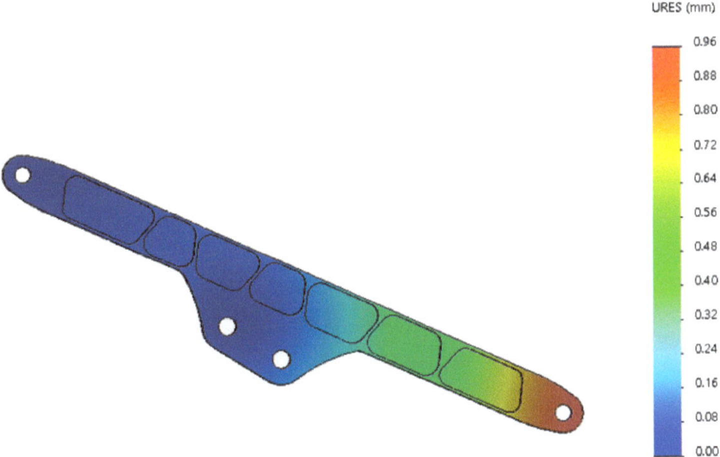

Figure 17. Strain cloud of second arm under excavation conditions.

The maximum stress of the third arm is 98.06 MPa, as shown in Figure 18, which is much less than the stress limit of 505 MPa. According to the stress cloud diagram in Figure 19, the middle reinforcement and the transition lug plate of the hinge hole of the three arms are under greater stress. The deformation cloud under excavation condition is shown in Figure 18, the maximum deformation is 3.27 mm at the end of the three arms and the swing-frame articulated shaft hole, and the remaining parts are relatively small.

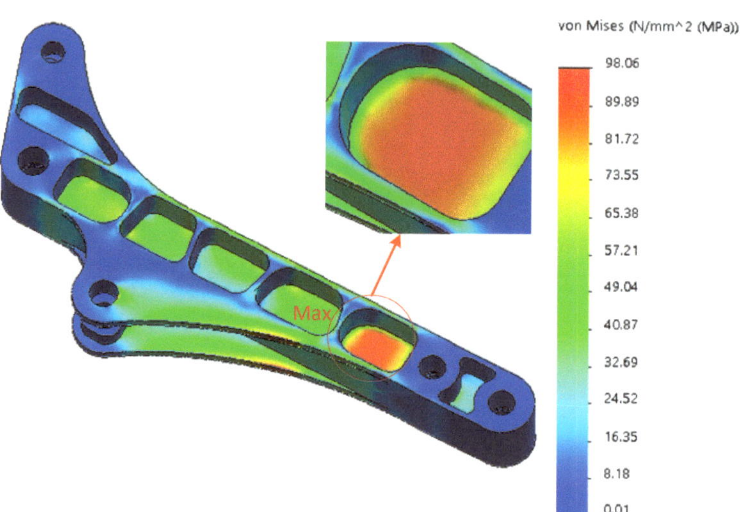

Figure 18. Stress cloud of third arm under excavation conditions.

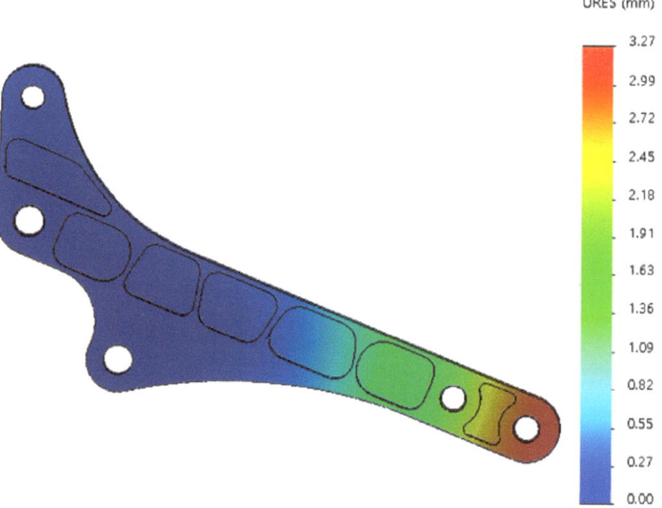

Figure 19. Strain cloud of third arm under excavation conditions.

4.3. Results of the Static Analysis of the Robot Arm under Handling Conditions

According to the dynamic simulation results of ADAMS under the handling condition of the robot arm, the working arm is adjusted to the attitude corresponding to the maximum force at the articulation point, and the pre-processing and displacement boundary conditions of the finite element analysis model of each working arm are completed according to the static force analysis process under the handling condition in the previous section.

The parts of the base with large stress variation and stress maximum are shown in Figure 20. The maximum stress appears at the bolt hole, and the stress maximum is 174.89 MPa, which is much smaller than the stress limit of 505 MPa. From the deformation cloud shown in Figure 21, it can be seen that the maximum deformation of the base is only 0.03 mm; the deformation is very small, which means that the base is relatively safe under stress.

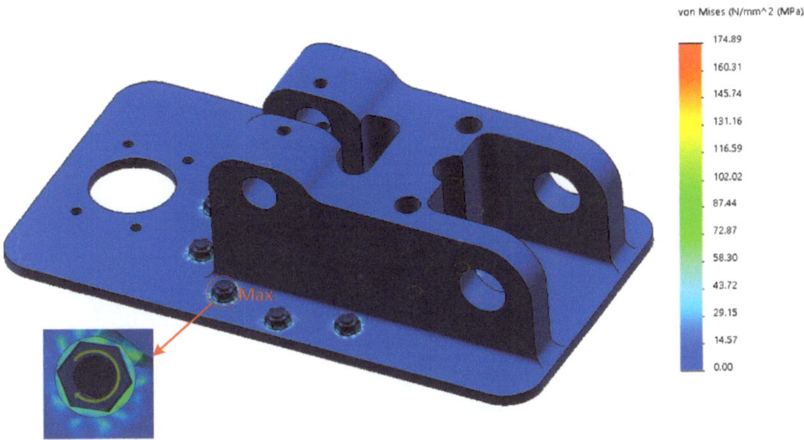

Figure 20. Stress cloud of base under handling conditions.

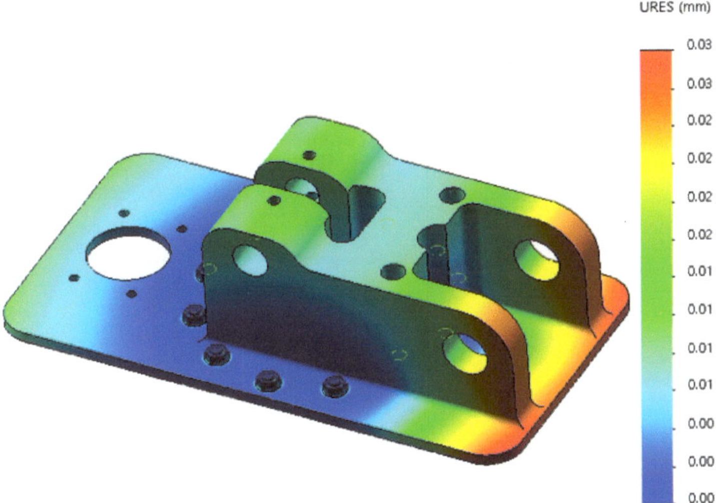

Figure 21. Strain cloud of base under handling conditions.

Big-arm stress changes and stress maximum parts as shown in Figure 22. The maximum stress appears in the open hole of the big arm, and the stress maximum value is 38.25 MPa, far less than the stress limit 505 MPa. According to the stress cloud diagram, the big arm in the middle of the digging position can be seen near the larger force. From the deformation cloud diagram shown in Figure 23, it can be seen that the maximum deformation occurs at the articulated shaft hole of the big arm and the cylinder barrel of the second arm, the maximum deformation is 0.01 mm, and the relative deformation is small.

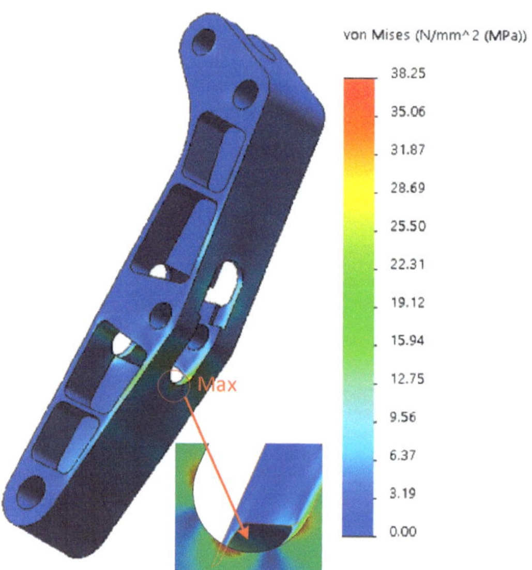

Figure 22. Stress cloud of first arm under handling conditions.

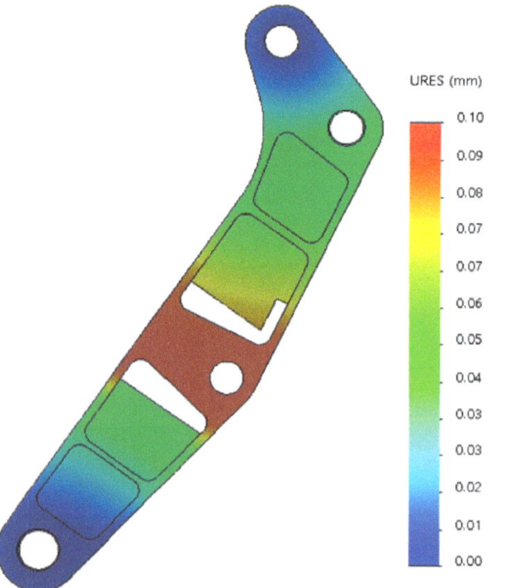

Figure 23. Strain cloud of first arm under handling conditions.

The parts of the second arm where the stress changes more and the stress maximum are shown in Figure 24. The maximum stress appears at the articulated hole linked with the big arm; the stress maximum is 48.35 MPa, much less than the stress limit 505 MPa. According to the stress cloud diagram, at the middle reinforcement of the second arm and the articulated hole, the stress is larger. The maximum position of deformation appears at the end of the second arm and the three arms' articulated shaft hole trunnion, as shown in Figure 25. The maximum value of deformation is 1.49 mm, and close to the direction of the base of the robot arm, the deformation is relatively small.

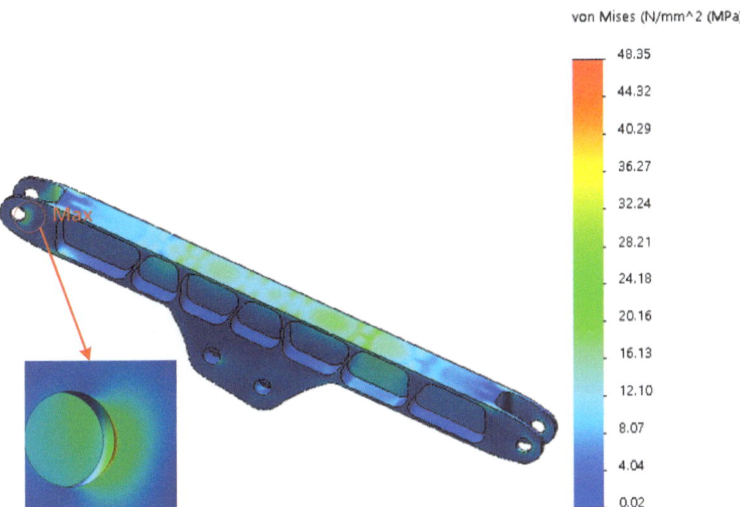

Figure 24. Stress cloud of second arm under handling conditions.

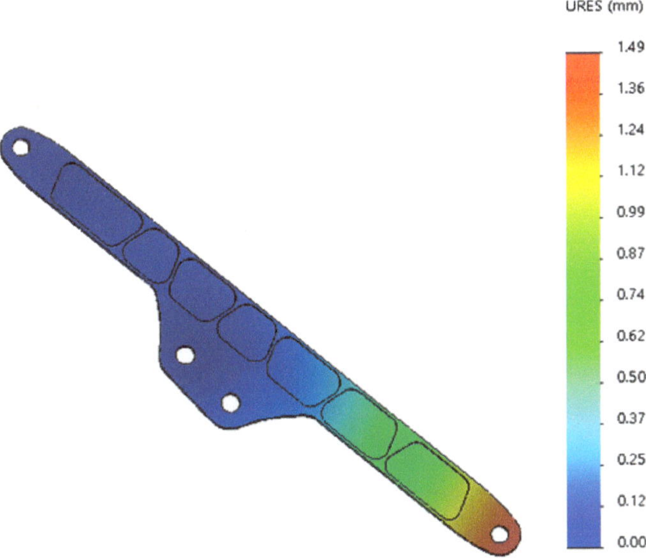

Figure 25. Strain cloud of second arm under handling conditions.

The maximum stress of the third arm is 20.12 MPa, as shown in Figure 26, which is much less than the stress limit of 505 MPa. According to the stress cloud diagram in Figure 26, the middle reinforcement bar and the transition lug plate of the hinge hole of the three arms are under greater stress. The deformation cloud under excavation condition is shown in Figure 27. The maximum deformation is 0.72 mm at the end of the three arms and the swing-frame articulated shaft hole. The remaining parts have a relatively small deformation.

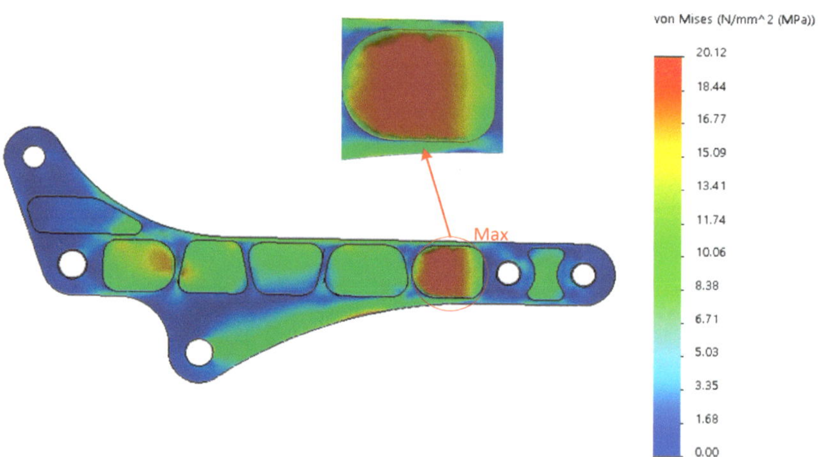

Figure 26. Stress cloud of third arm under handling conditions.

Figure 27. Strain cloud of third arm under handling conditions.

5. Prototype and Analysis of Hydraulic Machinery Arm

5.1. Prototype

It has been established through finite element analysis and dynamic simulation that the designed structure satisfies the necessary strength standards. Based on the planned robotic arm concept, a hydraulic robotic arm prototype is made and shown in Figure 28. The oil cylinder and hydraulic motor, which serve as the robotic arm's drive units, both have displacement sensors, setting the groundwork for eventual displacement closed-loop control. The hydraulic flow rate of each of the robotic arm's driving components is managed by a number of hydraulic proportional directional valves.

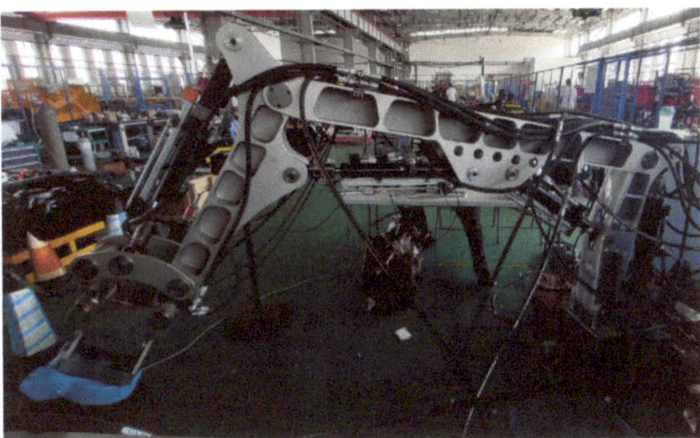

Figure 28. 7-DOFs rescue hydraulic robotic arm prototype.

5.2. Finite Element Experiments

The finite element method is widely used in structural design optimization, but the structural model is simplified for the sake of calculation simplicity, there are some differences with the actual structure and process, and the boundary conditions are set differently from the actual one. Therefore, the correctness of the finite element model is tested by the test results and, the smaller the error, the closer it is to the actual one. In this section, based on the completed FEM analysis, the stress–strain test is conducted on the robot arm to verify the feasibility of the finite element model.

5.2.1. Preliminary Preparation of the Experiment

(1) Experimental equipment: JINGYAN Stress Tester: SG04 type, as shown in Figure 29 below.

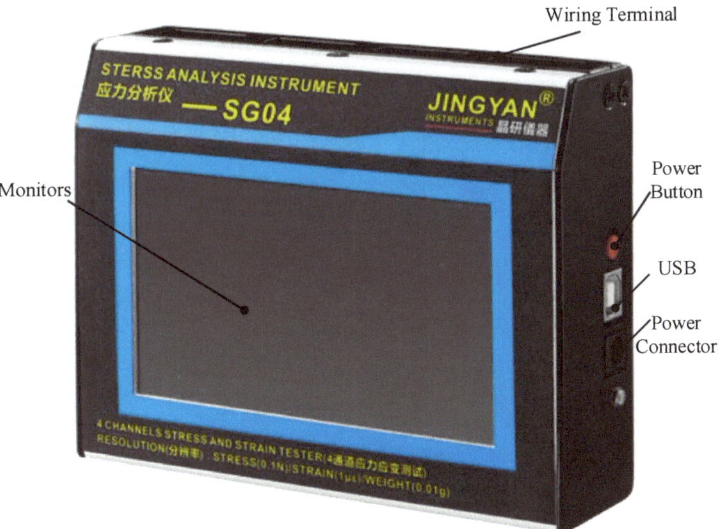

Figure 29. Jingyan SG04 stress tester.

The strain flower, with a resistance value 120Ω, sensitivity factor 2.0, Poisson's ratio 0.27, resolution 0.1, three axes at 45° to each other, can be used to test the strain in the direction of unknown principal stress. The strain flower schematic diagram and installation are shown in Figures 30 and 31.

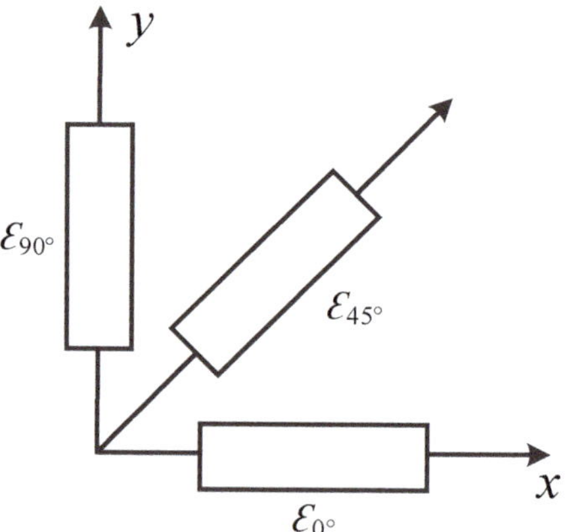

Figure 30. Strain flower schematic.

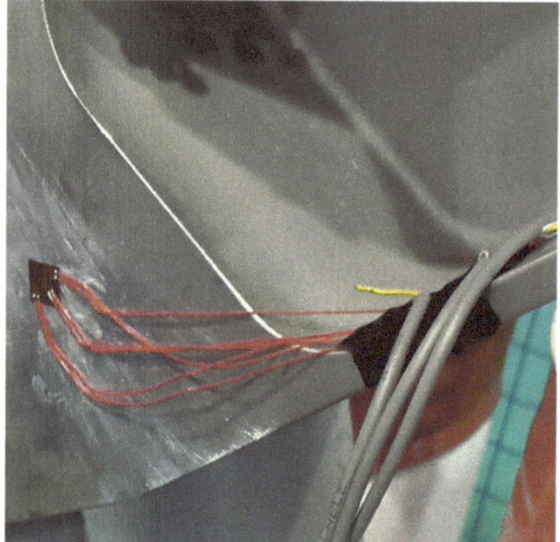

Figure 31. Strain relief flower installation diagram.

(2) Measuring bridge design and principle: A simple and reliable 1/4 bridge was used for the bridges, where each strain flower needs to be connected to three bridges, and the measuring bridge principle is shown in Figure 32 below.

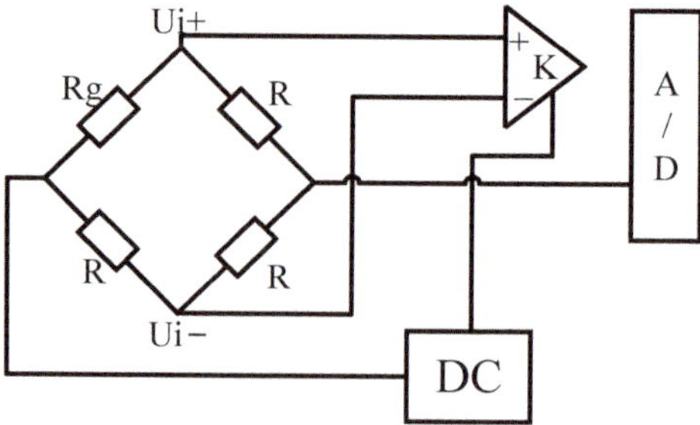

Rg-Measuring resistance R-Fixed Resistors
K-Amplifier gain

Figure 32. Schematic diagram of measuring bridge.

When the analyzer calculates the strain magnitude of the three bridge paths of the strain flower, the ultimate strain value of this experimental object can be found by the following equation:

$$\varepsilon_x = \varepsilon_{0°}, \varepsilon_y = \varepsilon_{90°}, \gamma_{xy} = \varepsilon_{0°} + \varepsilon_{90°} - 2\varepsilon_{45°} \tag{1}$$

$$\sigma_x = \frac{E}{(1-\mu^2)}(\varepsilon_x + \mu\varepsilon_y) \tag{2}$$

$$\sigma_y = \frac{E}{(1-\mu^2)}(\varepsilon_y + \mu\varepsilon_x) \tag{3}$$

$$\tau_x = \frac{E}{2(1+\mu)}\gamma_{xy} \tag{4}$$

where σ_x, σ_y, τ_x are the plane stresses; ε_x, ε_x, γ_{xy} are the corresponding strain values. This leads to:

$$\sigma_{max} = \frac{\sigma_x + \sigma_y}{2} + \sqrt{\left(\frac{\sigma_x - \sigma_y}{2}\right)^2 + \tau_x^2} \tag{5}$$

$$\sigma_{min} = \frac{\sigma_x + \sigma_y}{2} - \sqrt{\left(\frac{\sigma_x - \sigma_y}{2}\right)^2 + \tau_x^2} \tag{6}$$

$$\sigma_1 = \sigma_{max}, \sigma_2 = 0, \sigma_3 = \sigma_{min} \tag{7}$$

Then, the equivalent force value σ_s at this measurement location can be derived from the fourth strength theory as:

$$\sigma_s = \sqrt{\frac{1}{2}\left((\sigma_1 - \sigma_2)^2 + (\sigma_1 - \sigma_3)^2 + (\sigma_2 - \sigma_3)^2\right)} \tag{8}$$

(3) Cautions: In order to improve the accuracy of the experiment, the measurement locations should be polished before the patching, and, in addition, the stress concentrations appearing in the finite element analysis should be selected as much as possible when patching the strain gauges, such as axial holes and other easily damaged parts. The location of the patch point should be recorded. Some patch point locations are shown in Figure 33.

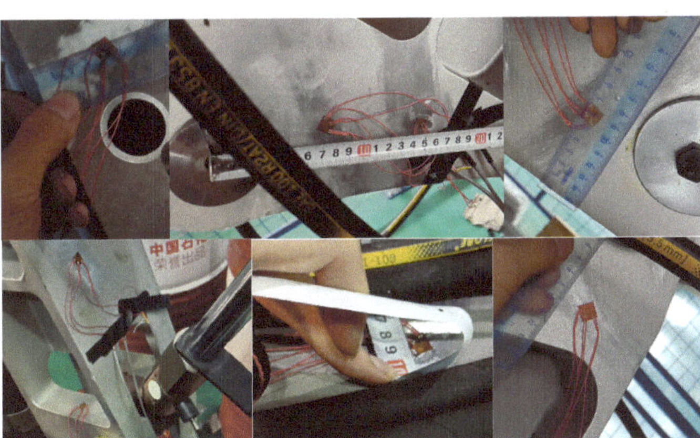

Figure 33. Strain gauge patch diagram.

(4) Test Method: The weight hanging at the end of the robotic arm is 300 kg. The robotic arm is adjusted to a specific attitude, the individual cylinder extensions are recorded, apply

strain gauges, and the channels are zeroed. The weight is weighed and gently suspended at the end of the robotic arm and, after the arm is balanced, the strain value at each test point is collected. The test arrangement is shown in Figure 34.

Figure 34. Schematic diagram of mechanical arm stress–strain test site layout.

After the test is completed, the robot arm is adjusted to the corresponding attitude, the corresponding load is applied, the load distribution at each point is simulated, the load distribution data is exported and loaded onto the finite element model of the robot arm, multiple points are taken for the experimental patch position, and the average value of the simulated value at that point is obtained as the simulated value of that position. The position of all measurement points on the robot arm is shown in Figure 35.

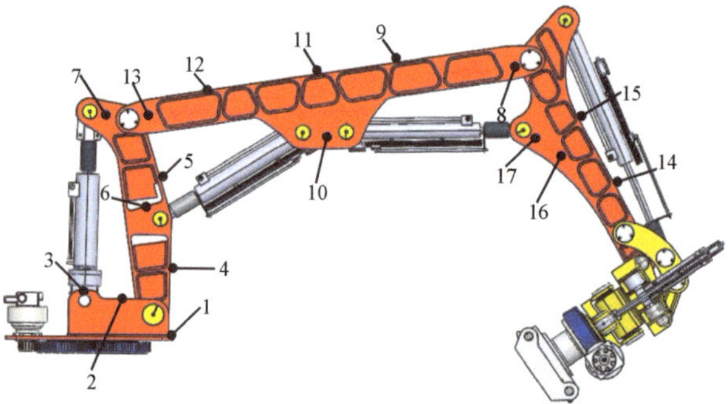

Figure 35. Schematic diagram of measurement point location.

5.2.2. Results and Analysis of the Experiment

The expansion and contraction of each hydraulic cylinder is shown in Table 5 below. The test point processing results are shown in Table 6. Referring to the above calculation and analysis process, we can see that the difference between the simulated value and the experimental measured stress value is not large, and the error between the experimental and simulated values may come from the model simplification and the error caused by the unevenness of the patch in the experiment. However, according to the above data, the simulation has good reference significance and can guarantee the structural and mechanical reliability of the robot arm, but still needs to leave enough safety margin in the design process, generally with a safety factor of 1.5–3 as a reference standard.

Table 5. Telescopic length table of the hydraulic cylinder.

The First Arm Cylinder (mm)	The Second Arm Cylinder (mm)	The Third Arm Cylinder (mm)	Rotary Hammer Cylinder (mm)
20	335	60	142

Table 6. Data processing results of stress and strain test points of prototypes.

Measurement Points	Experimental Value (MPa)	Simulation Value (MPa)	Absolute Error (MPa)
1	0.461	0.150	0.311
2	1.315	0.322	0.993
3	2.357	3.892	1.535
4	5.333	3.275	2.058
5	3.309	2.945	0.364
6	4.211	1.604	2.607
7	1.317	0.852	0.465
8	1.222	0.498	0.724
9	9.558	11.623	2.065
10	1.571	1.964	0.393
11	7.087	9.432	2.345
12	5.125	7.733	2.608
13	1.048	2.853	1.805
14	2.129	2.307	0.178
15	3.527	3.061	0.466
16	3.549	2.404	1.124
17	1.994	1.559	0.435

6. Discussion

Based on the aforementioned simulation and experimental findings, it is clear that the robotic arm used in this study, which was built using aluminum alloy 7075, has the advantages of low weight and high load. The mechanism of the research robotic arm is not, however, the best-designed structure. To further lower the weight of the robotic arm, topology optimization of the structure might be carried out. Stress–strain tests were performed to determine the robotic arm's strength, and the results show that while the finite element results are quite descriptive, the relative errors for points 1, 2, 6, 8, and 13 are relatively large. This suggests that the accuracy of the testing apparatus and the testing procedure can be further optimized.

7. Conclusions

In this paper, we started from the structural design of the seven-degree-of-freedom redundant rescue hydraulic robotic arm, completed the 3D model assembly in Solidworks, and then completed the finite element static analysis and structural optimization design of the robotic arm according to the actual working conditions of the rescue robotic arm, which is lightweight and has a high load and large working range. In ADAMS simulation software, a virtual prototype of the mechanical analysis of the rescue hydraulic robotic arm was built to simulate the two common working conditions of excavation and handling, and to analyze the force of the robotic arm, followed by finite element analysis of the key parts of the robotic arm. In order to verify the accuracy of the finite element static analysis method of the robotic arm in this paper, theoretical and practical research methods were used, and stress–strain experiments were conducted at important locations of the robotic arm according to the finite element analysis model, verifying the correctness of the finite element analysis method. In future research, different advanced control strategies will be applied to the robotic arm to further validate its capacity.

Author Contributions: Conceptualization, J.Z. and W.J.; methodology, W.J.; software, W.J. and Q.Z.; validation, W.J., Q.Z. and W.Z.; formal analysis, W.J. and Q.Z.; investigation, W.J.; resources, J.Z.; data curation, W.J. and W.Z.; writing—original draft preparation, W.J.; writing—review and editing, J.Z.; supervision, J.Z.; project administration, J.Z.; funding acquisition, J.Z. All authors have read and agreed to the published version of the manuscript.

Funding: This work is supported by National Natural Science Foundation of China (52275285).

Data Availability Statement: Data will be made available on request.

Conflicts of Interest: The authors declare no conflict of interest.

Abbreviations

SCARA—Selective Compliance Assembly Robot Arm; NSGA-II—Non-dominated sorting genetic algorithm II; 7-DOFs—seven degrees of freedom; FEM—Finite Element Method.

References

1. Takahashi, T.; Tadakuma, K.; Watanabe, M.; Takane, E.; Hookabe, N.; Kajiahara, H.; Yamasaki, T.; Konyo, M.; Tadokoro, S. Eversion Robotic Mechanism With Hydraulic Skeletonto Realize Steering Function. *IEEE Robot. Autom. Lett.* **2021**, *6*, 5413–5420. [CrossRef]
2. Han, S.; Chon, S.; Kim, J.; Seo, J.; Shin, D.G.; Park, S.; Kim, J.T.; Kim, J.; Jin, M.; Cho, J. Snake Robot Gripper Module for Search and Rescue in Narrow Spaces. *IEEE Robot. Autom. Lett.* **2022**, *7*, 1667–1673. [CrossRef]
3. Hagiwara, T.; Katou, Y. Trial Production of a Small Snake Type Robot That Can Be Mounted on a Disaster Rescue Robot. In Proceedings of the 2021 18th International Conference on Electrical Engineering/Electronics, Computer, Telecommunications and Information Technology (ECTI-CON), Chiang Mai, Thailand, 19–22 May 2021; pp. 1104–1107.
4. Yuan, J.; Wang, Z.; Zhang, Z.; Xing, Y.; Ji, A. Mechanism Design of a Transformable Crawling Robot and Feasibility Analysis for the Unstructured Environment. *Actuators* **2022**, *11*, 60. [CrossRef]
5. Voyles, R.M.; Larson, A.C. TerminatorBot: A Novel Robot With Dual-Use Mechanism for Locomotion and Manipulation. *IEEE/ASME Trans. Mechatron.* **2005**, *10*, 17–25. [CrossRef]
6. Cruz Ulloa, C.; Domínguez, D.; Del Cerro, J.; Barrientos, A. A Mixed-Reality Tele-Operation Method for High-Level Control of a Legged-Manipulator Robot. *Sensors* **2022**, *22*, 8146. [CrossRef]
7. Bai, L.; Li, X.; Sun, Y.; Zheng, J.; Chen, X. A Wheel-Legged Mobile Robot with Adjustable Body Length for Rescue and Search. In Proceedings of the 2021 6th IEEE International Conference on Advanced Robotics and Mechatronics (ICARM), Chongqing, China, 3–5 July 2021; pp. 340–345.
8. Wang, R.; Chen, Z.; Xu, K.; Wang, S.; Wang, J.; Li, B. Hybrid Obstacle-Surmounting Gait for Hexapod Wheel-Legged Robot in Special Terrain. In Proceedings of the 2021 6th IEEE International Conference on Advanced Robotics and Mechatronics (ICARM), Chongqing, China, 3–5 July 2021; pp. 1–6.
9. Lee, D.-H.; Huynh, T.; Kim, Y.-B.; Soumayya, C. Motion Control System Design for a Flying-Type Firefighting System with Water Jet Actuators. *Actuators* **2021**, *10*, 275. [CrossRef]
10. Liu, H.; Ge, J.; Wang, Y.; Li, J.; Ding, K.; Zhang, Z.; Guo, Z.; Li, W.; Lan, J. Multi-UAV Optimal Mission Assignment and Path Planning for Disaster Rescue Using Adaptive Genetic Algorithm and Improved Artificial Bee Colony Method. *Actuators* **2021**, *11*, 4. [CrossRef]
11. Focchi, M.; Bensaadallah, M.; Frego, M.; Peer, A.; Fontanelli, D.; Del Prete, A.; Palopoli, L. CLIO: A Novel Robotic Solution for Exploration and Rescue Missions in Hostile Mountain Environments. In Proceedings of the 2023 IEEE International Conference on Robotics and Automation (ICRA), London, UK, 29 May–2 June 2023; pp. 7742–7748.
12. Ye, C.; Su, Y.; Yu, S.; Wang, Y. Development of a Deformable Water-Mobile Robot. *Actuators* **2023**, *12*, 202. [CrossRef]
13. Yuguchi, Y.; Satoh, Y. Development of a Robotic System for Nuclear Facility Emergency Preparedness—Observing and Work-Assisting Robot System. *Adv. Robot.* **2002**, *16*, 481–484. [CrossRef]
14. Nishida, T.; Koga, H.; Fuchikawa, Y.; Kitayama, Y.; Kurogil, S.; Arimura, Y. Development of Pilot Assistance System with Stereo Vision for Robot Manipulation. In Proceedings of the 2006 SICE-ICASE International Joint Conference, Busan, Republic of Korea, 18–21 October 2006.
15. Wolf, A.; Brown, H.B.; Casciola, R.; Costa, A.; Schwerin, M.; Shamas, E.; Choset, H. A Mobile Hyper Redundant Mechanism for Search and Rescue Tasks. In Proceedings of the 2003 IEEE/RSJ International Conference on Intelligent Robots and Systems (IROS 2003) (Cat. No.03CH37453), Las Vegas, NV, USA, 27–31 October 2003; Volume 3, pp. 2889–2895.
16. Liang, X.; Wan, Y.; Zhang, C.; Kou, Y.; Xin, Q.; Yi, W. Robust Position Control of Hydraulic Manipulators Using Time Delay Estimation and Nonsingular Fast Terminal Sliding Mode. *Proc. Inst. Mech. Eng. Part I J. Syst. Control Eng.* **2018**, *232*, 50–61. [CrossRef]
17. Zhou, L.; Bai, S. A New Approach to Design of a Lightweight Anthropomorphic Arm for Service Applications. *J. Mech. Robot.* **2015**, *7*, 031001. [CrossRef]

18. Li, Z.; Milutinović, D.; Rosen, J. Design of a Multi-Arm Surgical Robotic System for Dexterous Manipulation. *J. Mech. Robot.* **2016**, *8*, 061017. [CrossRef]
19. Dämmer, G.; Gablenz, S.; Neumann, R.; Major, Z. Design, Topology Optimization, and Additive Manufacturing of a Pneumatically Actuated Lightweight Robot. *Actuators* **2023**, *12*, 266. [CrossRef]
20. Wu, Z.; Jiang, T. Property Analysis and Structure Optimization of Robotic Arm Based on Pro/E and ANSYS. *J. Chang. Univ. Sci. Technol. (Nat. Sci. Ed.)* **2011**, *34*, 3. [CrossRef]
21. Li, K.; Huo, Y.; Liu, Y.; Shi, Y.; He, Z.; Cui, Y. Design of a Lightweight Robotic Arm for Kiwifruit Pollination. *Comput. Electron. Agric.* **2022**, *198*, 107114. [CrossRef]
22. Evans, A.G. Lightweight Materials and Structures. *MRS Bull.* **2001**, *26*, 790–797. [CrossRef]
23. Yeo, S.J.; Oh, M.J.; Yoo, P.J. Structurally Controlled Cellular Architectures for High-Performance Ultra-Lightweight Materials. *Adv. Mater.* **2019**, *31*, 1803670. [CrossRef]
24. Sun, Q.; Zhou, G.; Meng, Z.; Jain, M.; Su, X. An Integrated Computational Materials Engineering Framework to Analyze the Failure Behaviors of Carbon Fiber Reinforced Polymer Composites for Lightweight Vehicle Applications. *Compos. Sci. Technol.* **2021**, *202*, 108560. [CrossRef]
25. Tam, K.-M.M.; Marshall, D.J.; Gu, M.; Kim, J.; Huang, Y.; Lavallee, J.; Mueller, C.T. Fabrication-Aware Structural Optimisation of Lattice Additive-Manufactured with Robot-Arm. *Int. J. Rapid Manuf.* **2018**, *7*, 120–168. [CrossRef]
26. Salvati, E.; Korsunsky, A.M. Micro-Scale Measurement & FEM Modelling of Residual Stresses in AA6082-T6 Al Alloy Generated by Wire EDM Cutting. *J. Mater. Process. Technol.* **2020**, *275*, 116373. [CrossRef]
27. Yanhui, Y.; Dong, L.; Ziyan, H.; Zijian, L. Optimization of Preform Shapes by RSM and FEM to Improve Deformation Homogeneity in Aerospace Forgings. *Chin. J. Aeronaut.* **2010**, *23*, 260–267. [CrossRef]
28. Liu, J.; Ma, Y. A Survey of Manufacturing Oriented Topology Optimization Methods. *Adv. Eng. Softw.* **2016**, *100*, 161–175. [CrossRef]
29. Mantovani, S.; Presti, I.L.; Cavazzoni, L.; Baldini, A. Influence of Manufacturing Constraints on the Topology Optimization of an Automotive Dashboard. *Procedia Manuf.* **2017**, *11*, 1700–1708. [CrossRef]
30. Zuo, K.-T.; Chen, L.-P.; Zhang, Y.-Q.; Yang, J. Manufacturing- and Machining-Based Topology Optimization. *Int. J. Adv. Manuf. Technol.* **2006**, *27*, 531–536. [CrossRef]
31. Rout, B.K.; Mittal, R.K. Optimal Design of Manipulator Parameter Using Evolutionary Optimization Techniques. *Robotica* **2010**, *28*, 381–395. [CrossRef]
32. Liu, W.; Zhu, H.; Wang, Y.; Zhou, S.; Bai, Y.; Zhao, C. Topology Optimization of Support Structure of Telescope Skin Based on Bit-Matrix Representation NSGA-II. *Chin. J. Aeronaut.* **2013**, *26*, 1422–1429. [CrossRef]

Article

Design and Experimental Testing of an Ankle Rehabilitation Robot

Ioan Doroftei [1,*] , **Cristina-Magda Cazacu** [1,*] **and Stelian Alaci** [2]

1   Mechanical Engineering, Mechatronics and Robotics Department, "Gheorghe Asachi" Technical University of Iasi, D. Mangeron Bvd. 43, 700050 Iasi, Romania
2   Mechanical Engineering Department, "Stefan cel Mare" University of Suceava, Universitatii Str. 13, 720229 Suceava, Romania; stelian.alaci@usm.ro
*   Correspondence: idorofte@mail.tuiasi.ro (I.D.); cristina-magda.cazacu@academic.tuiasi.ro (C.-M.C.)

Abstract: The ankle joint (AJ) is a crucial joint in daily life, responsible for providing stability, mobility, and support to the lower limbs during routine activities such as walking, jumping, and running. Ankle joint injuries can occur due to sudden twists or turns, leading to ligament sprains, strains, fractures, and dislocations that can cause pain, swelling, and limited mobility. When AJ trauma occurs, joint instability happens, causing mobility limitations or even a loss of joint mobility, and rehabilitation therapy is necessary. AJ rehabilitation is critical for those recovering from ankle injuries to regain strength, stability, and function. Common rehabilitation methods include rest, ice, compression, and elevation (RICE), physical therapy, ankle braces, and exercises to strengthen the surrounding muscles. Traditional rehabilitation therapies are limited and require constant presence from a therapist, but technological advancements offer new ways to fully recover from an injury. In recent decades there has been an upswing in research on robotics, specifically regarding rehabilitation. Robotic platforms (RbPs) offer several advantages for AJ rehabilitation assistance, including customized training programs, real-time feedback, improved performance monitoring, and increased patient engagement. These platforms use advanced technologies such as sensors, actuators, and virtual reality to help patients recover quicker and more efficiently. Furthermore, RbPs can provide a safe and controlled environment for patients who need to rebuild their strength and mobility. They can enable patients to focus on specific areas of weakness or instability and provide targeted training for faster recovery and reduced risk of re-injury. Unfortunately, high costs make it difficult to implement these systems in recuperative institutions, and the need for low-cost platforms is apparent. While different systems are currently being used, none of them fully satisfy patient needs or they lack technical problems. This paper addresses the conception, development, and implementation of rehabilitation platforms (RPs) that are adaptable to patients' needs by presenting different design solutions (DSs) of ankle RPs, mathematical modeling, and simulations of a selected rehabilitation platform (RP) currently under development. In addition, some results from practical tests of the first prototype of this RP are presented. One patient voluntarily agreed to use this platform for more rehabilitation sessions on her AJ (right leg). To counteract some drawbacks of the first prototype, some improvements in the RP design have been proposed. The results on testing the improved prototype will be the subject of future work.

Keywords: rehabilitation robot; ankle rehabilitation; robot design; modeling and simulation; experimental testing

Citation: Doroftei, I.; Cazacu, C.-M.; Alaci, S. Design and Experimental Testing of an Ankle Rehabilitation Robot. *Actuators* **2023**, *12*, 238. https://doi.org/10.3390/act12060238

Academic Editor: Nariman Sepehri

Received: 2 May 2023
Revised: 29 May 2023
Accepted: 6 June 2023
Published: 8 June 2023

1. Introduction

In recent years, research in the field of medical robotics has been intensified. Among the medical applications of robotics, high-precision surgical interventions or the recovery of motor functions following such interventions or accidents can be mentioned. As benefits of these systems, we can list: they eliminate overburdening of medical personnel, robots reduce waiting times, and physiotherapists can follow several patients at the same time. In addition, the optimization of tools used in surgery and microsurgery facilitates surgical interventions and reduces the incidence of errors and deaths in the scope of high-risk interventions. In conclusion, while doctors can control and monitor more patients with a reduced workload, patients receive better quality medical therapy thanks to modern and efficient equipment.

The structure of the ankle joint (AJ) plays a defining role in an individual's daily life, providing balance, stability, and the possibility of locomotion. Representing such an important structure in daily activities, the incidence of injuries at the level of this joint, including those of a neurological nature, caused by a stroke, is also high [1]. AJ traumas will immediately affect its stability and mobility. The recovery of the physically injured AJ involves, in the first stage, a reduction in edema: RICE and, possibly, anti-inflammatory drugs. Injured ligaments will form scar tissue, so patients will experience limited activity in the absence of proper rehabilitation. Traditional rehabilitation procedures are based on using simple and primitive devices such as elastic bands and foam rollers. These procedures need the permanent presence of the physiotherapist near the patient. In addition, the used exercises are time-consuming and repetitive, requiring effort both from the physiotherapist and from the patient. To counteract these aspects and to improve the quality of rehabilitation, in recent years, robots have been involved in rehabilitation therapies. The use of robots in such therapies is also based on the increasing number of elderly people who need assistance either due to aging or accidents.

Due to the multiple traumas that can appear at the lower limb, a large diversity of robotic systems have been designed for medical recovery, including AJ RPs. These devices are designed to improve the motor function of the ankle joint, which can be impaired due to various reasons, such as injury, disease, or aging. Typically, AJ RPs have two degrees of freedom (DOF) to enable dorsiflexion (DF)/plantar flexion (PF) and inversion (INV)/eversion (EV) movements. Some robots also include a third DOF to allow for abduction/adduction rotation. Examples include the ankle rehabilitation robot developed by Park et al. [2]. These robots can be classified into different categories based on their design, such as parallel devices, exoskeletons, or wearable devices.

Parallel devices are the most common type of AJ rehabilitation robots (RRs). They typically consist of two platforms, one fixed and one movable, connected by parallel linkages, providing multiple DOF for the ankle joint. Actuation is usually provided by electric motors or pneumatic actuators. Several studies have been conducted to evaluate the effectiveness of parallel AJ RRs in clinical settings. For example, a study by Kesar et al. [3] evaluated the Anklebot, a parallel robot for ankle rehabilitation, in stroke patients and found significant improvements in motor function. Another study by Lee et al. [4] developed an AJ rehabilitation robot (RR) using pneumatic artificial muscles and demonstrated improved ankle DF and PF in stroke patients.

Exoskeletons are another type of AJ RRs that provide support and assistance to the ankle joint through wearable robotic devices. They typically consist of rigid structures attached to the leg and foot, with actuators and sensors providing joint motion and feedback. A study by Lobo-Prat et al. [5] evaluated an ankle exoskeleton for gait rehabilitation in stroke patients and found improved ankle DF and reduced compensatory movements.

Wearable devices are a newer type of AJ RRs that are designed to be lightweight and portable, allowing patients to use them during daily activities. A study by Cheung et al. [6] developed a wearable AJ RR with self-aligning joints and found improved ankle DF and PF in healthy individuals.

Control strategies for AJ RRs play an important role in their effectiveness. Several studies have explored different control strategies, such as impedance control, admittance control, and trajectory tracking control. A study by Paradiso et al. [7] developed a robotic device for ankle rehabilitation with impedance control, while a study by Park et al. [8] developed a gait-enhancing mobile shoe using machine learning algorithms to adapt to each patient's unique gait. Table 1 summarizes some of the existing parallel AJ RRs, based on: actuation type, number of DOF, control strategies, and study with patients [9].

Table 1. Some existing parallel AJ RPs [9].

Reference	Actuation Type	Number of DOF	Control Strategies	Study with Patients
Girone et al. [10]	Pneumatic	6	Position control; force control	Yes
Yoon et al. [11]	Pneumatic	4	Position control	No
Dai et al. [12]	Electric	4	N/A	No
Liu et al. [13]	Electric	3	Position control; force control	No
Saglia et al. [14]	Electric	2	Position control; assistive control; admittance control	No
Malosio et al. [15]	Electric	3	Position control; admittance control	No
Ayas et al. [16]	Electric	2	Trajectory tracking; admittance adaptive control	No
Ai et al. [17]	Pneumatic	2	Adaptive backstepping sliding mode control	No
Jamwal et al. [18]	Pneumatic	3	Position control; adaptive control; adaptive impedance control	Yes
Zhang et al. [19]	Pneumatic	3	Position control; adaptive patient-cooperative control; adaptive trajectory tracking	Yes
Tsoi et al. [20]	Electric	3	Joint force control; impedance control	No
Wang et al. [21]	Electric	3	Position control	No
Valles et al. [22]	Electric	3	Position control; force control	No
Li et al. [23]	Electric	3	Position control; patient-passive compliant exercise; isotonic exercise; patient-active exercise	No

Unfortunately, robotic-assisted rehabilitation therapies at the level of recuperative institutions are highly costly. That is why research is needed for the development of robotic RPs that can allow therapy to be performed in the patient's home. Home-based AJ RRs is a growing field, as they offer patients more convenient and accessible options for rehabilitation. We may analyze these robots in terms of different aspects such as robot design, user interface, control strategies, clinical validation, and so on.

In terms of robot design, we may present some examples: Hong et al. [24] developed a portable and lightweight AJ RR that could be easily used at home by stroke patients; Liu et al. [25] designed a 3-DOF wearable AJ RR that could be easily worn by patients during their daily activities; and Zhou et al. [26] developed a modular AJ RR that could be customized based on the patient's needs and preferences.

Concerning the user interface, Aghaebrahimian et al. [27] developed a smartphone-based user interface for an AJ RR, allowing patients to monitor their progress and receive feedback on their rehabilitation; Kamal et al. [28] designed a gamified user interface for a home-based AJ RR, making rehabilitation more engaging and motivating for patients; and Wang et al. [29] developed a user-friendly interface for an AJ RR, allowing patients to easily adjust the robot's settings and track their rehabilitation progress.

Regarding clinical validation, Chen et al. [30] conducted a clinical study on a home-based AJ RR with 24 stroke patients and found that the robot was effective in improving the patients' ankle range of motion and gait performance; Li et al. [31] conducted a randomized controlled trial on a home-based AJ RR with 60 ankle sprain patients and found that the robot-assisted rehabilitation group had better ankle function than the control group; and Wang et al. [32] conducted a pilot study on a home-based AJ RR with 12 ankle sprain patients and found that the robot was effective in improving the patients' ankle strength and range of motion.

Overall, home-based AJ RRs have the potential to improve patient outcomes by providing more accessible and convenient rehabilitation options. However, more research is needed to validate their effectiveness and usability in clinical settings. Table 2 presents some AJ RPs suggested for domestic use, taking into account actuation type, number of DOF, and the main function of the developed rehabilitation systems [33].

Table 2. Some AJ RPs suggested for domestic use [33].

Reference	Actuation Type	Number of DOF	Function
Cioi et al. [34]	Pneumatic	6	Ankle rehabilitation for children with epilepsy
Girone et al. [10]	Pneumatic	6	AJ rehabilitation
Roy et al. [35]	Electric	3	Ankle training with a robotic device to improve hemiparetic gait after a stroke
Kim et al. [36]	Electric	2	Active ankle–foot orthosis for foot drop
Ward et al. [37]	Electric	2	Powered ankle–foot orthosis
Forrester et al. [38]	Electric	3	"AnkleBot" training on paretic ankle motor control in chronic stroke
Jamwal et al. [39]	Pneumatic	3	Treatment for an ankle sprain through physical therapy
Blanchette et al. [40]	Electro-hydraulic	2	Robotized ankle–foot orthosis
Takahashi et al. [41]	Pneumatic	2	An exoskeleton supplies plantar flexion assistance
Koller et al. [42]	Pneumatic	2	Powered ankle exoskeletons using neural measurements
Ren et al. [43]	Electric	2	Wearable AJ RR for in-bed acute stroke rehabilitation
Yeung et al. [44]	Electric	2	Robot-assisted ankle–foot orthosis to provide assistance post stroke
Awad et al. [45]	Electric	2	ReWalk ReStore dorsi flexor and plantar flexor

The analysis of the existing literature allows us to conclude that using robotic systems (RbSs) as an innovative substitute for traditional medical treatment is highly advantageous. Nonetheless, rehabilitation systems are faced with certain shortcomings which restrict their usage in recovery clinics, including poor interaction of the patient with the rehabilitation system, the need for additional safety measures during exercises, the inability to repeat sessions at home, limited real-time adjustment options, complex command interfaces, and difficulty handling the systems due to shape and dimension constraints. To overcome these limitations, specialized control techniques are necessary to ensure patient safety and recovery throughout system usage. Control algorithms aim to monitor RbPs used in rehabilitation exercises to enhance motor plasticity and improve motor function recovery. Despite the availability of numerous recovery systems in the literature, a device that satisfies all patient requirements and lacks technical issues does not yet exist. Hence, this study focuses on designing, developing, and implementing adaptable RPs tailored to patients' specific demands.

The remainder of this paper is organized as follows: Section 2 discusses the design of an ankle RP, including the motivation of the chosen design solution (DS), structural synthesis, design solutions (DSs), and the selection of a platform that will be practically realized. Section 3 presents mathematical modeling and simulation of the adopted DS, its dimensional synthesis, and experimental results during the test; in addition, this section presents some ethical and safety issues and a new proposed design, based on the conclusions that result from the practical test. In Section 4, some concluding remarks of this work are presented.

2. Materials and Methods

2.1. Design of the Ankle Rehabilitation Robot

2.1.1. Motivation of the Adopted Solutions

The AJ is very important both in a person's daily activities and in sports activities. Due to the fact that it is in great demand throughout the day, this joint can be subject to accidents. Although, in addition to classical therapies, many clinics use robotic systems for the rehabilitation of the AJ, these systems are either too complex, requiring the presence of a physiotherapist; boring and tiring, decreasing patients' motivation to use them; or too expensive to be purchased by patients in order to carry out recovery sessions at home. Starting from these aspects, there is the need to design robotic platforms that are as simple and friendly as possible, as well as at a low-cost price. In an attempt to design and create a robotic platform for the AJ, the authors of this paper have carried out various studies over the last few years [46–51].

2.1.2. Structural Synthesis

Taking into account the movements allowed by the AJ (Figure 1), we started the design of the robotic RP. This platform should be able to recover two of the three movements shown in Figure 1: DF/PF and EV/INV. The third movement allowed by the AJ (abduction/adduction) is not a key element in the rehabilitation of this joint, with it being a secondary movement. The ranges of the specified movements vary between 25° to 50° for PF, 20° to 30° for DF, 35° to 50° for INV, and 0° to 25° for EV [52]. We may conclude that the designed platform must be a spatial kinematic structure and it should allow two rotational movements around two perpendicular axes. This means that the driven link (DnL) of the RP must have two DOF. This link will be the plate supporting the sole of the foot (PSSF).

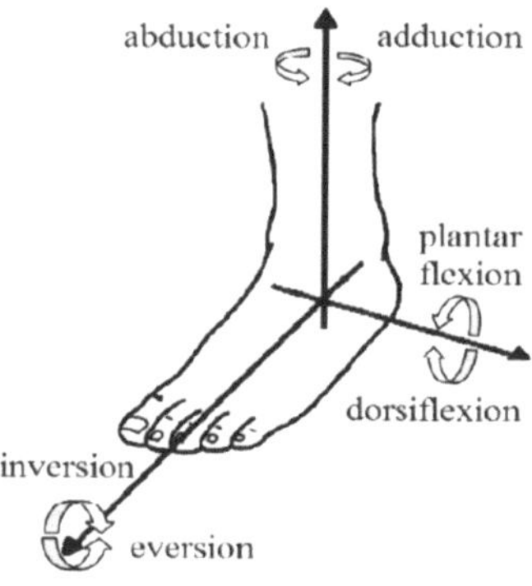

Figure 1. AJ rotational motions.

Based on the previous comments, some mechanisms with two DOF will be proposed for the AJ RP. "Fixed" RP versions with the base connected to the ground and also "portable" RP versions with the base connected to the calf will be proposed. For all of these RPs, the DnL will be the PSSF. None of the two types of RP may be used during walking. However, the "portable" versions could be displaced, before, during, or after the rehabilitation process. If we consider the DnL of a mechanism that represents the structure of an AJ RP, this link should have three DOF if all three movements of the AJ are intended to be recovered (Figure 2).

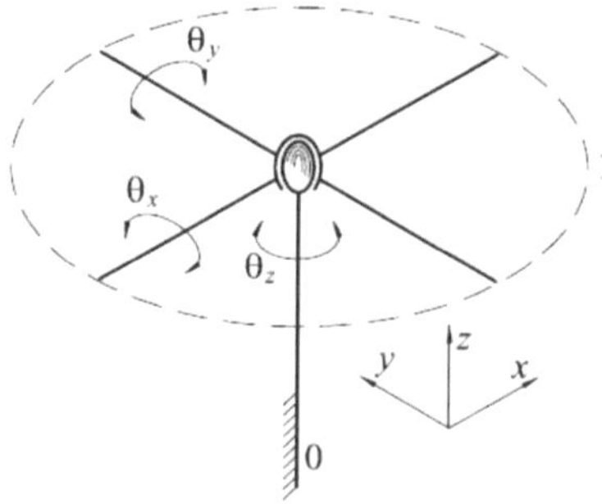

Figure 2. Principled schematics of an RP with three DOF.

We may find multiple actuating solutions of the DnL with three DOF. One of them could use a 3-SPS/S (spherical—prismatic—spherical/spherical) mechanism (see Figure 3a). The prismatic joints of this mechanism will be driving joints (DgJs). A second possible solution is using a 3-RSS/S (rotational—spherical—spherical/spherical) mechanism, shown in Figure 3b. In this case, DgJs are the revolute joints.

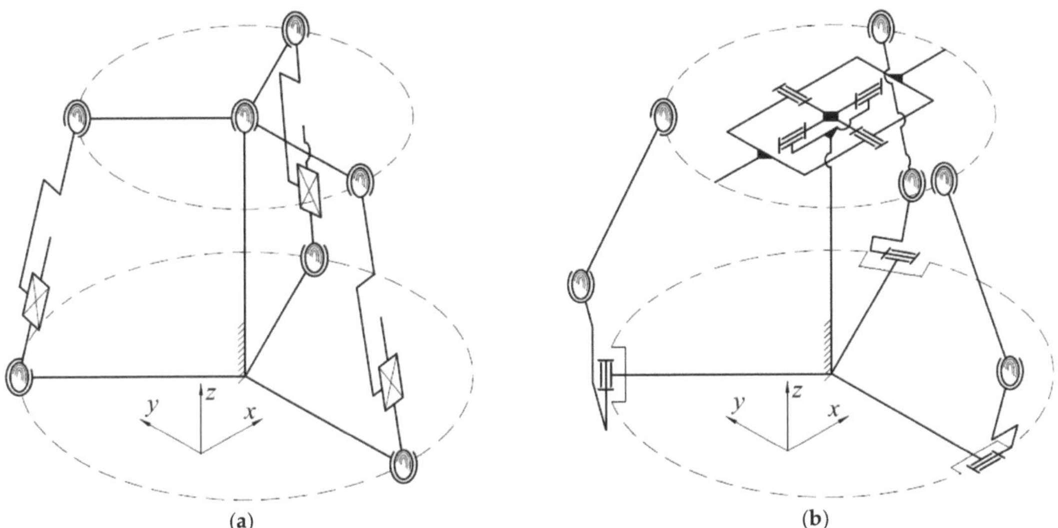

Figure 3. Kinematics of a RP with three DOF: (**a**) an RP using a 3-SPS/S mechanism and (**b**) an RP based on the 3-RSS/S mechanism.

Because we considered that the abduction/adduction movement of the AJ is a secondary one and we do not take it into account for recovery, the DnL (the PSSF) could only have two DOF (Figure 4).

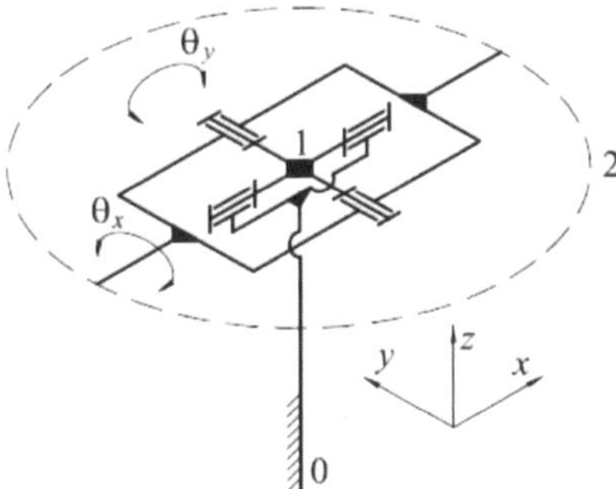

Figure 4. Principled schematics of an RP with two DOF.

We may use multiple actuating solutions of a DnL with two DOF, too. A first example is considering a mechanism such as 2-SPS/U (spherical—prismatic—spherical/universal), as shown in Figure 5a. To avoid rotation around its own axis of the SPS type kinematic chain, a 2-UPS/U (universal—prismatic—spherical/universal) mechanism may be utilized (Figure 5b). The actuation of prismatic joints usually requires linear actuators or rotary actuators and additional mechanical transmissions. For the linear actuators, we may use pneumatic or hydraulic actuators, but they require an external fluid source, which is not usually available at home. Moreover, pneumatic and hydraulic actuators exhibit nonlinearities in their operation. Mechanical transmissions are used to convert the rotational motion into a translational one (ball screw; rack and pinion). These transmissions are complicated and expensive. To counteract all the disadvantages mentioned above, we will propose an RP based on a Scotch–Yoke mechanism. This mechanism has advantages such as a high output torque and smooth operation. Two kinematic solutions (KSs) using the Scotch–Yoke mechanism will be proposed in the following. One of these solutions (KS-1), the "fixed kinematic solution", is shown in Figure 6. It has the base (link 0) fixed to the ground. The DnL 6 plays the role of the PSSF.

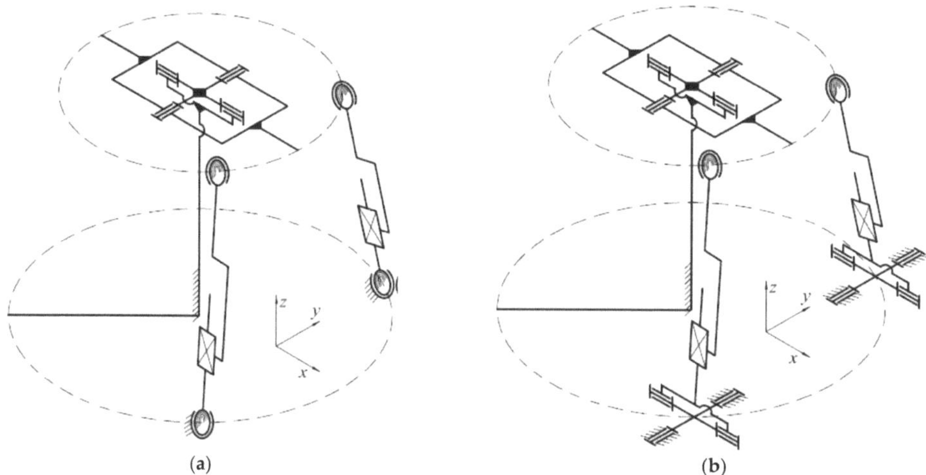

Figure 5. Kinematics of an RP with two DOF: (**a**) an RP using a 2-SPS/U mechanism and (**b**) an RP based on a 2-UPS/U mechanism.

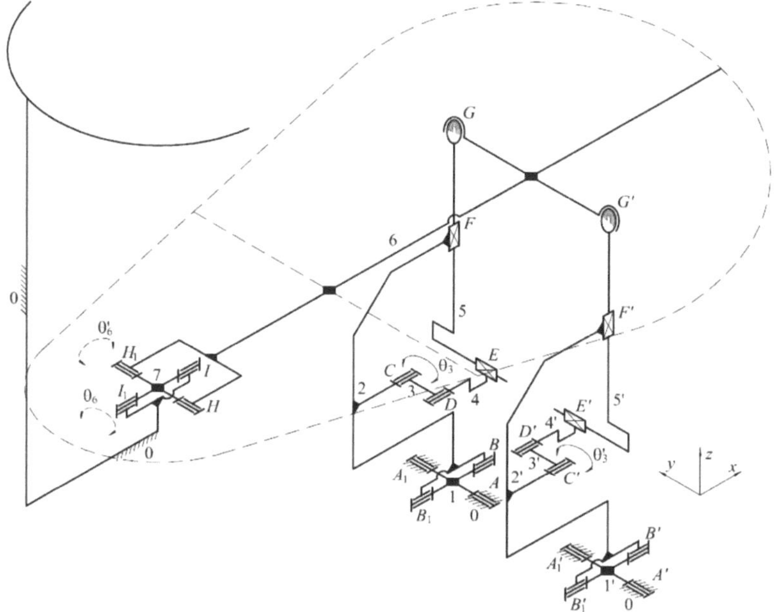

Figure 6. Kinematics of the RP based on the Scotch–Yoke mechanism, "fixed" version (KS-1).

The "portable kinematic solution" has the base (link 0) connected to the calf (KS-2; see Figure 7). The PSSF is, again, represented by the DnL 6. During the rehabilitation exercises, the leg should be suspended, with the calf resting on a solid surface (for example, a chair or a sofa). For both KSs mentioned above (KS-1 and KS-2), the driving links (DgLs) are the links 3 and 3′. To produce the INV/EV movement of the AJ, the DnL 6 must be rotated by the angle θ_6 around the x-axis. For performing this, the DgLs will be rotated by the same angle, $\theta_3 = \theta_{3'}$, in the same direction. The PF/DF movement of the DnL 6 (with the angle $\theta_{6'}$ around the y-axis) will be produced when the DgLs is rotated by the same angle but in opposite directions, $\theta_3 = -\theta_{3'}$.

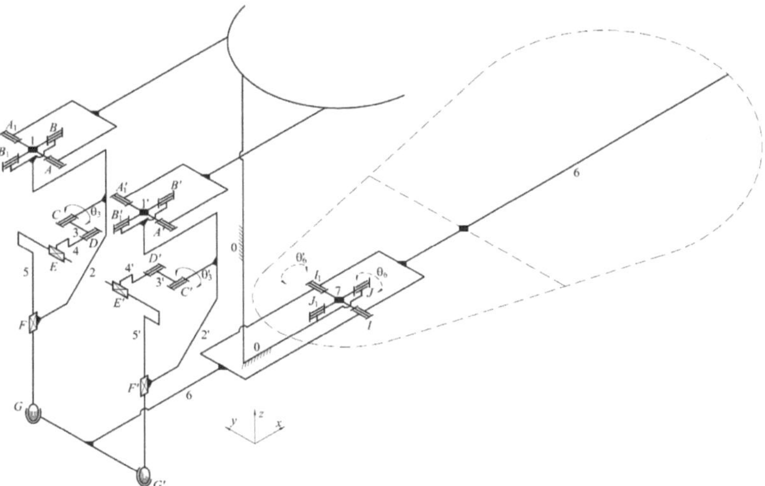

Figure 7. Kinematics of the RP based on the Scotch–Yoke mechanism, "portable" version (KS-2).

A second actuation solution of the DnL with two DOF (Figure 4) could use a 2-RSS/U (rotational—spherical—spherical/universal) mechanism, Figure 8. For this actuation solution, we may have two KSs, too. The first one (KS-3), called the "fixed kinematic solution", is shown in Figure 9 and the second one, the "portable kinematic solution" (KS-4), is represented in Figure 10. In these cases, the DnL 4 plays the role of the PSSF.

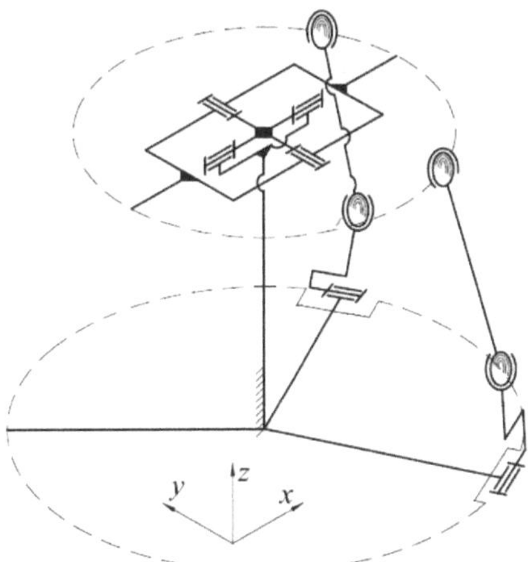

Figure 8. Kinematics of an RP with two DOF with the structure based on a 2-RSS/U mechanism.

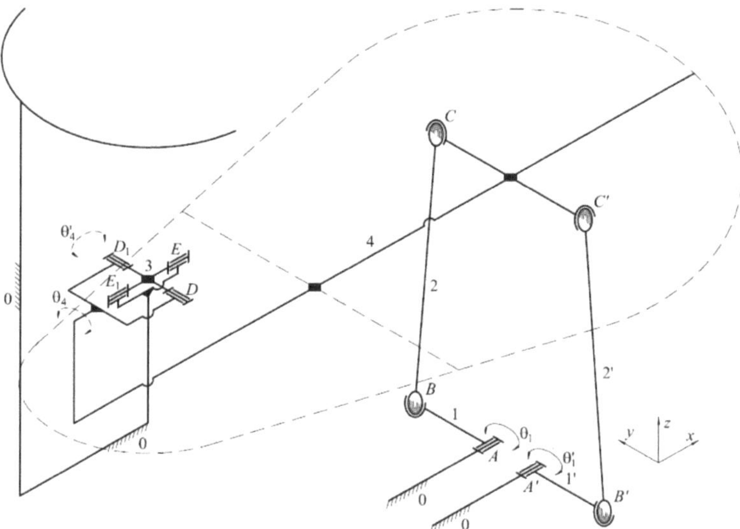

Figure 9. Kinematics of an RP based on the spatial four-bar linkage, "fixed" version (KS-3).

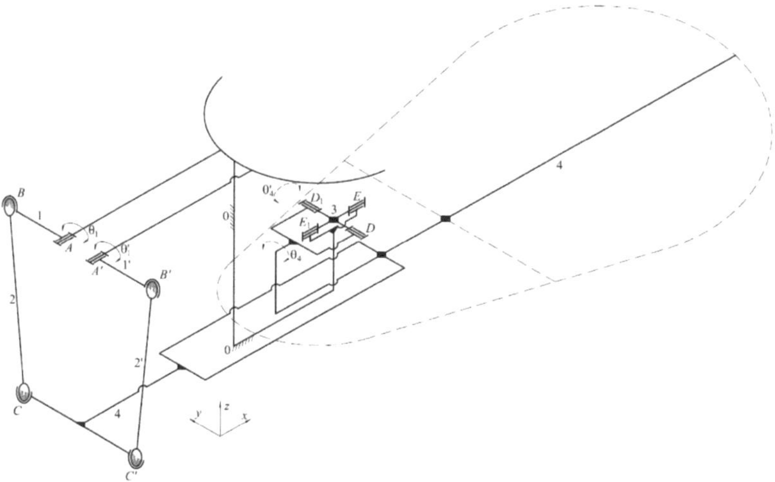

Figure 10. Kinematics of an RP based on the spatial four-bar linkage, "portable" version (KS-4).

For these last two mechanisms, the INV/EV movement of the AJ is produced when the DnL 4 is rotated by the angle θ_4 around the x-axis. We will determine whether the DgLs will be rotated by the same angle, $\theta_1 = \theta_{1'}$, in the same direction. If the DgLs is rotated by the same angle but in opposite directions, $\theta_1 = -\theta_{1'}$, the PF/DF movement of the DnL 4 (with the angle $\theta_{4'}$ around the y-axis) will be produced.

2.1.3. Design Solutions

Some DSs of "fixed" RPs based on the mechanisms discussed above have been analyzed. These designs could have collinear (DS-2, DS-4, DS-6, and DS-8) or parallel (DS-1, DS-3, DS-5, and DS-7) rotational axes of the two actuators. In addition, they could have the two DnL rotational axes coaxial (DS-1, DS-2, DS-5, and DS-6) with the AJ rotational axes or parallel (DS-3, DS-4, DS-7, and DS-8) to them. These DSs are summarized in Tables 3 and 4.

To select the design solution (DS) with the most advantages in use (Table 5), we have analyzed the proposed DSs based on several criteria:

- RP maintenance (C1)—these criteria take into account the actuator and mechanism type (joints type);
- Simplicity in use (C2)—ease of programming and use by the end user;
- RP cost (C3)—takes into account the cost prices of the components;
- RP overall dimensions (C4)—the overall dimensions and the mass of the platform are very important in choosing the technical solution;
- Minimum blocking probability (C5)—depends on the joints type;
- The DnL range of motion (C6)—the RP should cover the range of motion for both AJ movements considered for rehabilitation.

Table 3. DSs of robotic RPs with two DOF based on the spatial four-bar linkage (2-RSS/U).

The Center of the AJ Is Aligned with the Rotation Center of the Robot		The Center of the AJ and the Rotation Center of the Robot Are not Coincident	
Parallel Rotational Axes of DgLs	**Collinear Rotational Axes of DgLs**	**Parallel Rotational Axes of DgLs**	**Collinear Rotational Axes of DgLs**
DS-1	DS-2	DS-3	DS-4

Table 4. DSs of robotic RPs with two DOF based on the Scotch–Yoke mechanism (2-UPS/U).

The Center of the AJ Is Aligned with the Rotation Center of the Robot		The Center of the AJ and the Rotation Center of the Robot Are not Coincident	
Parallel Rotational Axes of DgLs	**Collinear Rotational Axes of DgLs**	**Parallel Rotational Axes of DgLs**	**Collinear Rotational Axes of DgLs**
DS-5	DS-6	DS-7	DS-8

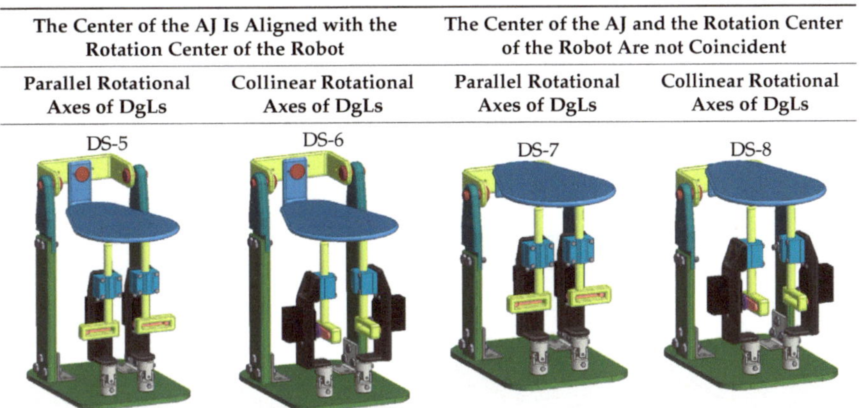

Points from 0 to 5 were awarded for each criterion and for each DS. RPs with the center of the AJ aligned with the rotation center of the robot (DS-1, DS-2, DS-5, and DS-6) have bigger overall dimensions than other platforms. In addition, RPs with collinear rotational axes of the DgLs (DS-2, DS-6, DS-4, and DS-8) have bigger dimensions compared with the platform with parallel rotational axes of these links (DS-1, DS-3, DS-5, and DS-7). DSs based on a 2-RSS/U mechanism (DS-1, DS-2, DS-3, and DS-4) have better maintenance than platforms based on a 2-UPS/U mechanism (DS-4, DS-5, DS-6, and DS-7). These last RPs (with a 2-UPS/U structure) are more expensive due to the ball nut screw transmissions that should be used for the prismatic joint. In addition, the blocking probability of the RPs based on a 2-UPS/U (DS-4, DS-5, DS-6, and DS-7) mechanism is higher. The range of motions for RPs with the center of the AJ not coincident with the rotation center of the robot (DS-3, DS-4, DS-7, and DS-8) is bigger. Among the discussed DSs, two solutions stand out that meet most of the requirements (DS-1 and DS-3). To select the DS which will be practically realized, dimensional synthesis, mathematical modeling, and simulations of the two mentioned DSs have to be carried out.

Table 5. Selection of the adopted DS.

	DS-1	DS-2	DS-3	DS-4	DS-5	DS-6	DS-7	DS-8
C1	5	5	5	5	4	4	4	4
C2	5	5	5	5	5	5	5	5
C3	5	5	5	5	4	4	4	4
C4	4	3	5	4	3	2	4	3
C5	4	4	5	4	3	3	3	3
C6	4	3	5	3	4	4	5	5
Total	27	25	30	26	23	22	25	24

3. Results

3.1. Mathematical Modeling and Simulation

Only two DSs based on the spatial four-bar mechanism (DS-1 and DS-3) were considered for mathematical modeling and simulation. The first one has two rotational axes coincident with the AJ axes. However, unfortunately, the simulation revealed that the DS-1 design solution, based on the kinematics shown in Figure 9, cannot assure the necessary movement ranges for AJ rehabilitation.

3.1.1. Mathematical Modeling of the DS-3 Design Solution

The kinematics of this DS of the RP is shown in Figure 11. The links 1 and 1′ are DgLs, while the plate 4 represents the DnL. To determine the relationship between the angular position of the DnL 4 with respect to the angular position of the driving link (DgL), the kinematic analysis of the mechanism is required.

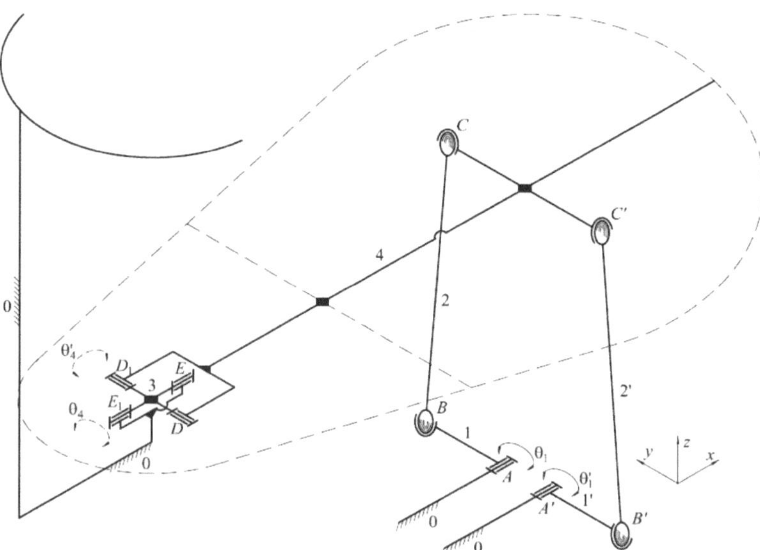

Figure 11. Kinematics of the DS-3 design solution.

The first considered case is that of the INV/EV movement, when both DgLs will be rotated with $\theta_1 = \theta_1'$. This movement is produced by the equivalent mechanism shown in Figure 12a. To write the kinematic equations, an equivalent mechanism driven by a single motor (shown in Figure 12b) is used.

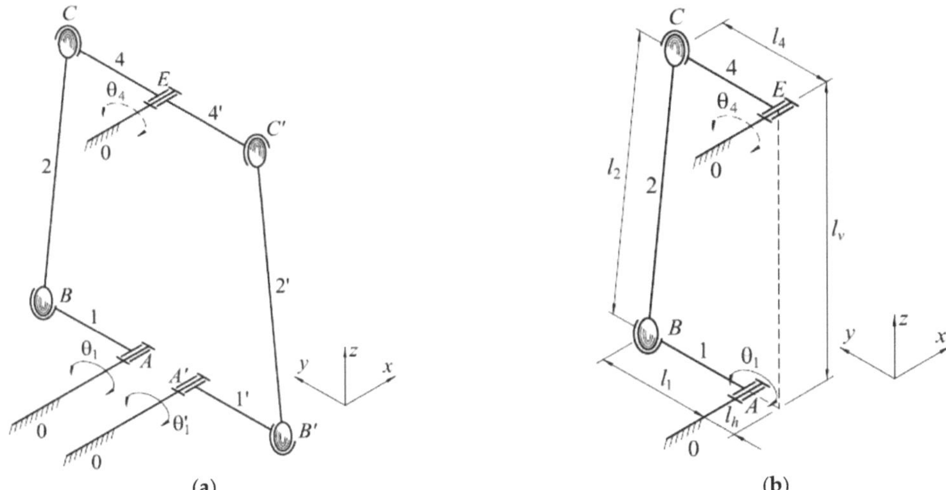

Figure 12. Equivalent mechanisms for EV/INV movement (DS-3 design solution): (**a**) mechanism with two DOF and (**b**) mechanism with one DOF.

The direct kinematics problem will lead to:

$$\theta_4 = 2 \cdot \mathrm{atan}\left(\frac{-B_1 \pm \sqrt{A_1{}^2 + B_1{}^2 - C_1{}^2}}{C_1 - A_1}\right), \tag{1}$$

where

$$\begin{cases} A_1 = -2 \cdot l_4 \cdot l_h - 2 \cdot l_1 \cdot l_4 \cdot \cos\theta_1 \\ B_1 = 2 \cdot l_4 \cdot l_v - 2 \cdot l_1 \cdot l_4 \cdot \sin\theta_1 \\ C_1 = 2 \cdot l_h \cdot l_1 \cdot \cos\theta_1 - 2 \cdot l_v \cdot l_1 \cdot \sin\theta_1 + l_1{}^2 - l_2^2 + l_4^2 + l_v^2 + l_h^2 \end{cases}, \tag{2}$$

while the inverse kinematics problem leads to:

$$\theta_1 = 2 \cdot \mathrm{atan}\left(\frac{-B_2 \pm \sqrt{A_2{}^2 + B_2{}^2 - C_2{}^2}}{C_2 - A_2}\right) \tag{3}$$

where

$$\begin{cases} A_1 = (2 \cdot l_h \cdot l_1 - 2 \cdot l_1 \cdot l_4 \cdot \cos\theta_4) \\ B_1 = (-2 \cdot l_v \cdot l_1 - 2 \cdot l_1 \cdot l_4 \cdot \sin\theta_4) \\ C_1 = l_1{}^2 - l_2^2 + l_4^2 + l_v^2 + l_h^2 + 2 \cdot l_4 \cdot (l_v \cdot \sin\theta_4 - l_h \cdot \cos\theta_4) \end{cases} \tag{4}$$

We will now consider a mechanism with two DOF (Figure 13a), responsible for PF/DF movement, when the DgLs 1 and 1′ are rotated with $\theta_1 = -\theta_1'$. To solve the direct kinematics problem, a mechanism with one DOF is used (Figure 13b). By solving the direct kinematics problem for this mechanism, we will obtain:

$$\theta_4' = 2 \cdot \mathrm{atan}\left(\frac{-B_3 \pm \sqrt{A_3{}^2 + B_3{}^2 - C_3{}^2}}{C_3 - A_3}\right), \tag{5}$$

where

$$\begin{cases} A_3 = -2 \cdot l_4'^2 \\ B_3 = -2 \cdot l_4' \cdot l_2 + 2 \cdot l_1 \cdot l_4' \cdot \sin \theta_1 \\ C_3 = l_1{}^2 - l_2^2 + l_4{}^2 + l_v{}^2 + l_h{}^2 + 2 \cdot l_4'^2 - 2 \cdot l_h \cdot l_4 - 2 \cdot l_1 \cdot l_v \cdot \sin \theta_1 + \\ \quad + 2 \cdot l_1 \cdot (l_h - l_4) \cdot \cos \theta_1 \end{cases} \quad \cdot \tag{6}$$

The inverse kinematics problem leads to:

$$\theta_1 = 2 \cdot \text{atan} \left(\frac{-B_4 \pm \sqrt{A_4{}^2 + B_4{}^2 - C_4{}^2}}{C_4 - A_4} \right), \tag{7}$$

where

$$\begin{cases} A_4 = 2 \cdot l_1 \cdot (l_h - l_4) \\ B_4 = 2 \cdot l_1 \cdot (l_4' \cdot \sin \theta_4' - l_v) \\ C_4 = l_1{}^2 - l_2{}^2 + l_4{}^2 + l_v{}^2 + l_h{}^2 + 2 \cdot l_4'^2 - 2 \cdot l_4' \cdot (l_2 \cdot \sin \theta_4' + l_4' \cdot \cos \theta_4') - 2 \cdot l_h \cdot l_4 \end{cases} \quad \cdot \tag{8}$$

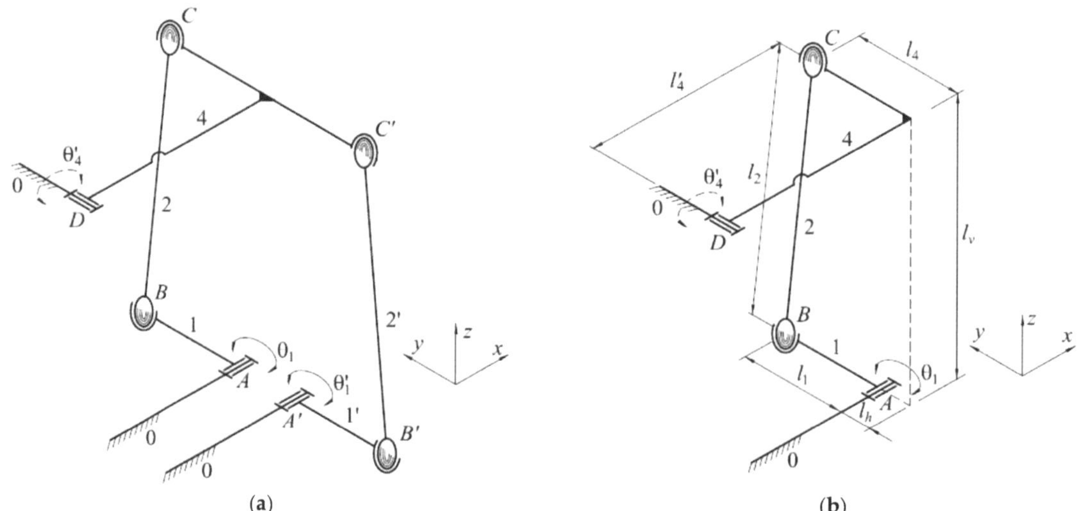

Figure 13. Equivalent mechanisms for PF/DF movement (DS-3 design solution): (**a**) real mechanism with two DOF [46] and (**b**) equivalent mechanism with one DOF.

Dimensional synthesis of the mechanisms and also a numerical simulation based on these equations must be further carried out.

3.1.2. Dimensional Synthesis and Simulation of the DS-3 Design Solution

For INV/EV movement, a simplified planar mechanism is used, represented by the extreme positions of the DnL (Figure 14a). For anatomical and dimensional reasons, we impose as known dimensions: l_h, l_v and l_4. The angle θ_4 is also known (θ_{41} for INV and θ_{42} for EV). The extreme positions of the crank are defined by θ_{11} and θ_{12} ($\theta_1 = \theta_{11} + \theta_{12}$). First, the dimensional synthesis of the mechanism responsible for INV/EV movement (Figure 14a) is solved, with the lengths l_1 and l_2 resulting in:

$$l_1 = \frac{b^2 - a^2}{2 \cdot b \cdot \cos(\psi + \theta_{12}) - 2 \cdot a \cdot \cos \beta'} \tag{9}$$

$$l_2 = \sqrt{l_1^2 + b^2 - 2 \cdot b \cdot l_1 \cdot \cos(\psi + \theta_{12})}, \tag{10}$$

where

$$a = \sqrt{l_h^2 + l_v^2 + l_4^2 - 2 \cdot \left(l_h^2 + l_v^2\right)^{1/2} \cdot l_4 \cdot \cos(\alpha' + \theta_{41} + \theta_{42})} \tag{11}$$

$$b = \sqrt{l_h^2 + l_v^2 + l_4^2 - 2 \cdot \left(l_h^2 + l_v^2\right)^{1/2} \cdot l_4 \cdot \cos \alpha'} \tag{12}$$

$$\alpha' = \frac{\pi}{2} - \theta_{42} - \mathrm{acos}\left(\frac{l_v}{\sqrt{l_h^2 + l_v^2}}\right) \tag{13}$$

$$\psi = \frac{\pi}{2} + \mathrm{acos}\left(\frac{l_v}{\sqrt{l_h^2 + l_v^2}}\right) - \mathrm{acos}\left(\frac{b^2 + l_h^2 + l_v^2 - l_4^2}{2 \cdot b \cdot \sqrt{l_h^2 + l_v^2}}\right) \tag{14}$$

$$\beta = \frac{\pi}{2} + \mathrm{acos}\left(\frac{l_v}{\sqrt{l_h^2 + l_v^2}}\right) - \mathrm{acos}\left(\frac{a^2 + l_h^2 + l_v^2 - l_4^2}{2 \cdot a \cdot \sqrt{l_h^2 + l_v^2}}\right) - \theta_{11} \tag{15}$$

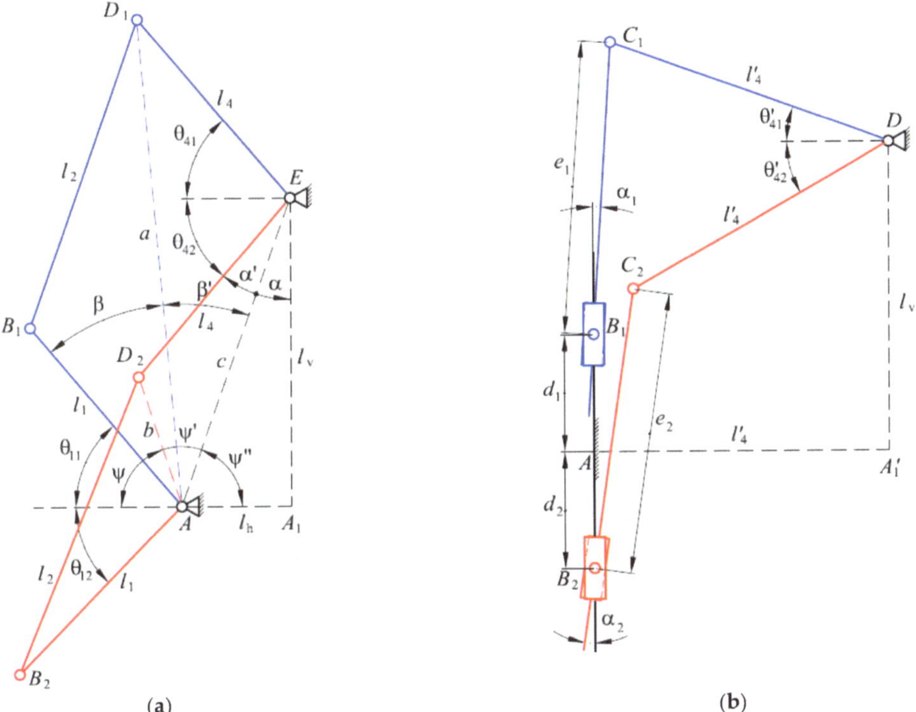

Figure 14. Simplified planar mechanism for geometric synthesis (DS-3 design solution): (**a**) for INV/EV movement and (**b**) for PF/DF movement [46]. In red and blue colors the extreme positions of the mechanisms are represented.

The geometric synthesis of the mechanism responsible for the PF/DF movement is solved using the simplified mechanism shown in Figure 14b. Based on the previous synthesis, d_1, e_1, d_2, e_2 and l_v are known, with:

$$d_1 = l_1 \cdot \sin \theta_{11}, \tag{16}$$

$$e_1 = l_2 \cdot \sin \theta_{21}, \tag{17}$$

$$d_2 = l_1 \cdot \sin \theta_{12}, \tag{18}$$

$$e_2 = l_2 \cdot \sin \theta_{22}. \tag{19}$$

The maximum values of the rehabilitation angle θ_4' are also known, considering that we know θ_{41}' and θ_{42}', which are the angular extreme positions of DgL. Following the analytical calculation, the last necessary unknown dimension is obtained, namely:

$$l_4' = \frac{(d_1 - l_v) \cdot \sin \theta_{41}' + \sqrt{(d_1 - l_v)^2 \cdot \sin^2 \theta_{41}' - 2 \cdot (1 - \cos \theta_{41}') \cdot \left[(d_1 - l_v)^2 - e_1^2\right]}}{2 \cdot (1 - \cos \theta_{41}')}, \tag{20}$$

or

$$l_4' = \frac{(d_2 + l_v) \cdot \sin \theta_{42}' + \sqrt{(d_2 + l_v)^2 \cdot \sin^2 \theta_{42}' - 2 \cdot (1 - \cos \theta_{42}') \cdot \left[(d_2 + l_v)^2 - e_2^2\right]}}{2 \cdot (1 - \cos \theta_{42}')}. \tag{21}$$

Based on the geometric synthesis and the kinematics problem, a prototype of the RP was designed and is shown in Figure 15, where 1—the base; 2 and 2'—electrical actuators (digital servos HD—1235 MG); 3 and 3'—driving links (the cranks of the two spatial four bar mechanisms); 4 and 4'—the rods of the two spatial four bar mechanisms; 5—driven link (the PSSF); 6 and 7—PSSF encoders; and 8 and 9—DgLs encoders.

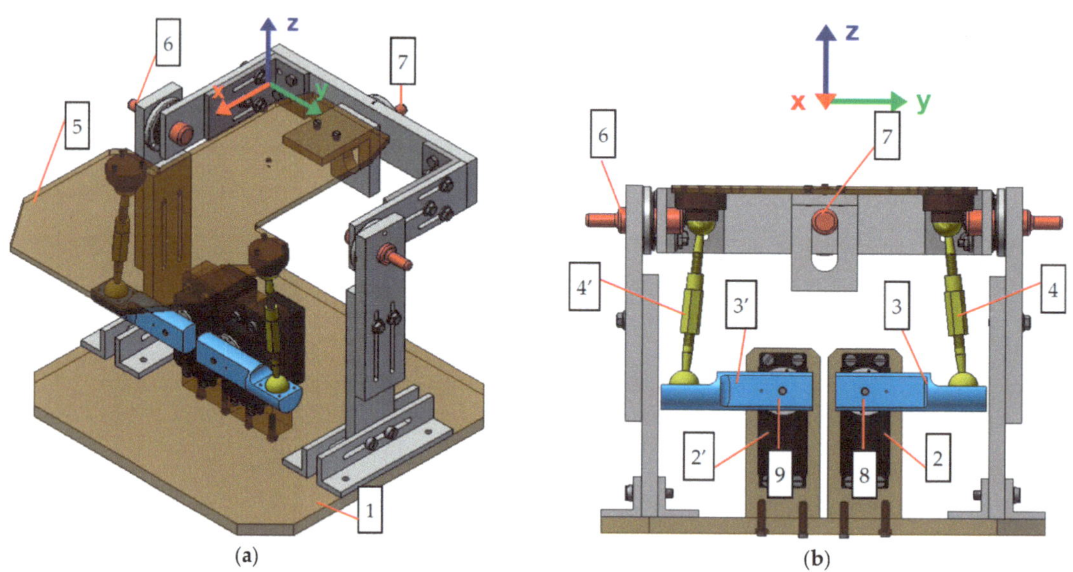

Figure 15. Three-dimensional CAD design of the RP (DS-3 design solution): (**a**) isometric view and (**b**) frontal view.

To prove that this design solution of the RP provides the DnL with the necessary movement ranges, a simulation of the virtual prototype was performed, using: $l_1 = 60$ mm; $l_2 = 103$ mm; $l_4 = 75$ mm; $l_h = 25$ mm; and $l_v = 101.5$ mm. A frame was attached to the CAD model with the origin at the center of the AJ, considered at 70 mm above the PSSF, Figure 16. When θ_1 and θ_1' angular positions of DgLs vary between the limits that ensure the two rehabilitated movements, the values of θ'_4 and θ_4 are those shown in Figure 17. The surfaces shown in Figure 17 could be considered the RP "workspace".

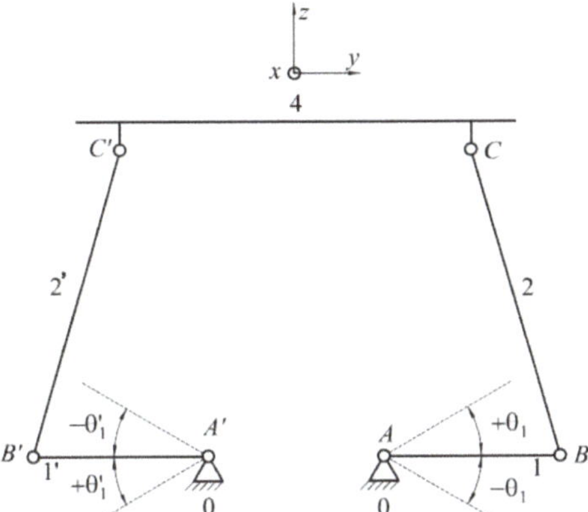

Figure 16. Attaching the frame to the DnL 4 (DS-3 design solution).

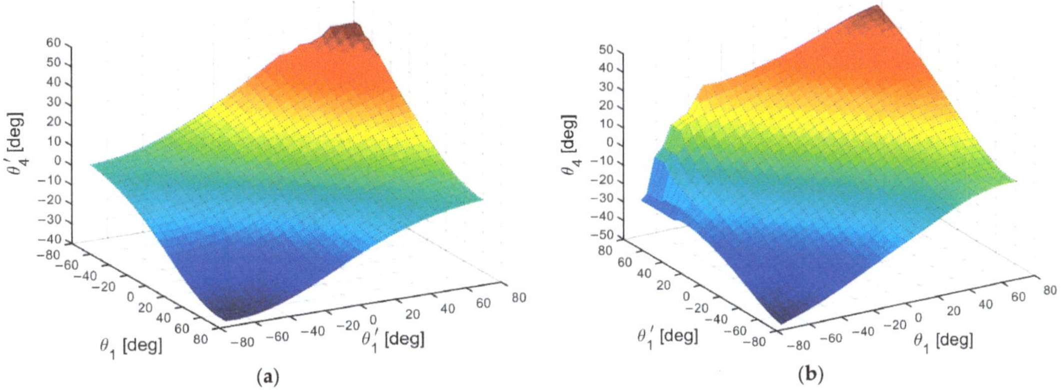

(a)

(b)

Figure 17. Maximum range of the PSSF angular positions according to the angular positions of the DgLs: (**a**) for PF/DF movement, $\theta_1 = -\theta'_1$, and (**b**) for INV/EV movement, $\theta_1 = \theta'_1$.

Considering that the angular position of the DnL should vary between $-25°$ and $0°$ for PF and between $0°$ and $50°$ for DF, the angular position of the DgLs will obtain values as follows: $\theta_1 = -32.4° \div 59°$ and $\theta'_1 = -59° \div 32.4°$, $\theta_1 = -\theta'_1$ (Figure 18a). For INV/EV movement, considering that the angular position of the DnL should vary between $-50°$ and $50°$, in order to assure the rehabilitation of both right and left leg AJs, the angular position of the DgLs should obtain values in the range $\theta_1 = \theta'_1 = -73.8° \div 73.8°$ (Figure 18b).

The curve in Figure 18a represents the diagonal of the surface in Figure 17a, when $\theta_1 = -\theta'_1$, and the curve in Figure 18b represents the diagonal of the surface in Figure 17b, when $\theta_1 = \theta'_1$.

The planar curves shown in Figure 19 represent the angular positions θ'_4 and θ_4 of the DnL according to the angular position θ_1 of the DgL for PF/DF movement (Figure 19a) and INV/EV movement (Figure 19b). As can be seen in these diagrams, the curves are identical, both for numerical simulation and for virtual RP simulation. The curves represented with a dashed red line represent the planar versions of the spatial curves presented in Figure 18a,b. In Figure 20, the angular positions of the DgLs and driven links (DnLs) are shown according to the time. Their linear variation can be observed, suggesting smooth operation and trouble-free performance of the proposed recovery exercises.

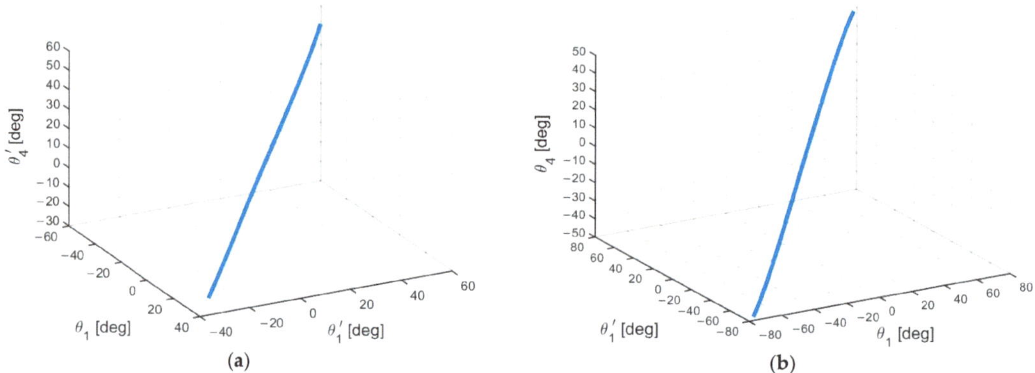

Figure 18. Angular position of the DnL according to the angular positions of the DgLs: (**a**) for the actual PF/DF movement, $\theta_1 = -\theta_1'$ and (**b**) for the actual INV/EV movement, $\theta_1 = \theta_1'$.

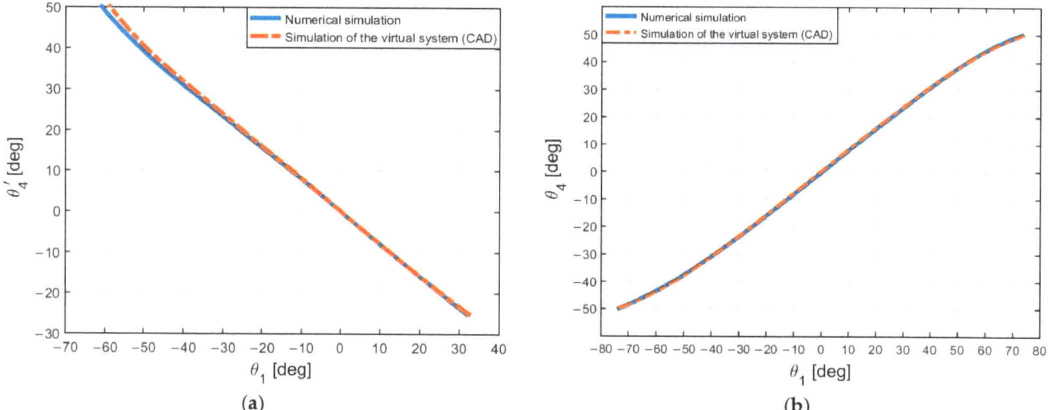

Figure 19. Numerical simulation results vs. results of the virtual RP simulation: (**a**) for PF/DF movement and (**b**) for INV/EV movement.

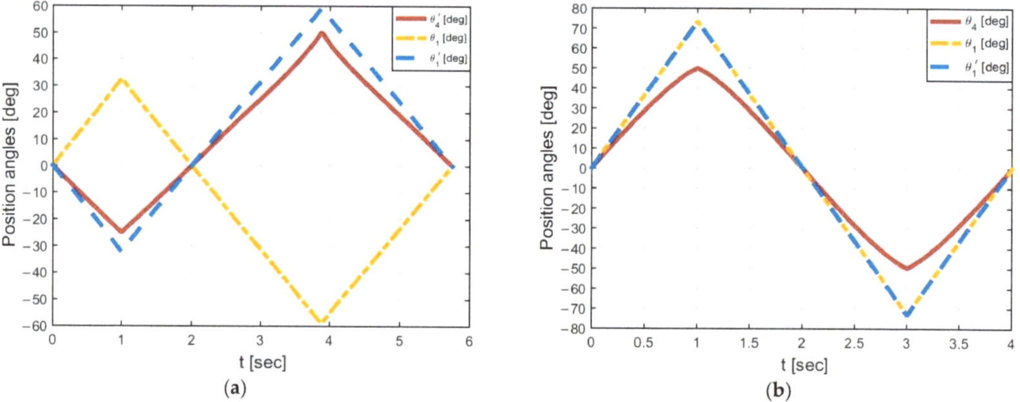

Figure 20. Angular positions of the DgLs and DnLs according to the time: (**a**) for PF/DF movement and (**b**) for INV/EV movement.

3.2. Experimental Results

3.2.1. Experimental Platform

The comparative study of the DSs discussed above led to the selection of optimal solutions, which are analyzed later. Thus, out of the two RPs based on the spatial four-bar linkage, only one remained under discussion (DS-3 design solution, Figure 11). This DS was practically realized (Figure 21). The general command and control architecture of the RP is presented in Figure 22. The therapist will set the extreme values of the angular positions for the PSSF through the graphical user interface (GUI), Figure 23. By doing so, the therapist will provide the input data for the microcontroller, according to the necessary rehabilitation exercises. Next, the microcontroller will send commands to the actuators to perform the required movements. Data collected from the encoders will be transmitted to the controller to be analyzed, resulting in visual feedback for both the patient and therapist.

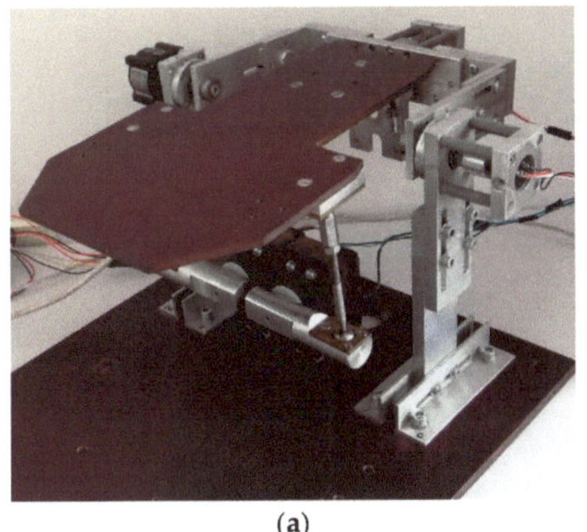

(a)

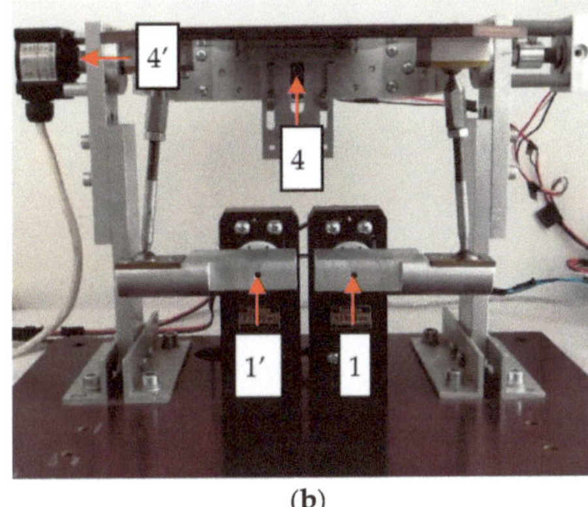

(b)

Figure 21. Prototype of the RP with two DOF (DS-3 design solution): (a) axonometric view and (b) frontal view.

Figure 22. General command and control architecture of the RP.

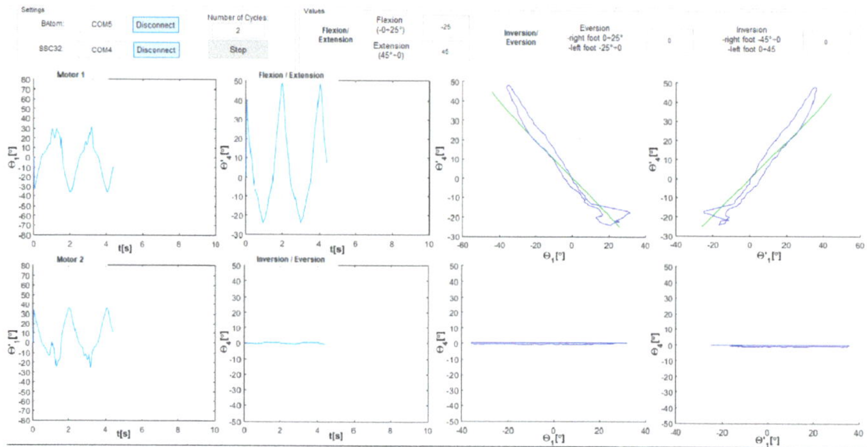

Figure 23. Graphical user interface.

As electric actuators, we have used two digital servos HD—1235 MG. Each one of these servos includes an encoder and its own electronics. These encoders (1 and $1'$ in Figure 21) are used to control the angular positions θ_1 and θ_1' of the DnLs, based on the information offered by the two encoders mounted on the rotational axes of the PSSF and also based on the mathematical modeling of the RP. The last two encoders (4 and $4'$ in Figure 21) are used to determine the angular positions θ_4 and θ_4' of the PSSF for each PF/DF or INV/EV movement. Through the GUI, the therapist will set the extreme values of θ_4 and θ_4'. To prevent supplementary injuries of the AJ during rehabilitation therapy, the current consumed by the motors is measured and certain limits for this current are imposed in the command/control program.

3.2.2. Experimental Tests and Results

Experimental tests on the RP were performed to evaluate if it may assure the range of angular strokes of the PSSF, according to PF/DF and INV/EV movement requirements. Complete angular strokes were performed for these movements, namely: 25° for PF, 45° for DF, 45° for INV (right leg), and 25° for EV (right leg). Figure 24 shows the values of θ_4' and θ_4 angular positions of the PSSF, relative to the θ_1 angular position of the DgL. These values are monitored during testing (dashed curves in blue). No patients were involved in these tests (the tests were performed without loading of the PSSF). In red, we may see the theoretical curves, generated with the numerical simulation results. Values above 0 for θ_4' (in Figure 24a) correspond to DF movement, while values on the negative axis correspond to PF movement. In Figure 24b, positive values of θ_4 correspond to INV movement, while negative values correspond to EV movement for the right leg. As we can see, the experimental curves are close to those obtained from numerical simulation for both movements that should be recovered.

Next, a volunteer patient with a fracture of the navicular bone in her right leg (Figure 25) was used to test the RP after a 30-day rest period. Due to this long period of immobilization, the patient suffered peripheral edema in the affected leg, walking difficulties and, also, difficulties in performing daily activities due to the increased stiffness of the AJ. Because of that, the physiotherapist recommended several types of PF/DF and INV/EV exercises to her with a frequency of at least once a day. The angular amplitudes of the AJ for the voluntary patient, before starting the therapy, were measured and noted (Figure 26).

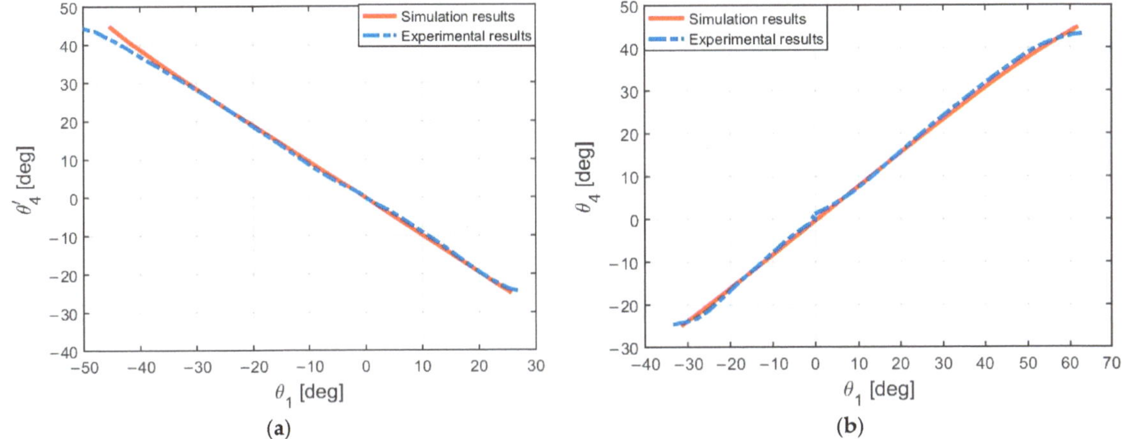

Figure 24. RP testing results [53]: (**a**) for PF/DF movement and (**b**) for INV/EV movement.

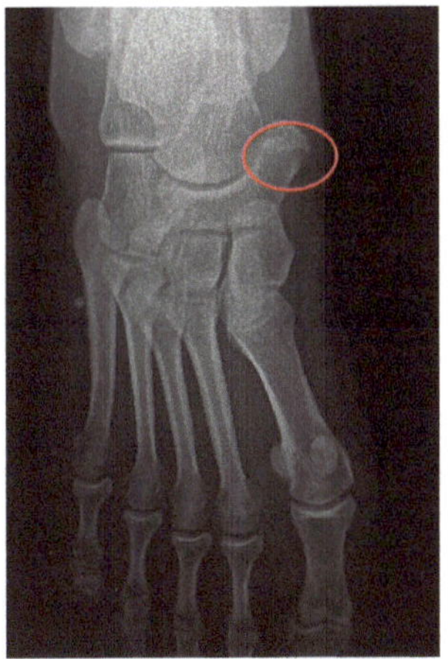

Figure 25. Fracture of the navicular bone, volunteer patient [53].

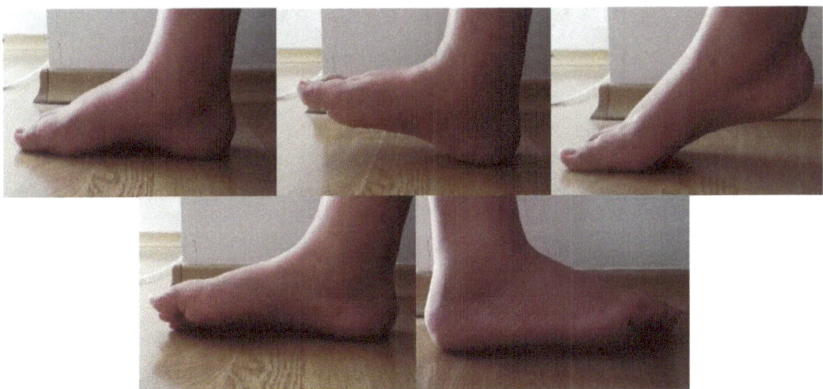

Figure 26. Angular amplitudes of the AJ for the voluntary patient before starting therapy.

Starting from this data, tests were carried out on the RP by progressively increasing the angular amplitudes of the PSSF, accurately finding the range of admissible values for rehabilitation exercises. The patient started the rehabilitation exercises for AJ of the right leg using the following as extreme angular positions: $\theta'_4 = -20° \div 25°$ for PF/DF movement and $\theta_4 = -15° \div 20°$ for EV/INV movement (Figure 27). Figure 28 denotes the results obtained during the first day of exercises. Deviations from the simulation results are due to the AJ's increased stiffness. Each recovery procedure was repeated at least 20 times daily.

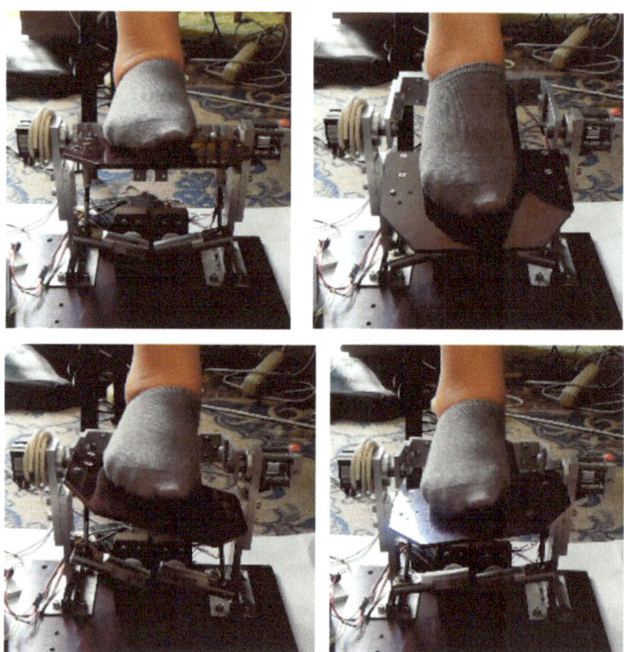

Figure 27. Images recorded during the first rehabilitation exercises of the voluntary patient.

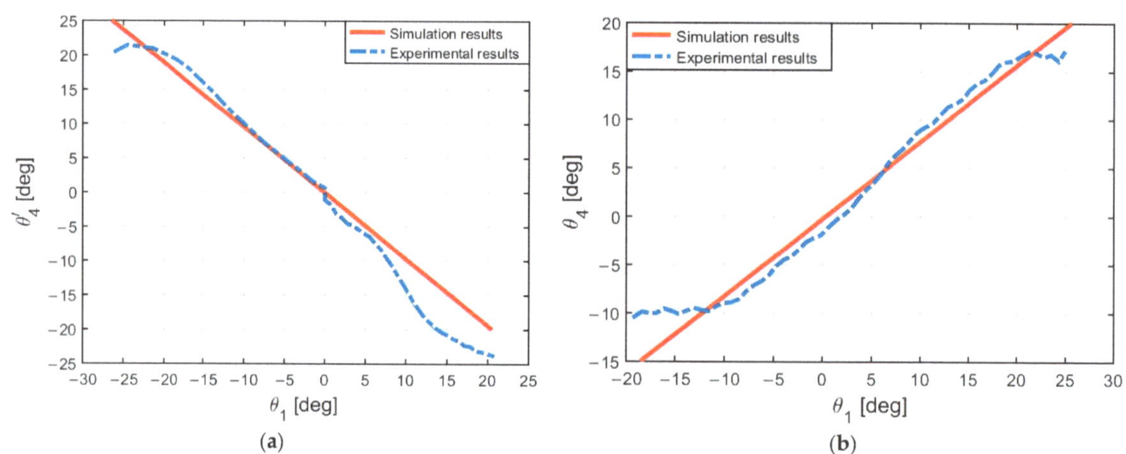

Figure 28. Results recorded during the rehabilitation exercises—day 1 [53]: (**a**) for PF/DF movement and (**b**) for INV/EV movement.

Before each session, in order to see if we were able to increase the intensity of the exercises, the patient's AJ was checked from a physical point of view to see the maximum angular ranges it can develop. After 10 days of therapy, which means half of the recommended recovery period, we were able to observe some improvements in the amplitude of PF/DF and INV/EV movements. More exactly, the range of θ'_4 varies between $-22° \div 30°$, and the range for θ_4 is $-23° \div 30°$ (Figure 29). We were also able to see that the deviations from the simulation results were much smaller. For the last rehabilitation exercises, we extended the angular extreme positions of the PSSF to $\theta'_4 = -25° \div 40°$ for PF/DF movement and $\theta_4 = -25° \div 35°$ for INV/EV movement (Figure 30). Comparing Figures 28 and 30, we can observe real progress in the AJ movement recovery, especially for INV/EV, where the stiffness of the joint was greatly increased.

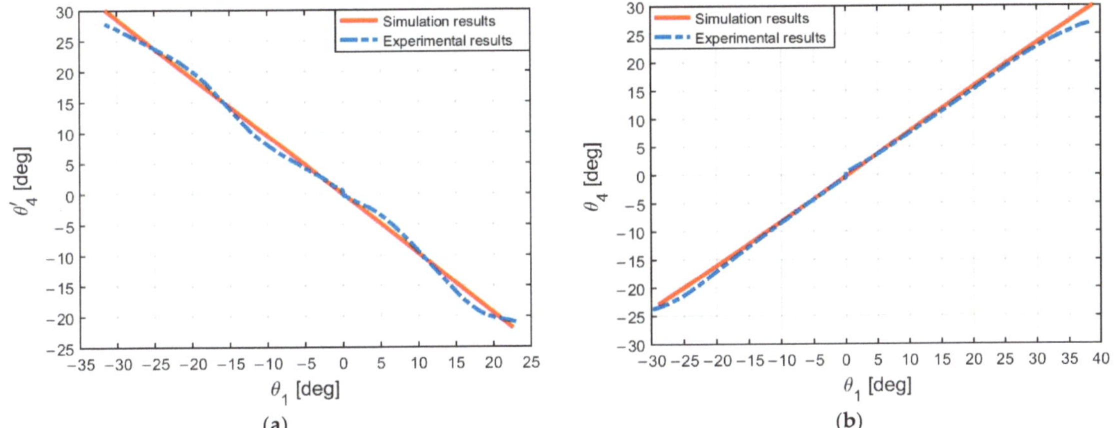

Figure 29. Results recorded during the rehabilitation exercises—day 10 [53]: (**a**) for PF/DF movement and (**b**) for INV/EV movement.

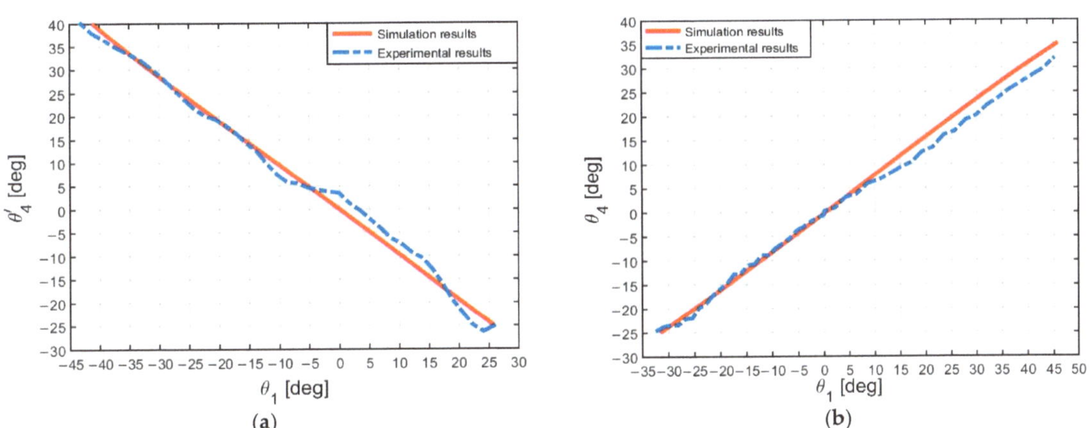

Figure 30. Results recorded during the rehabilitation exercises—day 20 [53]: (**a**) for PF/DF movement and (**b**) for INV/EV movement.

To better see the evolution in the range of motion for both PF/DF and INV/EV movements of the AJ, the curves shown in Figures 28–30 are represented together in Figure 31. As we can see, the ranges of θ'_4 and θ_4 increase for each new rehabilitation session (starting from the first day—continuous line in red—until the twentieth day—dashed line in green).

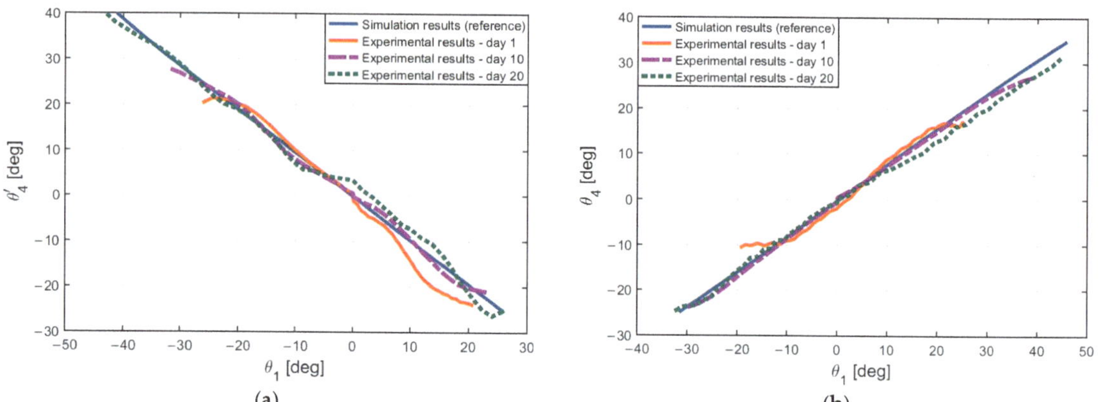

Figure 31. Results of the three rehabilitation sessions: (**a**) for PF/DF movement and (**b**) for INV/EV movement.

Figure 32 also shows the evolution of the AJ recovery results during rehabilitation therapy. The curves represented in this figure highlight the improvements in the AJ mobility at the end of the period, compared to the first day, for both PF/DF and INV/EV movements.

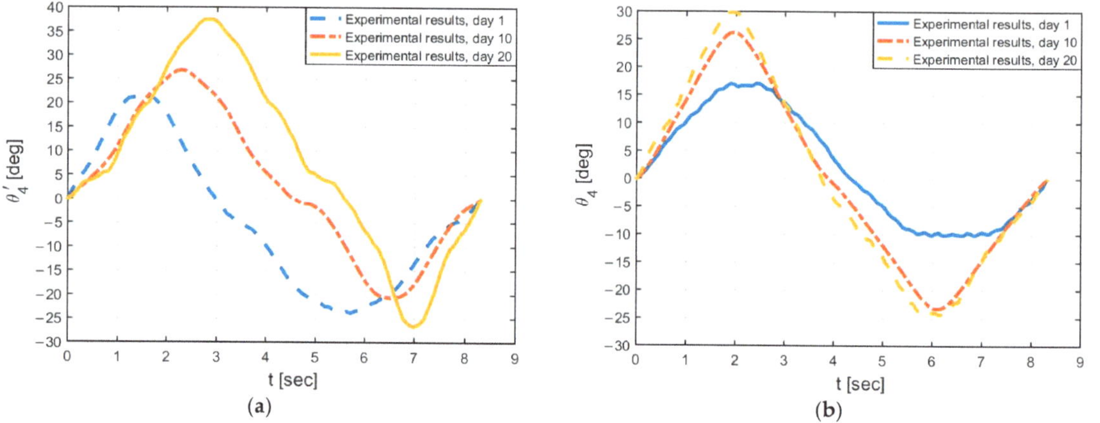

Figure 32. Results of the rehabilitation exercises over time during the therapy period [53]: (**a**) for PF/DF movement and (**b**) for INV/EV movement.

3.2.3. Ethical Issues

Ethical review and approval to use the RP for tests on human participants was not required. The single patient/participant in the study provided her written informed consent to participate. More than this, the patient is the coauthor of this study, and she suffered a fracture of the navicular bone in her right leg during her doctoral studies. She was a researcher and a patient at the same time. At the moment, the authors of this work do not intend to use the robotic platform in hospitals or to market this platform as a final product. Their intention is to improve the design according to the first prototype and its experimental tests to obtain a safe and effective RP.

3.2.4. Safety Issues

To prevent supplementary injuries of the AJ during rehabilitation therapy, the RP should be safe. In case the range of the PSSF angular positions exceeds the mobility range of the injured AJ or if the motors malfunction, they should stop. One possible solution is to measure the current consumed by the motors and to impose certain limits for this current in the command/control program. If the actual consumed current exceeds the limits imposed in the program, the motors should be shut down. This is the solution we are using at the moment. Another safety solution is to use compliant joints between the motors and the DgLs. If the encoders mounted on the output motors shafts send rotational values that exceed the rotational values sent by the encoders mounted on the DgLs, the motors should be shut down. This means that the AJ opposes resistance that exceeds the set value of the torque on the compliant joint. This second solution is intended to be further implemented.

3.3. New Proposed Design

Even though the robotic platform described here has demonstrated benefits for the rehabilitation of the human AJ, the center of the ankle suffers some displacement during exercise (see Figure 33). That is why the calf was not fixed during the rehabilitation exercises (this is the case for Rudgers Ankle [10] or ARBOT [14], etc., too). To avoid these displacements, an optimal RP should have a coincidence between the intersection point of the PF/DF and INV/EV rotational axes and the center of the AJ. To counteract this drawback of the realized RP, a new kinematics of the RP is proposed (Figure 34).

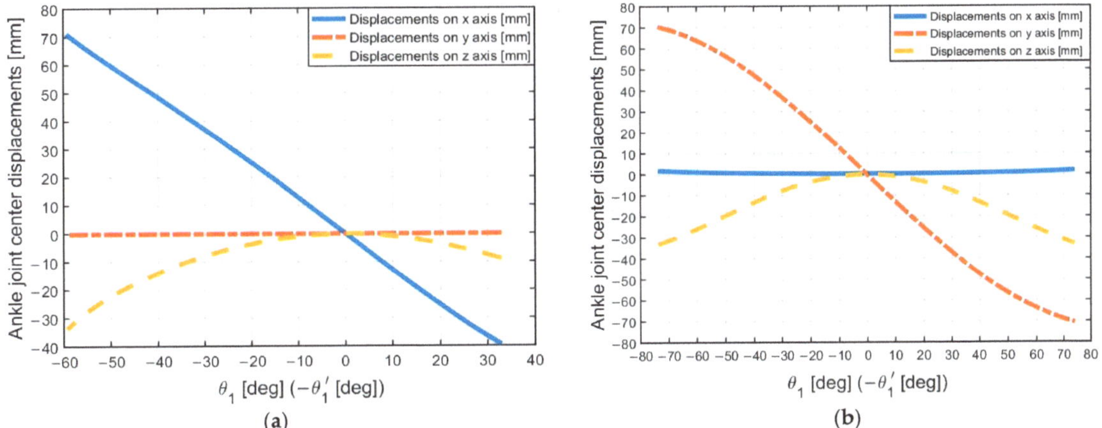

Figure 33. AJ center displacements during rehabilitation—simulation results (DS-3 design solution): (**a**) for PF/DF movement and (**b**) for INV/EV movement.

Based on the kinematic diagram shown in Figure 34, a new design for the RP was proposed. As we see in Figure 35, the center of AJ is aligned with the rotation center of the robot, which is the intersection point of the PF/DF and INV/EV rotational axes. In addition, the new robotic platform will enable its use on patients with different anthropomorphic dimensions thanks to its adjustable posture.

To underline the difference between the first prototype (Figure 15) and the new proposed RP (Figure 35), the two platforms are represented in Figure 36. The single difference between the first design and the new one is the shape of the PSSF, noted with 1 (in dark green). The new "U" shape of the PSSF has as an effect lowering of the AJ center of rotation at the intersection point of the two PSSF rotation axes.

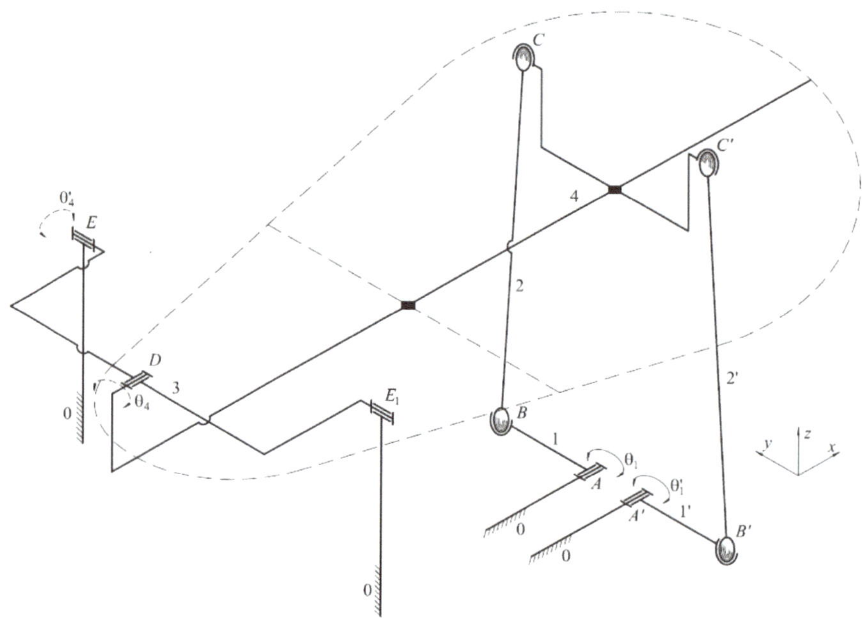

Figure 34. Kinematics of the new proposed RP.

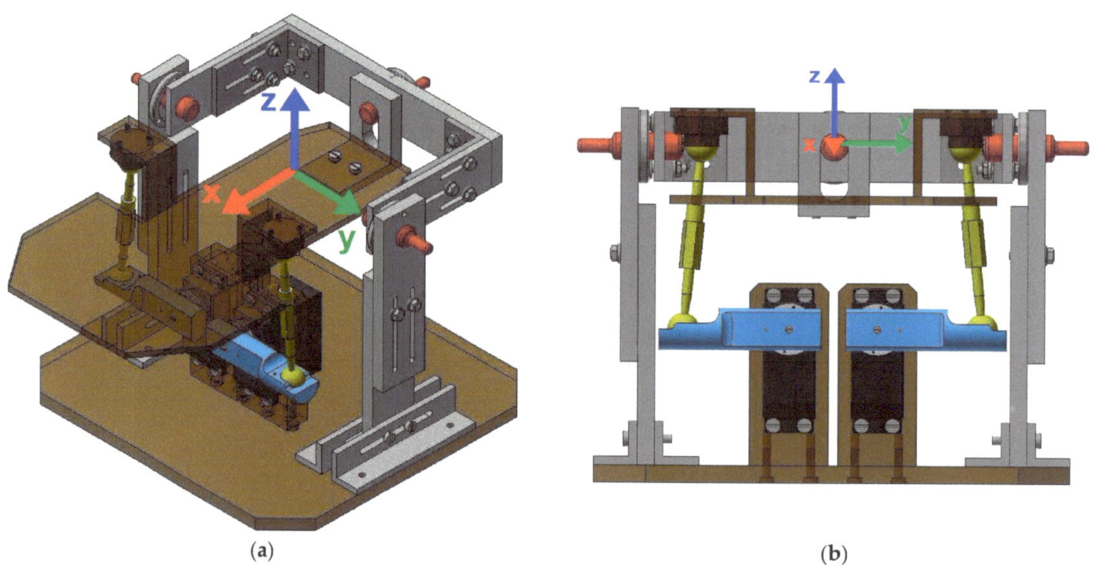

Figure 35. Three-dimensional CAD design of the new proposed RP: (**a**) isometric view and (**b**) frontal view.

The simulation of the new virtual robotic platform reveals that there is not any displacement of the AJ center during rehabilitation (Figure 37). The prototype of the new design is shown in Figure 38. The sole of the foot will be fixed on the plate representing the DnL while the shank will be connected to the robot base (Figure 38b).

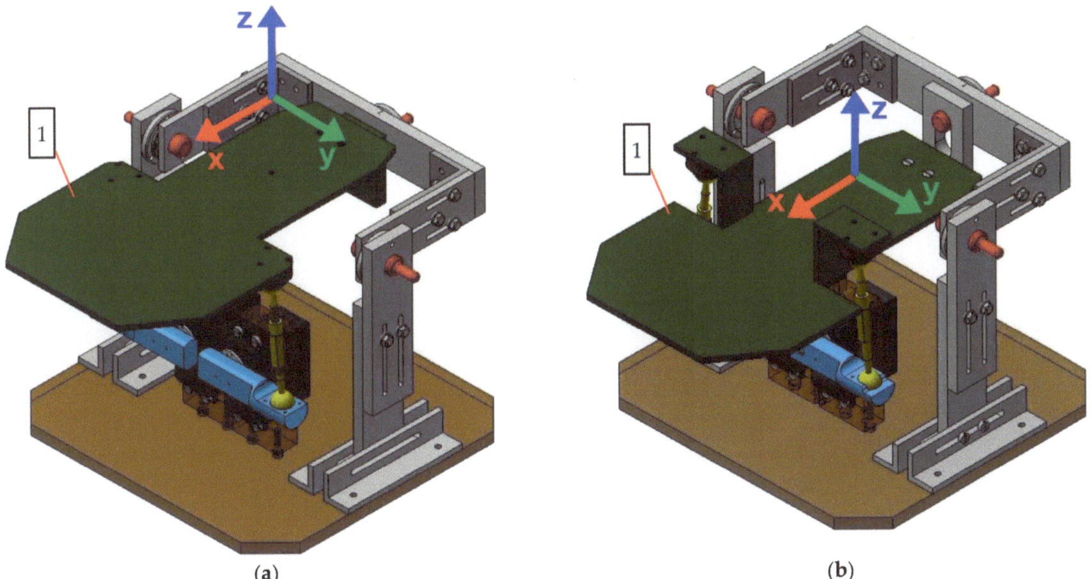

Figure 36. Three-dimensional CAD of both RPs: (**a**) the first design and (**b**) the new design.

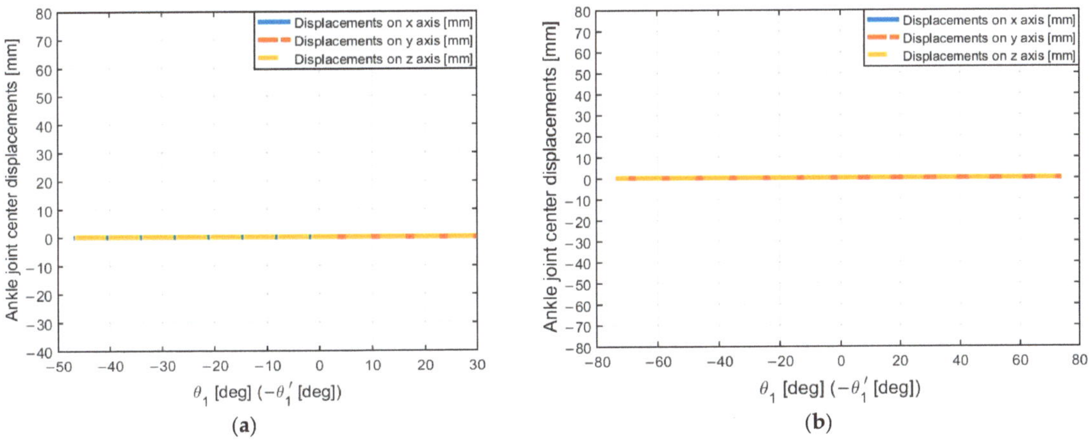

Figure 37. AJ center displacements during rehabilitation—simulation results (new RP design): (**a**) for PF/DF movement and (**b**) for INV/EV movement.

During the rehabilitation session, the patient will put their foot on the upper plate of the RP. The position of the PSSF can be adjusted, on z and x axes, so that the AJ rotation centre for different patients can coincide with the rotation centre of the mechanism. Results concerning the simulation of the new virtual robotic platform and also concerning its experimental tests will be the subject of future work.

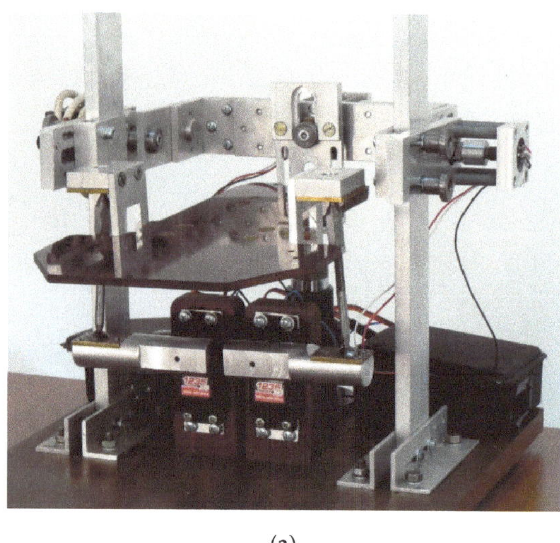
(a)

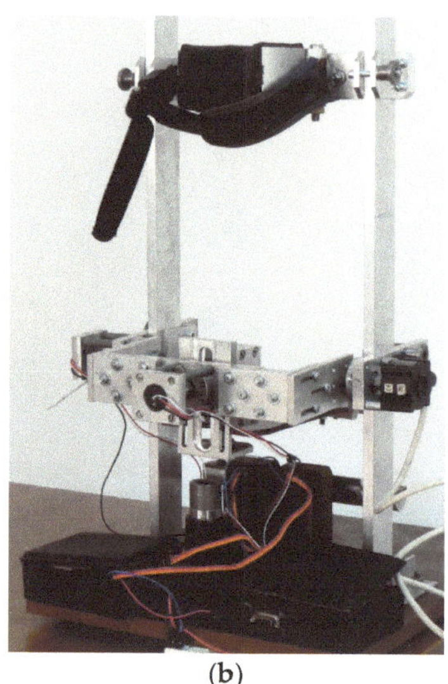
(b)

Figure 38. Improved prototype of the RP: (**a**) front isometric view and (**b**) back isometric view.

4. Discussion

When the AJ is injured, it can become unstable, and it can limit mobility or even cause loss of movement. Rehabilitation therapy is necessary to treat these traumas, and traditional therapies typically rely on elastic bands and foam rollers, which require constant assistance from a therapist and are often repetitive and time-consuming. Robotic systems have the potential to assist with AJ rehabilitation, but their high costs make them difficult to implement in recovery institutions. To address this issue, research has been carried out to develop low-cost, high-functionality robotic platforms for ankle rehabilitation and patient monitoring. This paper presents structural synthesis of AJ movements to identify new RP designs. Several solutions were proposed and compared based on a set of criteria, with two standing out as meeting most of the requirements. Dimensional synthesis, mathematical modeling, and simulation were used to select a final variant, with DS-3 based on the spatial four-bar mechanism chosen as the most practical solution.

The first experimental tests were carried out on a volunteer patient, who suffered from stiffness of the ankle joint, following rest from wearing the cast device. The assessment of the patient's recuperative progress was carried out through the monitoring of the angular strokes achieved by the ankle joint. The experimental results proved the efficiency of the system in patient recovery, as well as the validity of the mathematical model.

Even if the robotic platform described here demonstrated benefits on the rehabilitation of the human AJ, an optimal RP should have the center of the AJ aligned with the rotation center of the robot. This rotation center is the intersection point of the PF/DF and INV/EV rotational axes. Starting from that, a new RP design is proposed. The new RP will be further investigated in future work, including experimental tests. In addition, future studies are currently focused on the implementation of compliant joints between motors and DgLs, which could avoid overloading the ankle joint during rehabilitation exercises.

Author Contributions: Conceptualization, I.D. and C.-M.C.; methodology, I.D., C.-M.C. and S.A.; mathematical validation, I.D. and S.A.; experimental validation, C.-M.C.; investigation, C.-M.C.; resources, I.D. and C.-M.C.; writing—original draft preparation, I.D.; writing—review and editing, I.D.; supervision, I.D.; and project administration, I.D. All authors have read and agreed to the published version of the manuscript.

Funding: This research received no external funding.

Data Availability Statement: Not applicable.

Acknowledgments: This research was supported by the Doctoral School of "Gheorghe Asachi" Technical University of Iasi, Mechanical Engineering Faculty.

Conflicts of Interest: The authors declare no conflict of interest.

References

1. Valderrabano, V.; Barg, A.; Paul, J.; Pagenstert, G.; Wiewiorski, M. Foot and ankle injuries in professional soccer player. *Sport Orthop. Traumatol.* **2014**, *30*, 98–105. [CrossRef]
2. Park, S.; Hwang, Y.; Kim, H.; Kim, Y. Development of ankle rehabilitation robot with adjustable robotic mechanism. *J. Robot. Mechatron.* **2016**, *28*, 157–166.
3. Kesar, T.M.; Perumal, R.; Jancosko, A.; Reisman, D.S.; Rudolph, K.S.; Higginson, J.S.; Binder-Macleod, S.A. Novel patterns of functional electrical stimulation have an immediate effect on dorsiflexor muscle function during gait for people poststroke. *Phys. Ther.* **2011**, *91*, 1717–1728. [CrossRef] [PubMed]
4. Lee, H.J.; Kim, Y.; Cho, K.J. Development of a pneumatic artificial muscle-based ankle rehabilitation robot. *Int. J. Precis. Eng. Manuf.* **2016**, *17*, 1251–1259.
5. Lobo-Prat, J.; Siviy, C.; Reinkensmeyer, D.J. Design and validation of an ankle foot exoskeleton for running. *PLoS ONE* **2018**, *13*, e0201184.
6. Cheung, C.C.; Ng, S.K.; Chan, W.W. Design and control of a wearable ankle rehabilitation robot. *IEEE ASME Trans. Mechatron.* **2012**, *18*, 366–375.
7. Paradiso, R.; Marconi, L.; Farina, D. A novel robotic device for ankle rehabilitation. *IEEE Trans. Neural Syst. Rehabil. Eng.* **2009**, *17*, 287–295.
8. Park, H.; Jeong, G.W.; Kim, Y.J. Development of a gait enhancing mobile shoe using machine learning algorithms. *PLoS ONE* **2018**, *13*, e0201907.
9. Dong, M.; Zhou, Y.; Li, J.; Rong, X.; Fan, W.; Zhou, X.; Kong, Y. State of the art in parallel ankle rehabilitation robot: A systematic review. *J. NeuroEng. Rehabil.* **2021**, *18*, 1–15. [CrossRef]
10. Girone, M.; Burdea, G.; Bouzit, M. The Rutgers ankle orthopedic rehabilitation interface. In Proceedings of the ASME Dynamic Systems and Control Division, Nashville, TN, USA, 14–19 November 1999; pp. 305–312.
11. Yoon, J.; Ryu, J.; Lim, K.B. Reconfgurable ankle rehabilitation robot for various exercises: Research articles. *J. Robot. Syst.* **2005**, *22*, 15–33. [CrossRef]
12. Dai, J.S.; Zhao, T.; Nester, C. Sprained ankle physiotherapy based mechanism synthesis and stifness analysis of a robotic rehabilitation device. *Auton. Robot.* **2004**, *16*, 207–218. [CrossRef]
13. Liu, G.; Gao, J.; Yue, H.; Zhang, X.; Lu, G. Design and kinematics analysis of parallel robots for ankle rehabilitation. In Proceedings of the IEEE/RSJ International Conference on Intelligent Robots & Systems, Beijing, China, 9–15 October 2006; IEEE: Piscataway, NJ, USA, 2006; pp. 253–258.
14. Saglia, J.A.; Tsagarakis, N.G.; Dai, J.S.; Caldwell, D.G. A high-performance redundantly actuated parallel mechanism for ankle rehabilitation. *Int. J. Robot. Res.* **2009**, *28*, 1216–1227. [CrossRef]
15. Malosio, M.; Negri, S.P.; Pedrocchi, N.; Vicentini, F.; Caimmi, M.; Tosatti, L.M. A spherical parallel three degrees-of-freedom robot for ankle-foot neuro-rehabilitation. In Proceedings of the Annual International Conference of the IEEE Engineering in Medicine and Biology Society, San Diego, CA, USA, 28 August–September 2012; pp. 3356–3359.
16. Ayas, M.S.; Altas, I.H. Fuzzy logic based adaptive admittance control of a redundantly actuated ankle rehabilitation robot. *Control Eng. Pract.* **2017**, *59*, 44–54. [CrossRef]
17. Ai, Q.; Zhu, C.; Zuo, J.; Meng, W.; Liu, Q.; Xie, S.; Yang, M. Disturbance-estimated adaptive backstepping sliding mode control of a pneumatic muscles-driven ankle rehabilitation robot. *Sensors* **2018**, *18*, 66. [CrossRef] [PubMed]
18. Jamwal, P.K.; Hussain, S.; Ghayesh, M.H.; Rogozina, S.V. Impedance control of an intrinsically compliant parallel ankle rehabilitation robot. *IEEE Trans. Ind. Electron.* **2016**, *63*, 3638–3647. [CrossRef]
19. Zhang, M.; Cao, J.; Zhu, G.; Miao, Q.; Zeng, X.; Xie, S.Q. Reconfgurable workspace and torque capacity of a compliant ankle rehabilitation robot (CARR). *Robot. Auton. Syst.* **2017**, *98*, 213–221. [CrossRef]
20. Tsoi, Y.H.; Xie, S.Q.; Graham, A.E. Design, modeling and control of an ankle rehabilitation robot. In *Design and Control of Intelligent Robotic Systems*; Liu, D., Wang, L., Tan, K.C., Eds.; Springer: Berlin/Heidelberg, Germany, 2009; pp. 377–399.
21. Wang, C.; Fang, Y.; Guo, S.; Chen, Y. Design and kinematical performance analysis of a 3—RUS/RRR redundantly actuated parallel mechanism for ankle rehabilitation. *J. Mech. Robot.* **2013**, *5*, 041003. [CrossRef]
22. Valles, M.; Cazalilla, J.; Valera, A.; Mata, V.; Page, A.F.; Diaz-Rodriguez, M. A 3-PRS parallel manipulator for ankle rehabilitation: Towards a low-cost robotic rehabilitation. *Robotica* **2017**, *35*, 1939–1957. [CrossRef]
23. Li, J.; Zuo, S.; Zhang, L.; Dong, M. Mechanical design and performance analysis of a novel parallel robot for ankle rehabilitation. *J. Mech. Robot.* **2020**, *12*, 051007. [CrossRef]

24. Hong, D.; Gong, M.; Li, S.; Lu, W. Design of a Portable and Lightweight Ankle Rehabilitation Robot for Home Use by Stroke Patients. *J. Healthc. Eng.* **2020**, 1–9.
25. Liu, H.; Jin, R.; Xie, S.; Zhang, X.; Wang, F. Design and Development of a Wearable Ankle Rehabilitation Robot Based on Three-Dimensional Acceleration Sensor. *Appl. Sci.* **2019**, *9*, 4058.
26. Zhou, Z.; Jiang, Y.; Huang, H.; Wei, Q. Design and Control of a Modular Ankle Rehabilitation Robot. *J. Healthc. Eng.* **2019**, 1–12.
27. Aghaebrahimian, A.; Abdi, A.; Faraji, M. A Smartphone-Based User Interface for an Ankle Rehabilitation Robot. *J. Rehabil. Robot.* **2019**, *3*, 63–70.
28. Kamal, A.; Lemaire, E.D.; Sawan, M. Design and Evaluation of a Gamified User Interface for Home-Based Ankle Rehabilitation. *J. Med. Syst.* **2021**, *45*, 1–10.
29. Wang, L.; Chen, C.; Zheng, R.; Liu, H. Design of a User-Friendly Interface for an Ankle Rehabilitation Robot. *IEEE Access* **2020**, *8*, 135509–135517.
30. Chen, X.; Chen, S.; Chen, W.; Chen, Y.; Guo, X.; Liu, G. Clinical Efficacy and Safety of Home-Based Rehabilitation for Post-Stroke Ankle-Foot Dysfunction Using Ankle Rehabilitation Robot: A Randomized, Controlled Pilot Trial with 24 Weeks of Follow-Up. *Clin. Rehabil.* **2018**, *32*, 1174–1184.
31. Li, R.; Zhang, Y.; Xu, Z.; Li, M. Robot-Assisted Ankle Rehabilitation Training Improves Ankle Function in Patients with Acute Ankle Sprain: A Randomized Controlled Trial. *J. Orthop. Surg. Res.* **2020**, *15*, 1–9.
32. Wang, K.; Wang, Z.; Liu, H. Pilot Study on the Clinical Effect of a Home-Based Ankle Rehabilitation Robot. *J. Rehabil. Robot.* **2018**, *2*, 31–38.
33. Shah Nazar, P.; Pott, P.P. Ankle Rehabilitation Robotic Systems for domestic use—A systematic review. *Curr. Dir. Biomed. Eng.* **2022**, *8*, 65–68. [CrossRef]
34. Cioi, D.; Burdea, G.; Engsberg, J.R.; Janes, W. Ankle control and strength training for children with cerebral palsy using the Rutgers Ankle CP: A case study. In Proceedings of the IEEE International Conference on Rehabilitation Robotics, Zurich, Switzerland, 29 June–1 July 2011; p. 5975432.
35. Roy, A.; Forrester, L.W.; Macko, R. Short-term ankle motor performance with ankle robotics training in chronic hemiparetic stroke. *J. Rehabil. Res. Dev.* **2011**, *48*, 417–429. [CrossRef]
36. Kim, J.; Hwang, S.; Sohn, R.; Lee, Y.; Kim, Y. Development of an Active Ankle Foot Orthosis to Prevent Foot Drop and Toe Drag in Hemiplegic Patients: A Preliminary Study. *Appl. Bionics Biomech.* **2011**, *8*, 377–384. [CrossRef]
37. Ward, J.; Sugar, T.; Boehler, A.; Standeven, J.; Engsberg, J.R. Stroke Survivors' Gait Adaptations to a Powered Ankle Foot Orthosis. *Adv. Robot.* **2011**, *25*, 1879–1901. [CrossRef] [PubMed]
38. Forrester, L.W.; Roy, A.; Krebs, H.I.; Macko, R.F. Ankle training with a robotic device improves hemiparetic gait after a stroke. *Neurorehabilit. Neural Repair* **2011**, *25*, 369–377. [CrossRef] [PubMed]
39. Jamwal, P.K.; Sie, S.Q. An Adaptive Wearable Parallel Robot for the Treatment of Ankle Injuries. *IEEE/ASME Trans. Mechatron.* **2014**, *19*, 64–75. [CrossRef]
40. Blanchette, A.K.; Noel, M.; Richards, C.L.; Nadeau, S.; Bouyer, L.J. Modifications in ankle dorsiflexor activation by applying a torque perturbation during walking in persons post-stroke: A case series. *J. Neuroeng. Rehabil.* **2014**, *11*, 98. [CrossRef]
41. Takahashi, K.Z.; Lewek, M.D.; Sawicki, G.S. A neuromechanics-based powered ankle exoskeleton to assist walking post-stroke: A feasibility study. *J. Neuroeng. Rehabil.* **2015**, *12*, 23. [CrossRef]
42. Koller, J.R.; Remy, C.D.; Ferris, D.P. Comparing neural control and mechanically intrinsic control of powered ankle exoskeletons. In Proceedings of the IEEE International Conference on Rehabilitation Robotics, London, UK, 17–20 July 2017; pp. 294–299.
43. Ren, Y.; Wu, I.N.; Yang, C.Y.; Xu, T.; Harvey, R.L.; Zhang, L.Q. Developing a Wearable Ankle Rehabilitation Robotic Device for in-Bed Acute Stroke Rehabilitation. *IEEE Trans. Neural Syst. Rehabil. Eng.* **2017**, *25*, 589–596. [CrossRef]
44. Yeung, L.F.; Ockenfeld, C.; Pang, M.K.; Wai, H.W.; Soo, O.Y.; Li, S.W.; Tong, K.Y. Randomized controlled trial of robotassisted gait training with dorsiflexion assistance on chronic stroke patients wearing ankle-foot-orthosis. *J. Neuroeng. Rehabil.* **2018**, *15*, 51. [CrossRef]
45. Awad, L.N.; Esquenazi, A.; Francisco, G.E.; Nolan, K.J.; Jayaraman, A. The ReWalk ReStore™ soft robotic exosuit: A multi-site clinical trial of the safety, reliability, and feasibility of exosuit-augmented post-stroke gait rehabilitation. *J. Neuroeng. Rehabil.* **2020**, *17*, 80. [CrossRef]
46. Racu, C.M.; Doroftei, I. Design, modelling and simulation aspects of an ankle rehabilitation device. *IOP Conf. Ser. Mater. Sci. Eng.* **2016**, *145*, 052008. [CrossRef]
47. Racu, C.M.; Doroftei, I. Design Aspects of a New Device for Ankle Rehabilitation. *Appl. Mech. Mater.* **2015**, *809*, 986–991. [CrossRef]
48. Racu, C.M.; Doroftei, I. Ankle rehabilitation device with two degrees of freedom and compliant joint. *IOP Conf. Ser. Mater. Sci. Eng.* **2015**, *95*, 012054. [CrossRef]
49. Racu, C.M.; Doroftei, I. New Concepts of Ankle Rehabilitation Devices—Part I: Theoretical Aspects. In *New Advances in Mechanism and Machine Science*; Doroftei, I., Pisla, D., Lovasz, E., Eds.; Springer: Cham, Switzerland, 2018; pp. 223–231.
50. Racu, C.M.; Doroftei, I. New Concepts of Ankle Rehabilitation Devices—Part II: Design and Simulation. In *New Advances in Mechanism and Machine Science*; Doroftei, I., Pisla, D., Lovasz, E., Eds.; Springer: Cham, Switzerland, 2018; pp. 233–239.
51. Racu, C.M.; Doroftei, I. Preliminary results and evaluation of an ankle rehabilitation device. *IOP Conf. Ser. Mater. Sci. Eng.* **2020**, *997*, 012088. [CrossRef]

52. Parenteau, C.S.; Viano, D.C.; Petit, P.Y. Biomechanical properties of human cadaveric ankle-subtalar joints in quasi-static loading. *J. Biomech. Eng.* **1998**, *120*, 105–111. [CrossRef]
53. Doroftei, I.; Racu, C.M.; Baudoin, Y. Development of a Robotic Platform for Ankle Joint Rehabilitation. *Acta Tech. Napoc. Ser. Appl. Math. Mech. Eng.* **2021**, *64*, 301–310.